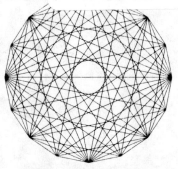

The DFT

An Owner's Manual for the
Discrete Fourier Transform

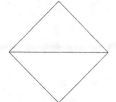

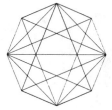

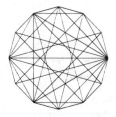

The DFT

An Owner's Manual for the
Discrete Fourier Transform

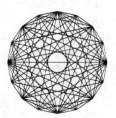

William L. Briggs
University of Colorado, Boulder

Van Emden Henson
Naval Postgraduate School

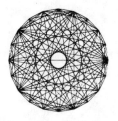

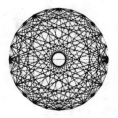

siam

Society for Industrial and Applied Mathematics
Philadelphia

Library of Congress Cataloging-in-Publication Data

Briggs, William L.
    The DFT: an owner's manual for the discrete Fourier transform /
William L. Briggs, Van Emden Henson.
        p.   cm.
    Includes bibliographical references (p.   -   ) and index.
    ISBN 0-89871-342-0 (pbk.)
    1. Fourier transformations.   I. Henson, Van Emden.   II. Title.
QA403.5.B75   1995
515'.723—dc20   95-3232

to our parents

....

Muriel and Bill
and
Louise and Phill

Contents

Preface

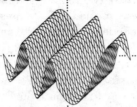

Let's begin by describing what this book is and what it is not. **Fourier analysis** is the study of how functions defined on a continuum (that is, at *all* points of an interval) can be represented and analyzed in terms of periodic functions like sines and cosines. While this is an immensely elegant and important subject, many practical problems involve doing Fourier analysis on a computer or doing Fourier analysis on *samples* of functions, or both. When carried out in these modes, Fourier analysis becomes **discrete Fourier analysis** (also called **practical, computational,** and **finite Fourier analysis**). All of these names convey a sense of usefulness and tangibility that is certainly one of the hallmarks of the subject. So the first claim is that this book is about discrete Fourier analysis, with the emphasis on discrete.

Fourier analysis in the continuous setting involves both Fourier series and Fourier transforms, but its discrete version has the good fortune of requiring only a single process which has come to be called the **discrete Fourier transform (DFT)**. One might argue that **discrete Fourier series** or **Fourier sums** are more appropriate names, but this is a decision that we will concede to generations of accepted usage.

There are several aspects of DFTs that make the subject so compelling. One is its immediate applicability to a bewildering variety of problems, some of which we shall survey during the course of the book. Another tantalizing quality of the subject

is the fact that it appears quite distinct from continuous Fourier analysis and at the
same time is so closely related to it. Much of the book is devoted to exploring the
constant interplay among the DFT, Fourier series, and Fourier transforms. For this
reason, the subject can feel like a delicate dance on the borders between the continuous
and the discrete, and the passage between these two realms never fails to be exciting.
Therefore, in a few words, this book is a study of the DFT: its uses, its properties, its
applications, and its relationship to Fourier series and Fourier transforms.

Equally important is what this book is *not*! Undoubtedly, one reason for the
spectacular rise in the popularity of practical Fourier analysis is the invention of the
algorithm known as the **fast Fourier transform (FFT)**. For all of its magic and
power, the FFT is *only* an extremely efficient way to compute the DFT. It must be said
clearly at the outset that this book is *not* about the FFT, which has a vast literature
of its own. We will devote only one chapter to the FFT, and it is a whirlwind survey
of that complex enterprise. For the purposes of this book, we will assume that the
FFT is the computational realization of the DFT; it is a black box that can be found,
already programmed, in nearly every computer library in the world. For that reason,
whenever the DFT is mentioned or used, we will assume that there is always a very
efficient way to evaluate it, namely the FFT.

A reader might ask: why another book on Fourier analysis, discrete or otherwise?
And that is the first question that the authors must answer. It is not as if this
venerable subject has been neglected in its two centuries of existence. To the contrary,
the subject in all of its guises and dialects undoubtedly receives more use and attention
every year. It is difficult to imagine that one collection of mathematical ideas could
bear in such a fundamental way on problems from crystallography to geophysical
signal processing, from statistical analysis to remote sensing, from prime factorization
to partial differential equations. Surely it is a statement, if not about the structure
of the world around us, then about the structure that we impose upon that world.
And therein lies one justification of the book: a field in which there is so much change
and application deserves to be revisited periodically and reviewed, particularly at an
introductory level.

These remarks should not imply that outstanding treatments of Fourier analysis
do not already exist. Definitive books on Fourier analysis, such as those written by
Carslaw [30], Lighthill [93], and Bracewell [13], have been followed by more recent
books on discrete Fourier analysis, such as those by Brigham [20], [21], Čižek [36], and
Van Loan [155]. The present book owes much to all of its predecessors; and indeed
nearly every fact and result in the following pages appears *somewhere* in the literature
of engineering and applied mathematics. This leads to a second reason for writing,
namely to collect the many scattered and essential results about DFTs and place them
in a single introductory book in which the DFT itself is on center stage.

Underlying this book is the realization that Fourier analysis has an enormous and
increasingly diverse audience of students and users. That audience is the intended
readership of the book: it has been written for students and practitioners, and every
attempt has been made to write it as a tutorial. Hopefully it is already evident
that the style is informal; it will also be detailed when necessary. Motivation and
interpretation generally take precedence over rigor, although an occasional proof does
appear when it might enhance understanding. The text is laced with numerical and
graphical examples that illuminate the discussion at hand. Each chapter is supported
by a healthy dose of instructive problems of both an analytical and computational

nature. In short, the book is meant to be used for teaching and learning.

Although there are some mathematical prerequisites for this book, it has been made as self-contained as possible. A minimum prerequisite is a standard three-semester calculus course, provided it includes complex exponentials, power series, and geometric series. The book is not meant to provide a thorough exposition of Fourier series and Fourier transforms; therefore, some prior exposure to these topics would be beneficial. On occasion a particularly "heavy" theorem will be used without proof, and this practice seems justified when it allows a useful result to be reached.

The DFT Book

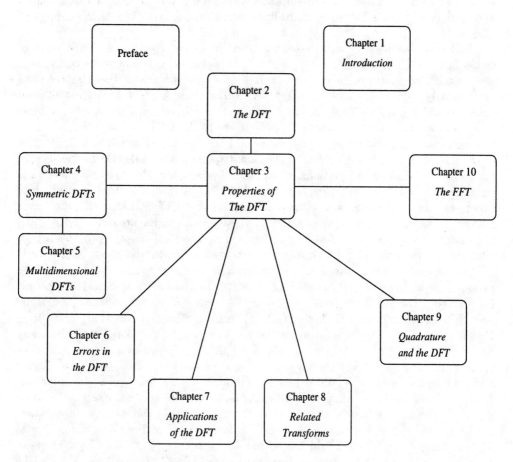

FIG. 0.1. *This is the first of 99 figures in this book. It shows the interdependence of its chapters and suggests some pathways through the book.*

Unlike the DFT, this book is not entirely linear. The map in Figure 0.1 suggests possible pathways through the book. Chapter 1 uses two examples to provide some historical and practical motivation for the DFT. Readers who disdain appetizers before nine-course meals or overtures before operas can move directly to Chapter 2 with minimal loss.

The DFT edifice is built in Chapter 2 from several different perspectives, each of which has some historical or mathematical significance: first as an approximation to Fourier transforms, then as an approximation to Fourier series coefficients, then as a form of interpolation, and finally with the use of spike (or delta) functions. Any and all of these four paths lead to the same end: the definition of the DFT. The absolutely fundamental **reciprocity relations** which underlie all uses of the DFT are developed early in this chapter and must not be overlooked. Chapter 3 is the requisite chapter that deals with the many properties of the DFT which are used throughout the remainder of the book.

At this point the path splits and there are several options. In Chapter 4 special **symmetric** forms of the DFT are considered in detail. This collection of ideas and techniques is tremendously important in computations. Equally important is the subject of DFTs in more than one dimension since many computationally intensive applications of the DFT involve multidimensional transforms. This is the subject of Chapter 5.

The toughest passage may be encountered in Chapter 6, in which the issue of errors in the DFT is treated. Be forewarned: this chapter is meant to be methodical and exhaustive—perhaps to a fault. While all of the results in this chapter have undoubtedly appeared before, we know of no other place in which they have all been collected and organized. This chapter is both the most theoretical chapter of the book *and* the most important for practical applications of the DFT.

The final four chapters offer some choice and can be sampled in any order. The first of these chapters deals with a few of the many applications of the DFT, specifically, the solution of difference equations, digital signal processing, FK migration of seismic data, and the x-ray tomography problem. The next of these chapters is a survey of other transforms that are related in some way to the DFT. Another chapter delves into a host of issues surrounding the DFT, viewed as a quadrature rule. Not wishing to neglect *or* emphasize the fast Fourier transform, the final chapter provides a high altitude survey of the FFT for those who wish to look "under the hood" of the DFT.

A book that began as a 200-page tutorial could not have grown to these proportions without some help. We are grateful to the following SIAM staffpeople for bringing the book to fruition: Nancy Abbott (designer), Susan Ciambrano (acquisitions editor), Beth Gallagher (editor), J. Corey Gray (production), and Vickie Kearn (publisher). Early versions of the book were read by Bob Palais, Ridgway Scott, Gil Strang, Roland Sweet, and Charles Van Loan. We appreciate the time they spent in this task, as well as the suggestions and the encouragement that they offered. Cleve Moler, using *Matlab*, showed us the mandala figures that grace the cover and frontmatter of the book (they should be displayed in color for best effect). Scanning of one of the book's figures was done by David Canright, and we also owe thanks to the writers of the freeware package SU (Seismic Unix) at the Colorado School of Mines. One of us (WLB) would like to thank the people at Dublin City University for nine months of Irish hospitality during the early stages of the writing. Finally, the completion of this book would have been impossible without the love and forbearance of our wives and daughters, who provided encouragement, chili dinners, waltzes with snowmen, and warm blankets in spare bedrooms: we thank Julie and Katie, Teri and Jennifer.

You probably agree that this preamble has gone on long enough; it's time for the fun (and the learning) to begin.

List of (Frequently and Consistently Used) Notation

$a(-n)$	$a \times 10^{-n}$
A, B	length of spatial domain
$\mathbf{A}, \mathbf{B}, \mathbf{C}$	generic matrices (**bold face**)
c_k	Fourier coefficients
$\mathcal{C}$	discrete cosine transform operator
$C^m[a, b]$	space of m times continuously differentiable functions on $[a, b]$
$\mathcal{D}$ or $\mathcal{D}_N$	DFT operator
$\mathcal{F}$	Fourier transform operator
f, g, h, u	generic functions
$f * g$	convolution of two functions
$\hat{f}, \hat{g}, \hat{h}$	Fourier transforms of f, g, h
f_n, g_n, h_n	generic sequences or sequences of samples of functions f, g, h
$f_n * g_n$	convolution of two sequences
f_{mn}	array of samples of a function $f(x, y)$
$\tilde{f}_n$	output of inverse DFT
F_k	sequence of DFT coefficients
F_{jk}	array of DFT coefficients
$\overline{F}jk$	auxiliary array (not complex conjugate)
$\mathcal{I}$	a spatial domain, usually $[-A/2, A/2]$
N	number of points in a DFT
$\mathcal{N}$	the set of integers $\{-N/2 + 1, \ldots, N/2\}$
$n = P : Q$	the consecutive integers between P and Q, inclusive
p_a	square pulse or boxcar function of width a
$\mathcal{R}$	replication operator
$\text{Re}\{\}, \text{Im}\{\}$	real, imaginary part
S	discrete sine transform operator
T	length of the input (time) domain, Chapter 7
$\mathbf{W}_N$	matrix of the N-point DFT
x_n, y_n	grid points in the spatial domain
$\mathbf{x}, \mathbf{y}, \mathbf{z}$	generic vectors (**bold face**)
$\delta(x)$	Dirac delta function, x a real number
$\delta(k)$	Kronecker delta sequence, k an integer
$\hat{\delta}_N(k)$	modular Kronecker delta sequence, k an integer
Δx or h	grid spacing (sampling interval) in spatial domain
$\Delta\omega$	grid spacing (sampling interval) in frequency domain
η	wavelength of two-dimensional wave in y-direction
η_n	extrema of an orthogonal polynomial
λ	wavelength of two-dimensional wave
μ	wavelength of two-dimensional wave in x-direction
ν	length of frequency vector of two-dimensional wave
ω	variable in frequency domain
ω_c	cut-off frequency
ω_k	grid point in frequency domain
ω_N	$e^{i2\pi/N}$
Ω, Λ	length of frequency domain
σ	second variable in frequency domain
Σ'	sum in which first term is weighted by $1/2$
Σ''	sum in which first and last terms are weighted by $1/2$
ξ_n	zeros of an orthogonal polynomial

Chapter **1**

Introduction

The material essential for a student's mathematical laboratory is very simple. Each student should have a copy of Barlow's tables of squares, etc., a copy of Crelle's Calculation Tables, *and a seven-place table of logarithms. Further it is necessary to provide a stock of computing paper, . . . and lastly, a stock of computing forms for practical Fourier analysis. With this modest apparatus nearly all of the computations hereafter described may be performed, although time and labour may often be saved by the use of multiplying and adding machines when these are available.*
— E. T. Whittaker and G. Robinson
The Calculus of Observations, 1924

1

1.1.   A Bit of History

Rather than beginning with the definition of the discrete Fourier transform on the first page of this book, we thought that a few pages of historical and practical introduction might be useful. Some valuable perspective comes with the realization that the DFT was not discovered ten years ago, nor was it invented with the fast Fourier transform (FFT) thirty years ago. It has a fascinating history, spanning over two centuries, that is closely associated with the development of applied mathematics and numerical analysis. Therefore, this chapter will be devoted, in part, to the history of the DFT. However, before diving into the technicalities of the subject, there is insight to be gained from seeing the DFT in action on a specific problem. Therefore, a few pages of this chapter will also be spent extracting as much understanding as possible from a very practical example. With these excursions into history and applications behind us, the path to the DFT should be straight and clear.

Let's begin with some history. Fourier analysis is over 200 years old, and its history is filled with both controversy and prodigious feats. Interwoven throughout it is the thread of discrete or practical Fourier analysis which is most pertinent to this book. In order to appreciate the complete history, one must retreat some 60 years prior to the moment in 1807 when Jean Baptiste Joseph Fourier presented the first version of his paper on the theory of heat conduction to the Paris Academy. The year 1750 is a good starting point: George II was king of England and the American colonies were in the midst of the French and Indian War; Voltaire, Rousseau, and Kant were writing in Europe; Bach had just died, Mozart was soon to be born; and the calculus of Newton and Leibnitz, published 75 years earlier, was enabling the creation of powerful new theories of celestial and continuum mechanics.

There were two outstanding problems of the day that focused considerable mathematical energy, and formed the seeds that ultimately became Fourier analysis. The first problem was to describe the vibration of a taut string anchored at both ends (or equivalently the propagation of sound in an elastic medium). Remarkably, the wave equation as we know it today had already been formulated, and the mathematicians Jean d'Alembert, Leonhard Euler, Daniel Bernoulli, and Joseph-Louis Lagrange had proposed methods of solution around 1750. Bernoulli's solution took the form of a trigonometric series

$$y = A \sin x \cos at + B \sin 2x \cos 2at + \cdots,$$

in which x is the spatial coordinate and t is the time variable. This solution already anticipated the continuous form of a Fourier series. It appears that both Euler and Lagrange actually *discretized* the vibrating string problem by imagining the string to consist of a finite number of connected particles. The solution of this discrete problem required finding *samples* of the function that describes the displacement of the string. A page from Lagrange's work on this problem (see Figure 1.1), published in 1759 [90], contains all of the ingredients of what we would today call a discrete Fourier sine series.

The second problem which nourished the roots of Fourier analysis, particularly in its discrete form, was that of determining the orbits of celestial bodies. Euler, Lagrange, and Alexis Claude Clairaut made fundamental contributions by proposing that data taken from observations be approximated by linear combinations of periodic functions. The calculation of the coefficients in these trigonometric expansions led to a computation that we would call a discrete Fourier transform. In fact, a paper

ET LA PROPAGATION DU SON.   81

ment, pour arriver à une qui ne contienne plus qu'une seule de ces variables; mais il est facile de voir qu'en s'y prenant de cette façon on tomberait dans des calculs impraticables à cause du nombre indéterminé d'équations et d'inconnues; il est donc nécessaire de suivre une autre route : voici celle qui m'a paru la plus propre.

24. Je multiplie d'abord chacune de ces équations par un des coefficients indéterminés $D_1, D_2, D_3, D_4, \ldots$, en supposant que le premier D_1 soit égal à 1; ensuite je les ajoute toutes ensemble : j'ai

$$y_1\left[D_1\sin\frac{\varpi}{2m} + D_2\sin\frac{2\varpi}{2m} + D_3\sin\frac{3\varpi}{2m} + \ldots + D_{m-1}\sin\frac{(m-1)\varpi}{2m}\right]$$
$$+ y_2\left[D_1\sin\frac{2\varpi}{2m} + D_2\sin\frac{4\varpi}{2m} + D_3\sin\frac{6\varpi}{2m} + \ldots + D_{m-1}\sin\frac{2(m-1)\varpi}{2m}\right]$$
$$+ y_3\left[D_1\sin\frac{3\varpi}{2m} + D_2\sin\frac{6\varpi}{2m} + D_3\sin\frac{9\varpi}{2m} + \ldots + D_{m-1}\sin\frac{3(m-1)\varpi}{2m}\right]$$
$$\cdots\cdots\cdots\cdots\cdots\cdots\cdots\cdots$$
$$+ y_{m-1}\left[D_1\sin\frac{(m-1)\varpi}{2m} + D_2\sin\frac{2(m-1)\varpi}{2m} + \ldots + D_{m-1}\sin\frac{(m-1)^2\varpi}{2m}\right]$$
$$= D_1 S_1 + D_2 S_2 + D_3 S_3 + \ldots + D_{m-1}S_{m-1}.$$

Qu'on veuille à présent la valeur d'un y quelconque, par exemple de y_μ, on fera évanouir les coefficients des autres y, et l'on obtiendra l'équation simple

$$y_\mu\left[D_1\sin\frac{\mu\varpi}{2m} + D_2\sin\frac{2\mu\varpi}{2m} + D_3\sin\frac{3\mu\varpi}{2m} + \ldots + D_{m-1}\sin\frac{(m-1)\mu\varpi}{2m}\right]$$
$$= D_1 S_1 + D_2 S_2 + D_3 S_3 + \ldots + D_{m-1}S_{m-1}.$$

On déterminera ensuite les valeurs des quantités $D_2, D_3, D_4, \ldots$, qui sont en nombre de $m-2$, par les équations particulières qu'on aura en supposant égaux à zéro les coefficients de tous les autres y: on aura par là l'équation générale

$$D_1\sin\frac{\lambda\varpi}{2m} + D_2\sin\frac{2\lambda\varpi}{2m} + D_3\sin\frac{3\lambda\varpi}{2m} + \ldots + D_{m-1}\sin\frac{(m-1)\lambda\varpi}{2m} = 0,$$

I.   11

FIG. 1.1.   *This page from* The Nature and Propagation of Sound, *written by Lagrange in 1759* [90], *dealt with the solution of the vibrating string problem. Lagrange assumed that the string consists of* $m-1$ *particles whose displacements,* y_n, *are sums of sine functions of various frequencies. This representation is essentially a discrete Fourier sine transform. Note that* ϖ *means* π.

published in 1754 by Clairaut contains what has been described as the first explicit formula for the DFT [74].

The story follows two paths at the beginning of the nineteenth century. Not surprisingly, we might call one path *continuous* and the other *discrete*. On the continuous path, in 1807 Fourier presented his paper before the Paris Academy, in which he asserted that an arbitrary function can be represented as an infinite series of sines and cosines. The paper elicited only mild encouragement from the judges and the suggestion that Fourier refine his work and submit it for the grand prize in 1812. The Academy's panel of judges for the grand prize included Lagrange, Laplace, and Legendre, and they did award Fourier the grand prize in 1812, but not without reservations. Despite the fact that Euler and Bernoulli had introduced trigonometric representations of functions, and that Lagrange had already produced what we would call a Fourier series solution to the wave equation, Fourier's broader claim that an *arbitrary function* could be given such a representation aroused skepticism, if not outrage. The grand prize came with the deflating assessment that

> the way in which the author arrives at his equations is not exempt from difficulties, and his analysis still leaves something to be desired, be it in generality, or be it even in rigor.

Historians are divided over how much credit is due to Lagrange for the discovery of Fourier series. One historian [154] has remarked that

> certainly Lagrange could find nothing new in Fourier's theorem except the sweeping generality of its statement and the preposterous legerdemain advanced as a proof.

Fourier's work on heat conduction and the supporting theory of trigonometric series culminated in the *Théorie analytique de la chaleur* [60], which was published in 1822. Since that year, the work of Fourier has been given more generous assessments. Clerk Maxwell called it a "great mathematical poem." The entire career of William Thompson (Lord Kelvin) was impacted by Fourier's theory of heat, and he proclaimed that

> it is difficult to say whether their uniquely original quality, or their transcendently intense mathematical interest, or their perennially important instructiveness for the physical sciences, is most to be praised.

Regardless of the originality and rigor of the work when it was first presented, there can be little doubt that for almost 200 years, the subject of Fourier analysis has changed the entire landscape of mathematics and its applications [156], [162].

The continuous path did not end with Fourier's work. The remainder of the nineteenth century was an incubator of mathematical thought in Europe. Some of the greatest mathematicians of the period such as Poisson, Dirichlet, and Riemann advanced the theory of trigonometric series and addressed the challenging questions of convergence. The campaign continued into the twentieth century when Lebesgue, armed with his new theory of integration, was able to produce even more general statements about the convergence of trigonometric series.

Let's return to the beginning of the nineteenth century and follow the second path, with all of its intrigue. As mentioned earlier, Clairaut and Lagrange had considered the problem of fitting astronomical data, and because those data had periodic patterns, it was natural to use approximating functions consisting of sines and cosines. Since the data represented *discrete* samples of an unknown function, and since the approximating functions were *finite* sums of trigonometric functions, this work also led to some of the earliest expressions of the discrete Fourier transform.

The work of Lagrange on interpolation was undoubtedly known to the German mathematician Carl Friedrich Gauss, whose prolific stream of mathematics originated in Göttingen. Almost a footnote to Gauss' vast output was his own contribution to trigonometric interpolation, which also contained the discrete Fourier transform. Equally significant is a small calculation buried in his treatise on interpolation [61] that appeared posthumously in 1866 as an unpublished paper. This work has been dated to 1805, and it contains the first clear and indisputable use of the **fast Fourier transform (FFT)**, which is generally attributed to Cooley and Tukey in 1965. Ironically, Gauss' calculation was cited in 1904 in the mathematical encyclopedia of Burkhardt [25] and again in 1977 by Goldstine [67]. The entire history of the FFT was recorded yet again in 1985 in a fascinating piece of mathematical sleuthing by Heideman, Johnson, and Burrus [74], who remark that "Burkhardt's and Goldstine's works went almost as unnoticed as Gauss' work itself."

To introduce the discrete Fourier transform and to recognize a historical landmark, it would be worthwhile to have a brief look at the problem that Gauss was working on when he resorted to his own fast Fourier transform. Around 1800 Gauss became interested in astronomy because of the discovery of the asteroid Ceres, the orbit of which he was able to determine with great accuracy. At about the same time, Gauss obtained the data presented in Table 1.1 for the position of the asteroid Pallas.

TABLE 1.1

Data on the asteroid Pallas used by Gauss, reproduced by Goldstine from tables of Baron von Zach. Reprinted here, by permission, from [H. H. Goldstine, A History of Numerical Analysis from the 16th through the 19th Century, Springer–Verlag, Berlin, 1977]. ©1977, Springer–Verlag.

Ascension, θ (degrees)	0	30	60	90	120	150
Declination, X (minutes)	408	89	−66	10	338	807
Ascension, θ (degrees)	180	210	240	270	300	330
Declination, X (minutes)	1238	1511	1583	1462	1183	804

Departing from Gauss' notation, we will let θ and X represent the ascension (still measured in degrees) and the declination, respectively. We will also let (θ_n, X_n) denote the actual data pairs where $n = 0, \ldots, 11$. Since the values of X appear to have a periodic pattern as they vary with θ, the goal is to fit these twelve data points with a trigonometric expression of the form

$$X = f(\theta) = a_0 + \sum_{k=1}^{5} \left[\left(a_k \cos\left(\frac{2\pi k\theta}{360}\right) + b_k \sin\left(\frac{2\pi k\theta}{360}\right) \right) \right] + a_6 \cos\left(\frac{2\pi \cdot 6 \cdot \theta}{360}\right).$$

Notice that there are twelve unknown coefficients (the a_k's and b_k's), each of which multiplies a sine or cosine function (called **modes**) with a certain period or frequency. The fundamental period (represented by the coefficients a_1 and b_1) is 360 degrees; the modes with $k > 1$ have smaller periods of $360/k$ degrees.

Twelve conditions are needed to determine the twelve coefficients. Those conditions are simply that the function f match the data; that is,

$$f(\theta_n) = X_n$$

for $n = 0 : 11$ (throughout the book the notation $n = N_1 : N_2$ will be used to indicate that the integer index n runs through all consecutive integers between and including the integers N_1 and N_2). These conditions amount to a system of twelve linear equations that even in Gauss' day could have been solved without undue exertion. Nevertheless Gauss, either in search of a shortcut or intrigued by the symmetries of the sine and cosine functions, discovered a way to collect the coefficients and equations in three subproblems that were easily solved. The solutions to these subproblems were then combined into a final solution. Goldstine [67] has reproduced this remarkable computation, and it is evident that the splitting process discovered by Gauss lies at the heart of the modern fast Fourier transform method.

At the moment we are most interested in the outcome and interpretation of Gauss' calculation. Either by solving a 12×12 linear system or by doing a small FFT, the coefficients of the interpolation function are given in Table 1.2.

TABLE 1.2
Coefficients in Gauss' trigonometric interpolating function

k	0	1	2	3	4	5	6
a_k	780.6	−411.0	43.4	−4.3	−1.1	.3	.1
b_k	–	−720.2	-2.2	5.5	−1.0	−.3	–

A picture is far more illuminating than the numbers. Figure 1.2 shows the data (marked as ×'s) and the function f plotted as a smooth curve. Notice that the interpolation function does its job: it passes through the data points. Furthermore, the function and its coefficients exhibit the "frequency structure" of the data. The coefficients with the largest magnitude belong to the constant mode (corresponding to a_0), and to the fundamental mode (corresponding to a_1 and b_1). However, contributions from the higher frequency modes are needed to represent the data exactly.

His use of the FFT aside, Gauss' fitting of the data with a sum of trigonometric functions is an example of a DFT and is a prototype problem for this book. As we will see shortly, the "input" for such a problem may originate as data points (as in this example) or as a function defined on an interval. In the latter case, the process of finding the coefficients in the interpolating function is equivalent to approximating either the Fourier series coefficients or the Fourier transform of the given function. We will explore all of these uses of the DFT in the remainder of this book.

1.2. An Application

We now turn to a practical application of the DFT to a problem of **data analysis** or, more specifically, **time series analysis**. Even with many details omitted, the example offers a vivid demonstration of some of the fundamental uses of the DFT, namely spectral decomposition and filtering. On a recent research expedition to the Trapridge Glacier in the Yukon Territory, Canada, data were collected by sensors in the

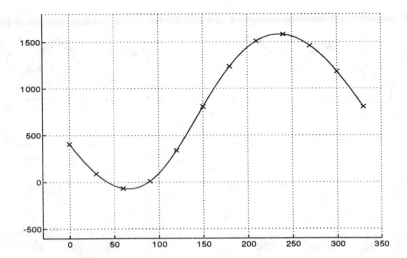

FIG. 1.2. *The 12 data points (marked by ×'s) describe the position of the asteroid Pallas. The declination (in minutes) of Pallas is plotted along the vertical axis against the ascension (in degrees). A weighted sum of 12 sine and cosine functions with different periods is used to fit the data (smooth curve). The coefficients in that weighted sum are determined by a discrete Fourier transform.*

bed of the glacier, 80 meters below the surface [132]. In particular, measurements of the turbidity (amount of suspended material) of the subglacial water were taken every $\Delta t = 10$ minutes $\approx .0069$ days. When plotted, these data produce the jagged curve shown in the upper left graph of Figure 1.3 (time increases to the right and turbidity values are plotted on the vertical axis). The curve actually consists of $N = 368$ data points, which represent $N\Delta t = 2.55$ days of data.

Notice that the data exhibit both patterns *and* irregularities. On the largest scale we see a wave-like pattern with a period of approximately one day; this is an explicable diurnal variation in the water quality variables of the glacier that has a relatively *low frequency*. On a smaller time scale the data appear to be infected with a *high frequency* oscillation which is attributable partly to instrument noise. Between these high and low frequency patterns, there may be other effects that occur on intermediate time scales.

One of the most revealing analyses that can be done on a set of data, or any "signal" for that matter, is to decompose the data according to its various frequencies. This procedure is often called **spectral analysis** or **spectral decomposition**, and it provides an alternative picture of the data in the **frequency domain**. This frequency picture tells how much of the variability of the data is comprised of low frequency waves and how much is due to high frequency patterns. Using the glacier data, here is how it works. The first question that needs to be answered concerns the range of frequencies that appear in a data set such as that shown in the top left graph of Figure 1.3. Just a few physical arguments provide the answer. We will let the total time interval spanned by the data set be denoted A; for the glacier data, we have $A = 2.55$ days. As mentioned earlier, we will let the spacing between data points be denoted Δt. These two parameters, the length of the time interval A, and the so-called **sampling rate** Δt are absolutely crucial in the following arguments. Here are the key observations that give us the range of frequencies.

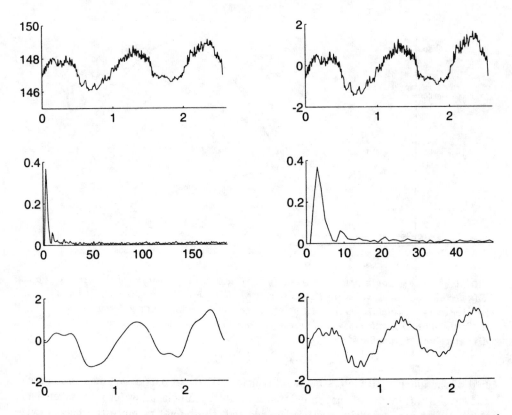

FIG. 1.3. *The above graphs show how the DFT can be used to carry out a spectral analysis of a data set. The original data set consisting of $N = 368$ subglacial water turbidity measurements taken every $\Delta t = 10$ minutes is shown in the top left figure. The same data set with the "mean subtracted out" is shown in the top right figure. The center left figure shows the spectrum of the data after applying the DFT. The horizontal axis now represents frequency with values between $\omega_1 = .39$ and $\omega_{184} = 72$ days^{-1}. The relative weight of each frequency component in the overall structure of the turbidity data is plotted on the vertical axis. The center right figure is a close-up of the spectrum showing the low frequency contributions. By filtering the high frequency components, the data can be smoothed. The lower left and right figures show reconstructions of the data using the lowest 10 and lowest 50 frequencies, respectively.*

If we collect data for 2.55 days, the longest complete oscillation that we can resolve lasts for precisely 2.55 days. Patterns with periods greater than 2.55 days cannot be fully resolved within a 2.55 day time frame. What is the frequency of this longest wave? Using the units of periods per day for frequency, we see that this wave has a frequency of

$$\omega_1 = \frac{1}{A} = \frac{1 \text{ period}}{2.55 \text{ days}} = .39 \text{ days}^{-1}.$$

The subscript 1 on ω_1 means that this is the first (and lowest) of an entire set of frequencies. Therefore, the length of the time interval A days determines the lowest frequency in the system.

We now consider periodic patterns with higher frequencies. A wave with two full periods over A days has a frequency of $\omega_2 = 2/A$ days^{-1}. Three full periods over the interval corresponds to a frequency of $\omega_3 = 3/A$ days^{-1}. Clearly, the full set of frequencies is given by

$$\omega_k = \frac{k}{2.55} = \frac{k}{A} = \text{days}^{-1}.$$

Having determined where the range of frequencies begins, an even more important consideration is where it ends. Not surprisingly, the sampling rate Δt determines the maximum frequency. One might argue that there should be no maximum frequency, since oscillations continue on ever-decreasing time scales all the way to molecular and atomic levels. However, when we take samples every $\Delta t = 10$ minutes, there *is* a limit to the time scales that can be resolved. In pages to come, we will demonstrate in many ways the following fundamental fact: if a signal (or phenomenon of any kind) is sampled once every Δt units, then a wave with a period less than $2\Delta t$ cannot be resolved accurately. The most detail that we can see with a "shutter speed" of Δt is a wave that has a peak at one sample time, a valley at the next, and a peak at the next. This "up-down-up" pattern has a period of $2\Delta t$ units, and a frequency of

$$\omega_{max} = 1/(2\Delta t) \quad \text{units}^{-1}.$$

Where does this maximum frequency fit into the set of frequencies that we denoted $\omega_k = k/A$? Notice that $\Delta t = A/N$, where N is the total number of data points. Therefore

$$\omega_{max} = \frac{1}{2\Delta t} = \frac{1}{2 \cdot A/N} = \frac{N}{2} \cdot \frac{1}{A} = \omega_{\frac{N}{2}}.$$

In other words, the highest frequency that can be resolved with a sample rate of $\Delta t = A/N$ is the $N/2$ frequency in the set of frequencies. (If N is odd, a modified argument can be made, as shown in Chapter 3.) In summary, the $N/2$ frequencies that can be resolved over an interval of A units with a sample rate of Δt units are given by

$$\omega_1 = \frac{1}{A}, \ldots, \omega_k = \frac{k}{A}, \ldots, \omega_{\frac{N}{2}} = \frac{1}{2\Delta t} = \frac{N}{2A}.$$

For the glacier data set, the numerical values of these frequencies that can be resolved are

$$\omega_1 = .39, \quad \omega_2 = .78, \ldots, \omega_k = .39 \, k, \ldots, \omega_{184} = 72 \quad \text{days}^{-1}.$$

We can now turn to an analysis of the turbidity data. A common first step in data or signal analysis is to subtract out the mean. This maneuver amounts to computing the average of the turbidity values and subtracting this average from all of the turbidity values. The result of this rescaling is shown in the top right plot of Figure 1.3; the data values appear as fluctuations about the mean value of zero. With this adjusted data set, we can perform the frequency or spectral decomposition of the data. The details of this calculation will be omitted, since it involves the discrete Fourier transform, which is the subject of the remainder of the book! However, we can examine the output of this calculation and appreciate the result. The center left graph of Figure 1.3 shows what is commonly called the **spectrum** or **power spectrum** of the data set. The horizontal axis of this graph is no longer *time*, but rather *frequency*. Notice that $N/2 = 184$ frequencies are represented on this axis. On the vertical axis is a measure of the relative weight of each of the frequencies in the overall structure of the turbidity data. Clearly, most of the "energy" in the data resides in the lower frequencies, that is, the ω_k's with $k < 10$. The center right graph of Figure 1.3 is a close-up of the spectrum for the lower frequencies. In this picture it is evident that the dominant frequencies are ω_2 and ω_3, which correspond to periods of 1.25 days and 0.83 days. These periods are the nearest to the prominent diurnal (daily) oscillation that we observed in the data at the outset. Thus the spectral decomposition does capture both the obvious patterns in the data, as well as other hidden frequency patterns that are not so obvious visually.

What about the experimental "noise" that also appears in the data? This noise shows up as all of the high frequency ($k >> 10$) contributions in the spectrum plot. This observation brings us to another strategy. *If* we believe that the high frequency oscillations really are spurious and contain no useful physical information, we might be tempted to **filter** these oscillations out of the data. Using the original data set, this could be a delicate task, since the high frequency part of the signal is spread throughout the data set. However, with the spectral decomposition plot in front of us, it is an easy task, since the high frequency noise is localized in the high frequency part of the plot.

The process of filtering is a science and an art that goes far beyond the meat-cleaver approach that we are about to adopt. Readers who wish to see the more refined aspects of filtering should consult one of many excellent sources [20], [70], [108]. With apologies aside, the idea of filtering can be demonstrated by simply removing all of the high frequency contributions in the spectrum above a chosen frequency. This method corresponds to using a sharp low-pass filter (low frequencies are allowed to pass) that truncates the spectrum.

With this new truncated spectrum, it is now possible to reconstruct the data set in an inverse operation that also requires the use of the DFT. Omitting details, we can observe the outcome of this reconstruction in the lower plots of Figure 1.3. The first of these two figures shows the result of removing all frequencies above ω_{10}. Notice that all of the noise in the data has been eliminated, and indeed even the low frequency oscillations have been smoothed considerably. The second of the reconstructions uses all frequencies below ω_{50}; this filter results in a slightly sharper replica of the low frequency oscillations, with more of the high frequency noise also evident.

This discussion of spectral decomposition and filtering has admittedly been rather sparse in details and technicalities. Hopefully it has illustrated a few fundamental concepts that will recur endlessly in the pages to come. In summary, given a function or data set defined in a **time** or **spatial domain**, the DFT serves as a tool that gives a frequency picture of that function or data set. The frequency picture depends on the length of the interval A on which the original data set is defined, and it also depends on the rate Δt that is used to sample the data. Once the data set is viewed in the frequency domain, there are operations such as filtering that can be applied if necessary. Equally important is the fact that if the problem calls for it, there is always a path back to the original domain; this passage also uses the DFT. All of these ideas will be elaborated and explored in the pages to come. It is now time to dispense with generalities and history, and look at the DFT in all of its power and glory.

1.3. Problems

1. Gauss' problem revisited.   Consider the following simplified version of Gauss' data fitting problem. Assume that the following $N = 4$ data pairs (x_n, y_n)

are collected:

$$(0, y_0), \quad \left(\frac{\pi}{2}, y_1\right), \quad (\pi, y_2), \quad \left(\frac{3\pi}{2}, y_3\right),$$

where the values y_0, y_1, y_2, and y_3 are known, but will be left unspecified. The data set is also assumed to be periodic with period of 2π, so one could imagine another data point $(2\pi, y_4) = (2\pi, y_0)$ if necessary. This data set is to be fit with a function of the form

$$y = f(x) = a_0 + a_1 \cos x + b_1 \sin x + a_2 \cos 2x, \qquad (1.1)$$

where the four coefficients a_0, a_1, b_1, a_2 are unknown, but will be determined from the data.

(a) Verify that each of the individual functions in the representation (1.1), namely, $f_0(x) = 1$, $f_1(x) = \cos x$, $f_2(x) = \sin x$, and $f_3(x) = \cos(2x)$ has a period of 2π (that is, it satisfies $f(x) = f(x + 2\pi)$ for all x).

(b) Conclude that the entire function f also has a period of 2π.

(c) Plot each of the individual functions f_0, f_1, f_2, and f_3, and note the period and frequency of each function.

(d) To find the coefficients a_0, a_1, b_1, and a_2 in the representation (1.1), impose the **interpolating conditions**

$$f(x_n) = y_n$$

for $n = 0, 1, 2, 3$, where $x_n = n\pi/2$. Write down the resulting system of linear equations for the unknown coefficients (recalling that y_0, y_1, y_2, and y_3 are assumed to be known).

(e) Use any available means (pencil and paper will suffice!) to solve for the coefficients a_0, a_1, b_1, and a_2 given the data values

$$y_0 = 2, \quad y_1 = 1, \quad y_2 = 3, \quad y_3 = 2.$$

(f) Plot the resulting function f and verify that it does pass through each of the four data points.

2. The complex exponential. Given the Euler relation

$$e^{i\theta} = \cos\theta + i\sin\theta,$$

verify that

(a) $e^{im\theta} = \cos(m\theta) + i\sin(m\theta)$,

(b) $e^{i\theta}e^{i\phi} = e^{i(\theta+\phi)} = \cos(\theta + \phi) + i\sin(\theta + \phi)$,

(c) $\cos\theta = \dfrac{e^{i\theta} + e^{-i\theta}}{2}$ and $\sin\theta = \dfrac{e^{i\theta} - e^{-i\theta}}{2i}$ (also called Euler relations),

(d) $e^{i2\pi m/N} = 1$ if m is any integer multiple of the integer N,

(e) $e^{i2\pi nk/N} = e^{i2\pi(n+mN)k/N}$ where m and N are any integers.

3. Modes, frequencies, and periods. Consider the functions (or modes) $v_k(x) = \cos(\pi kx/2)$ and $w_k(x) = \sin(\pi kx/2)$ on the interval $[-2, 2]$.

(a) Plot on separate sets of axes (hand sketches will do) the functions v_k and w_k for $k = 0, 1, 2, 3, 4$. In each case, note the period of the mode and the frequency of the mode. How are the period and frequency of a given mode related to the index k?

(b) Now consider the grid with grid spacing (or sampling rate) $\Delta x = 1/2$ consisting of the points $x_n = n/2$ where $n = -4 : 4$. Mark this grid on the plots of part (a) and indicate the sampled values of the mode.

(c) What are the values of v_0 and w_0 at the grid points?

(d) What are the values of v_4 and w_4 at the grid points?

(e) Plot v_5 on the interval $[-2, 2]$. What is the effect of sampling v_5 at the grid points? Compare the values of v_5 and the values of v_3 at the sample points.

4.   Sampling and frequencies.   Throughout this book, we have, somewhat arbitrarily, used Δx to denote a grid spacing or sampling rate. This notation suggests that the sampling takes place on a *spatial* domain. However, sampling can also take place on *temporal* domains, as illustrated by the glacier data example in the text. In this case Δt is a more appropriate notation for the grid spacing.

(a) Imagine that you have collected $N = 140$ equally spaced data values (for example, temperatures or tidal heights) over a time interval of duration $A = 14$ days. What is the grid spacing (or sampling rate) Δt? What is the minimum frequency ω_1 that is resolved by this sampling? What is the maximum frequency ω_{70} that is resolved by this sampling? What are the units of these frequencies?

(b) Imagine that you have collected $N = 120$ equally spaced data values (for example, material densities or beam displacements) over a spatial interval of length $A = 30$ centimeters. What is the grid spacing Δx? What is the minimum frequency ω_1 that is resolved by this sampling? What is the maximum frequency ω_{60} that is resolved by this sampling? What are the units of these frequencies? Note that in spatial problems frequencies are often referred to as *wavenumbers*.

5. Complex forms.   It will be extremely useful to generalize problem 1 and obtain a sneak preview of the DFT. There are some advantages to formulating the above interpolating conditions in terms of the complex exponential. Show that with $N = 4$ the real form of the interpolating conditions

$$y_n = a_0 + a_1 \cos\left(\frac{2\pi n}{N}\right) + b_1 \sin\left(\frac{2\pi n}{N}\right) + a_2 \cos\left(\frac{4\pi n}{N}\right)$$

for $n = 0, 1, 2, 3$ may also be written in the form

$$y_n = \sum_{k=0}^{3} c_k e^{i2\pi nk/N}$$

for $n = 0, 1, 2, 3$, where the coefficients c_0, c_1, c_2, and c_3 may be complex-valued. Assuming that all of the data values are real, show that c_0 and c_2 are real-valued, and that $\text{Re}\{c_1\} = \text{Re}\{c_3\}$ and $\text{Im}\{c_1\} = -\text{Im}\{c_3\}$. Then the following relations between the coefficients $\{a_0, a_1, b_1, a_2\}$ and the coefficients $\{c_0, c_1, c_2, c_3\}$ result:

$$a_0 = c_0, \quad a_1 = 2\text{Re}\{c_1\}, \quad b_1 = -2\text{Im}\{c_1\}, \quad a_2 = c_2.$$

6. More general complex forms. To sharpen your skills in working with complex exponentials, carry out the calculation of the previous problem for a data set with N points in the following two cases.

(a) Assume that the real data pairs (x_n, y_n) are given for $n = 0, \ldots, N - 1$ where $x_n = 2\pi n/N$. Show that the interpolating conditions take the real form

$$y_n = a_0 + \sum_{k=1}^{\frac{N}{2}-1} \left[a_k \cos\left(\frac{2\pi n k}{N}\right) + b_k \sin\left(\frac{2\pi n k}{N}\right) \right] + a_{\frac{N}{2}} \cos(\pi n),$$

for $n = 0 : N - 1$. Show that these conditions are equivalent to the conditions in complex form

$$y_n = \sum_{k=0}^{N-1} c_k e^{i2\pi n k/N}$$

for $n = 0 : N - 1$, where c_0 and $c_{\frac{N}{2}}$ are real, $\text{Re}\{c_k\} = \text{Re}\{c_{N-k}\}$, and $\text{Im}\{c_k\} = -\text{Im}\{c_{N-k}\}$. The relations between the real coefficients $\{a_k, b_k\}$ and the complex coefficients c_k are $a_0 = c_0$, $a_{\frac{N}{2}} = c_{\frac{N}{2}}$, and

$$a_k = 2\text{Re}\{c_k\}, \quad b_k = -2\text{Im}\{c_k\}$$

for $k = 1 : N/2 - 1$.

(b) Now assume the N real data pairs (x_n, y_n) are given for $n = -N/2 + 1, \ldots, N/2$, where $x_n = 2\pi n/N$. Notice that the sampling interval is now $[-\pi, \pi]$. The data fitting conditions take the real form

$$y_n = a_0 + \sum_{k=1}^{\frac{N}{2}-1} \left[a_k \cos\left(\frac{2\pi n k}{N}\right) + b_k \sin\left(\frac{2\pi n k}{N}\right) \right] + a_{\frac{N}{2}} \cos(\pi n),$$

where $n = -N/2 + 1 : N/2$. Show that these conditions are equivalent to the conditions in complex form

$$y_n = \sum_{k=-\frac{N}{2}+1}^{\frac{N}{2}} c_k e^{i2\pi n k/N}$$

for $n = -N/2+1 : N/2$, where c_0 and $c_{\frac{N}{2}}$ are real, $\text{Re}\{c_k\} = \text{Re}\{c_{-k}\}$, and $\text{Im}\{c_k\} = -\text{Im}\{c_{-k}\}$. The relations between the real coefficients $\{a_k, b_k\}$ and the complex coefficients c_k are $a_0 = c_0$, $a_{\frac{N}{2}} = c_{\frac{N}{2}}$, and

$$a_k = 2\text{Re}\{c_k\}, \quad b_k = -2\text{Im}\{c_k\}$$

for $k = 1 : N/2 - 1$.

We will see shortly that all of the above procedures for determining the coefficients c_k from the data values y_n (or vice-versa) are versions of the DFT.

7. Explain those figures.   The cover and frontmatter of this book display several polygonal, mandala-shaped figures which were generated using the DFT. With just a small preview of things to come, we can explain these figures. As will be shown early in the next chapter, given a sequence $\{f_n\}_{n=0}^{N-1}$, one form of the DFT of this sequence is the sequence $\{F_k\}_{k=0}^{N-1}$, where

$$F_k = \frac{1}{N} \sum_{n=0}^{N-1} f_n e^{-i2\pi nk/N}.$$

(a) Let $N = 4$ for the moment and compute the DFT of the unit vectors $(1,0,0,0)$, $(0,1,0,0)$, $(0,0,1,0)$, and $(0,0,0,1)$. Verify that the corresponding DFTs are $\frac{1}{4}(1,1,1,1)$, $\frac{1}{4}(1,-i,1,i)$, $\frac{1}{4}(1,-1,1,-1)$, and $\frac{1}{4}(1,i,-1,-i)$.

(b) Plot each of these DFT vectors by regarding each component of a given vector as a point in the complex plane and draw a connecting segment from each component to its predecessor, cyclically (i.e., $F_0 \to F_1 \to F_2 \to F_3 \to F_1$).

(c) Repeat this process with the eight unit vectors $\hat{e}_j = (0,0,\ldots,1,0,0,\ldots,0)$, where the 1 is in the jth component. Show that the 8-point DFT of $\hat{e}_j$ is

$$F_k = \frac{1}{8} \left(\cos\left(\frac{\pi jk}{4}\right) - i\sin\left(\frac{\pi jk}{4}\right) \right),$$

which holds for $k = 0 : 7$.

(d) Verify that plotting each of these eight DFTs in the complex plane (connecting end-to-end as described above) produces a figure similar to the ones shown on the cover and elsewhere.

Chapter **2**

The Discrete Fourier Transform

Our life is an apprenticeship to the truth that around every circle another can be drawn.
— Ralph Waldo Emerson

15

2.1.   Introduction

We must first agree on what the words **discrete Fourier**[1] **transform** mean. What, exactly, is the DFT? The simple response is to give a formula, such as

$$F_k = \frac{1}{N} \sum_{n=-\frac{N}{2}+1}^{\frac{N}{2}} f_n\, e^{-i2\pi nk/N}, \tag{2.1}$$

and state that this holds for k equal to any N consecutive integers. Equation (2.1) is, in fact, a definition that we shall use, but such a response sheds no light on the original question. What, then, is the DFT? Is it a Fourier transform, as its name might imply? If it is not a Fourier transform, does it approximate one? The adjective *discrete* suggests that it may be more closely related to the Fourier series than to the continuous Fourier transform. Is this the case? There are no simple answers to these questions. Viewed from certain perspectives, the DFT is each of these things. Yet from other vantage points it presents different faces altogether. Our intent is to arrive at an answer, but we shall not do so in the span of two or three pages. In fact, the remainder of this book will be devoted to formulating an answer.

We will begin with a derivation of a DFT formula. Notice that we did not say *the* derivation of *the* DFT. There are almost as many ways to derive DFTs as there are applications of them. In the course of this chapter, we will arrive at the DFT along the following four paths:

- approximation to the Fourier transform of a function,

- approximation to the Fourier coefficients of a function,

- trigonometric approximation, and

- the Fourier transform of a spike train.

This approach may seem a bit overzealous, but it does suggest the remarkable generality of the DFT and the way in which it underlies all of Fourier analysis. The first topic will be the DFT as an approximation to the Fourier transform. This choice of a starting point is somewhat arbitrary, but it does have the advantage that the absolutely fundamental **reciprocity relations** can be revealed as early as possible. This first derivation of the DFT is followed by the official presentation of the DFT, its orthogonality properties, and the essential inverse DFT relation. We then present three alternative, but illuminating derivations of the DFT within the frameworks of Fourier series, trigonometric approximation, and distribution theory. Those who feel that one derivation of the DFT suffices can pass over the remaining three and proceed directly to the last section of the chapter. However, we feel that all four paths provide important and complimentary views of the DFT and deserve at least a quick reading. While there is certainly technical material throughout this chapter, our intent is to present some fairly qualitative features of the DFT, and return to technical matters, deeper properties, and applications in later chapters.

---

[1] Born in Auxerre, France in 1768, orphaned at the age of eight, JEAN BAPTISTE JOSEPH FOURIER was denied entrance to the artillery because he was not of noble birth, "although he were a second Newton." Fourier was an active supporter of the French Revolution and accompanied Napoleon to Egypt, where he served as secretary of the Institute of Egypt. He died of an aneurism on May 16, 1830.

2.2.   DFT Approximation to the Fourier Transform

A natural problem to examine first is the approximation of the Fourier transform of a (possibly complex-valued) function f of a real variable x. We should recognize that, in practice, f may not appear explicitly as a function, but may be given as a set of discrete data values. However, for the moment let's assume that f is defined on the interval $(-\infty, \infty)$ and has some known properties, one of which is that it is absolutely integrable on the real line. This means that

$$\int_{-\infty}^{\infty} |f(x)| \, dx \; < \; \infty.$$

Then we may define a function $\hat{f}(\omega)$ by

$$\hat{f}(\omega) = \int_{-\infty}^{\infty} f(x)e^{-i2\pi\omega x} dx \, , \tag{2.2}$$

where $-\infty < \omega < \infty$ and $i = \sqrt{-1}$. We have chosen to offend half of our audience by letting i (the mathematician's choice), rather than j (the engineer's preference), be the imaginary unit! The function $\hat{f}$ is called the **Fourier transform** of f and is uniquely determined by (2.2). The transform $\hat{f}$ is said to be defined in the **frequency domain** or **transform domain**, while the input function f is said to be defined in the **spatial domain** if x is a spatial coordinate, or in the **time domain** if f is a time-dependent function. Of tremendous importance is the fact that there is also an inverse relationship between f and $\hat{f}$ [13], [55], [88], [111], [160] given by

$$f(x) = \int_{-\infty}^{\infty} \hat{f}(\omega)e^{i2\pi\omega x} d\omega. \tag{2.3}$$

This relationship gives f as the **inverse Fourier transform** of $\hat{f}(\omega)$.

Uninitiated readers should not feel disadvantaged with this sudden exposure to Fourier transforms. In fact, it can be argued that an understanding of the DFT can lead to a better understanding of the Fourier transform. Let's pause to give a brief physical interpretation of the Fourier transform, which will be applicable to the DFT. It all begins with a look at the **kernel** of the Fourier transform, which is the term $e^{-i2\pi\omega x}$. Similarly, the kernel of the inverse Fourier transform is $e^{i2\pi\omega x}$. Using the Euler[2] formula, these kernels may be written

$$e^{\pm i2\pi\omega x} \; = \; \cos(2\pi\omega x) \pm i\sin(2\pi\omega x).$$

We see that for a fixed value of ω, the kernel consists of waves (sines and cosines) with a **period** (or **wavelength**) of $1/\omega$, measured in the units of x (either length or time). These waves are called **modes**. Equivalently, the modes corresponding to a fixed value of ω have a **frequency** of ω periods per unit length or cycles per unit

---

[2]LEONHARD EULER (1707–1783), among the most prolific and accomplished mathematicians of his or any other time, has lent his name to equations, relations, theorems, methods, formulas, numbers, and curiosities in every branch of mathematics. Within the 886 books and papers that make up Euler's work (nearly half published posthumously) are found what we know today as Euler's relation, the Euler differential equation, Euler's method for solving quartics, the Euler method for numerical solution of ODEs; the list seems endless.

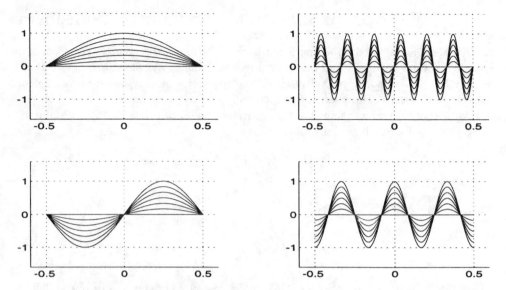

FIG. 2.1. *All of Fourier analysis (the DFT, Fourier series, and Fourier transforms) concerns the representation of functions or data in terms of modes consisting of sines and cosines with various frequencies. The figure shows some typical modes of the form $A\cos(2\pi\omega x)$ or $A\sin(2\pi\omega x)$. The amplitude, A, varies between 0 and 1. Counterclockwise from the top left are $\cos(\pi x)$, with frequency $\omega = 1/2$ cycles/unit and period (wavelength) 2 units; $\sin(2\pi x)$, with frequency $\omega = 1$ and wavelength 1; $\cos(6\pi x)$, with frequency $\omega = 3$ and a wavelength $1/3$; and $\sin(12\pi x)$ with frequency $\omega = 6$ and wavelength $1/6$.*

time (the combination cycles per second is often called **Hertz**[3]). Figure 2.1 shows several typical modes of the form $\cos(2\pi\omega x)$ or $\sin(2\pi\omega x)$ that might be used in the representation of a function.

The inverse Fourier transform relationship (2.3) can be regarded as a recipe for assembling the function f as a combination of modes of all frequencies $-\infty < \omega < \infty$. The mode associated with a particular frequency ω has a certain weight in this combination, and this weight is given by $\hat{f}(\omega)$. This process of assembling a function from all of the various modes is often called **synthesis**: given the weights of the modes $\hat{f}(\omega)$, the function f can be constructed. The complete set of values of $\hat{f}$ is also called the **spectrum** of f since it gives the entire "frequency content" of the function or signal f. Equally important is the opposite process, often called **analysis**: given the function f, we can find the amount, $\hat{f}(\omega)$, of the mode with frequency ω that is present in f. Analysis can be done by applying the forward transform (2.2). This interpretation is illustrated in Figure 2.2, in which a real-valued function

$$f(x) = e^{-|x|}\cos(\pi x)$$

is shown together with its (in this case) real-valued Fourier transform,

$$\hat{f}(\omega) = \frac{1}{4\pi^2(\omega^2 - \omega) + \pi + 1} + \frac{1}{4\pi^2(\omega^2 + \omega) + \pi + 1}.$$

---

[3]HEINRICH HERTZ was born in 1857 and studied under Helmholtz and Kirchhoff. His experimental work in electrodynamics led to the discovery of Hertzian waves, which Marconi later used in his design of the radio. Hertz was a professor of physics at the University of Bonn and died a month before his 37th birthday.

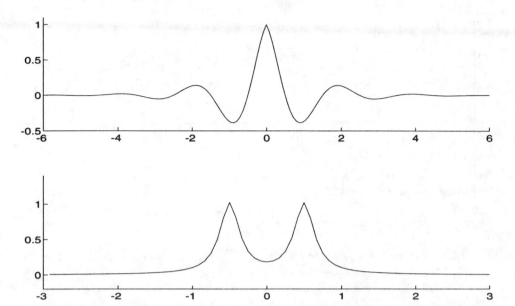

FIG. 2.2. *A real-valued function* $f(x) = e^{-|x|}\cos(\pi x)$, *defined on* $(-\infty, \infty)$, *is shown (for* $x \in [-6, 6]$) *in the top figure. Its Fourier transform* $\hat{f}$ *is also real-valued, and is shown (for* $\omega \in [-3, 3]$) *in the bottom figure. The value of the Fourier transform* $\hat{f}(\omega)$ *gives the amount by which the modes* $\cos(2\pi\omega x)$ *and* $\sin(2\pi\omega x)$ *are weighted in the representation of* f.

With this capsule account of the Fourier transform, let's see how the DFT emerges as a natural approximation. First a practical observation is needed: when a function is given it is either already limited in extent (for example, f might represent an image that has well-defined boundaries) *or*, for the sake of computation, f must be assumed to be zero outside some finite interval. Therefore, for the moment we will assume that $f(x) = 0$ for $|x| > A/2$. The Fourier transform of such a function with limited extent is given by

$$\hat{f}(\omega) = \int_{-\infty}^{\infty} f(x)e^{-i2\pi\omega x}\,dx = \int_{-\frac{A}{2}}^{\frac{A}{2}} f(x)e^{-i2\pi\omega x}\,dx. \qquad (2.4)$$

It is this integral that we wish to approximate numerically.

To devise a method of approximation, the interval of integration $[-A/2, A/2]$ is divided into N subintervals of length $\Delta x = A/N$. Assuming for the moment that N is even, a grid with $N + 1$ equally spaced points is defined by the points $x_n = n\Delta x$ for $n = -N/2 : N/2$. Thus the set of grid points is

$$x_{-\frac{N}{2}} = -\frac{A}{2}, \ldots, x_0 = 0, \ldots, x_{\frac{N}{2}} = \frac{A}{2}.$$

We now assume that the function f is known at the grid points (in fact, f might be known *only* at these points). Letting the integrand be

$$g(x) = f(x)e^{-i2\pi\omega x},$$

we may apply the trapezoid rule [22], [80], [166] (see Figure 2.3) to this integral. This

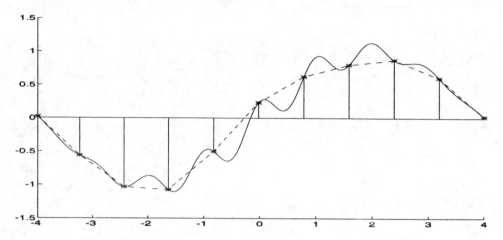

FIG. 2.3. *In order to approximate the integral $\int_{-A/2}^{A/2} g(x)dx$, a grid is established on the interval $[-A/2, A/2]$ consisting of $N+1$ equally spaced points $x_n = n\Delta x$, where $\Delta x = A/N$ and $n = -N/2 : N/2$. Here $A = 8$ while $N = 10$. The trapezoid rule results if the integrand is replaced by straight line segments over each subinterval, and the area of the region under the curve is approximated by the sum of the areas of the trapezoids.*

leads to the approximation

$$\int_{-\frac{A}{2}}^{\frac{A}{2}} g(x)dx \approx \frac{\Delta x}{2}\left\{ g\left(-\frac{A}{2}\right) + 2 \sum_{n=-\frac{N}{2}+1}^{\frac{N}{2}-1} g(x_n) + g\left(\frac{A}{2}\right)\right\}.$$

For now we will add the requirement that $g(-A/2) = g(A/2)$, an assumption that will be the subject of great scrutiny in the pages that follow. With this assumption the trapezoid rule approximation may be written

$$\hat{f}(\omega) = \int_{-\frac{A}{2}}^{\frac{A}{2}} g(x)dx \approx \Delta x \sum_{n=-\frac{N}{2}+1}^{\frac{N}{2}} g(x_n)$$

$$= \frac{A}{N} \sum_{n=-\frac{N}{2}+1}^{\frac{N}{2}} f(x_n)e^{-i2\pi\omega x_n}.$$

At the moment, this approximation can be evaluated for *any* value of ω. Looking ahead a bit, we anticipate approximating $\hat{f}$ only at selected values of ω. Therefore, we must determine how many and which values of ω to use. For the purposes of the DFT, we need the sampled values $f(x_n)$ to determine the approximations to $\hat{f}(\omega)$ uniquely, and vice versa. Since N values of $f(x_n)$ are used in the trapezoid rule approximation, it stands to reason that we should choose N values for ω at which to approximate $\hat{f}$. That is the easier question to answer!

The question of *which* frequency values to use requires a discussion of fundamental importance to the DFT, because it leads to the **reciprocity relations**. It is impossible to overstate the significance of these relations, and indeed they will reappear often in the remainder of this book. The reciprocity relations are the keystone of the DFT that holds its entire structure in place.

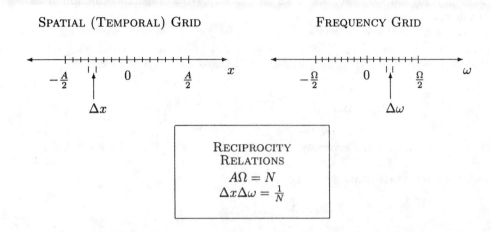

Fig. 2.4. *The spatial (or temporal) grid of the DFT is associated with the frequency grid through the reciprocity relations. With the grid spacings and grids lengths as shown in the figure, these relations state that $\Delta x \Delta \omega = 1/N$ and $A\Omega = N$.*

Ironically, for all of their importance, the reciprocity relations can be exposed with very little effort and stated quite succinctly. We are presently working on a spatial or temporal domain $[-A/2, A/2]$ with grid spacing Δx and grid points $x_n = n\Delta x$. Closely associated with that domain is a frequency domain that we will denote $[-\Omega/2, \Omega/2]$. This frequency domain will also be equipped with a grid consisting of N equally spaced points separated by a distance $\Delta \omega$. We will denote these grid points $\omega_k = k\Delta \omega$, where $k = -N/2+1 : N/2$. The task is to relate the four grid parameters Δx, $\Delta \omega$, A, Ω, assuming that both the spatial and frequency grids consist of N points. The reciprocity relations serve precisely this purpose. Figure 2.4 shows the two grids and their relevant parameters.

Imagine all modes (sines and cosines) that have an integer number of periods on $[-A/2, A/2]$ and fit exactly on the interval. Of these waves, consider the wave with the largest possible period. This wave is often called the **one-mode**, or **fundamental mode**. Clearly, it has one full period on the interval $[-A/2, A/2]$ or a period of A units. What is the frequency of this wave? This wave has a frequency of $1/A$ periods per unit length. This frequency is the *lowest* frequency associated with the interval $[-A/2, A/2]$. Therefore, we will denote this fundamental unit of frequency

$$\Delta \omega = \frac{1}{A},$$

and it will be the grid spacing in the frequency domain. All other frequencies recognized by the DFT will be integer multiples of $\Delta \omega$ corresponding to modes with an integer number of periods on $[-A/2, A/2]$. Since there are N grid points on the frequency interval $[-\Omega/2, \Omega/2]$, and the grid points are separated by $\Delta \omega$, it follows that $\Omega = N\Delta \omega$. Combining these two expressions, we have the first reciprocity relation:

$$\Omega = N\Delta \omega = \frac{N}{A} \quad \text{or} \quad A\Omega = N.$$

This relation asserts that the lengths of the spatial (or temporal) domain and the frequency domain vary inversely with each other. We will examine the implications of this relation in a moment.

A second reciprocity relation can now be found quickly. Since the interval $[-A/2, A/2]$ is covered by N equally spaced grid points separated by Δx, we know that $N\Delta x = A$. Combining this with the fact that $\Delta\omega = 1/A$, we see that

$$\frac{1}{\Delta\omega} = A = N\Delta x \quad \text{or} \quad \Delta x \Delta\omega = \frac{1}{N}.$$

As in the first reciprocity relation, we conclude that the grid spacings in the two domains are also related inversely. Let us summarize with the following definition.

▶ **Reciprocity Relations** ◀

$$A\Omega = N \quad \text{and} \quad \Delta x \Delta\omega = \frac{1}{N}. \tag{2.5}$$

The two reciprocity relations are not independent, but they are both useful. The first relation tells us that if the number of grid points N is held fixed, an increase in the length of the spatial domain comes at the expense of a decrease in the length of the frequency domain. (Remember how the first reciprocity relation was derived: if A is increased, it means that longer periods are allowed on the spatial grid, which means that the fundamental frequency $\Delta\omega$ decreases, which means that the length of the frequency domain $\Omega = N\Delta\omega$ must decrease.) The second reciprocity relation can be interpreted in a similar way. Halving Δx with N fixed also halves the length of the spatial domain. The fundamental mode on the original grid has a frequency of $1/A$ cycles per unit length, while on the new shorter grid it has a frequency of $1/(A/2)$ or $2/A$ cycles per unit length. Thus $\Delta\omega$ is doubled in the process. All sorts of variations on these arguments, allowing N to vary also, can be formulated. They are useful thought experiments that lead to a better understanding of the reciprocity relations (see problems 14 and 15).

With the reciprocity relations established, we can now return to the trapezoid rule approximation and extract the DFT in short order. First we use f_n to denote the sampled values $f(x_n)$ for $n = -N/2 + 1 : N/2$. Then, agreeing to approximate $\hat{f}$ at the frequency grid points $\omega_k = k\Delta\omega = k/A$, we note that

$$x_n\omega_k = (n\Delta x)(k\Delta\omega) = \frac{nA}{N}\frac{k}{A} = \frac{nk}{N}.$$

The sum in the trapezoid rule becomes

$$\hat{f}(\omega_k) \approx \frac{A}{N} \sum_{n=-\frac{N}{2}+1}^{\frac{N}{2}} f(x_n)e^{-i2\pi\omega_k x_n} = \frac{A}{N} \sum_{n=-\frac{N}{2}+1}^{\frac{N}{2}} f_n e^{-i2\pi nk/N}.$$

Therefore, our approximations to the Fourier transform $\hat{f}$ at the frequency grid points $\omega_k = k/A$ are given by

$$\hat{f}(\omega_k) = \hat{f}\left(\frac{k}{A}\right) = \int_{-\frac{A}{2}}^{\frac{A}{2}} f(x)e^{-i2\pi kx/A}dx \approx A\underbrace{\frac{1}{N} \sum_{n=-\frac{N}{2}+1}^{\frac{N}{2}} f_n e^{-i2\pi nk/N}}_{F_k},$$

for $k = -N/2 + 1 : N/2$. The expression on the right above the brace is our chosen definition of the DFT. Given the set of N sample values f_n, the DFT consists of the N coefficients

$$F_k = \frac{1}{N} \sum_{n=-\frac{N}{2}+1}^{\frac{N}{2}} f_n e^{-i2\pi nk/N}$$

for $k = -N/2 + 1 : N/2$. In addition to identifying the DFT, we can conclude that approximations to the Fourier transform $\hat{f}(\omega_k)$ are given by $\hat{f}(\omega_k) \approx A F_k$. This approximation and the errors it entails will be investigated in later chapters. Let us now officially introduce the DFT.

2.3.   The DFT–IDFT Pair

For convenience, we will adopt the notation

$$\omega_N = e^{i2\pi/N} = \cos\left(\frac{2\pi}{N}\right) + i\sin\left(\frac{2\pi}{N}\right),$$

so that

$$\omega_N^{-nk} = e^{-i2\pi nk/N} \quad \text{and} \quad \omega_N^{nk} = e^{i2\pi nk/N}.$$

With this bit of notation in hand, we may define the DFT in the following fashion.

▶ **Discrete Fourier Transform**   ◀

Let N be an even positive integer and let f_n be a sequence[4] of N complex numbers where $n = -N/2 + 1 : N/2$. Then its discrete Fourier transform is another sequence of N complex numbers given by

$$F_k = \frac{1}{N} \sum_{n=-\frac{N}{2}+1}^{\frac{N}{2}} f_n \omega_N^{-nk} \tag{2.6}$$

for $k = -N/2 + 1 : N/2$.

Much of our work will be done with this form of the DFT in which N is assumed to be even. However, there is an analogous version that applies when N is odd and it is worth stating here.

If N is an odd positive integer and f_n is a sequence of N complex numbers where $n = -(N-1)/2 : (N-1)/2$, then its discrete Fourier transform is another sequence of N complex numbers given by

$$F_k = \frac{1}{N} \sum_{n=-\frac{N-1}{2}}^{\frac{N-1}{2}} f_n \omega_N^{-nk} \tag{2.7}$$

for $k = -(N-1)/2 : (N-1)/2$.

---

[4] The terms *sequence* and *vector* will be used interchangeably when referring to the input to the DFT. Although *vector* may seem the more accurate term, *sequence* is also appropriate because the input is often viewed as an infinite set obtained by extending the original set periodically.

The implications of using even and odd values of N will be explored in Chapter 3. We shall often employ the operator notation $\mathcal{D}\{f_n\}$ to mean the DFT of the sequence f_n, and $\mathcal{D}\{f_n\}_k$ to indicate the kth element of the transform, so that $\mathcal{D}\{f_n\}_k = F_k$.

Those readers who are familiar with the literature of DFTs and FFTs will recognize that our definition differs from those that use indices running from 0 to $N-1$. We have not made this choice lightly. There are good arguments to be made in favor of (and against) the use of either definition. Certainly those who view the input to their DFTs as causal (time-dependent) sequences will be more comfortable with the indices $0:N-1$. For other (often spatial) applications, such as image reconstruction from projections, it is more convenient to place the origin in the center of the image space, leading to definitions such as (2.6) and (2.7). Our choice is motivated by the fact that many theoretical explanations and considerations are simpler or more natural with indices running between $-N/2$ and $N/2$. It is important to note, however, that *anything* that can be done using one index set can also be done using the other, and usually with little alteration. For this reason we now present an alternate definition, and will occasionally use it.

▶ **Discrete Fourier Transform (Alternate Form)** ◀

Let N be a positive integer and let f_n be a sequence of N complex numbers where $n = 0:N-1$. Then its discrete Fourier transform is another sequence of N complex numbers given by

$$F_k = \frac{1}{N} \sum_{n=0}^{N-1} f_n \omega_N^{-nk} \qquad (2.8)$$

for $k = 0:N-1$.

As shown in problem 17, the alternate form of the DFT (2.8) is entirely equivalent to the original form (2.6). One obvious advantage of the alternate definition is that it is independent of the parity (even/oddness) of N. There are many other alternate forms of the DFT which will be presented in detail in Chapter 3.

In general, the output of the DFT, F_k, is a complex-valued sequence. Of impending significance is the fact that it is also an N-periodic sequence satisfying $F_k = F_{k\pm N}$ (problem 9). It is frequently convenient to examine the real and imaginary parts of the DFT independently. These two sequences are denoted by $\mathrm{Re}\{F_k\}$ and $\mathrm{Im}\{F_k\}$, respectively. Using the same notation to denote the real and imaginary parts of the input sequence f_n, we observe that the real and imaginary parts of the DFT are given by

$$
\begin{aligned}
F_k &= \frac{1}{N} \sum_{n=-\frac{N}{2}+1}^{\frac{N}{2}} f_n \omega_N^{-nk} \\
&= \frac{1}{N} \sum_{n=-\frac{N}{2}+1}^{\frac{N}{2}} (\mathrm{Re}\{f_n\} + i\mathrm{Im}\{f_n\}) \left(\cos\left(\frac{2\pi nk}{N}\right) - i \sin\left(\frac{2\pi nk}{N}\right) \right) \\
&= \frac{1}{N} \sum_{n=-\frac{N}{2}+1}^{\frac{N}{2}} \left(\mathrm{Re}\{f_n\} \cos\left(\frac{2\pi nk}{N}\right) + \mathrm{Im}\{f_n\} \sin\left(\frac{2\pi nk}{N}\right) \right)
\end{aligned}
$$

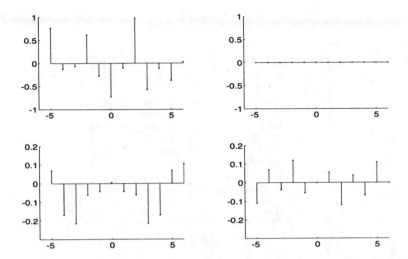

FIG. 2.5. *The real (top left) and imaginary (top right) parts of the 12-point input sequence f_n of Table 2.1 are shown. The real part of F_k is shown on the bottom left, while the imaginary part of F_k is shown on the bottom right. Observe that the real part of F_k is an even sequence satisfying* $\operatorname{Re}\{F_k\} = \operatorname{Re}\{F_{-k}\}$, *while the imaginary part is odd, satisfying* $\operatorname{Im}\{F_k\} = -\operatorname{Im}\{F_{-k}\}$.

$$+ \frac{i}{N} \sum_{n=-\frac{N}{2}+1}^{\frac{N}{2}} \left(\operatorname{Im}\{f_n\} \cos\left(\frac{2\pi nk}{N}\right) - \operatorname{Re}\{f_n\} \sin\left(\frac{2\pi nk}{N}\right) \right).$$

Therefore, the real and imaginary parts of the output F_k are defined as follows.

▶ **Real Form of the DFT** ◀

$$\operatorname{Re}\{F_k\} = \frac{1}{N} \sum_{n=-\frac{N}{2}+1}^{\frac{N}{2}} \operatorname{Re}\{f_n\} \cos\left(\frac{2\pi nk}{N}\right) + \operatorname{Im}\{f_n\} \sin\left(\frac{2\pi nk}{N}\right),$$

$$\operatorname{Im}\{F_k\} = \frac{1}{N} \sum_{n=-\frac{N}{2}+1}^{\frac{N}{2}} \operatorname{Im}\{f_n\} \cos\left(\frac{2\pi nk}{N}\right) - \operatorname{Re}\{f_n\} \sin\left(\frac{2\pi nk}{N}\right).$$

Example: Numerical evaluation of a DFT. Before proceeding further, it is useful to solidify these concepts with a simple example. Consider the 12-point real-valued sequence f_n given in Table 2.1. Its DFT F_k has been computed explicitly from the definition and is also shown in the table. Figure 2.5 shows both the input sequence f_n and the DFT F_k graphically.

The interpretation of the DFT coefficients is essentially the same as that given for the Fourier transform itself. The kth DFT coefficient F_k gives the "amount" of the kth mode (with a frequency ω_k) that is present in the input sequence f_n. The N modes distinguished by the DFT, that is, the set of functions $e^{i2\pi\omega_k x}$, for $k = -N/2+1 : N/2$ (as well as their real-valued components $\cos(2\pi\omega_k x)$ and $\sin(2\pi\omega_k x)$), we refer to as the **basic modes** of the DFT.

TABLE 2.1
Numerical 12-point DFT.

n,k	Re $\{f_n\}$	Im $\{f_n\}$	Re $\{F_k\}$	Im $\{F_k\}$
-5	0.7630	0	0.0684	-0.1093
-4	-0.1205	0	-0.1684	0.0685
-3	-0.0649	0	-0.2143	-0.0381
-2	0.6133	0	-0.0606	0.1194
-1	-0.2697	0	-0.0418	-0.0548
0	-0.7216	0	0.0052	0
1	-0.0993	0	-0.0418	0.0548
2	0.9787	0	-0.0606	-0.1194
3	-0.5689	0	-0.2143	0.0381
4	-0.1080	0	-0.1684	-0.0685
5	-0.3685	0	0.0684	0.1093
6	0.0293	0	0.1066	0.0000

Let's look at these modes more closely. In sharp distinction to the Fourier transform which uses modes of all frequencies, there are only N distinct modes in an N-point DFT, with roughly $N/2$ different frequencies. The modes can be labeled by the frequency index k, and each mode has a value at each grid point x_n where $n = -N/2 : N/2$. Therefore, we will denote the nth component of the kth DFT mode as

$$v_n^k = \omega_N^{-nk} = e^{-i2\pi nk/N} = \cos\left(\frac{2\pi nk}{N}\right) - i\sin\left(\frac{2\pi nk}{N}\right),$$

for $n, k = -N/2 + 1 : N/2$. For the case at hand, the $N = 12$ modes can be grouped by their frequencies as follows:

- $k = 0:\ v_n^0 = 1,$

- $k = \pm 1:\ v_n^{\pm 1} = \cos(\pi n/6) \mp i\sin(\pi n/6),$

- $k = \pm 2:\ v_n^{\pm 2} = \cos(\pi n/3) \mp i\sin(\pi n/3),$

- $k = \pm 3:\ v_n^{\pm 3} = \cos(\pi n/2) \mp i\sin(\pi n/2),$

- $k = \pm 4:\ v_n^{\pm 4} = \cos(2\pi n/3) \mp i\sin(2\pi n/3),$

- $k = \pm 5:\ v_n^{\pm 5} = \cos(5\pi n/6) \mp i\sin(5\pi n/6),$

- $k = 6:\ v_n^6 = \cos(\pi n) = (-1)^n.$

The real (cosine) part of each of these modes is plotted in Figure 2.6, which clearly shows the distinct frequencies that are represented by the DFT. Notice that in counting the DFT modes we can list either

N complex modes v_n^k for $k = -N/2 + 1 : N/2$, *or*

a sine and cosine mode for $k = 1 : N/2 - 1$ plus a cosine mode for $k = 0$ and $k = N/2$.

Either way there are exactly N distinct modes (see problem 10).

Returning to Table 2.1, a brief examination of the data serves to introduce a number of the themes that will appear throughout the book. First, observe that the

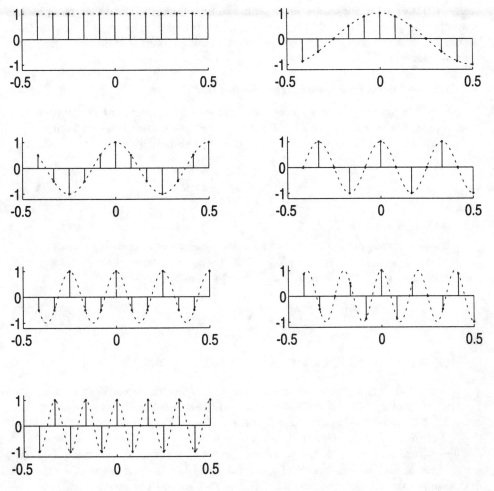

FIG. 2.6.   *An N-point DFT uses exactly N distinct modes with roughly N/2 different frequencies. The figure shows the cosine modes* $\cos(2\pi nk/N)$ *for the DFT with* $N = 12$. *The mode with frequency index k has* $|k|$ *periods on the domain where* $k = -N/2 + 1 : N/2$. *Shown are modes* $k = 0, \pm 1$ (*top row*); $k = \pm 2, \pm 3$ (*second row*), $k = \pm 4, \pm 5$ (*third row*), *and* $k = 6$ (*bottom*). *The "spikes" indicate the discrete values* $\cos(2\pi nk/N)$, *while the dotted curves indicate the corresponding continuum functions* $\cos(2\pi kx)$ *for* $x \in (-1/2, 1/2)$.

input is a real-valued sequence, while the output is complex-valued with the exception of F_0 and F_6, which are real-valued. The real part of the output is an even sequence with the property $\mathrm{Re}\,\{F_k\} = \mathrm{Re}\,\{F_{-k}\}$, while the imaginary part is an odd sequence with the property that $\mathrm{Im}\,\{F_k\} = -\mathrm{Im}\,\{F_{-k}\}$. The value of F_0 (often called the **DC component**) also equals the sum of the elements of the input sequence (problem 19). These observations constitute our first look at the symmetries of the DFT, a topic which constitutes Chapter 4. At this point our intent is merely to observe that there often exist lovely and important relationships between the input and output sequences, and that by examining these relationships thoroughly we may develop a powerful set of tools for discrete Fourier analysis.

The utility of the DFT, like that of any transform, arises because a difficult

problem in the spatial (or temporal) domain can be transformed into a simpler problem
in another domain. Ultimately the solution in the second domain must be transformed
back to the original domain. To accomplish this task, an inverse transform is required.
We now define it.

▶ **Inverse Discrete Fourier Transform (IDFT)** ◀

Let N be an even positive integer and let F_k be a sequence of N complex
numbers where $k = -N/2 + 1 : N/2$. Then its inverse discrete Fourier
transform is another sequence of N complex numbers given by

$$f_n = \sum_{k=-N/2+1}^{N/2} F_k \omega_N^{nk} \qquad (2.9)$$

for $n = -N/2 + 1 : N/2$.

If N is an odd positive integer and F_k is a sequence of N complex numbers,
where $k = -(N-1)/2 : (N-1)/2$, then its inverse discrete Fourier
transform is another sequence of N complex numbers given by

$$f_n = \sum_{k=-\frac{N-1}{2}}^{\frac{N-1}{2}} F_k \omega_N^{nk} \qquad (2.10)$$

for $n = -(N-1)/2 : (N-1)/2$.

Notice that this definition confers periodicity on the sequence f_n; it satisfies
$f_n = f_{n+N}$ (problem 9). As in the case of the DFT, we shall often employ an operator
notation, $\mathcal{D}^{-1}\{F_k\}$, to mean the IDFT of the sequence F_k, and $\mathcal{D}^{-1}\{F_k\}_n$ to indicate
the nth element of the inverse transform; therefore $\mathcal{D}^{-1}\{F_k\}_n = f_n$.

The notation and the discussion above suggest that the DFT and the IDFT really
are inverses of each other, but this fact certainly has not been shown! Therefore, the
next task is to show that the operators $\mathcal{D}$ and $\mathcal{D}^{-1}$ satisfy the inverse relations

$$\mathcal{D}^{-1}\{\mathcal{D}\{f_n\}\}_n = f_n \quad \text{and} \quad \mathcal{D}\{\mathcal{D}^{-1}\{F_k\}\}_k = F_k.$$

To verify the inversion formula, we must first develop the **orthogonality property**
of the complex exponential, a property that is crucial to the whole business of Fourier
series, Fourier transforms, and DFTs.

Let's begin with the discrete orthogonality property for the complex exponential.
To facilitate the discussion, we will introduce some notation known as the **modular
Kronecker[5] delta**. Letting k be any integer, we define $\hat{\delta}_N(k)$ by

$$\hat{\delta}_N(k) = \begin{cases} 1 & \text{if } k = 0 \text{ or a multiple of } N, \\ 0 & \text{otherwise.} \end{cases}$$

For example, $\hat{\delta}_4(0) = 1, \hat{\delta}_4(1) = 0, \hat{\delta}_4(8) = 1, \hat{\delta}_4(-3) = 0$. With this notation we can
state the orthogonality property; it is so important that we exalt it as a theorem (with
a proof!).

---

[5]LEOPOLD KRONECKER (1823–1891) specialized in the theory of equations, elliptic function theory,
and algebraic number theory while at the University of Berlin. Kronecker held firm in his belief that
all legitimate mathematics be based upon finite methods applied to integers, and to him is attributed
the quote "God made the whole numbers, all the rest is the work of man."

THEOREM 2.1. ORTHOGONALITY. *Let j and k be integers and let N be a positive integer. Then*

$$\sum_{n=0}^{N-1} e^{i2\pi nj/N} e^{-i2\pi nk/N} = N\,\hat{\delta}_N(j-k). \tag{2.11}$$

Proof: As before, we let $\omega_N = e^{i2\pi/N}$. Consider the N complex numbers ω_N^k, for $k = 0 : N-1$. They are called the Nth **roots of unity** because they satisfy

$$(\omega_N^k)^N = (e^{i2\pi k/N})^N = e^{i2\pi k} = 1,$$

and therefore are zeros of the polynomial $z^N - 1$. In fact, ω_N^k is one of the Nth roots of unity for *any* integer k, but it is easy to show that the sequence $\{\omega_N^k\}_{k=-\infty}^{\infty}$ is N-periodic (problem 9), so that the complete set of these roots may be specified by ω_N^{-k} for any N consecutive integers k. We first factor the polynomial $z^N - 1$ as

$$\begin{aligned}
z^N - 1 &= (z-1)(z^{N-1} + z^{N-2} + \ldots + z + 1) \\
&= (z-1)\sum_{n=0}^{N-1} z^n.
\end{aligned}$$

Noting that ω_N^k is a root of $z^n - 1 = 0$, there are two cases to consider. If we let $z = \omega_N^{j-k}$ where $j - k$ is not a multiple of N, then $z \neq 1$, and we have

$$\sum_{n=0}^{N-1} z^n = \sum_{n=0}^{N-1} \omega_N^{(j-k)n} = 0.$$

On the other hand, if $j - k$ is a multiple of N then $\omega_N^{j-k} = 1$ and

$$\sum_{n=0}^{N-1} \omega_N^{(j-k)n} = \sum_{n=0}^{N-1} 1 = N.$$

The orthogonality property follows from these two cases. ∎

Another proof of the orthogonality property can be constructed using the geometric series (see problem 12). Notice that since the sequence ω_N^k is N-periodic, the orthogonality property holds when the sum in (2.11) is computed over any N consecutive values of n; in other words,

$$\sum_{n=P}^{P+N-1} e^{i2\pi nj/N} e^{-i2\pi nk/N} = N\,\hat{\delta}_N(j-k)$$

for any integer P.

The orthogonality property of the DFT is a relationship between vectors. Orthogonality of two vectors $\mathbf{x}$ and $\mathbf{y}$ in any vector space means that the inner product of the two vectors is zero; that is,

$$\langle \mathbf{x}, \mathbf{y} \rangle = 0.$$

For real vectors the inner product is simply

$$\langle \mathbf{x}, \mathbf{y} \rangle = \sum_n x_n y_n,$$

while for complex-valued vectors the inner product is

$$\langle \mathbf{x}, \mathbf{y} \rangle = \sum_n x_n y_n^*,$$

where $*$ denotes complex conjugation, that is, $(a + ib)^* = a - ib$. Thus, if we define the complex N-vector

$$\mathbf{w}^k = \begin{pmatrix} \omega_N^0 \\ \omega_N^k \\ \omega_N^{2k} \\ \vdots \\ \omega_N^{(N-1)k} \end{pmatrix},$$

then

$$\langle \mathbf{w}^j, \mathbf{w}^k \rangle = N \hat{\delta}_N (j - k)$$

and any N consecutive values of k yield an orthogonal set of vectors.

With the puissance of orthogonality, we are ready to examine the inverse relation between the DFT and the IDFT. Here is the fundamental result.

THEOREM 2.2. INVERSION. *Let f_n be a sequence of N complex numbers and let $F_k = \mathcal{D}\{f_n\}_k$ be the DFT of this sequence. Then $\mathcal{D}^{-1}\{\mathcal{D}\{f_n\}_k\}_n = f_n$.*

Proof: Combining the definitions of $\mathcal{D}$ and $\mathcal{D}^{-1}$, we can write

$$\mathcal{D}^{-1}\{\mathcal{D}\{f_n\}_k\}_n = \sum_{k=-\frac{N}{2}+1}^{\frac{N}{2}} F_k \omega_N^{nk}$$

$$= \sum_{k=-\frac{N}{2}+1}^{\frac{N}{2}} \left(\frac{1}{N} \sum_{j=-\frac{N}{2}+1}^{\frac{N}{2}} f_j \omega_N^{-jk} \right) \omega_N^{nk}$$

$$= \frac{1}{N} \sum_{j=-\frac{N}{2}+1}^{\frac{N}{2}} f_j \underbrace{\sum_{k=-\frac{N}{2}+1}^{\frac{N}{2}} \omega_N^{k(n-j)}}_{N\hat{\delta}_N(n-j)},$$

where we have applied the orthogonality property. The inner sum in this equation is nonzero only when $j = n$ in the outer sum, yielding

$$\mathcal{D}^{-1}\{\mathcal{D}\{f_n\}_k\}_n = \frac{1}{N} N f_n = f_n. \qquad \blacksquare$$

It is equally easy to show that $\mathcal{D}\{\mathcal{D}^{-1}\{F_k\}_n\}_k = F_k$, so that indeed we have a pair of operators that are inverses of each other (problem 16).

Example revisited. Let's return to the example of Table 2.1 and Figure 2.5 in which a 12-point DFT of a real-valued sequence was computed, and now interpret it in light of the inverse DFT. We will begin by writing the inverse DFT using only real quantities. Recall that $F_k = \text{Re}\{F_k\} + i\text{Im}\{F_k\}$, and that in this case, in which f_n is real-valued, $\text{Re}\{F_k\}$ is an even sequence and $\text{Im}\{F_k\}$ is an odd sequence. Furthermore, we saw that the DFT coefficients F_0 and $F_{N/2}$ are real. The IDFT now looks like

$$f_n = \sum_{k=-\frac{N}{2}+1}^{\frac{N}{2}} F_k \omega_N^{nk}$$

$$= \quad F_0 + \sum_{k=1}^{\frac{N}{2}-1} (F_k \omega_N^{nk} + F_{-k} \omega_N^{-nk}) + (-1)^n F_{\frac{N}{2}}.$$

Let's pause for an explanation of a few maneuvers which we have used for the first of many times. As noted earlier, when the input sequence is real, the DFT coefficients F_0 and $F_{N/2}$ are real. Therefore, we have pulled these terms out of the IDFT sum, and used the fact that $\omega_N^{nk} = \cos(\pi n) = (-1)^n$, when $k = N/2$. We have also folded the sum so it runs over the indices $k = 1 : N/2 - 1$.

We may now continue and write

$$
\begin{aligned}
f_n &= \quad F_0 + \sum_{k=1}^{\frac{N}{2}-1} (\text{Re}\{F_k\} + i\text{Im}\{F_k\}) \omega_N^{nk} \\
&\quad + \sum_{k=1}^{\frac{N}{2}-1} (\text{Re}\{F_{-k}\} + i\text{Im}\{F_{-k}\}) \omega_N^{-nk} + (-1)^n F_{\frac{N}{2}} \\
&= \quad F_0 + \sum_{k=1}^{\frac{N}{2}-1} \text{Re}\{F_k\} \underbrace{(\omega_N^{nk} + \omega_N^{-nk})}_{2\cos(2\pi nk/N)} \\
&\quad +i \sum_{k=1}^{\frac{N}{2}-1} \text{Im}\{F_k\} \underbrace{(\omega_N^{nk}) - \omega_N^{-nk})}_{2i\sin(2\pi nk/N)} + (-1)^n F_{\frac{N}{2}}.
\end{aligned}
$$

A few more crucial facts have been used to get this far. Since the real part of F_k is an even sequence, $\text{Re}\{F_k\} = \text{Re}\{F_{-k}\}$; and since the imaginary part of F_k is an odd sequence, $\text{Im}\{F_k\} = -\text{Im}\{F_{-k}\}$ (see the DFT coefficients in Table 2.1). Furthermore, the Euler relations have been used to collect the complex exponentials and form sines and cosines as indicated. We are now almost there. The above argument continues to its completion as

$$f_n = F_0 + 2 \sum_{k=1}^{\frac{N}{2}-1} \left(\text{Re}\{F_k\} \cos\left(\frac{2\pi nk}{N}\right) - \text{Im}\{F_k\} \sin\left(\frac{2\pi nk}{N}\right) \right) + (-1)^n F_{\frac{N}{2}},$$

where $n = -N/2 + 1 : N/2$. We see clearly that the sequence f_n consists entirely of real quantities and is thus real-valued, as shown in Table 2.1. Furthermore, the meaning of the DFT coefficients is reiterated: the real part of F_k is the weight of the cosine part of the kth mode, and the imaginary part of F_k is the weight of the sine part of the kth mode in the recipe for the input sequence f_n.

Before closing this section, we will have a look at the DFT in the powerful and elegant language of matrices. The DFT maps a set of N input values f_n into N output values F_k. We have shown that it is an invertible operator, and we will soon show that it is a linear operator. Therefore, the DFT can be represented as the product of an N-vector and an $N \times N$ matrix. In this setting it is more convenient to use the alternate definition of the DFT

$$F_k = \frac{1}{N} \sum_{n=0}^{N-1} f_n \omega_N^{-nk}$$

for $k = 0 : N - 1$. If $\mathbf{f}$ represents the vector of input data,

$$\mathbf{f} = (f_0, f_1, f_2, \ldots, f_{N-1})^T,$$

and $\mathbf{F}$ represents the vector of output values,

$$\mathbf{F} = (F_0, F_1, F_2, \ldots, F_{N-1})^T,$$

then the DFT can be written as

$$\mathbf{F} = \mathbf{W}\mathbf{f}.$$

The DFT matrix $\mathbf{W}$ is the square, nonsingular matrix

$$\mathbf{W} = \frac{1}{N} \begin{pmatrix} \omega_N^0 & \omega_N^0 & \omega_N^0 & \cdots & \omega_N^0 \\ \omega_N^0 & \omega_N^{-1} & \omega_N^{-2} & \cdots & \omega_N^{-(N-1)} \\ \omega_N^0 & \omega_N^{-2} & \omega_N^{-4} & \cdots & \omega_N^{-2(N-1)} \\ \vdots & \vdots & \vdots & & \vdots \\ \omega_N^0 & \omega_N^{-(N-1)} & \omega_N^{-2(N-1)} & \cdots & \omega_N^{-(N-1)(N-1)} \end{pmatrix}.$$

This matrix has many important properties [36], [99], some of which we list below.

- Since the DFT is invertible, we observe that a matrix representation for the IDFT exists and is $\mathbf{W}^{-1}$.

- The matrix $\mathbf{W}$ is symmetric, so that $\mathbf{W}^T = \mathbf{W}$.

- The inverse of $\mathbf{W}$ is a multiple of its complex conjugate:

$$\mathbf{W}^{-1} = N\mathbf{W}^*.$$

Therefore, $N\mathbf{W}\mathbf{W}^H = \mathbf{I}$ (where H denotes the conjugate transpose and $\mathbf{I}$ is the identity matrix), and $\mathbf{W}$ is unitary up to a factor of N. The factor of N can be included in the definitions of both the DFT and the IDFT, in which case $\mathbf{W}\mathbf{W}^H = \mathbf{I}$ (problem 18).

- For $N > 4$ the matrix $\mathbf{W}$ has four distinct eigenvalues, namely

$$\lambda_1 = \sqrt{N}, \quad \lambda_2 = -\sqrt{N}, \quad \lambda_3 = 0 - i\sqrt{N}, \quad \lambda_4 = 0 + i\sqrt{N}.$$

The multiplicities of the eigenvalues are m_1, m_2, m_3, and m_4, respectively, and are related to the order N of the matrix as shown in Table 2.2 [6], [36], [99]. The value of the determinant of $\mathbf{W}$ is also related to N, and is included in the table.

While we will not spend a great deal of time considering the matrix formulation of the DFT, there are certain applications for which this is an extremely useful approach. For example, the multitude of FFT algorithms can be expressed in terms of factorizations of the matrix $\mathbf{W}$ [155]. For those interested in far deeper and more abstract algebraic and number theoretic properties of the DFT, we recommend the challenging paper by Auslander and Tolimieri [6].

N	m_1	m_2	m_3	m_4	det $\mathbf{W}$
$4n$	$n+1$	n	n	$n-1$	$-i(-1)^n N^{N/2}$
$4n+1$	$n+1$	n	n	n	$(-1)^n N^{N/2}$
$4n+2$	$n+1$	$n+1$	n	n	$-N^{N/2}$
$4n+3$	$n+1$	$n+1$	$n+1$	n	$i(-1)^n N^{N/2}$

2.4. DFT Approximations to Fourier Series Coefficients

As closely as the DFT is related to the Fourier transform, it may be argued that it holds even more kinship to the coefficients of the Fourier series. It is a simple matter to use the Fourier series to derive the DFT formula, and we will do so shortly. But first, it is worthwhile to spend a few pages highlighting some of the important features of the theory of Fourier series. We begin with the following definition.

▶ **Fourier Series** ◀

Let f be a function that is periodic with period A (also called A-**periodic**). Then the **Fourier series associated with f is the trigonometric series**

$$f(x) \sim \sum_{k=-\infty}^{\infty} c_k e^{i2\pi kx/A}, \qquad (2.12)$$

where the coefficients c_k are given by

$$c_k = \frac{1}{A} \int_{-\frac{A}{2}}^{\frac{A}{2}} f(x) e^{-i2\pi kx/A} dx. \qquad (2.13)$$

The symbol $\sim$ means that the Fourier series is *associated* with the function f. We would prefer to make the stronger statement that the series *equals* the function at every point, but without imposing additional conditions on f, this cannot be said. The conditions that are required for convergence of the series to f will be outlined shortly. For the moment, we will assume that the function is sufficiently well behaved to insure convergence, and will now show that expression (2.13) for the coefficients c_k is correct. As with the DFT, our ability to find the coefficients c_k depends on an orthogonality property.

THEOREM 2.3. ORTHOGONALITY OF COMPLEX EXPONENTIALS. *Let j and k be any integers. Then the set of complex exponential functions $e^{i2\pi kx/A}$ satisfies the* *orthogonality relation*

$$\int_{-\frac{A}{2}}^{\frac{A}{2}} e^{i2\pi jx/A} e^{-2\pi kx/A} dx = A\delta(j-k), \qquad (2.14)$$

where we have used the ordinary Kronecker delta

$$\delta(k) = \begin{cases} 1 & \text{if } k = 0, \\ 0 & \text{if } k \neq 0. \end{cases}$$

Proof: The orthogonality property follows from a direct integration, the details of which are explored in problem 22. ∎

As in the discrete case, continuous orthogonality can be viewed as orthogonality of "vectors" in a space of functions defined on an interval $\mathcal{I}$. The inner product of f and g on $\mathcal{I}$ is defined as

$$\langle f, g \rangle = \int_{\mathcal{I}} f(x)g^*(x) dx.$$

For a vector space of complex-valued functions on the interval $[-A/2, A/2]$, the functions

$$\mathbf{w}^k(x) = e^{i2\pi kx/A}$$

are orthogonal since

$$\langle \mathbf{w}^j, \mathbf{w}^k \rangle = A\delta(j-k).$$

To find the coefficients c_k, we **assume** that the A-periodic function f is the sum of its Fourier series, so that

$$f(x) = \sum_{j=-\infty}^{\infty} c_j e^{i2\pi jx/A}. \qquad (2.15)$$

Multiplying both sides of this equation by $(1/A)e^{-i2\pi kx/A}$ and assuming that term-by-term integration over $[-A/2, A/2]$ is permitted, we find that

$$\frac{1}{A}\int_{-\frac{A}{2}}^{\frac{A}{2}} f(x)e^{-i2\pi kx/A} dx = \frac{1}{A}\int_{-\frac{A}{2}}^{\frac{A}{2}} \sum_{j=-\infty}^{\infty} c_j e^{i2\pi(j-k)x/A} dx$$

$$= \sum_{j=-\infty}^{\infty} c_j \underbrace{\frac{1}{A}\int_{-\frac{A}{2}}^{\frac{A}{2}} e^{i2\pi(j-k)x/A} dx}_{\delta(j-k)}$$

$$= c_k.$$

By the orthogonality property, $1/A$ times the integral vanishes unless $k = j$, in which case it is unity. The only term that survives in the series on the right is the $k = j$ term, which gives

$$c_k = \frac{1}{A}\int_{-\frac{A}{2}}^{\frac{A}{2}} f(x)e^{-i2\pi kx/A} dx.$$

Since many applications involve real-valued functions f, it is often convenient to write the Fourier series in a form that involves no complex-valued quantities. For these situations we give the following definition.

▶ **Fourier Series for Real-Valued f** ◀

Let f be a real-valued function that is A-periodic. Then the Fourier series associated with f is the trigonometric series

$$f(x) \sim \frac{a_0}{2} + \sum_{k=1}^{\infty} a_k \cos\left(\frac{2\pi kx}{A}\right) + \sum_{k=1}^{\infty} b_k \sin\left(\frac{2\pi kx}{A}\right), \qquad (2.16)$$

where the coefficients are given by

$$a_k = \frac{2}{A} \int_{-\frac{A}{2}}^{\frac{A}{2}} f(x) \cos\left(\frac{2\pi kx}{A}\right) dx$$

for $k = 0, 1, 2, \ldots$, and by

$$b_k = \frac{2}{A} \int_{-\frac{A}{2}}^{\frac{A}{2}} f(x) \sin\left(\frac{2\pi kx}{A}\right) dx$$

for $k = 1, 2, \ldots$.

The equivalence of the real and complex forms of the Fourier series for real-valued functions is easily established using the Euler relations, and is the subject of problem 23. The real form of the Fourier series, like the complex form, depends on an orthogonality property. In this case, we use the space of real-valued functions on the interval $[-A/2, A/2]$ with the inner product

$$\langle f, g \rangle = \int_{-\frac{A}{2}}^{\frac{A}{2}} f(x)g(x)\, dx.$$

It can be shown that the functions

$$\mathbf{w}^k(x) = \cos\left(\frac{2\pi kx}{A}\right) \quad \text{and} \quad \mathbf{v}^k(x) = \sin\left(\frac{2\pi kx}{A}\right),$$

where $j, k = 0, \pm 1, \pm 2, \pm 3, \ldots$, form orthogonal sets (problem 24).

But now we must ask about when a Fourier series converges. What conditions are sufficient for the Fourier series associated with f to converge to f? The answer to this question is not simple. In fact, astonishing as it may seem considering the venerability of the topic, there are still open questions concerning Fourier series convergence. There is a vast literature on the issue of convergence ([30], [34], [36], [153], [158], among many others) and a treatment of this topic in any detail is far beyond the scope or purpose of this book. We will state here a widely used criterion for the convergence of the Fourier series.

First we need some definitions. A function f is said to be **piecewise continuous on an interval** $[a, b]$ if it is continuous at all points in the interval except at a finite number of points x_i at which it is either not defined or discontinuous, but at which the one-sided limits

$$f(x_i^+) = \lim_{h \to 0^+} f(x_i + h) \quad \text{and} \quad f(x_i^-) = \lim_{h \to 0^-} f(x_i + h)$$

exist and are finite. An alternative view is that f is piecewise continuous on $[a, b]$ if that interval can be divided into a finite number of subintervals, on each of which f is continuous. More generally, a function f is said to be **piecewise continuous** (everywhere) if it is piecewise continuous on every finite interval. As an example, the function

$$f(x) = \begin{cases} x^2 & \text{for} \quad x < 2, \\ \frac{x^2-9}{x-3} & \text{for} \quad 2 \le x < 3 \text{ or } x > 3, \end{cases}$$

is piecewise continuous, despite being undefined at $x = 3$ and having a jump discontinuity at $x = 2$.

A function f is said to be **piecewise smooth on an interval** $[a, b]$ if both f and f' are piecewise continuous on the interval, and is said to be **piecewise smooth** (everywhere) if it is piecewise smooth on every finite interval. Piecewise smoothness of a function implies the existence of the one-sided derivatives

$$f'(x_i^+) = \lim_{h \to 0^+} \frac{f(x_i + h) - f(x_i)}{h} \quad \text{and} \quad f'(x^-) = \lim_{h \to 0^-} \frac{f(x_i + h) - f(x_i)}{h}$$

at every point of discontinuity. Figure 2.7 illustrates these "piecewise properties." Shown in the upper two figures are the function $f(x)$ and its derivative $f'(x)$, which are defined by

$$f(x) = \begin{cases} \dfrac{x}{2} + \dfrac{1}{2} & \text{for} \quad x \le 0, \\[2ex] x \ln x & \text{for} \quad 0 < x \le 1, \\[2ex] (x-1)\ln(x-1) + \dfrac{1}{5} & \text{for} \quad 1 < x, \end{cases}$$

and

$$f'(x) = \begin{cases} \dfrac{1}{2} & \text{for} \quad x \le 0, \\[2ex] \ln x + 1 & \text{for} \quad 0 < x \le 1, \\[2ex] \ln(x-1) + 1 & \text{for} \quad 1 < x. \end{cases}$$

The function $f(x)$ is piecewise continuous since it is continuous on each of the subintervals $[-1, 0]$, $(0, 1]$, and $(1, 2]$ and has both left- and right-hand limits at $x = 0$ and $x = 1$. The derivative $f'(x)$, however, does *not* have left-hand limits at either $x = 0$ or $x = 1$, but rather approaches $-\infty$. Hence $f(x)$ is piecewise continuous but not piecewise smooth. Shown in the lower figures are the function $g(x)$ and its derivative $g'(x)$, defined by

$$g(x) = \begin{cases} \dfrac{x}{2} + \dfrac{1}{2} & \text{for} \quad x \le 0, \\[2ex] x^2 \ln x & \text{for} \quad 0 < x \le 1, \\[2ex] (x-1)^2 \ln(x-1) + \dfrac{1}{5} & \text{for} \quad 1 < x, \end{cases}$$

and

$$g'(x) = \begin{cases} \dfrac{1}{2} & \text{for} \quad x \le 0, \\[2mm] 2x \ln x + x & \text{for} \quad 0 < x \le 1, \\[2mm] 2(x-1)\ln(x-1) + (x-1) & \text{for} \quad 1 < x. \end{cases}$$

Unlike $f(x)$, the function $g(x)$ has a piecewise continuous derivative, and therefore is piecewise smooth. We may now state the main convergence result.

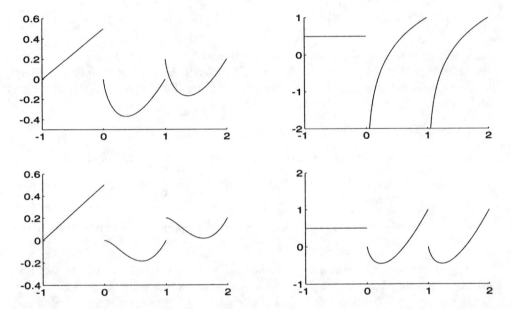

FIG. 2.7. *The upper left figure shows the graph of the piecewise continuous function $f(x)$ on the interval $[-1, 2]$. The derivative $f'(x)$ is shown in the upper right. The function $f(x)$ is continuous on the subintervals $(-1, 0)$, $(0, 1)$, and $(1, 2)$ and has one-sided limits at 0 and 1. However, $f'(x)$, while continuous on the subintervals $(-1, 0)$, $(0, 1)$, and $(1, 2)$, has no left-hand limits at either 0 or 1. Hence $f(x)$ is not piecewise smooth. The lower left figure shows the graph of a function $g(x)$ that is piecewise smooth on $[-1, 2]$. The function is piecewise continuous, and one-sided derivatives exist at each of the points 0 and 1.*

THEOREM 2.4. CONVERGENCE OF FOURIER SERIES. *Let f be a piecewise smooth A-periodic function. Then the Fourier series for f*

$$\sum_{k=-\infty}^{\infty} c_k e^{i2\pi kx/A} \quad where \quad c_k = \frac{1}{A} \int_{-\frac{A}{2}}^{\frac{A}{2}} f(x) e^{-i2\pi kx/A} dx$$

converges (pointwise) for every x to the value

$$\frac{f(x^+) + f(x^-)}{2}.$$

We shall not prove this theorem here, but the interested reader can find a good discussion of the proof in [158]. However, some important observations about this result should be made. Since at a point of continuity the right- and left-hand limits of a function must be equal, and equal to the function value, it follows that at any point of continuity, the Fourier series converges to $f(x)$. At any point of discontinuity, the series converges to the average value of the right- and left-hand limits.

So far, the Fourier series has been defined only for periodic functions. However, an important case that arises often is that in which f is defined and piecewise smooth only on the interval $[-A/2, A/2]$; perhaps f is not defined outside of that interval, or perhaps it is not a periodic function at all. In order to handle this situation we need to know about the **periodic extension** of f, the function h defined by

$$h(x + sA) = f(x), \qquad x \in \left(-\frac{A}{2}, \frac{A}{2}\right), \qquad s = 0, \pm 1, \pm 2, \ldots.$$

The periodic extension of f is simply the repetition of f every A units on both sides of the interval $[-A/2, A/2]$. It should be verified that h *is* an A-periodic function. Here is the important role of the periodic extension h: if the Fourier series for f converges on $[-A/2, A/2]^6$, then it converges

- to the value of f at points of continuity on $(-A/2, A/2)$,

- to the average value of f at points of discontinuity on $(-A/2, A/2)$,

- to the value of the periodic extension of f at points of continuity outside of $(-A/2, A/2)$, and

- to the average value of the periodic extension at points of discontinuity outside of $(-A/2, A/2)$.

Figure 2.8 shows a function f defined on an interval $[-1, 1]$ and its periodic extension beyond that interval. The periodic extension is the function to which the Fourier series of f converges for all x provided we use average values at points of discontinuity. In particular, if $f(-A/2) \neq f(A/2)$ then the Fourier series converges to the average of the function values at the right and left endpoints

$$\frac{1}{2}\left[f\left(-\frac{A}{2}^+\right) + f\left(\frac{A}{2}^-\right)\right].$$

These facts must be observed scrupulously when a function is sampled for input to the DFT.

With this prelude to Fourier series, we are now in a position to derive the DFT as an approximation to the integral that gives the Fourier series coefficients c_k. The process is similar to that used to derive the DFT as an approximation to the Fourier transform. Now we consider approximations to the integral

$$c_k = \frac{1}{A} \int_{-\frac{A}{2}}^{\frac{A}{2}} f(x)e^{-i2\pi kx/A}dx. \tag{2.17}$$

---

[6]There seems to be no agreement in the literature about whether the interval for defining Fourier series should be the closed interval $[-A/2, A/2]$, a half-open interval $(-A/2, A/2]$, or the open interval $(-A/2, A/2)$. Arguments can be made for or against any of these choices. We will use the closed interval $[-A/2, A/2]$ throughout the book to emphasize the point (the subject of sermons to come!) that in defining the input to the DFT, values of the sampled function at *both* endpoints contribute to the input.

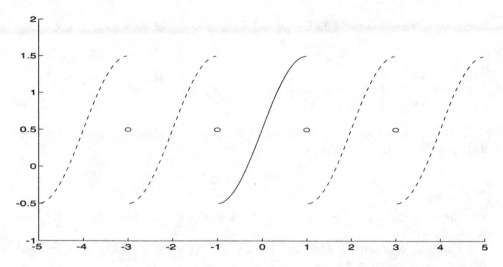

FIG. 2.8.   *The piecewise smooth function f shown here is defined on the interval $[-1,1]$ (solid curve). Its Fourier series on that interval converges to the periodic extension of f (solid and dashed curve) at all points x. Notice that at points at which the periodic extension is discontinuous (such as $x = \pm 1$) the Fourier series converges to the average value of the function at the discontinuity (marked by the small ovals).*

As before, the interval of integration is divided into N subintervals of equal length, and let the grid spacing be $\Delta x = A/N$. A grid with $N + 1$ equally spaced points over the interval $[-A/2, A/2]$ is defined by the points $x_n = n\Delta x$ for $n = -N/2 : N/2$. Furthermore, we let

$$g(x) = f(x)e^{-i2\pi kx/A}$$

be the integrand in this expression. Applying the trapezoid rule gives the approximations

$$c_k = \frac{1}{A} \int_{-\frac{A}{2}}^{\frac{A}{2}} g(x)dx \approx \frac{1}{A}\frac{\Delta x}{2}\left\{ g\left(-\frac{A}{2}\right) + 2\sum_{n=-\frac{N}{2}+1}^{\frac{N}{2}-1} g(x_n) + g\left(\frac{A}{2}\right)\right\}.$$

Now the question of endpoint values enters in a critical way. Recall that in approximating the Fourier transform, we were able to associate the trapezoid rule with the DFT because of the assumption that $g(-A/2) = g(A/2)$. Now we must reason differently and actually avoid making an unnecessary assumption.

We have already seen that if the periodic extension of f is discontinuous at the endpoints $x = \pm A/2$, then, when its Fourier series converges, it converges to the average value

$$\frac{1}{2}\left[f\left(-\frac{A^+}{2}\right) + f\left(\frac{A^-}{2}\right)\right].$$

Therefore, it is the average value of f at the endpoints that must be used in the trapezoid rule. Noting that the kernel $e^{-i2\pi kx/A}$ has the value $(-1)^k$ at $x = \pm A/2$, we see that function g that must be used for the trapezoid rule is

$$g(x) = \begin{cases} f(x)e^{-i2\pi kx/A} & \text{for} \quad x \neq \pm\frac{A}{2}, \\ \frac{(-1)^k}{2}\left[f\left(-\frac{A}{2}^+\right) + f\left(\frac{A}{2}^-\right)\right] & \text{for} \quad x = \pm\frac{A}{2}. \end{cases}$$

It should be verified that this choice of g, dictated by the convergence properties of the Fourier series, guarantees that

$$g\left(-\frac{A}{2}\right) = g\left(\frac{A}{2}\right).$$

In a similar way, an average value must be used at any grid points at which f has discontinuities.

Using this definition of g, along with the observation that $2\pi kx_n/A = 2\pi nk/N$, reduces the trapezoid rule to

$$c_k = \frac{1}{A}\int_{-\frac{A}{2}}^{\frac{A}{2}} g(x)dx \quad \approx \quad \frac{1}{A}\Delta x \sum_{n=-\frac{N}{2}+1}^{\frac{N}{2}} g(x_n)$$

$$= \quad \frac{1}{A}\frac{A}{N} \sum_{n=-\frac{N}{2}+1}^{\frac{N}{2}} f(x_n)e^{-i2\pi nk/N}.$$

Letting $f_n = f(x_n)$, we see that an approximation to the Fourier series coefficient c_k is given by

$$c_k \approx \sum_{n=-\frac{N}{2}+1}^{\frac{N}{2}} f_n e^{-i2\pi kn/N} = \mathcal{D}\{f_n\}_k$$

which, for $k = -N/2 + 1 : N/2$, is precisely the definition of the DFT. Thus we see that the DFT gives approximations to the first N Fourier coefficients of a function f on a given interval $[-A/2, A/2]$ in a very natural way. There are subtleties concerning the use of average values at the endpoints and discontinuites, but the importance of this issue will be emphasized many times in hopes of removing the subtlety!

We have now shown that the DFT provides approximations to both the Fourier transform and the Fourier coefficients. The errors in these approximations will be investigated thoroughly in Chapter 6. Having related the DFT to both the Fourier transform and Fourier coefficients, we close this section by completing the circle and establishing a simple but important connection between Fourier transforms and Fourier series. When a function f is spatially limited, meaning that $f(x) = 0$ for $|x| > A/2$, we see that the Fourier transform evaluated at the frequency $\omega_k = k/A$ is given by

$$\hat{f}(\omega_k) = \int_{-\frac{A}{2}}^{\frac{A}{2}} f(x)e^{-i2\pi kx/A}dx.$$

Now compare this expression to the integral for the Fourier coefficient of f on $[-A/2, A/2]$:

$$c_k = \frac{1}{A}\int_{-\frac{A}{2}}^{\frac{A}{2}} f(x)e^{-i2\pi kx/A}dx.$$

The relationship is evident. In the case of a function that is zero outside of the interval $[-A/2, A/2]$, the Fourier transform and the Fourier coefficients are related by

$$\hat{f}(\omega_k) = \hat{f}\left(\frac{k}{A}\right) = Ac_k$$

for $-\infty < k < \infty$. This relationship will prove to be quite useful in the pages to come.

2.5. The DFT from Trigonometric Approximation

The derivations shown so far have evolved from the problem of approximating either the Fourier series coefficients or the Fourier transform of a particular function. Another way to uncover the DFT follows by considering the problem of approximating (or fitting) a set of data with a function known as a trigonometric polynomial. The goal is to find a linear combination of sines and cosines that "best" approximates a given data set. One of the beautiful connections of mathematics is that the solution to this problem leads to the DFT. We have already seen a preview of this development in Chapter 1, with Gauss'[7] interpolation of the orbit of Ceres.

Suppose that we are given N data pairs that we will denote (x_n, f_n), where $n = -(N-1)/2 : (N-1)/2$. (This derivation is one instance in which it is more convenient to work with an odd number of samples; everything said here can be done with minor modifications for even N.) The x_n's are real and are assumed to be equally spaced points on an interval $[-A/2, A/2]$; that is, $x_n = n\Delta x$ where $\Delta x = A/N$. The f_n's may be complex-valued. The data pairs may be viewed as samples of a continuous function f that have been gathered at the grid points x_n. But equally likely is the instance in which the pairs originate as a discrete set of collected data.

We seek the best possible approximation to the data using the N-term trigonometric polynomial ψ_N, given by

$$\psi_N(x) = \sum_{k=-\frac{N-1}{2}}^{\frac{N-1}{2}} \alpha_k e^{i2\pi kx/A}.$$

We describe the function ψ_N as a **trigonometric polynomial** because it is a polynomial in the quantity $e^{i2\pi x/A}$. There are many conditions that might be imposed to determine the "best" approximation to a data set. We will use the least squares criterion and require that the sum of the squares of the differences between the data values and the approximation function ψ_N at the points x_n be minimized. Said a little more concisely, we seek to choose the coefficients α_k to minimize the **discrete least squares error**

$$E = \sum_{n=-\frac{N-1}{2}}^{\frac{N-1}{2}} |f_n - \psi_N(x_n)|^2.$$

---

[7]Born in 1777, in Brunswick, Germany, to uneducated parents, CARL FRIEDRICH GAUSS is universally regarded as one of the three greatest mathematicians of all time, the other two being Archimedes, Newton, Euler, Hilbert, or Euclid (pick two). He completed his doctoral thesis at age 20, and became Professor of Mathematics at the University of Göttingen in 1807, where he lived until his death in 1855. Gauss made lasting contributions to algebra, astronomy, geodesy, number theory, and physics.

The least squares error is a real-valued, nonnegative function of the N coefficients

$$\alpha_{-(N-1)/2}, \; \alpha_{-(N-1)/2+1}, \; \cdots, \; \alpha_{-1}, \; \alpha_0, \; \alpha_1, \ldots, \; \alpha_{(N-1)/2}.$$

Therefore, a necessary condition for the minimization of E is that the first partial derivatives of E with respect to each of the N coefficients vanish. Observing that

$$|f_n - \psi_N(x_n)|^2 = (f_n - \psi_N(x_n))(f_n - \psi_N(x_n))^*,$$

we arrive at the so-called **normal equations** for the problem (see problem 26):

$$\frac{\partial E}{\partial \alpha_k} = \sum_{n=-\frac{N-1}{2}}^{\frac{N-1}{2}} \left[e^{-i2\pi nk/N} \left(f_n - \sum_{p=-\frac{N-1}{2}}^{\frac{N-1}{2}} \alpha_p \, e^{i2\pi np/N} \right) \right] = 0, \qquad (2.18)$$

where $k = -(N-1)/2 : (N-1)/2$. Rearranging the terms gives us the set of N equations

$$\sum_{n=-\frac{N-1}{2}}^{\frac{N-1}{2}} f_n e^{-i2\pi nk/N} = \sum_{n=-\frac{N-1}{2}}^{\frac{N-1}{2}} \sum_{p=-\frac{N-1}{2}}^{\frac{N-1}{2}} \alpha_p \, e^{i2\pi np/N} \, e^{-i2\pi nk/N}. \qquad (2.19)$$

These expressions can be further simplified if we use our conventional notation that $\omega_N = e^{i2\pi/N}$. The normal equations now appear as

$$\sum_{n=-\frac{N-1}{2}}^{\frac{N-1}{2}} f_n \omega_N^{-nk} = \sum_{p=-\frac{N-1}{2}}^{\frac{N-1}{2}} \alpha_p \underbrace{\sum_{n=-\frac{N-1}{2}}^{\frac{N-1}{2}} \omega_N^{(p-k)n}}_{N\hat{\delta}_N(p-k)},$$

where again $k = -(N-1)/2 : (N-1)/2$. As indicated, the inner sum on the right side begs to be treated by the orthogonality relation. Doing so, we find that

$$\sum_{n=-\frac{N-1}{2}}^{\frac{N-1}{2}} f_n \omega_N^{-nk} = N \alpha_k \qquad (2.20)$$

for $k = -(N-1)/2 : (N-1)/2$. Notice the minor miracle that has occurred! The rather dense set of normal equations that linked the coefficients α_p in an obscure way has been separated or "diagonalized" so that there is a single equation for each of the N coefficients. From this last expression (2.20) it is evident that the least squares error is minimized when the coefficients in the approximating polynomial are given by the DFT of the data,

$$\alpha_k = \frac{1}{N} \sum_{n=-\frac{N-1}{2}}^{\frac{N-1}{2}} f_n \omega_N^{-nk} \qquad (2.21)$$

for $k = -(N-1)/2 : (N-1)/2$.

Having determined that the DFT coefficients give the trigonometric polynomial with minimum least squares error, it is natural to ask just how good this "best" polynomial is. What is the size of the error? Recall that the least squares error E is given by

$$E = \sum_{n=-\frac{N-1}{2}}^{\frac{N-1}{2}} |f_n - \psi_N(x_n)|^2$$

$$= \sum_{n=-\frac{N-1}{2}}^{\frac{N-1}{2}} (f_n - \psi_N(x_n))(f_n - \psi_N(x_n))^*$$

$$= \sum_{n=-\frac{N-1}{2}}^{\frac{N-1}{2}} |f_n|^2 - \sum_{n=-\frac{N-1}{2}}^{\frac{N-1}{2}} f_n \psi_N^*(x_n)$$

$$- \sum_{n=-\frac{N-1}{2}}^{\frac{N-1}{2}} f_n^* \psi_N(x_n) + \sum_{n=-\frac{N-1}{2}}^{\frac{N-1}{2}} |\psi_N(x_n)|^2.$$

A direct calculation shows (problem 27) that each of the last three sums in this expression has the value $N \sum_n |\alpha_n|^2$, so that

$$E = \sum_{n=-\frac{N-1}{2}}^{\frac{N-1}{2}} |f_n|^2 - N \sum_{n=-\frac{N-1}{2}}^{\frac{N-1}{2}} |\alpha_n|^2. \qquad (2.22)$$

Observing that $|\alpha_n|^2 = \alpha_n \alpha_n^*$, we may use (2.21) to obtain

$$E = \sum_{n=-\frac{N-1}{2}}^{\frac{N-1}{2}} |f_n|^2 - \frac{N}{N^2} \sum_{n=-\frac{N-1}{2}}^{\frac{N-1}{2}} \sum_{p=-\frac{N-1}{2}}^{\frac{N-1}{2}} \sum_{m=-\frac{N-1}{2}}^{\frac{N-1}{2}} \left(f_p f_m^* e^{i2\pi n(m-p)/N} \right)$$

$$= \sum_{n=-\frac{N-1}{2}}^{\frac{N-1}{2}} |f_n|^2 - \frac{1}{N} \sum_{p=-\frac{N-1}{2}}^{\frac{N-1}{2}} \sum_{m=-\frac{N-1}{2}}^{\frac{N-1}{2}} f_p f_m^* \underbrace{\sum_{n=-\frac{N-1}{2}}^{\frac{N-1}{2}} e^{i2\pi n(m-p)/N}}_{N\hat{\delta}_N(m-p)}.$$

Since the last sum involves N terms, we may invoke orthogonality and conclude that it is zero except when $p = m$, in which case it has the value N. Thus we observe, perhaps unexpectedly, that

$$E = \sum_{n=-\frac{N-1}{2}}^{\frac{N-1}{2}} |f_n|^2 - \frac{1}{N} \sum_{p=-\frac{N-1}{2}}^{\frac{N-1}{2}} \sum_{m=-\frac{N-1}{2}}^{\frac{N-1}{2}} f_p f_m^* N\hat{\delta}_N(m-p)$$

$$= \sum_{n=-\frac{N-1}{2}}^{\frac{N-1}{2}} |f_n|^2 - \sum_{m=-\frac{N-1}{2}}^{\frac{N-1}{2}} |f_m|^2 = 0.$$

Since the sum of the squares of the individual errors at the grid points is zero, it necessarily follows that the individual errors themselves must be zero. In other words, the approximation function ψ_N must pass through each of the data points or $\psi_N(x_n) = f_n$ at each grid point. This means that ψ_N is both a least squares approximation and an **interpolating** function for the data. We have arrived at the fact that the least squares approximation is an interpolant in a rather circuitious manner. Indeed, the DFT can be derived directly by requiring that the polynomial ψ_N interpolate the data; this approach is considered in problem 28.

One immediate consequence of this discovery is that with $E = 0$, (2.22) may now be written as

$$\sum_{n=-\frac{N-1}{2}}^{\frac{N-1}{2}} |f_n|^2 = N \sum_{n=-\frac{N-1}{2}}^{\frac{N-1}{2}} |\alpha_n|^2, \tag{2.23}$$

a fundamental property of the DFT known as **Parseval's**[8] **relation.**

Example: A least squares approximation. We would like to present an example of trigonometric approximation using the DFT. In order to provide some continuity with the earlier sections of the chapter, why not use the data set given in Table 2.1? We assume that those data are collected on the interval $[-1/2, 1/2]$, and tabulate them again in Table 2.3.

Notice that the first variable consists of equally spaced points x_n on the interval $[-1/2, 1/2]$; therefore, we can take $A = 1$ in the above formulation. We have discovered that the coefficients in the least squares approximation polynomial

$$\psi_{12}(x) = \sum_{k=-5}^{6} \alpha_k e^{i2\pi k x}$$

are given by $\alpha_k = \mathcal{D}\{f_n\}_k$, and these values are also shown in Table 2.3. Recall that in this case, since the data f_n are real, the coefficients α_k have the property

$$\mathrm{Re}\{\alpha_k\} = \mathrm{Re}\{\alpha_{-k}\} \quad \text{and} \quad \mathrm{Im}\{\alpha_k\} = -\mathrm{Im}\{\alpha_{-k}\}.$$

If we take the values of α_k given in Table 2.3 together with the symmetry in their real and imaginary parts, it can be shown (see problem 25) that

$$\psi_{12}(x) = \alpha_0 + 2\sum_{k=1}^{5} \left(\mathrm{Re}\{\alpha_k\} \cos(2\pi k x) - \mathrm{Im}\{\alpha_k\} \sin(2\pi k x) \right) + \alpha_6 \cos(12\pi x)$$

on the interval $[-1/2, 1/2]$. Note that this expression is essentially the real form of the inverse DFT. This trigonometric polynomial is plotted in Figure 2.9 together with the 12 data points. Clearly, the approximation function does its job: it passes through each of the data points.

2.6. Transforming a Spike Train

We have now derived the DFT as an approximation to the Fourier transform, as an approximation to the Fourier coefficients, and as the coefficients in an interpolating

---

[8]Little is known about the life of MARC ANTOINE PARSEVAL DES CHÊNES. He was born in 1755 and was forced to flee France in 1792 after writing poems critical of Napoleon. Nominated for membership in the Paris Academy of Sciences five times, he was never elected. The only enduring mathematical result published in his five memoirs is what we now call Parseval's relation.

TABLE 2.3
Coefficients of a least squares trigonometric polynomial.

n, k	x_n	f_n	Re $\{\alpha_k\}$	Im $\{\alpha_k\}$
-5	$-5/12$	0.7630	0.0684	-0.1093
-4	$-1/3$	-0.1205	-0.1684	0.0685
-3	$-1/4$	-0.0649	-0.2143	-0.0381
-2	$-1/6$	0.6133	-0.0606	0.1194
-1	$-1/12$	-0.2697	-0.0418	-0.0548
0	0	-0.7216	0.0052	0
1	1/12	-0.0993	-0.0418	0.0548
2	1/6	0.9787	-0.0606	-0.1194
3	1/4	-0.5689	-0.2143	0.0381
4	1/3	-0.1080	-0.1684	-0.0685
5	5/12	-0.3685	0.0684	0.1093
6	1/2	0.0293	0.1066	0.0000

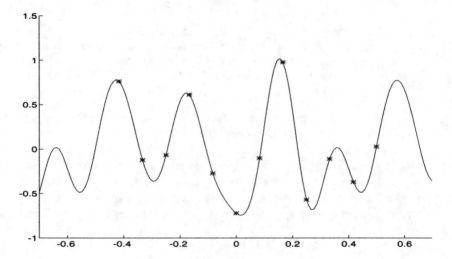

FIG. 2.9. *The figure shows $N = 12$ data points at equally spaced points of the interval $[-1/2, 1/2]$. Using the DFT, the least squares trigonometric polynomial for this data set can be found. This polynomial actually interpolates the data, meaning that it passes through each data point.*

function. While this may seem like an exhaustive exhibition of the DFT, we will now take another approach by computing the Fourier transform of a sampled waveform, and showing that the DFT appears once again. This is a very natural and complementary approach since it requires applying the Fourier transform to a discretized function rather than approximating a Fourier integral. While this approach is stimulating, and also provides a quick survey of delta distributions, readers who feel that the DFT has already had sufficient introduction can move ahead to Chapter 3.

In order to proceed, we must quickly review the **Dirac**[9] **delta function**, often called the **impulse function**. This object, denoted $\delta(t)$, is not, in fact, a function;

---

[9]PAUL ADRIEN MAURICE DIRAC (1902–1984) was one of the greatest theoretical physicists of the twentieth century. His groundbreaking work in quantum mechanics led to the discovery of quantum electrodynamics, the relativistic equations of the electron, and a Noble Prize in 1933.

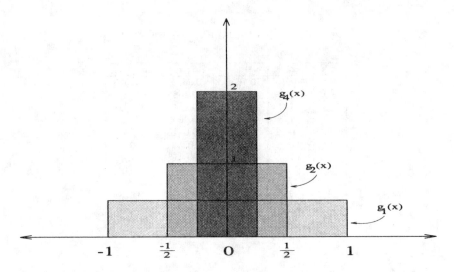

FIG. 2.10. *Several members of the sequence $g_n(x)$ are shown, where $g_n(x) = n/2$ if $|x| \le 1/n$, and $g_n(x) = 0$ otherwise. Note that each member of the sequence is nonzero over a smaller interval and has a greater amplitude than its predecessors. Furthermore, the area under the curve is unity for every member of the sequence.*

it is a **distribution** or a **generalized function** (these two terms actually have different meanings, but for our purposes, they are essentially the same [62], [93], [128]). However, we will abide by common practice and call δ the delta function. The first task is to avoid confusing the delta *function* with the Kronecker delta *sequences* $\hat{\delta}_N(n)$ and $\delta(n)$. Hopefully the notation and context will make the meaning clear.

The delta function can be defined as the limit of a sequence of functions $\{g_n(x)\}$ having certain properties. A sequence $\{g_n(x)\}$ that leads to the δ function must satisfy

$$\lim_{n \to \infty} g_n(x) = 0 \quad \text{if} \quad x \ne 0$$

and

$$\lim_{n \to \infty} \int_{-\infty}^{\infty} g_n(x)dx = 1.$$

A simple example of a sequence possessing these desired properties is

$$g_n(x) = \begin{cases} \dfrac{n}{2} & \text{if} \quad |x| \le \dfrac{1}{n}, \\[2ex] 0 & \text{otherwise.} \end{cases}$$

Successive members of this sequence have increasing amplitude over smaller and smaller intervals about the origin (see Figure 2.10): they get taller and skinnier, while maintaining unit area. For this reason the delta function is often said to have zero width, infinite height, with unit area under the curve. We may view the δ function as a **spike**.

It is tempting to define $\delta(x) = \lim_{n \to \infty} g_n(x)$, but this is not technically correct, since the sequence does not converge at $x = 0$. Nevertheless, we do take this limit as an informal definition, valid except at $x = 0$. The function is officially defined by its action when integrated against other regular functions. That is, we define properties of the delta function implicitly by requiring that, for any suitable function $f(x)$,

$$\int_{-\infty}^{\infty} f(x)\delta(x)dx = \lim_{n\to\infty} \int_{-\infty}^{\infty} f(x)g_n(x)dx.$$

It should be verified that the sequence $\{g_n(x)\}$ (plus several more given in problem 29) satisfies the following properties.

1. **Zero value.** The δ function is zero almost everywhere; that is,

$$\delta(x) = 0 \quad \text{for all} \quad x \neq 0. \tag{2.24}$$

2. **Scaling property.** The δ function has the property that for α a real number,

$$\delta(\alpha x) = \frac{1}{|\alpha|}\delta(x). \tag{2.25}$$

3. **Unit area.** The area under the δ function is unity, or

$$\int_{-\infty}^{\infty} \delta(x)dx = 1. \tag{2.26}$$

4. **Product with regular functions.** For every real number y the product of a continuous function f with the δ function satisfies

$$f(x)\,\delta(x-y) = f(y)\,\delta(x-y). \tag{2.27}$$

By this we mean that

$$\lim_{n\to\infty} \int_{-\infty}^{\infty} f(x)g_n(x-y)dx = \lim_{n\to\infty} \int_{-\infty}^{\infty} f(y)g_n(x-y)dx.$$

5. **Sifting property.** Integrating a function against the δ function has the effect of sifting (or testing for) the value of the function at the origin; that is,

$$\int_{-\infty}^{\infty} f(x)\delta(x)dx = f(0). \tag{2.28}$$

6. **General sifting property.** Generalizing the previous property, the δ function has the property that

$$\int_{-\infty}^{\infty} f(x)\delta(x-y)dx = f(y). \tag{2.29}$$

The sifting property is extremely useful. Suppose, for example, that one wishes to shift a function, but not alter it in any other way (think of introducing a time delay in a signal). Then, by the sifting property, we may conclude that

$$f(x - x_0) = \int_{-\infty}^{\infty} f(y)\delta(y - x - x_0)dy.$$

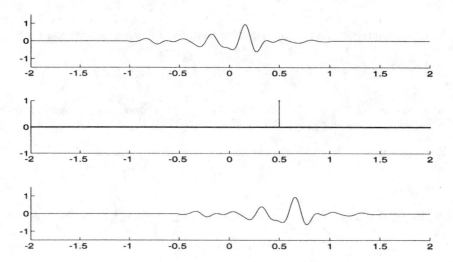

FIG. 2.11. *The figure shows that the convolution of a function f (top) with a spike $\delta(x - x_0)$ located at x_0 (middle) applies a shift to $f(x)$ so that $f(x) * \delta(x - x_0) = f(x - x_0)$ (bottom).*

This integral is a special example of an operation called **convolution**, which we will meet again. We see that the convolution of a function f with a shifted δ function has the effect of translating the function to the location of the δ function (see Figure 2.11).

We will use the δ function to give us yet another derivation of the DFT shortly. Before we do, however, it is worthwhile to apply these properties to develop a few more important Fourier transform pairs.

Example: Fourier transform of a spike. In engineering and physics, we often idealize the response of a system to a brief pulse input by using a δ function which is a spike of infinitesmal duration and infinite amplitude (integrating to unity). What frequencies make up a spike? Consider the Fourier transform of a spike. By the sifting property, we see that

$$\int_{-\infty}^{\infty} \delta(x)e^{-i2\pi\omega x}\,dx \;=\; e^0 \;=\; 1.$$

The transform of a spike in the spatial domain is a flat spectrum; that is, all frequencies are present at unit strength.

One might ask what the presence of a unit spike at the origin in the frequency domain implies about its spatial domain partner. In other words, what is the inverse Fourier transform of a delta function? Intuitively, a spike at zero frequency should correspond to a spatial function that is constant. Indeed, the inverse Fourier transform of a δ distribution is easily obtained, again by the sifting property:

$$\int_{-\infty}^{\infty} \delta(\omega)e^{i2\pi\omega x}\,d\omega \;=\; e^0 \;=\; 1.$$

Example: Fourier transform of the exponential function. What does the presence of a single spike in the frequency domain mean if it is *not* at the origin? Again, arguing by intuition, we expect a single spike at the frequency $\omega = \omega_0$ to represent a function that oscillates at precisely the frequency $\omega = \omega_0$. Consider the inverse transform of the shifted δ function. Once again the sifting property says that

$$\mathcal{F}^{-1}\{\delta(\omega - \omega_0)\} = \int_{-\infty}^{\infty} \delta(\omega - \omega_0)e^{i2\pi\omega x} d\omega$$

$$= e^{i2\pi\omega_0 x} = \cos(2\pi\omega_0 x) + i\,\sin(2\pi\omega_0 x).$$

This result leads easily to the Fourier transform of a cosine, since by a similar argument

$$\mathcal{F}^{-1}\{\delta(\omega + \omega_0)\} = \cos(-2\pi\omega_0 x) + i\,\sin(-2\pi\omega_0 x).$$

Adding and subtracting the inverse transforms of $\delta(\omega - \omega_0)$ and $\delta(\omega + \omega_0)$, we find that

$$\mathcal{F}^{-1}\left\{\frac{1}{2}\delta(\omega - \omega_0) + \frac{1}{2}\delta(\omega + \omega_0)\right\} = \cos(2\pi\omega_0 x)$$

and

$$\mathcal{F}^{-1}\left\{\frac{1}{2i}\delta(\omega - \omega_0) - \frac{1}{2i}\delta(\omega + \omega_0)\right\} = \sin(2\pi\omega_0 x).$$

We now know that the Fourier transform of an exponential $e^{i2\pi\omega_0 x}$ is a single spike located at $\omega = \omega_0$, and that $\cos(2\pi\omega_0 x)$ has a strictly real Fourier transform consisting of a pair of spikes at $\omega = \pm\omega_0$, while the Fourier transform of $\sin(2\pi\omega_0 x)$ is purely imaginary, and consists of a pair of spikes at $\omega = \pm\omega_0$. Figure 2.12 displays the Fourier transform pairs developed using the delta function.

We can now use these results to derive the DFT from a different perspective. Instead of using a sampled version of a continuous function (as in previous approaches), this derivation will apply a continuous transform to a sampled function. Consider a "function" that consists of a finite sequence of regularly spaced spikes, each with some amplitude; we will refer to such a sequence as a **spike train**. (This object is sometimes referred to as a **comb function**, or a **shah function**.) Continuing to use the function notation, a spike train can be represented by a sum of shifted δ functions. If the spikes are separated by a distance Δx and are located at the grid points $x_n = n\Delta x$, then the spike train can be written as

$$h(x) = \sum_n f_n \delta(x - x_n),$$

where f_n is a set of amplitudes for the spikes. Note that $h(x)$ has no meaning in the ordinary sense of a function, being almost everywhere equal to zero.

Notice that we may regard this spike train as samples of an underlying continuum function f where $f_n = f(x_n)$. This can be seen by using property 4 above and writing

$$h(x) = \sum_n f_n \delta(x - x_n) = \sum_n f(x_n)\delta(x - x_n)$$

$$= \sum_n f(x)\delta(x - x_n).$$

In the last line we see that the delta functions $\delta(x - x_n)$ isolate the values of the function f at the grid points $x = x_n$.

Suppose now that the spike train consists of precisely N spikes located at the points $x = x_n = nA/N$ of the interval $[-A/2, A/2]$, where $n = -N/2 + 1 : N/2$. Let's

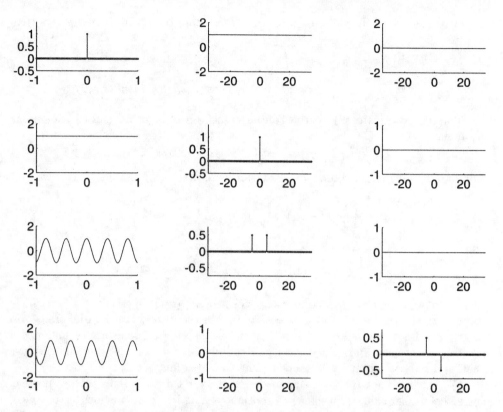

FIG. 2.12. *Several Fourier transform pairs derived with the help of the delta function are shown here. Each row of figures consists of a real-valued function in the spatial domain (left), the real part of its Fourier transform in the frequency domain (center), and the imaginary part of its transform in the frequency domain (right). Shown are: the transform of a spike or impulse (top row), the inverse transform of a spike or impulse (second row), the transform of $\cos(2\pi\omega_0 x)$ (third row), and the transform of $\sin(2\pi\omega_0 x)$ (bottom row).*

see what happens if we formally take the Fourier transform of this spike train. It is given by

$$
\begin{aligned}
\mathcal{F}\{h(x)\}(\omega) &= \int_{-\infty}^{\infty} \left(\sum_{n=-\frac{N}{2}+1}^{\frac{N}{2}} f_n \delta(x - x_n) \right) e^{-i2\pi\omega x}\, dx \\[2mm]
&= \sum_{n=-\frac{N}{2}+1}^{\frac{N}{2}} f_n \int_{-\infty}^{\infty} \delta(x - x_n) e^{-i2\pi\omega x}\, dx \\[2mm]
&= \sum_{n=-\frac{N}{2}+1}^{\frac{N}{2}} f_n e^{-i2\pi\omega x_n}.
\end{aligned}
$$

We have used the sifting property to arrive at the last line which tells us that the Fourier transform of a spike train is just a linear combination of exponentials weighted by the amplitudes that appear in the spike train.

At this point there are two ways to proceed. Having determined the Fourier transform of the spike train, we wish to sample it, so that we have a discrete set of data in both domains. We are at liberty to choose any values of ω at which to sample the Fourier transform. We can rely on our previous experience with the reciprocity relations and choose the N sample points $\omega_k = k\Delta\omega = k/A$, where $k = -N/2 + 1 : N/2$. The sampled version of the Fourier transform is then

$$
\begin{aligned}
\mathcal{F}\{h(x)\}(\omega_k) &= \sum_{n=-\frac{N}{2}+1}^{\frac{N}{2}} f_n e^{-i2\pi\omega_k x_n} \\
&= \sum_{n=-\frac{N}{2}+1}^{\frac{N}{2}} f_n e^{-i2\pi nk/N} \\
&= N\mathcal{D}\{f_n\}_k
\end{aligned}
$$

for $k = -N/2 + 1 : N/2$. In other words, if we are willing to appeal to the reciprocity relations, the DFT appears as the samples of the Fourier transform of the spike train.

However, there is a slightly more adventurous and independent way to capture the DFT. If we assume no familiarity with the reciprocity relations, then the choice of the sample points ω_k is not obvious. They *can* be determined if we require that the inverse Fourier transform applied to the spike train of samples $\mathcal{F}\{h(x)\}(\omega_k)$ brings us back to the input values f_n with which we began. It is an interesting argument and is worth investigating. Before we present it, let's provide some notational relief by agreeing to let the set of indices

$$
\left\{ -\frac{N}{2}+1,\ -\frac{N}{2}+2, \ldots,\ -1, 0, 1, \ldots,\ \frac{N}{2} \right\}
$$

be denoted $\mathcal{N}$.

Let $\omega_k = k\Delta\omega$, where $\Delta\omega$ is as yet unspecified, and $k \in \mathcal{N}$. The sampled version of the Fourier transform can be written as the spike train

$$
\mathcal{F}\{h(x)\}(\omega) = \sum_{k\in\mathcal{N}} \sum_{n\in\mathcal{N}} f_n e^{-i2\pi\omega x_n} \delta(\omega - \omega_k). \tag{2.30}
$$

Another use of property 4 allows us to write

$$
\mathcal{F}\{h(x)\}(\omega) = \sum_{k\in\mathcal{N}} \sum_{n\in\mathcal{N}} f_n e^{-i2\pi\omega_k x_n} \delta(\omega - \omega_k). \tag{2.31}
$$

Notice that (2.31), the continuum Fourier transform of a spike train, is itself a spike train. The amplitudes of the spikes, that is, the coefficients of the $\delta(\omega - \omega_k)$ terms, are given by

$$
\sum_{n\in\mathcal{N}} f_n e^{-i2\pi\omega_k x_n}. \tag{2.32}
$$

Forming the inverse Fourier transform of the spike train (2.31), we have

$$
\mathcal{F}^{-1}\left\{\mathcal{F}\left\{h(x)\right\}(\omega)\right\}(x) = \int_{-\infty}^{\infty} \sum_{k \in \mathcal{N}} \sum_{n \in \mathcal{N}} f_n e^{-i2\pi\omega_k x_n} \delta(\omega - \omega_k) e^{i2\pi\omega x} d\omega
$$

$$
= \sum_{k \in \mathcal{N}} \sum_{n \in \mathcal{N}} f_n e^{-i2\pi\omega_k x_n} \int_{-\infty}^{\infty} \delta(\omega - \omega_k) e^{i2\pi\omega x} d\omega
$$

$$
= \sum_{k \in \mathcal{N}} \sum_{n \in \mathcal{N}} f_n e^{-i2\pi\omega_k(x_n - x)},
$$

which is a continuous function of x. At the peril of becoming strangled in our notation, we now sample this function by evaluating it at all of the original grid points $x_p = pA/N$, where $p \in \mathcal{N}$. This yields the sampled inverse Fourier transform

$$
\mathcal{F}^{-1}\left\{\mathcal{F}\left\{h(x)\right\}(\omega)\right\}(x_p) = \sum_{k \in \mathcal{N}} \sum_{n \in \mathcal{N}} f_n e^{-i2\pi\omega_k(x_n - x_p)}.
$$

Rearranging the order of summation, replacing x_n and x_p by their definitions, and combining the exponentials we may write

$$
\mathcal{F}^{-1}\left\{\mathcal{F}\left\{h(x)\right\}(\omega)\right\}(x_p) = \sum_{n \in \mathcal{N}} f_n \sum_{k \in \mathcal{N}} e^{i2\pi\omega_k(p-n)A/N}.
$$

We now see precisely which value for $\Delta\omega$ we must select to recover the original values of the input f_n. If we select $\Delta\omega = 1/A$, then $\omega_k = k/A$, and the previous equation becomes

$$
\mathcal{F}^{-1}\left\{\mathcal{F}\left\{h(x)\right\}(\omega)\right\}(x_p) = \sum_{n \in \mathcal{N}} f_n \underbrace{\sum_{k \in \mathcal{N}} e^{i2\pi k(p-n)/N}}_{N\hat{\delta}_N(p-n)}.
$$

The key to this passage is the orthogonality property. The summation over k collapses to a single term, namely $N\hat{\delta}_N(p-n)$, and we obtain

$$
\mathcal{F}^{-1}\left\{\mathcal{F}\left\{f(x)\right\}(\omega)\right\}(x_p) = N \sum_{n \in \mathcal{N}} f_n \hat{\delta}_N(p-n) = f_p.
$$

Hence, with the choice $\Delta\omega = 1/A$, we find that the inverse Fourier transform of the spike train of Fourier transform values (2.31) returns the original input data, up to a multiple of N. For this reason, it seems natural to adopt $1/N$ times the sequence of amplitudes (2.32) of the spike train $\mathcal{F}\left\{h(x)\right\}(\omega)$ as the DFT. The DFT then becomes

$$
\mathcal{D}\left\{f_n\right\}_k = \frac{1}{N} \sum_{n \in \mathcal{N}} f_n e^{-i2\pi nk/N},
$$

which is precisely the formula obtained in previous derivations.

The upshot of this argument (either using the reciprocity relations or orthogonality) is that samples of a continuous function f can be expressed as a spike train of delta functions. The (continuous) Fourier transform of that spike train when sampled at the appropriate frequencies is precisely the DFT up to the scaling factor N. Although the demonstration of this fact relies on the use of the delta function, it does offer an alternative pathway to the DFT.

2.7.   Limiting Forms of the DFT–IDFT Pair

The remainder of this chapter will be devoted to a qualitative exploration of the relationships among the DFT, the Fourier series, and the Fourier transform. The apparent similarities among this trio are tantalizing; at the same time, there are differences that must be understood and reconciled. Of particular interest is the question: *exactly what does the DFT approximate?* We will try to answer this question conceptually by looking at the DFT in various limits, most notably when $N \to \infty$. While the discussion in this chapter will be incomplete, it does reach some salient qualitative conclusions which are valuable in their own right. It also sets the stage for the full analysis which will take place in Chapter 6.

Limiting Forms of the Forward DFT

Let's begin with a process that is hopefully becoming familiar. A function f is sampled on the interval $[-A/2, A/2]$ at the N grid points $x_n = n\Delta x$, where $\Delta x = A/N$. We will denote the sampled values of the function $f_n = f(x_n)$. Now consider the DFT as it has already been defined. It looks like

$$\mathcal{D}\{f_n\}_k = \frac{1}{N} \sum_{n=-\frac{N}{2}+1}^{\frac{N}{2}} f_n e^{-i2\pi nk/N}$$

for $k = -N/2 + 1 : N/2$. The discussion can be divided quite concisely into three situations for the forward DFT.

1. Let's assume first that f is a function with period A and that the goal is to approximate its Fourier coefficients. Since $x_n = n\Delta x = nA/N$, we can replace n/N by x_n/A in the DFT definition and write

$$AD\{f_n\}_k = \frac{A}{N} \sum_{n=-\frac{N}{2}+1}^{\frac{N}{2}} f_n e^{-i2\pi x_n k/A} = \Delta x \sum_{n=-\frac{N}{2}+1}^{\frac{N}{2}} f_n e^{-i2\pi x_n k/A}$$

for $k = -N/2 + 1 : N/2$. We now ask what happens if we hold A and k fixed and let $N \to \infty$. Since $\Delta x = A/N$, $\Delta x \to 0$ as $N \to \infty$, and the DFT sum approaches an integral. In fact,

$$\lim_{\substack{N\to\infty \\ \Delta x\to 0}} AD\{f_n\}_k = \lim_{\substack{N\to\infty \\ \Delta x\to 0}} \Delta x \sum_{n=-\frac{N}{2}+1}^{\frac{N}{2}} f(x_n) e^{-i2\pi x_n k/A}$$

$$= \int_{-\frac{A}{2}}^{\frac{A}{2}} f(x) e^{-i2\pi kx/A} dx$$

$$= Ac_k,$$

or simply

$$\lim_{\substack{N\to\infty \\ \Delta x\to 0}} \mathcal{D}\{f_n\}_k = c_k.$$

We see that as $N \to \infty$ and $\Delta x \to 0$, the DFT approaches the Fourier coefficients of f on the interval $[-A/2, A/2]$. For finite values of N the DFT only approximates the c_k's, and the error in this approximation is a **discretization error** that arises because Δx is small, but nonzero. Letting $N \to \infty$ and $\Delta x \to 0$ reduces this error.

2. Consider the case of a spatially limited function f which satisfies $f(x) = 0$ for $|x| > A/2$. Now the goal is to approximate the Fourier transform of f. We have already observed that in this case, the value of the Fourier transform at the frequency grid point $\omega_k = k/A$ is a multiple of the kth Fourier coefficient; that is,

$$c_k = \frac{1}{A}\hat{f}\left(\frac{k}{A}\right) = \frac{1}{A}\hat{f}(\omega_k).$$

Therefore, if we hold A fixed (which means $\Delta\omega = 1/A$ is also fixed) and hold k fixed (which means that $\omega_k = k\Delta\omega$ is also fixed), then we have that

$$\lim_{\substack{N \to \infty \\ \Delta x \to 0}} A\mathcal{D}\{f_n\}_k = \hat{f}(\omega_k).$$

It is interesting to note that by the reciprocity relation $A\Omega = N$, the length of the frequency domain Ω increases in this limit. In other words, as $N \to \infty$ and $\Delta x \to 0$, higher frequencies can be resolved by the grid which is reflected in an increase in the length of the frequency domain. As in the previous case, the error in the DFT approximation to $\hat{f}(\omega_k)$ is a discretization error that decreases as $N \to \infty$.

It is misleading to think that the DFT becomes the entire Fourier transform as the number of data samples becomes infinite. Letting $N \to \infty$ allows the DFT sum to converge to the Fourier integral, but the transform is approximated only at certain isolated points in the frequency domain since $\Delta\omega$ remains constant as $N \to \infty$. The DFT does *not* provide any information whatsoever about $\hat{f}(\omega)$ if $\omega \neq k/A$. This distinction will prove to be of crucial importance in understanding the relationship between the DFT and the Fourier transform.

3. The third case is that in which f does not vanish outside of a finite interval. As in the previous case we can let $N \to \infty$ and $\Delta x \to 0$ with k held constant in the DFT. This limit gives us

$$\lim_{\substack{N \to \infty \\ \Delta x \to 0}} A\mathcal{D}\{f_n\}_k = \lim_{\substack{N \to \infty \\ \Delta x \to 0}} \Delta x \sum_{n=-\frac{N}{2}+1}^{\frac{N}{2}} f_n e^{-i2\pi\omega_k x_n k}$$

$$= \int_{-\frac{A}{2}}^{\frac{A}{2}} f(x)e^{-i2\pi\omega_k x}dx.$$

However, since f is nonzero outside of the interval $[-A/2, A/2]$, the integral in the previous line is necessarily an approximation to the Fourier transform

$$\hat{f}(\omega_k) = \int_{-\infty}^{\infty} f(x)e^{-i2\pi\omega_k x}dx.$$

The error in this approximation is a **truncation error** due to the fact that the interval of integration $(-\infty, \infty)$ has been truncated. Therefore, a second limit is required to eliminate this error and recover the exact value of $\hat{f}(\omega_k)$. Consider the effect of letting $A \to \infty$. By the reciprocity relations, this limit will also force $\Delta\omega = 1/A$ to zero. At this point we are involved in a very formal and conceptual exercise. However, the conclusion is quite instructive. Imagine that as $A \to \infty$ and $\Delta\omega \to 0$, we hold the combination $\omega_k = k\Delta\omega$ fixed. Then we can write formally that

$$\lim_{\substack{A \to \infty \\ \Delta\omega \to 0}} \int_{-\frac{A}{2}}^{\frac{A}{2}} f(x)e^{-i2\pi\omega_k x}\,dx = \hat{f}(\omega_k) \quad \text{for} \quad \omega_k \text{ fixed.}$$

We can now summarize this two-step limit process. The combined effect of the two limits might be summarized in the following manner:

$$\lim_{\substack{A \to \infty \\ \Delta\omega \to 0}} \lim_{\substack{N \to \infty \\ \Delta x \to 0}} A\mathcal{D}\{f_n\}_k = \hat{f}(\omega_k) \quad \text{for} \quad \omega_k \text{ fixed.}$$

The outer limit eliminates the truncation error due to the fact that the DFT approximates an integral on the finite interval $[-A/2, A/2]$, while the inner limit eliminates the discretization error that arises because the DFT uses a discrete grid in the spatial domain. Of course, these limits cannot be carried out computationally, but they do tell us how DFT approximations can be improved by changing various parameters. These limits will be revisited and carried out explicitly in Chapter 6.

Refining the Frequency Grid

We have seen that with the interval $[-A/2, A/2]$ fixed, the quantity $A\mathcal{D}\{f_n\}_k$ approximates the Fourier transform at the discrete frequencies $\omega_k = k/A$. A common question is: what can be done if values of the Fourier transform are needed at points other than k/A? An approach to approximating $\hat{f}(\omega)$ at intermediate values of ω can be developed from the reciprocity relations. By way of example, suppose that values of $\hat{f}(\omega)$ are desired on a grid with spacing $\Delta\omega^{\text{new}} = 1/(2A)$ rather than $\Delta\omega = 1/A$. From the reciprocity relation (2.5),

$$\Delta x^{\text{new}} \Delta\omega^{\text{new}} = \frac{1}{N}.$$

Therefore,

$$\Delta\omega^{\text{new}} = \frac{1}{2A} \quad \text{implies that} \quad \Delta x^{\text{new}} = \frac{2A}{N}.$$

We see that if the number of points N is held fixed and the grid spacing Δx is doubled, then the grid spacing in the frequency domain $\Delta\omega$ is halved. Notice that this process also doubles the length of the spatial domain $(N\Delta x)$ and halves the length of the frequency domain $(N\Delta\omega)$. This operation clearly provides a refinement of the frequency grid, as shown in Figure 2.13.

If we are willing to double the number of grid points, then it is possible to refine the frequency grid without increasing Δx or reducing the length of the frequency domain Ω. The new grid lengths must also obey the reciprocity relations. Therefore, increasing N to $2N$ with Ω constant has the effect of doubling the length of the spatial domain since the reciprocity relation with the new grid parameters,

$$(2A)\Omega = 2N,$$

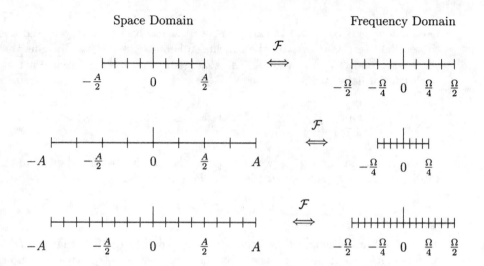

FIG. 2.13. *The reciprocity relations determine how changes in the spatial grid affect the frequency grid. If the number of grid points N is held fixed and the grid spacing Δx is doubled, then the length of the spatial domain is doubled, and the frequency grid spacing is halved, as is the length of the frequency domain (top → middle). If the number of grid points N is doubled with Δx fixed, then $\Delta \omega$ is halved while the length of the frequency domain remains constant (top → bottom). Both of these processes result in a refinement of the frequency grid.*

must still be obeyed. At the same time, the other reciprocity relation decrees that the frequency grid spacing must decrease if the reciprocity relations are to be satisfied:

$$\Delta x \left(\frac{\Delta \omega}{2} \right) = \frac{1}{2N}.$$

This manipulation provides the desired refinement in the frequency domain without losing representation of the high frequencies as shown in Figure 2.13. In either case, doubling A has the effect of halving $\Delta \omega$, which refines the frequency grid. How does this refinement affect the sampling of the given function f, which must now be sampled on the interval $[-A, A]$? If f is spatially limited and zero outside the interval $[-A/2, A/2]$, then it must be extended with zeros on the intervals $[-A, -A/2]$ and $[A/2, A]$. If f is not zero outside the interval $[-A/2, A/2]$, then we simply sample f at the appropriate grid points on the interval $[-A, A]$. We will return to the practice of padding input sequences with zeros in Chapter 3.

Limiting Forms of the IDFT

We will close this section by asking: *what does the inverse DFT approximate?* The question will be answered by looking at the limiting forms of the IDFT. As with the forward DFT the account can be given in three parts.

1. We first consider the case in which coefficients F_k are given. These are presumed to be the Fourier coefficients of a function f on an interval $[-A/2, A/2]$. The meaning of the coefficients F_k determines how we interpret the output of the

IDFT. The IDFT appears in its usual form as

$$\mathcal{D}^{-1}\{F_k\}_n = \sum_{k=-\frac{N}{2}+1}^{\frac{N}{2}} F_k e^{i2\pi nk/N}$$

for $n = -N/2 + 1 : N/2$. We anticipate using a spatial grid on the interval $[-A/2, A/2]$ which has grid points $x_n = n\Delta x = nA/N$. Setting $n/N = x_n/A$ in the IDFT we have

$$\mathcal{D}^{-1}\{F_k\}_n = \sum_{k=-\frac{N}{2}+1}^{\frac{N}{2}} F_k e^{i2\pi k x_n/A}$$

for $n = -N/2 + 1 : N/2$. We ask about the effect of letting $N \to \infty$ in this IDFT expression. Holding x_n and A fixed, we see that the IDFT becomes the Fourier series representation for f evaluated at the point $x = x_n$:

$$\lim_{N\to\infty} \mathcal{D}^{-1}\{F_k\}_n = \lim_{N\to\infty} \sum_{k=-\frac{N}{2}+1}^{\frac{N}{2}} F_k e^{i2\pi k x_n/A} = f(x_n).$$

So this is the first interpretation of the IDFT: if the input sequence F_k is regarded as a set of Fourier coefficients for a function f on a known interval $[-A/2, A/2]$, the IDFT approximates the Fourier series synthesis of f at isolated points x_n. The error that arises in the IDFT approximations to the Fourier series is a truncation error, since the IDFT is a truncated Fourier series representation.

2. Now consider the case in which the input F_k to the IDFT is regarded as a set of samples of the Fourier transform $\hat{f}$ of a function f. Assume furthermore that the function f is **band-limited**, which means that $\hat{f}(\omega) = 0$ for $|\omega| > \Omega/2$ for some cut-off frequency $\Omega/2$. The samples of the Fourier transform, $F_k = \hat{f}(\omega_k)$, are taken at N equally spaced points $\omega_k = k\Delta\omega$ of the interval $[-\Omega/2, \Omega/2]$, where $\Delta\omega = \Omega/N$. The grid that has been created in the frequency domain induces a grid in the spatial domain, and not surprisingly, the reciprocity relations come into play. As we have seen, the grid points in the spatial domain are $x_n = n\Delta x = nA/N$, where $\Delta x = 1/\Omega$.

Now let's have a look at the IDFT. It can be written

$$\Delta\omega \mathcal{D}^{-1}\{F_k\}_n = \Delta\omega \sum_{k=-\frac{N}{2}+1}^{\frac{N}{2}} F_k e^{i2\pi nk/N} = \Delta\omega \sum_{k=-\frac{N}{2}+1}^{\frac{N}{2}} F_k e^{i2\pi x_n \omega_k}$$

for $n = -N/2 + 1 : N/2$. We have used the relationships between the two grids to argue that

$$\frac{nk}{N} = \frac{nA}{N}\frac{k}{A} = x_n \omega_k.$$

We can now see the effect of letting $N \to \infty$. Recall that the length of the frequency domain Ω is fixed, as is the spatial grid point x_n. Therefore, as

$N \to \infty$, it follows that $\Delta\omega = \Omega/N \to 0$, and the length of the spatial domain A increases. Therefore, in this limit the IDFT approaches an integral; specifically,

$$\lim_{\substack{N\to\infty \\ \Delta\omega\to 0}} \Delta\omega \mathcal{D}^{-1}\{F_k\}_n = \lim_{\substack{N\to\infty \\ \Delta\omega\to 0}} \Delta\omega \sum_{k=-\frac{N}{2}+1}^{\frac{N}{2}} F_k e^{i2\pi nk/N}$$

$$= \lim_{\substack{N\to\infty \\ \Delta\omega\to 0}} \Delta\omega \sum_{k=-\frac{N}{2}+1}^{\frac{N}{2}} \hat{f}(\omega_k) e^{i2\pi x_n \omega_k}$$

$$= \int_{-\frac{\Omega}{2}}^{\frac{\Omega}{2}} \hat{f}(\omega) e^{i2\pi \omega x_n} d\omega$$

$$= \mathcal{F}^{-1}\{\hat{f}(\omega)\}(x_n)$$

$$= f(x_n).$$

Note that the integral over the interval $[-\Omega/2, \Omega/2]$ is the inverse Fourier transform, since $\hat{f}$ is assumed to be zero outside of this interval. The interpretation of this chain of thought is that the IDFT approximates the inverse Fourier transform evaluated at $x = x_n$, which is just $f(x_n)$. The error that is reduced by letting $N \to \infty$ is a discretization error, since a discrete grid is used in the frequency domain.

3. The assumption of band-limiting is a rather special case. More typically, the Fourier transform does not vanish outside of a finite interval. In this more general case, we might still expect the IDFT to approximate the inverse Fourier transform, and hence samples of f; however, there are both discretization errors and truncation errors to overcome. We may begin by assuming that Ω (and hence, by reciprocity, Δx) are held fixed while $N \to \infty$ and $\Delta\omega \to 0$. As in the previous case we have

$$\lim_{\substack{N\to\infty \\ \Delta\omega\to 0}} \Delta\omega \mathcal{D}^{-1}\{F_k\}_n = \int_{-\frac{\Omega}{2}}^{\frac{\Omega}{2}} \hat{f}(\omega) e^{i2\pi \omega x_n} d\omega.$$

However, this last integral is not the inverse Fourier transform

$$f(x_n) = \int_{-\infty}^{\infty} \hat{f}(\omega) e^{i2\pi \omega x_n} d\omega$$

since $\hat{f}$ no longer vanishes outside $[-\Omega/2, \Omega/2]$. In order to recover the inverse Fourier transform, we must also (quite formally) let $\Omega \to \infty$. Note that by the reciprocity relations this also means that $\Delta x \to 0$, and so we must delicately keep the grid point x_n fixed. The limit takes the form

$$\lim_{\substack{\Omega\to\infty \\ \Delta x\to 0}} \int_{-\frac{\Omega}{2}}^{-\frac{\Omega}{2}} \hat{f}(\omega) e^{i2\pi \omega x_n} d\omega = \mathcal{F}^{-1}\{\hat{f}(\omega)\}(x_n) = f(x_n) \quad \text{for} \quad x_n \text{ fixed.}$$

Combining the two limit steps we may write collectively that for a fixed value of x_n

$$\lim_{\substack{\Omega\to\infty \\ \Delta x\to 0}} \lim_{\substack{N\to\infty \\ \Delta\omega\to 0}} \Delta\omega \mathcal{D}^{-1}\{F_k\}_n = f(x_n).$$

The outer limit accounts for the fact that the IDFT approximations use a finite interval of integration in the frequency domain. The inner limit recognizes the fact that the IDFT uses a finite grid to approximate an integral. This two-step limit process will also be examined more carefully and carried out explicity in Chapter 6.

This concludes an extremely heuristic analysis of the relationships among the DFT/IDFT, Fourier series, and Fourier transforms. Hopefully even this qualitative treatment illustrates a few fundamental lessons. First, it is essential to know what the input sequence to either the DFT or the IDFT represents, since this will determine what the output sequence approximates. We have seen that the DFT may provide approximations to either Fourier coefficients or samples of the Fourier transform, depending upon the origins of the input sequence. We also demonstrated that the IDFT can approximate either the value of a Fourier series or the inverse Fourier transform at selected grid points, again depending upon the interpretation of the input data. Beyond these observations, it is crucial to understand how the DFT/IDFT approximations can be improved computationally. While the limiting processes just presented are formal, they can be simulated in computation by varying the grid parameters in appropriate ways. Above all, we wish to emphasize that there are subtleties when it comes to interpreting the DFT and the IDFT. We will most assuredly return to these subtleties in the pages to come.

2.8. Problems

8. The roots of unity.   Show that ω_N^{-k} satisfies the equation $z^N - 1 = 0$ for $k = 0 : N - 1$. Show that ω_N^{-k} satisfies the equation $z^N - 1 = 0$ for $k = P : P + N - 1$, where P is any integer.

9. Periodicity of ω_N^{nk}.

(a) Show that the sequence $\{\omega_N^k\}_{k=-\infty}^{\infty}$ is periodic of length N.

(b) Show that the sequence $\{\omega_N^{nk}\}_{n,k=-\infty}^{\infty}$ is periodic in both n and k with period N.

(c) Show also that $\omega_N^k \neq \omega_N^p$ if $0 < |k - p| < N$.

(d) Show f_n and F_k as given by the forward (2.6) and inverse (2.9) DFT are N-periodic sequences.

10. Modes of the DFT.   Write out and sketch the real modes (cosine and sine modes) of the DFT in the case $N = 8$. For each mode note its period and frequency.

11. General orthogonality.   Show that the discrete orthogonality relation may be defined on any set of N consecutive integers; that is, for any integer P,

$$\sum_{n=P}^{P+N-1} e^{i2\pi nj/N} e^{-i2\pi nk/N} = N \hat{\delta}_N(j - k).$$

12. Another proof of orthogonality.   An entirely different approach may be used to prove orthogonality without considering the polynomial $z^N - 1$. Use the partial sum of the geometric series

$$\sum_{n=P}^{P+N-1} a^n = \frac{a^P - a^{P+N}}{1-a}$$

to prove the orthogonality property of problem 11.

13. Matrix representation.   In the text, the matrix, **W**, for the alternate DFT

$$F_k = \frac{1}{N} \sum_{n=0}^{N-1} f_n \omega_N^{-nk},$$

for $k = 0 : N-1$, was presented. Find the DFT matrix that corresponds to the DFT definition

$$F_k = \frac{1}{N} \sum_{n=-\frac{N}{2}+1}^{\frac{N}{2}} f_n \omega_N^{-nk},$$

for $k = -N/2 + 1 : N/2$, when $N = 8$.

14. The reciprocity relations.   In each of the following cases assume that a function f is sampled at N points of the given interval in the spatial domain. Give the corresponding values of $\Delta\omega$ and Ω in the frequency domain. Then give the values of all grid parameters $(A, \Delta\omega, \text{and } \Omega)$ if (i) N is doubled and Δx is halved, and (ii) N is doubled leaving Δx unchanged. Sketches of the spatial and frequency grids before and after each change would be most informative.

   (a) $[-1, 1]$ with $N = 8$,

   (b) $[-\pi, \pi]$ with $N = 16$,

   (c) $[-1/2, 1/2]$ with $N = 32$.

15. Using the reciprocity relations to design grids.

   (a) The function f has a period of two and is sampled with $N = 64$ grid points on an interval of length $A = 2$. These samples are used as input to the DFT in order to approximate the Fourier coefficients of f. What are the minimum and maximum frequencies that are represented by the DFT? Do these minimum and maximum frequencies change if the same function is sampled, with $N = 64$, on an interval of length $A = 4$?

   (b) The function f is zero for $|x| > 5$. The DFT will be used to approximate the Fourier transform of f at discrete frequencies up to a maximum frequency of 100 periods per unit length. Use the reciprocity relations to select a grid in the spatial domain that will provide this resolution. Specify the minimum N and the maximum Δx that will do the job.

   (c) The function f is available for sampling on the interval $(-\infty, \infty)$. The DFT will be used to approximate the Fourier transform of f at discrete frequencies up to a maximum frequency of 500 periods per unit length. Use the reciprocity relations to select a grid in the spatial domain that will provide this resolution. Specify the minimum N and the maximum Δx that will do the job.

16. Inverse relation. Follow the proof of Theorem 2.2 to show that

$$\mathcal{D}\left\{\mathcal{D}^{-1}\left\{F_k\right\}_n\right\}_k = F_k.$$

17. Equivalence of DFT forms. Assuming that the input sequence f_n and the output sequence F_k are periodic, show that the N-point DFT can be defined on any set of N consecutive integers,

$$F_k = \frac{1}{N} \sum_{n=P}^{P+N-1} f_n \omega_N^{-nk}$$

for $k = Q : Q + N - 1$, where P and Q are any integers.

18. Matrix properties. Show that the DFT matrix $\mathbf{W}$ is symmetric and that $\mathbf{W}^{-1} = N\mathbf{W}^*$. Furthermore, show that $N\mathbf{W}\mathbf{W}^H = \mathbf{I}$, where H denotes the conjugate transpose. Hence $\mathbf{W}$ is a unitary matrix up to a factor of N.

19. Average value property. Show that the zeroth DFT coefficient is the average value of the input sequence

$$F_0 = \frac{1}{N} \sum_{N=-\frac{N}{2}+1}^{\frac{N}{2}} f_n.$$

20. Modes of the alternate DFT. Consider the alternate form of the DFT

$$F_k = \sum_{n=0}^{N-1} f_n \omega_N^{-nk}.$$

Now the modes ω_N^{-nk} are indexed by $k = 0 : N - 1$. For each value of k in this set, find the period and frequency of the corresponding mode. In particular, verify that the modes with $k = 0$ and $k = N/2$ (for N even) are real. Show that the modes with indices k and $N - k$ have the same period, which means that the high frequency modes are clustered near $k = N/2$, while the low frequency modes are near $k = 0$ and $k = N$.

21. Sampling input for the DFT. Consider the functions

$$\text{(a) } f(x) = x, \qquad \text{(b) } f(x) = |x|, \qquad \text{(c) } f(x) = 1 - |x|$$

on the interval $[-1, 1]$. Sketch each function on $[-1, 1]$ and sketch its periodic extension on $[-4, 4]$. Show how each function should be sampled if the DFT is to be used to approximate its Fourier coefficients on $[-1, 1]$.

22. Continuous orthogonality. Show that the continuous Fourier modes on the interval $[-A/2, A/2]$ satisfy the orthogonality relation

$$\frac{1}{A} \int_{-\frac{A}{2}}^{\frac{A}{2}} e^{i2\pi jx/A} e^{-i2\pi kx/A} dx = \delta(j - k)$$

for integers j and k. (Hint: Consider $j = k$ separately and integrate directly when $j \neq k$.)

23. Real form of the Fourier Series. Use the Euler relations

$$\cos \phi = \frac{e^{i\phi} + e^{-i\phi}}{2}, \qquad \sin \phi = \frac{e^{i\phi} - e^{-i\phi}}{2i}$$

to show that if f is real-valued, then the real form of the Fourier series

$$f(x) \sim \frac{a_0}{2} + \sum_{k=1}^{\infty} a_k \cos(2\pi kx/A) + \sum_{k=1}^{\infty} b_k \sin(2\pi kx/A)$$

with coefficients given by

$$a_k = \frac{2}{A} \int_{-\frac{A}{2}}^{\frac{A}{2}} f(x) \cos\left(\frac{2\pi kx}{A}\right) dx$$

for $k = 0, 1, 2, \ldots$, and by

$$b_k = \frac{2}{A} \int_{-\frac{A}{2}}^{\frac{A}{2}} f(x) \sin\left(\frac{2\pi kx}{A}\right) dx$$

for $k = 1, 2, \ldots$, is equivalent to the complex form

$$f(x) \sim \sum_{k=-\infty}^{\infty} c_k e^{i2\pi k, x/A},$$

where the coefficients c_k are given by

$$c_k = \frac{1}{A} \int_{-\frac{A}{2}}^{\frac{A}{2}} f(x) e^{-i2\pi kx/A} dx$$

for $k = -\infty : \infty$. Find the relationship between c_k and $\{a_k, b_k\}$.

24. Orthogonality of sines and cosines. Show that the functions $\mathbf{w}^k$ and $\mathbf{v}^j$, defined by

$$\mathbf{w}^k(x) = \cos(2\pi kx/A) \quad \text{and} \quad \mathbf{v}^j(x) = \sin(2\pi jx/A),$$

satisfy the following orthogonality properties on $[-A/2, A/2]$:

$$\langle \mathbf{w}^k, \mathbf{v}^j \rangle = 0 \quad \text{for all } j, k,$$

$$\langle \mathbf{w}^k, \mathbf{w}^j \rangle = \begin{cases} 0 & \text{if} \quad k \neq j, \\ A/2 & \text{if} \quad k = j, \end{cases}$$

$$\langle \mathbf{v}^k, \mathbf{v}^j \rangle = \begin{cases} 0 & \text{if} \quad k \neq j, \\ A/2 & \text{if} \quad k = j \neq 0, \end{cases}$$

where

$$\langle \mathbf{v}, \mathbf{w} \rangle = \int_{-A/2}^{A/2} \mathbf{v}(x) \mathbf{w}(x) dx.$$

25. Real form of the trigonometric polynomial. Assuming N is odd, show that if the coefficients α_k in the N-term trigonometric polynomial

$$\psi_N(x) = \sum_{k=-\frac{N-1}{2}}^{\frac{N-1}{2}} \alpha_k e^{i2\pi kx/A}$$

satisfy

$$\mathrm{Re}\{\alpha_k\} = \mathrm{Re}\{\alpha_{-k}\} \quad \text{and} \quad \mathrm{Im}\{\alpha_k\} = -\mathrm{Im}\{\alpha_{-k}\}$$

(which will be the case if the data f_n are real-valued), then the polynomial can be written in the form

$$\psi_N(x) = \alpha_0 + 2\sum_{k=1}^{\frac{N-1}{2}} \left(\mathrm{Re}\{\alpha_k\} \cos\left(\frac{2\pi kx}{A}\right) - \mathrm{Im}\{\alpha_k\} \sin\left(\frac{2\pi kx}{A}\right) \right) + \alpha_{\frac{N}{2}} \cos(\pi Nx).$$

26. The normal equations. Show that the necessary conditions that the least squares error be minimized,

$$\frac{\partial E}{\partial \alpha_k} = 0,$$

for $k = -(N-1)/2 : (N-1)/2$, yield the system of normal equations (2.18)

$$\frac{\partial E}{\partial \alpha_k} = \sum_{n=-\frac{N-1}{2}}^{\frac{N-1}{2}} \left[e^{-i2\pi nk/N} \left(f_n - \sum_{p=-\frac{(N-1)}{2}}^{\frac{(N-1)}{2}} \alpha_p\, e^{i2\pi np/N} \right) \right] = 0,$$

where $k = -(N-1)/2 : (N-1)/2$.

27. Properties of the least squares approximation. Verify the following facts that were used to relate the DFT to the least squares trigonometric approximation. The least squares approximation to the N data points (x_n, f_n) is denoted ψ_N.

(a) Use the orthogonality property and the fact that $\alpha_k = \mathcal{D}\{f_n\}_k$ to show that

$$\sum_{n=-\frac{N-1}{2}}^{\frac{N-1}{2}} |\psi_N(x_n)|^2 = \sum_{n=-\frac{N-1}{2}}^{\frac{N-1}{2}} f(x_n)\psi_N^*(x_n)$$

$$= \sum_{n=-\frac{N-1}{2}}^{\frac{N-1}{2}} f^*(x_n)\psi_N(x_n)$$

$$= N \sum_{n=-\frac{N-1}{2}}^{\frac{N-1}{2}} |\alpha_n|^2.$$

(b) Show that with the coefficients α_k given by the DFT, the error in the least squares approximations is zero; that is,

$$E = \sum_{n=-\frac{N-1}{2}}^{\frac{N-1}{2}} |f_n|^2 - N \sum_{n=-\frac{N-1}{2}}^{\frac{N-1}{2}} |\alpha_n|^2 = 0.$$

28. Trigonometric interpolation.   The result of the least squares approximation question was that the trigonometric polynomial ψ_N actually interpolates the given data. Show this conclusion directly: given a data set (x_n, f_n) for $n = -N/2+1 : N/2$, where $x_n = nA/N$, assume an interpolating polynomial of the form

$$\psi_N(x) = \sum_{k=-\frac{N}{2}+1}^{\frac{N}{2}} \alpha_k e^{i2\pi kx/A}.$$

Then impose the interpolation conditions $\psi_N(x_n) = f_n$ for $n = -N/2 + 1 : N/2$ to show that the coefficients are given by $\alpha_k = \mathcal{D}\{f_n\}_k$.

29. The Dirac δ function.   Sketch and show that each of the following sequences of functions $\{g_n(x)\}_{n=1}^{\infty}$ satisfies the three properties

$$\lim_{n\to\infty} g_n(x) = 0, \quad \text{for } x \neq 0,$$

$$\lim_{n\to\infty} \int_{-\infty}^{\infty} g_n(x)dx = 1,$$

$$\lim_{n\to\infty} \int_{-\infty}^{\infty} g_n(x)h(x)dx = h(0),$$

and that each sequence could thus be used as a defining sequence for the Dirac δ function.

(a) $g_n(x) = \begin{cases} n/2 & \text{for } |x| \leq 1/n, \\ 0 & \text{for } |x| > 1/n, \end{cases}$ (b) $g_n(x) = \dfrac{n}{2}e^{-n|x|},$

(c) $g_n(x) = \begin{cases} n(1 - n|x|) & \text{for } |x| \leq 1/n, \\ 0 & \text{for } |x| > 1/n, \end{cases}$ (d) $g_n(x) = \dfrac{n}{\sqrt{\pi}}e^{-n^2 x^2}.$

Thus each of these sequences could be used to define the Dirac δ function.

Chapter **3**

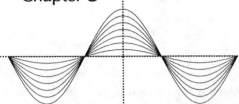

Properties of the DFT

*All intelligent
conversation is playing
on words. The rest is
definition or instruction.*
— Herman Wouk
The Caine Mutiny

3.1. Alternate Forms for the DFT

The DFT arises in so many different settings and is used by practitioners in so many different fields that, not surprisingly, it appears in many different disguises. In defining the DFT in the previous chapter, we issued the proviso that the definition used primarily in this book, namely

$$F_k = \frac{1}{N} \sum_{n=-\frac{N}{2}+1}^{\frac{N}{2}} f_n \omega_N^{-nk},$$

for $k = -N/2 + 1 : N/2$, is only one of many that appear in the literature. To underscore this point we will occasionally use different forms of the DFT even in *this* book. Given this state of affairs, it seems reasonable to devote just a few moments to other forms of the DFT that might be encountered in practice. Knowing the protracted deliberations that led to our choice of a DFT definition, it would be foolish to suggest that one form is superior among all others. There is no single DFT that has a clear advantage over all others. The best attitude is to accept the DFT's multiple personalities and to deal with them however they appear. This approach is particularly valuable in working with DFT (or FFT) software, an issue we will also discuss briefly.

Many of the variations on the DFT are truly superficial. Other authors have chosen virtually every combination of the following options:

1. placing the scaling factor $1/N$ on the forward or inverse transform (the option of using $1/\sqrt{N}$ on both transforms also occurs),

2. using i or j for $\sqrt{-1}$,

3. using $\pm$ in the exponent of the kernel on the forward or inverse transform, and

4. including *in the notation* the left-hand or right-hand endpoint of the sampling interval (we have seen that the endpoint value actually used must be the average of the endpoint values).

Having listed these options we will dispense with them as notational differences. There are other choices that are more substantial.

Much of the latitude in defining the DFT comes from the implied periodicity of both the input sequence f_n and the transform sequence F_k. An N-point DFT can be defined on *any* N consecutive terms of the periodic sequence f_n, and can be used to define any N consecutive terms of the periodic transform sequence F_k. In other words, a general form of the forward N-point DFT is

$$F_k = \frac{1}{N} \sum_{n=P+1}^{P+N} f_n \omega_N^{-nk}$$

for $k = Q+1 : Q+N$, where P and Q are any integers and N is any positive integer. Of this infinitude of DFTs, only a few are useful and have been committed to practice. We will consider these select few in this section.

There appear to be three factors distinguishing the DFTs that are extant in the literature. We will base our DFT taxonomy upon these three factors:

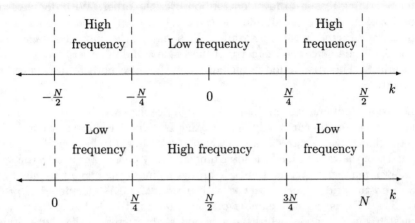

FIG. 3.1. *Using centered indices (top figure), the low frequencies correspond to $|k| < N/4$ for a single set of sample points, while the high frequencies correspond to $|k| > N/4$. With noncentered indices, the low frequencies correspond to $0 \leq k < N/4$ and $3N/4 < k < N$, and the high frequencies correspond to $N/4 < k < 3N/4$ (for a single set).*

1. Is the sampling interval *centered about the origin or not*? If the sampling takes place on an interval of length A, this is the choice between the intervals $[-A/2, A/2]$ and $[0, A]$. The centered forms of the DFT generally occur when the sample interval is a spatial domain (for example, an image or an object), whereas the noncentered forms are generally used for time-dependent (causal) sequences. Centered indices (for example, $n, k = -N/2 : N/2$) have the appeal that in the frequency domain, the low frequency indices are clustered about $k = 0$, while the high frequency indices are near the two ends of the index set for $|k| \approx N/2$. In the noncentered cases (for example, $n, k = 0 : N - 1$), the low frequency indices are near the ends of the index set ($k \approx 0$ and $k \approx N$), and the high frequency indices are near the center ($k \approx N/2$). Some may find it disconcerting that for the noncentered cases, the low frequencies are associated with both low- and high-valued indices. The distribution of high and low frequencies in various cases is summarized in Figure 3.1.

2. Does the DFT use an *even or odd number of sample points*? This distinction is different than the issue of whether N itself is even or odd, and some confusion is inevitable. For example, with N even it is possible to define a DFT on N points (an even number of points) by using the index ranges $n, k = -N/2 + 1 : N/2$ or on $N + 1$ points (an odd number of points) by using the index ranges $n, k = -N/2 : N/2$. As another example, with N even *or* odd, it is possible to define a DFT on $2N$ points (an even number of points) using the index ranges $n, k = 0 : 2N - 1$ or on $2N + 1$ points (an odd number of points) using the index ranges $n, k = 0 : 2N$. Examples of these various combinations will be given shortly.

3. Is the DFT defined on a *single or double set of points*? If the input and output sequences consist of either $N, N - 1$, or $N + 1$ sample points, we will say that the DFT is defined on a **single set** of points. If the DFT is defined on either $2N$ or $2N + 1$ points, we will use the term **double set** of points. There seems to be

no overriding reason to favor one option over the other. With double sets, the parity of N is immaterial, and the choice of an even or odd number of sample points is reflected in the index range. This may be the only advantage, and it is not significant since the geometry of the single and double set modes is virtually identical. This distinction is included only because both forms appear in the literature.

The various combinations of these three binary choices are shown in Table 3.1, and account for most of the commonly occuring DFTs. The DFT used primarily in this book is the single/even/centered case in which N is assumed to be even. The geometry of the modes of the various DFTs is intriguing and may be a factor in selecting a DFT. With an even number of points, the DFT always includes the highest frequency mode resolvable on the grid, namely $\cos(\pi n)$ with period $2\Delta x$. As elucidated in problems 32 and 33, a DFT on an odd number of points does not include the $\cos(\pi n)$ mode; its highest frequency modes will have a period slightly greater than $2\Delta x$ (typically something like $2(1 + 1/N)\Delta x$).

We will offer two examples of how Table 3.1 can be used to construct DFTs. The first is quite straightforward; the second presents some unexpected subtleties.

Example: Double/even/noncentered DFT. Assume that a function f is sampled at $2N$ points of the interval $[0, A]$. Since $\Delta x = A/(2N)$, those grid points are given by $x_n = nA/(2N)$, where $n = 0 : 2N - 1$. Notice that the resulting DFT is defined on an even number of grid points. Denoting the samples of f by $f_n = f(x_n)$, the forward transform is given by

$$F_k = \frac{1}{2N} \sum_{n=0}^{2N-1} f_n e^{-i\pi nk/N}$$

for $k = 0 : 2N - 1$. Orthogonality relations can be established on this set of grid points, and the inverse DFT is easily found to be

$$f_n = \sum_{k=0}^{2N-1} F_k e^{i\pi nk/N}$$

for $n = 0 : 2N - 1$. A slightly modified set of reciprocity relations also holds for this DFT. Since the length of the spatial domain is A, it follows that $\Delta \omega = 1/A$; therefore, $\Delta x \Delta \omega = 1/(2N)$. The length of the frequency domain is $\Omega = 2N/A$, which leads to the second reciprocity relation, $A\Omega = 2N$. One would discover that this alternate DFT changes very few, if any, of the DFT properties.

Example: Single/odd/centered DFT. Now assume that the function f is sampled at $N + 1$ equally spaced points of the interval $[-A/2, A/2]$ (including the origin), where N itself is even. This means that the grid points for the DFT are given by $x_n = nA/(N + 1)$, where $n = -N/2 : N/2$. Letting $f_n = f(x_n)$, the resulting DFT is given by

$$F_k = \frac{1}{N + 1} \sum_{n=-\frac{N}{2}}^{\frac{N}{2}} f_n \omega_{N+1}^{-nk}$$

for $k = -N/2 : N/2$. It can be shown (problem 34) that the kernel ω_{N+1}^{-nk} satisfies an orthogonality property on the indices $n = -N/2 : N/2$, and an inverse DFT can be defined in the expected manner. However, there are some curious properties of this

TABLE 3.1
Various forms of the DFT.

Type	Comments	Index sets n, k	Mode	Highest frequency indices
single even centered	N even N points	$-\frac{N}{2}+1:\frac{N}{2}$	$e^{i2\pi nk/N}$	$N/2$
single even centered	N odd $N-1$ points	$-\frac{N-1}{2}+1:\frac{N-1}{2}$	$e^{i2\pi nk/N-1}$	$(N-1)/2$
single even noncentered	N even N points	$0:N-1$	$e^{i2\pi nk/N}$	$N/2$
single odd centered	N even $N+1$ points	$-\frac{N}{2}:\frac{N}{2}$	$e^{i2\pi nk/N+1}$	$\pm N/2$
single odd centered	N odd N points	$-\frac{N-1}{2}:\frac{N-1}{2}$	$e^{i2\pi nk/N}$	$\pm(N-1)/2$
single odd noncentered	N odd N points	$0:N-1$	$e^{i2\pi nk/N}$	$(N\pm1)/2$
double even centered	N even or odd $2N$ points	$-N+1:N$	$e^{i\pi nk/N}$	N
double even noncentered	N even or odd $2N$ points	$0:2N-1$	$e^{i\pi nk/N}$	N
double odd centered	N even or odd $2N+1$ points	$-N:N$	$e^{i2\pi nk/2N+1}$	$\pm N$
double odd noncentered	N even or odd $2N+1$ points	$0:2N$	$e^{i2\pi nk/2N+1}$	$N, N+1$

DFT (and the double/odd/centered DFT) that are not shared by the other forms. Notice that the DFT points $x_n = nA/(N+1)$, while uniformly distributed on the interval $[-A/2, A/2]$ do *not* include either endpoint. However, the zeros of highest frequency sine mode occur at the points $\xi_n = nA/N$ for $n = -N/2 : N/2$, which *do* include the endpoints $\pm A/2$. There is a mismatch between these two sets of points (as illustrated in Figure 3.2) which does not occur in the other DFTs of Table 3.1. The discrepancy between the x_n's and the ξ_n's decreases as N increases. This quirk does not reduce the legitimacy of the odd/centered forms as full-fledged DFTs, but they should carry the warning that they sample the input differently than the other DFTs. There is still the question of how well the odd/centered DFTs approximate Fourier coefficients and Fourier integrals. While this is the subject of the next chapter, we

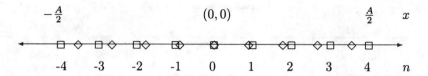

$$-\tfrac{A}{2} \qquad\qquad (0,0) \qquad\qquad \tfrac{A}{2} \quad x$$

FIG. 3.2. When a DFT is defined on a centered interval with an odd number of sample points (single/odd/centered), the sample points $x_n = nA/(N+1)$ do not coincide with the zeros of the sine modes, $\xi_n = nA/N$. The two sets of points do coincide in the other forms of the DFT shown in Table 3.1. The figure shows the mismatch for a nine-point DFT ($N = 8$); the x_n's are marked by $\diamond$, and the ξ_n's are marked by $\square$.

should state here for the sake of completeness that these forms of the DFT appear to have the same error properties as the other DFTs. To the best of our knowledge, this question has not been fully addressed or resolved in the literature. One could verify that the odd/centered forms do possess the usual DFT properties that will be discussed in this chapter.

To conclude the discussion of the single/odd/centered DFT, we note that it carries its own reciprocity relations. Since the sampling interval has length A, the grid spacing in the frequency domain is $\Delta\omega = 1/A$, and hence $\Delta x \Delta\omega = 1/(N+1)$. The DFT is defined on $N+1$ points, so the extent of the frequency grid is $\Omega = (N+1)\Delta\omega = (N+1)/A$. This implies the second reciprocity relation, $A\Omega = N+1$. It should be noted that the frequency grid points are given by $\omega_k = k/A$. This says that the highest frequencies,

$$\omega_{\pm\frac{N}{2}} = \frac{N}{2A} = \frac{\Omega}{2}\frac{N}{N+1},$$

do not (quite) coincide with either endpoint of the frequency domain which have the values $\pm\Omega/2$. Thus there is a similar anomaly in the grid points of the frequency domain which should not present accuracy problems as long as one realizes precisely which frequencies are represented by the DFT.

The question of different forms of the DFT arises in a critical way when it comes to using software. Care and ingenuity must be used to insure that the input is in the proper form and the output is interpreted correctly. We proceed by example and list several popular software packages that use different forms of the DFT; inclusion does not represent author endorsement!

1. *Mathematica* [165] uses the N-point DFT

$$F_k = \frac{1}{\sqrt{N}} \sum_{n=1}^{N} f_n e^{i2\pi(n-1)(k-1)/N}$$

   for $k = 1 : N$, with the single proviso that "the zero frequency term appears at position 1 in the resulting list." This is essentially the single/even/noncentered or single/odd/noncentered cases of Table 3.1, except that the input and output sequences are defined for $n, k = 1 : N$.

2. The *IMSL* (International Mathematical and Statistical Libraries) routine

FFTCC uses the following DFT:

$$F_{k+1} = \sum_{n=0}^{N-1} f_{n+1} e^{i2\pi nk/N}$$

for $k = 0 : N - 1$. This is essentially the single/even/noncentered or single/odd/noncentered form of Table 3.1, depending on whether N is even or odd. Again, the input and output sequences must be defined with positive (nonzero) indices.

3. The package *Matlab* [97] uses the DFT definition

$$F_{k+1} = \sum_{n=0}^{N-1} f_{n+1} e^{i2\pi nk/N},$$

for $k = 0 : N - 1$, in its FFT routine, which agrees with the IMSL definition.

4. A widely used mainframe DFT software package *FFTPACK* [140] uses the definition

$$F_k = \sum_{n=1}^{N} f_n e^{-i2\pi(n-1)(k-1)/N}$$

for $k = 1 : N$. Except for scaling factors, and the appearance of a minus sign in the exponential, this definition matches that of *Mathematica*.

5. The mathematical environment *MAPLE* [31] computes the DFT in the form

$$F_{k+1} = \frac{1}{N} \sum_{n=0}^{N-1} f_n \omega_N^{-nk},$$

for $k = 0 : N - 1$, which has the effect of using a different index set for the input and output.

6. *MATHCAD* [96] offers both real and complex DFTs. The real version has the form

$$F_k = \frac{1}{\sqrt{N}} \sum_{n=0}^{N-1} f_n \omega_N^{nk},$$

for $k = 0 : N - 1$, where f_n is a real sequence and F_k is complex.

This small software survey confirms the assertion made earlier that DFTs appear in all sorts of costumes. Anyone who works regularly with the DFT will eventually encounter it in more than one form. Hopefully this section has helped prepare the reader for that eventuality.

3.2. Basic Properties of the DFT

At this stage, the DFT is like a town that we know by name, but have never visited. We have a definition of the DFT, derived in several ways, and we have discussed some alternate forms in which the DFT may appear. However, we still know very

little about it. Now is the time to become acquainted with the DFT by exploring
its properties. As with its cousins, the Fourier series and the Fourier transform, the
DFT is useful precisely because it possesses special properties that enable it to solve
many problems easily. Such simplification often occurs because a problem specified
originally in one domain (a spatial or time domain) can be reformulated in a simpler
form in the frequency domain. The bridge between these two domains is the DFT,
and the properties of the DFT tell us how a given problem is modified when we pass
from one domain to another.

Before launching this exploration, one cover statement might be made. In
the previous section we observed that the DFT may take a bewildering variety of
(ultimately equivalent) forms. The form of the DFT pair that will be used throughout
this chapter is

$$\mathcal{D}\{f_n\}_k = F_k = \frac{1}{N} \sum_{n=-\frac{N}{2}+1}^{\frac{N}{2}} f_n \omega_N^{-nk}, \tag{3.1}$$

for $k = -N/2 + 1 : N/2$, which is the forward transform, and

$$\mathcal{D}^{-1}\{F_k\}_n = f_n = \sum_{k=-\frac{N}{2}+1}^{\frac{N}{2}} F_k \omega_N^{nk}, \tag{3.2}$$

for $n = -N/2 + 1 : N/2$, which is the inverse transform. However, every property that
is discussed inevitably holds for any legitimate form of the DFT that one chooses to
use. With that sweeping unproven statement underlying our thoughts, let's discuss
properties of the DFT.

Periodicity

A natural point of departure is a property that we have already mentioned and used.
The complex sequences f_n and F_k defined by the N-point DFT pair (3.1) and (3.2)
have the property that they are N-periodic, which means that

$$f_{n+N} = f_n \quad \text{and} \quad F_{k+N} = F_k \quad \text{for} \quad \text{all integers } n \text{ and } k.$$

This property follows immediately from the fact that

$$\omega_N^{-n(k+N)} = \omega_N^{-nN} \quad \text{and} \quad \omega_N^{(n+N)k} = \omega_N^{nk}.$$

Among the ramifications of this property is the fact that by extending either sequence
f_n or F_k periodically beyond the set of indices $n, k = -N/2 + 1 : N/2$, the DFT can
be defined on any two sets of N consecutive integers. Another crucial consequence is
that in sampling a function for input to the DFT, the sequence of samples must be
periodic and satisfy $f_{-\frac{N}{2}} = f_{\frac{N}{2}}$.

Linearity

One of the fundamental properties of the DFT is linearity: the DFT of a linear
combination of input sequences is the same as a linear combination of their DFTs.
Linearity enables us to separate signals into various components (the basis for spectral
analysis) and to keep those of interest while discarding those of no interest (the

principle underlying filtering theory). Linearity can be expressed in the following way. If f_n and g_n are two complex-valued sequences of length N, and α and β are complex numbers, then

$$\mathcal{D}\{\alpha f_n + \beta g_n\}_k = \alpha \mathcal{D}\{f_n\}_k + \beta \mathcal{D}\{g_n\}_k.$$

This property follows from the following argument:

$$
\begin{aligned}
\mathcal{D}\{\alpha f_n + \beta g_n\}_k &= \frac{1}{N} \sum_{n=-\frac{N}{2}+1}^{\frac{N}{2}} (\alpha f_n + \beta g_n)\,\omega_N^{-nk} \\
&= \alpha \frac{1}{N} \sum_{n=-\frac{N}{2}+1}^{\frac{N}{2}} f_n \omega_N^{-nk} + \beta \frac{1}{N} \sum_{n=-\frac{N}{2}+1}^{\frac{N}{2}} g_n \omega_N^{-nk} \\
&= \alpha \mathcal{D}\{f_n\}_k + \beta \mathcal{D}\{g_n\}_k.
\end{aligned}
$$

Precisely the same argument can be used to show that the inverse DFT is also a linear operator.

Shift and Modulation

Two closely related properties that have important implications are the shift and modulation properties. The **shift** property tells us the effect of taking the DFT of a sequence that has been shifted (or translated). A brief calculation using the DFT definition (3.1) can be used directly to derive the shift property (problem 43). Here is a slightly different approach. Consider the sequence f_n that has been shifted j units to the right. Using the IDFT, it can be written

$$
\begin{aligned}
f_{n-j} &= \sum_{k=-\frac{N}{2}+1}^{\frac{N}{2}} F_k\,\omega_N^{(n-j)k} \\
&= \sum_{k=-\frac{N}{2}+1}^{\frac{N}{2}} \left[F_k\,\omega_N^{-jk} \right] \omega_N^{nk} \\
&= \mathcal{D}^{-1}\left\{ F_k\,\omega_N^{-jk} \right\}_n.
\end{aligned}
$$

From this last statement it follows that

$$\mathcal{D}\{f_{n-j}\}_k = F_k\,\omega_N^{-jk}. \tag{3.3}$$

In words, transforming a sequence that has been shifted j units to the right has the effect of *rotating* the DFT coefficients of the original sequence in the complex plane; in fact, the original coefficient F_k is rotated by an angle $-2\pi jk/N$. The magnitude of the DFT coefficients remains unchanged under these rotations. This effect is illustrated in Figure 3.3. A popular special case of this property is that in which the original sequence is shifted by half of a period ($N/2$ units). The resulting sequence of DFT coefficients has the property that

$$\mathcal{D}\left\{ f_{n-\frac{N}{2}} \right\}_k = (-1)^k\,F_k,$$

where F_k is the DFT of the unshifted sequence.

The **modulation** or **frequency shift** property gives the effect of modulating the input sequence, that is, multiplying the elements of the input sequence f_n by ω_N^{nj}, where j is a fixed integer. In a somewhat symmetrical and predictable way, it results in a DFT sequence that is shifted relative to the DFT of the unmodulated sequence f_n. A brief argument demonstrates the effect of modulation:

$$\mathcal{D}\left\{f_n \omega_N^{nj}\right\}_k = \frac{1}{N} \sum_{n=-\frac{N}{2}+1}^{\frac{N}{2}} \left(f_n \omega_N^{nj}\right) \omega_N^{-nk}$$

$$= \frac{1}{N} \sum_{n=-\frac{N}{2}+1}^{\frac{N}{2}} f_n \omega_N^{-n(k-j)}$$

$$= \mathcal{D}\{f_n\}_{k-j} = F_{k-j}. \tag{3.4}$$

The property says that if an input sequence is modulated by modes of the form $\cos(2\pi n j/N)$ or $\sin(2\pi n j/N)$ with a frequency of j cycles on the domain, the resulting DFT is shifted by j units. This property can be visualized best when f_n is real-valued and the modulation is done by a real mode (problem 61).

Hermitian Symmetry

The DFT and the IDFT are similar in form, as shown by their respective definitions, (3.1) and (3.2). Indeed, the relationship is sufficiently close that either transform can be used to compute the other, with a simple alteration to the input data. For example, the inverse transform $f_n = \mathcal{D}^{-1}\{F_k\}_n$ can be computed by taking the conjugate of the *forward* transform of a mildly modified version of the sequence F_k. A similar maneuver can be used to compute $\mathcal{D}\{f_n\}_k$. The two properties that capture the similarity between the DFT and the IDFT are usually called **Hermitian**[1] **symmetry** properties, and are given by

$$f_n = \mathcal{D}\{N F_k^*\}^* \quad \text{and} \quad F_k = \mathcal{D}^{-1}\left\{\frac{1}{N} f_n^*\right\}^*, \tag{3.5}$$

where $*$ indicates complex conjugation.

These relations are easy to establish using the definitions of the DFT and its inverse, which is best left as an exercise for the reader (see problem 42). In principle, the Hermitian symmetry properties suggest that only one algorithm is needed to compute both the forward and inverse DFTs. In practice, however, the FFT (discussed in Chapter 10) is used to compute DFTs, and literally hundreds of specialized FFT algorithms exist, each tuned to operate on a specific form of the input.

DFT of a Reversed Sequence

Suppose we are given the input sequence f_n, where as usual $n = -N/2 + 1 : N/2$. Consider the input sequence f_{-n} formed by reversing the terms of f_n. The new

---

[1]Mathematically gifted from childhood, CHARLES HERMITE (1822–1901) did not receive a professorship at the Sorbonne until the age of 54. He made fundamental contributions to number theory and the theory of functions (particularly elliptic and theta functions).

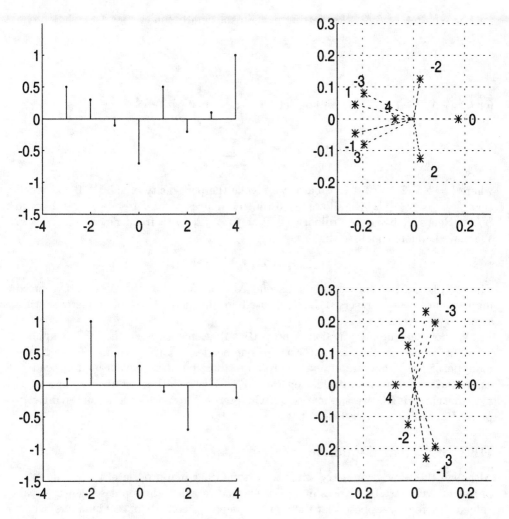

FIG. 3.3. *According to the DFT shift property, if an input sequence f_n of length $N = 8$ (upper left) with DFT coefficients F_k (upper right) is shifted to the right by two units to produce the sequence f_{n-2} (lower left), the resulting DFT coefficients F'_k (lower right) are rotated in the complex plane through angles of $-\pi k/2$ where $k = -3 : 4$. Note that the magnitude of individual DFT coefficients remains unchanged under the rotation. Note also that since both input sequences are real, both DFT sequences are conjugate symmetric.*

sequence is

$$f_{\frac{N}{2}}, \; f_{\frac{N}{2}-1}, \; \cdots, \; f_1, \; f_0, \; f_{-1}, \; \cdots, \; f_{-\frac{N}{2}+1}.$$

What can we say about the DFT of the altered sequence $\mathcal{D}\{f_{-n}\}_k$? Substituting the new sequence into the definition DFT (3.1), we obtain

$$\mathcal{D}\{f_{-n}\}_k = \frac{1}{N} \sum_{n=-\frac{N}{2}+1}^{\frac{N}{2}} f_{-n} \omega_N^{-nk}$$

for $k = -N/2 + 1 : N/2$. Now letting $p = -n$ the sum becomes

$$\mathcal{D}\{f_{-n}\}_k = \frac{1}{N} \sum_{p=-\frac{N}{2}}^{\frac{N}{2}-1} f_p \omega_N^{pk}$$

for $k = -N/2 + 1 : N/2$. If we now let $k = -l$, we may write

$$\mathcal{D}\{f_{-n}\}_{-l} = \frac{1}{N} \sum_{p=-\frac{N}{2}}^{\frac{N}{2}-1} f_p \omega_N^{-pl}, \tag{3.6}$$

where $l = -N/2 : N/2 - 1$. Finally, we invoke the periodicity of the DFT and recall that $f_{-N/2} = f_{N/2}$, which allows us to run the indices p and l from $-N/2 + 1$ up to $N/2$. Hence, (3.6) is the ordinary DFT of $f_p = f_{-n}$ with frequency index $l = -k$. We may therefore conclude that

$$\mathcal{D}\{f_{-n}\}_k = \mathcal{D}\{f_n\}_{-k} = F_{-k}, \tag{3.7}$$

where F_k is the DFT of the original sequence. The fact that the DFT of a reversed input sequence is a reversed DFT is used in the derivations of numerous other symmetry properties.

We now take up a discussion of properties that emerge when the DFT is applied to input sequences that exhibit certain symmetries. These symmetries can take many forms, but they are all extremely important, because they ultimately lead to computational savings in evaluating the DFT. The resulting algorithms, known as **symmetric DFTs**, are the subject of Chapter 4. However, the actual symmetry properties can be discussed right now.

DFT of a Real Sequence

Although the term "symmetry" may seem somewhat inappropriate, by far the most prevalent symmetry that arises in practice is that in which the input sequence is real-valued. Therefore, suppose that the input sequence f_n consists of real numbers, which means that $f_n^* = f_n$. Then we find that

$$F_k^* = \left\{ \frac{1}{N} \sum_{n=-\frac{N}{2}+1}^{\frac{N}{2}} f_n \omega_N^{-nk} \right\}^* = \frac{1}{N} \sum_{n=-\frac{N}{2}+1}^{\frac{N}{2}} f_n^* (\omega_N^{-nk})^*$$

$$= \frac{1}{N} \sum_{n=-\frac{N}{2}+1}^{\frac{N}{2}} f_n \omega_N^{-n(-k)}$$

$$= F_{-k}.$$

We have used the fact that $f_n = f_n^*$ and $(\omega_N^{-p})^* = \omega_N^p$ in reaching this conclusion. Any complex-valued sequence possessing the property $F_k^* = F_{-k}$ is said to be **conjugate symmetric** or **conjugate even**. Therefore, the DFT of a real sequence is conjugate symmetric. Notice that if we write F_k in the form $F_k = \text{Re}\{F_k\} + i\,\text{Im}\{F_k\}$, then conjugate symmetry implies that the real part of the transform is an even sequence while the imaginary part of the transform is an odd sequence:

$$\text{Re}\{F_{-k}\} = \text{Re}\{F_k\} \quad \text{and} \quad \text{Im}\{F_{-k}\} = -\text{Im}\{F_k\}. \tag{3.8}$$

The fact that the DFT of a real sequence is conjugate symmetric also implies that the IDFT of a conjugate symmetric sequence is real (problem 46).

There are some important implications of this property, especially since real-valued input sequences are extremely common in applications. The first consequence is economy in storage. Clearly, a real-valued input sequence of length N can be stored in N real storage locations. The output sequence F_k is complex-valued and might appear to require $2N$ real storage locations. However, since $F_k^* = F_{-k}$, it follows that F_0 and $F_{N/2}$ are real. Furthermore, we observe that the real parts of F_k are needed only for $k = 0 : N/2$, and the imaginary parts are needed only for $k = 1 : N/2 - 1$, since the remaining values can be obtained from (3.8). A quick count shows that there are precisely N independent real quantities in the output sequence F_k. If N is odd, then the relations (3.8) hold for $k = 0 : (N-1)/2$, so knowledge of F_k for these values of k is sufficient to determine the entire sequence. This property (and analogous properties for other symmetric input sequences) can be used to achieve savings in both computation and storage in the DFT.

DFT of a Conjugate Symmetric Sequence

Given the previous property, it should be no surprise that the DFT of a conjugate symmetric sequence is real. In fact, if $f_n^* = f_{-n}$, then

$$F_k^* = \left\{ \frac{1}{N} \sum_{n=-\frac{N}{2}+1}^{\frac{N}{2}} f_n \omega_N^{-nk} \right\}^* = \frac{1}{N} \sum_{n=-\frac{N}{2}+1}^{\frac{N}{2}} f_{-n} \omega_N^{nk} = \frac{1}{N} \sum_{p=-\frac{N}{2}+1}^{\frac{N}{2}} f_p \omega_N^{-pk} = F_k,$$

where we have used the periodicity of f_n to write the last sum. Since $F_k = F_k^*$, the DFT is real-valued.

DFTs of Even or Odd Sequences

We now consider the transform of complex sequences with either even or odd symmetry. By even symmetry, we mean a sequence in which $f_{-n} = f_n$. Notice that the reversal of an even sequence is the original sequence. Having already established that the DFT of a reversed input sequence f_{-n} is the reversed output sequence F_{-k}, it follows that the DFT of an even sequence is also even. Therefore, we conclude that $F_{-k} = F_k$. This property may also be shown directly using the definition of the DFT (3.1) (problem 44).

We define an odd sequence as one in which $f_{-n} = -f_n$. By the reversed sequence property and the linearity property we may conclude that the DFT of an odd sequence satisfies $F_{-k} = -F_k$; thus the DFT of an odd sequence is itself an odd sequence.

DFTs of Real Even and Real Odd Sequences

The majority of DFT applications involve real-valued input sequences, and these sequences often possess other symmetries. For example, a sequence that is both real and even, so that $f_n = f_{-n}$ and $f_n = f_n^*$, must have a DFT that is conjugate symmetric (because f_n is real) *and* even (because f_n is even). These two properties imply that $\text{Re}\{F_k\}$ is an even sequence and $\text{Im}\{F_k\} = 0$. Thus we conclude that the DFT of a real, even sequence is also real and even. We can analyze the DFT of an input sequence that is both real and odd in a similar way. Recall that the DFT of an odd sequence is odd $(F_{-k} = -F_k)$, and the DFT of a real sequence is conjugate symmetric $(F_{-k} = F_k^*)$. These facts together imply that $\mathrm{Re}\,\{F_k\} = 0$ and $\mathrm{Im}\,\{F_k\}$ is an odd sequence; or, stated in words, the DFT of a real, odd sequence is a purely imaginary, odd sequence. These properties can also be shown directly from the definition of the DFT (problem 44).

There are many other symmetries that could be considered, and almost any symmetry will give a new property. Those we have outlined above are the most common and useful. Among the symmetries omitted (until Chapter 4) are those called **quarter-wave symmetries**, which are extremely important in the numerical solution of partial differential equations. The symmetry properties which were included in this discussion are summarized graphically in Figures 3.4 and 3.5.

Before leaving the topic of symmetries entirely, we mention the fact that, while some sequences exhibit symmetries and others do not, **any** sequence can be decomposed into a pair of symmetric sequences. We briefly outline how this is done, and relate the DFTs of the symmetric component sequences to the DFT of the original sequence.

Waveform Decomposition

An arbitrary sequence f_n can always be decomposed into the sum of two sequences, one of which is even and the other odd. This is accomplished by defining

$$f_n^{\mathrm{even}} = \frac{f_n + f_{-n}}{2} \quad \text{and} \quad f_n^{\mathrm{odd}} = \frac{f_n - f_{-n}}{2}$$

and noting that

$$f_n = f_n^{\mathrm{even}} + f_n^{\mathrm{odd}}.$$

But since $\mathcal{D}\,\{f_{-n}\}_k = F_{-k}$ (reversal property), we may use the linearity of the DFT to show that

$$\mathcal{D}\,\{f_n^{\mathrm{even}}\}_k = \frac{F_k + F_{-k}}{2} \quad \text{and} \quad \mathcal{D}\,\{f_n^{\mathrm{odd}}\}_k = \frac{F_k - F_{-k}}{2}.$$

This property is called the **waveform decomposition** property and may be verified carefully in problem 45. Notice the consistency in these relations:

$$\mathcal{D}\,\{f_n\}_k = \mathcal{D}\,\{f_n^{\mathrm{even}} + f_n^{\mathrm{odd}}\}_k = F_k.$$

At this point we turn our attention from the symmetry properties of the DFT, and examine what could be termed operational properties. That is, how does the DFT behave under the action of certain operations? While many operations could be of interest, we will restrict our attention to convolution and correlation, two of the most useful such creatures.

Discrete (Cyclic) Convolution

The discrete convolution theorem is among the most important properties of the DFT. It underlies essentially all of the Fourier transform-based signal processing done today, and by itself accounts for much of the utility of the DFT. We begin by defining and discussing the notion of discrete convolution. Given two N-periodic sequences f_n and

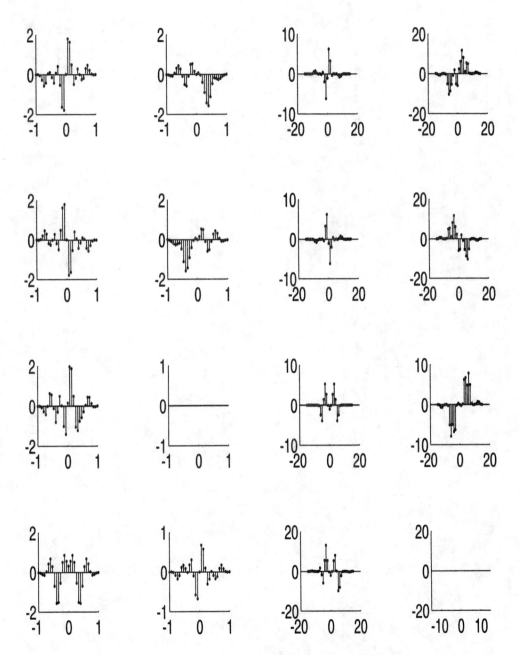

FIG. 3.4. *Various symmetries of the DFT are summarized in this figure. In the left two columns, the real and imaginary parts of the input are shown, while the right two columns show the corresponding real and imaginary parts of the DFT. The symmetries shown are: arbitrary input (top row), reversed arbitrary input, i.e., reversed input sequence of top row (second row), real input (third row), and conjugate symmetric input (bottom row).*

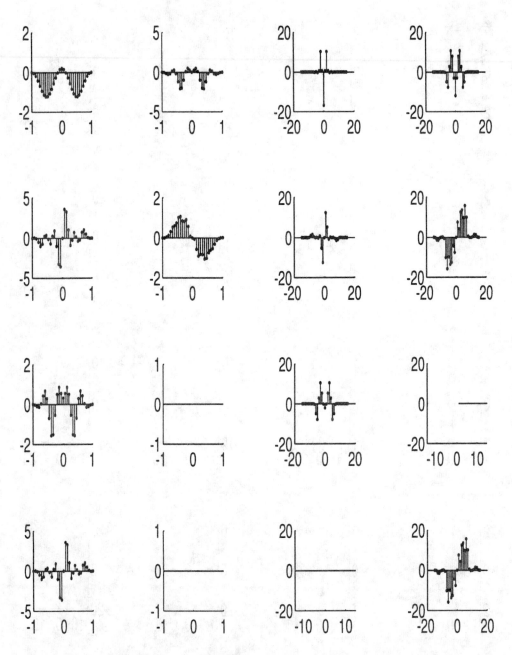

FIG. 3.5. *More of the various symmetries of the DFT are summarized in this figure. In the left two columns, the real and imaginary parts of the input are shown, while the right two columns show the corresponding real and imaginary parts of the DFT. The symmetries shown are: even input (top row), odd input (second row), real even input (third row), and real odd input (bottom row).*

g_n defined for the indices $n = -N/2 + 1 : N/2$, their **discrete (cyclic) convolution**, denoted $f_n * g_n$, is another sequence h_n defined by

$$h_n = f_n * g_n = \sum_{j=-\frac{N}{2}+1}^{\frac{N}{2}} f_j g_{n-j} \tag{3.9}$$

for $n = -N/2 + 1 : N/2$. Notice that h_n is also an N-periodic sequence.

It is never enough merely to define the convolution operator; some interpretation is always needed. Convolution may be viewed in several different ways. Because of its importance, we will highlight some of its properties and try to instill a sense of its significance. We note first that the order of the sequences in convolution is immaterial, since if we let $p = n - j$ and invoke the periodicity of f_n and g_n we find that

$$f_n * g_n = \sum_{j=-\frac{N}{2}+1}^{\frac{N}{2}} f_j g_{n-j} = \sum_{p=-\frac{N}{2}+1}^{\frac{N}{2}} f_{n-p} g_p = g_n * f_n.$$

In addition, a scalar multiple can be "passed through" the convolution; that is,

$$\alpha(f_n * g_n) = (\alpha f_n) * g_n = f_n * (\alpha g_n).$$

The convolution operator is developed graphically in Figure 3.6 and may be described in the following way. The nth term of the convolution h_n results from reversing the order of the convolving sequence g_j (forming the sequence g_{-j}), shifting this sequence to the left an amount n (forming the sequence g_{n-j}), and then forming the scalar product of that sequence with the sequence f_j. This process is repeated for each $n = -N/2 + 1 : N/2$.

Further insight into discrete convolution might be offered by considering a sifting property of the δ sequence. Let's look at the effect of convolving an arbitrary sequence f_n with the sequence $\delta(n - n_0)$, where n_0 is a fixed integer. (Recall that $\delta(n - n_0)$ is zero unless $n = n_0$.) A short calculation reveals that

$$f_n * \delta(n - n_0) = \sum_{j=-\frac{N}{2}+1}^{\frac{N}{2}} f_j \delta(n - n_0 - j) = f_{n-n_0}.$$

We see that if a sequence is convolved with a δ sequence centered at the index n_0, the effect is to shift the sequence n_0 units to the right (see problem 60).

We will illustrate convolution with one of its most frequent uses. As shown in Figure 3.7, the sequence f_n is obtained by sampling a superposition of cosine waves with various frequencies, including several with high frequencies. The sequence

$$g_n = \{0, 0, 0, \ldots, g_{-10}, g_{-9} \ldots, g_{-1}, g_0, g_1, \ldots, g_9, g_{10}, \ldots, 0, 0, 0\}$$

includes only 21 nonzero (positive) entries, centered about g_0, which sum to unity. Convolution of f with g may be thought of as a running 21-point weighted average of f. As seen in the figure, the resulting convolution is a smoothed version of f_n, in which the higher frequency components have been removed. In this light, we may view convolution of two sequences as a filtering operation in which one of the sequences is input data and the other is a filter. Having developed the view of convolution as a

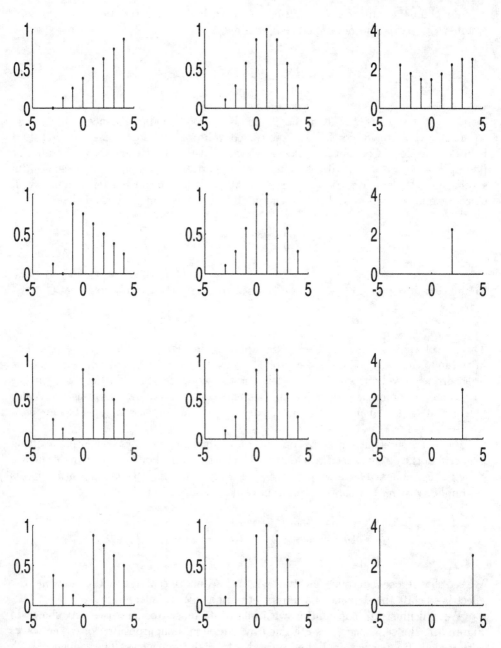

FIG. 3.6. *Each term of the convolution sequence h_n is the scalar product of one sequence f_n with a shifted version of a second sequence g_n in reversed order. Each h_n may be viewed as a weighted sum of the input f_n, the weighting given by g_{-n}. The top row shows, left to right, the sequences g_n, f_n, and h_n. The second row shows the shifted, reversed sequence g_{2-n} (left), the sequence f_n (middle), and the output entry h_2 (right), which is the scalar product of $g_{(2-n)}$ and f_n. The third row shows the shifted, reversed sequence g_{3-n} (left), the sequence f_n (middle), and the output entry h_3 (right), the scalar product of $g_{(3-n)}$ and f_n. The bottom row shows the shifted, reversed sequence g_{4-n} (left), the sequence f_n (middle), and the output entry h_4 (right), the scalar product of $g_{(4-n)}$ and f_n.*

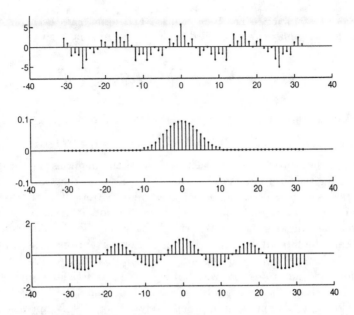

FIG. 3.7.   *The convolution operator acts as a filter. In the case shown, the input f_n (top) is an oscillatory sequence of length $N = 64$, while the convolving sequence g_n (middle) is a 21-point weighted average operator (whose entries sum to unity). The output $h_n = f_n * g_n$ (bottom) is a smoothed version of the input f_n, that also has damped amplitude.*

filtering operator, we now arrive at one of the most important of the DFT properties. We will dignify it with theorem status and state the convolution theorem.

THEOREM 3.1. DISCRETE CONVOLUTION THEOREM. *Let f_n and g_n be periodic sequences of length N whose DFTs are F_k and G_k. Then the DFT of the convolution $h_n = f_n * g_n$ is*

$$H_k = \mathcal{D}\{f_n * g_n\}_k = N F_k G_k.$$

That is, the DFT of the convolution is the pointwise product of the DFTs. This is often expressed by saying that convolution in the time (or spatial) domain corresponds to multiplication in the frequency domain.

Proof: We begin by writing f_n and g_{n-j} as the IDFTs of their DFTs. By the modulation property we know that $g_{n-j} = \mathcal{D}^{-1}\left\{G_k \omega_N^{-jk}\right\}$. Hence we can write

$$f_n * g_n = \sum_{n=-\frac{N}{2}+1}^{\frac{N}{2}} f_n g_{n-j}$$

$$= \sum_{j=-\frac{N}{2}+1}^{\frac{N}{2}} \left[\underbrace{\left(\sum_{p=-\frac{N}{2}+1}^{\frac{N}{2}} F_p \omega_N^{jp} \right)}_{f_n} \underbrace{\left(\sum_{k=-\frac{N}{2}+1}^{\frac{N}{2}} \left[G_k \omega_N^{-jk} \right] \omega_N^{nk} \right)}_{g_{n-j}} \right]$$

$$= \sum_{k=-\frac{N}{2}+1}^{\frac{N}{2}} G_k \omega_N^{nk} \sum_{p=-\frac{N}{2}+1}^{\frac{N}{2}} F_p \underbrace{\left(\sum_{j=-\frac{N}{2}+1}^{\frac{N}{2}} \omega_N^{j(p-k)} \right)}_{N\hat{\delta}_N(p-k)},$$

where the order of summation has been changed and terms have been regrouped. The orthogonality property may now be applied to the innermost sum to give

$$f_n * g_n = N \sum_{k=-\frac{N}{2}+1}^{\frac{N}{2}} F_k G_k \omega_N^{nk}.$$

This shows that $f_n * g_n$ is the IDFT of $N F_k G_k$. Stated differently,,

$$f_n * g_n = N\mathcal{D}^{-1}\{F_k G_k\}_n \quad \text{or} \quad \mathcal{D}\{f_n * g_n\}_k = N F_k G_k. \qquad \blacksquare$$

We are now in a position to understand one of the mightiest applications of the DFT, that of digital filtering. Assume that f_n is a signal that will be filtered, and g_n represents a selected filter. Essentially, we may perform filtering by computing the DFT of the input signal ($f_n \rightarrow F_k$), multiplying it term-by-term with the (precomputed) set of weights G_k to form the sequence $F_k G_k$, and computing the IDFT to obtain the filtered sequence. Here is the important observation. Computing a convolution of two N-point sequences by its definition requires on the order of N^2 multiplications and additions. However, if the DFTs and IDFTs are done using an FFT (which, as shown in Chapter 10, requires approximately $N \log N$ operations), the convolution can be done with approximately $2N \log N + N$ operations ($N \log N$ operations for each of the two FFTs and N operations for the pointwise products of the DFTs). Particularly for large values of N, the use of the FFT *and* the convolution theorem results in tremendous computational savings.

There is one other curious perspective on convolution that should be mentioned, since it has some important practical consequences. Consider the familiar exercise of multiplying two polynomials

$$a(x) = a_0 + a_1 x + \cdots + a_n x^n$$

and

$$b(x) = b_0 + b_1 x + \cdots + b_m x^m.$$

The product of these two polynomials looks like

$$
\begin{aligned}
c(x) &= a_0 b_0 + (a_0 b_1 + a_1 b_0)x + (a_0 b_2 + a_1 b_1 + a_2 b_0)x^2 + \cdots + a_n b_m x^{n+m} \\
&= \sum_{k=0}^{m+n} c_k x^k \quad \text{where} \quad c_k = \sum_{j=\max(k-m,0)}^{\min(k,n)} a_j b_{k-j}.
\end{aligned}
$$

A close inspection of the coefficients of the product polynomial reveals that the coefficient c_k has the form of a convolution. We can make this more precise as follows.

We will let $N = m + n + 1$ and then extend the two sets of coefficients $\{a_k\}$ and $\{b_k\}$ with zeros in the following way:

$$
\hat{a}_k = \begin{cases} 0 & \text{if } -N+1 \leq k < 0, \\ a_k & \text{if } 0 \leq k \leq n, \\ 0 & \text{if } n < k \leq N, \end{cases}
\quad \text{and} \quad
\hat{b}_k = \begin{cases} 0 & \text{if } -N+1 \leq k < 0, \\ b_k & \text{if } 0 \leq k \leq m, \\ 0 & \text{if } m < k \leq N. \end{cases}
$$

The auxiliary sequences $\hat{a}_k$ and $\hat{b}_k$ have length $2N$, and now we can write

$$c_k = \hat{a}_k * \hat{b}_k = \sum_{j=-N+1}^{N-1} \hat{a}_j \hat{b}_{k-j}.$$

In other words, the coefficients of the product polynomial can be obtained by taking the $2N$-point convolution of the auxiliary sequences. Notice that, of the entire convolution sequence c_k, only the coefficients

$$c_0, c_1, \ldots, c_{m+n} = c_{N-1}$$

are needed to form the product polynomial. This association between convolution and products of polynomials has important implications in algorithms for high precision arithmetic, as outlined in problem 50.

We have scarcely touched upon the subject of convolution with all of its applications and algorithmic implications. For a far more detailed account of related issues (for example, noncyclic convolution, convolution of long sequences, and FFT/convolution algorithms) the interested reader is referred to any signal processing text or to [28], [73], or [107].

Frequency Convolution

Just as the DFT of a convolution of two sequences is the product of their DFTs, it can be shown that the DFT of the product of two sequences is the convolution of their DFTs; that is,

$$\mathcal{D}\{f_n g_n\}_k = F_k * G_k.$$

To show this, we consider the inverse transform of the convolution and write

$$
\begin{aligned}
\mathcal{D}^{-1}\{F_k * G_k\}_n &= \sum_{k=-\frac{N}{2}+1}^{\frac{N}{2}} \left(\sum_{j=-\frac{N}{2}+1}^{\frac{N}{2}} F_j G_{k-j} \right) \omega_N^{nk} \\
&= \sum_{j=-\frac{N}{2}+1}^{\frac{N}{2}} F_j \underbrace{\left(\sum_{k=-\frac{N}{2}+1}^{\frac{N}{2}} G_{k-j} \omega_N^{nk} \right)}_{g_n \omega_N^{-nj}} \\
&= \sum_{j=-\frac{N}{2}+1}^{\frac{N}{2}} F_j g_n \omega_N^{nj} \\
&= g_n \underbrace{\sum_{j=-\frac{N}{2}+1}^{\frac{N}{2}} F_j \omega_N^{nj}}_{f_n} \\
&= g_n f_n.
\end{aligned}
$$

The modulation property has been used in the third line of this argument. The frequency convolution theorem, $\mathcal{D}\{f_n g_n\} = F_k * G_k$, now follows immediately by applying $\mathcal{D}$ to each side of this equation.

Discrete Correlation

Closely related to the convolution of two sequences is an operation known as **correlation**. Correlation is (perhaps) more intuitive than convolution, because its name describes what it does: it determines how much one sequence resembles (correlates with) another. Consider two real sequences that need to be compared. Intuitively, one might multiply the sequences pointwise and sum the results. If the sequences are identical then the result will be large and positive. If they are identical except for being of opposite sign, then the result will be large and negative. If they are entirely dissimilar, then we expect that the agreement or disagreement of the signs will be random, and the resulting sum should be small in magnitude. The discrete correlation operator embodies this idea, except that it may be applied to complex, as well as real-valued sequences, and it also computes a correlation for the two sequences when one has been retarded, or lagged, relative to the other.

Here is how discrete correlation works. Let f_n and g_n be N-periodic sequences. The discrete correlation of the sequences f_n and g_n, denoted $f_n \otimes g_n$, is an N-periodic sequence h_n defined by

$$h_n = f_n \otimes g_n = \sum_{j=-\frac{N}{2}+1}^{\frac{N}{2}} f_j g_{n+j} \qquad (3.10)$$

for $n = -N/2 + 1 : N/2$. With this definition in hand, we state the **discrete correlation theorem**, namely that

$$\mathcal{D}\{f_n \otimes g_n\}_k = N F_k^* G_k. \qquad (3.11)$$

As with discrete convolution, the discrete correlation of two sequences can be done efficiently by performing the pointwise product of their DFTs, and then transforming back to the original domain with an inverse DFT. There is also a corresponding relationship for correlation of two sequences in the frequency domain. It is given by

$$\mathcal{D}\{f_n^* g_n\}_k = F_k \otimes G_k. \qquad (3.12)$$

We leave the verification of these properties to problem 47.

Parseval's Relation

A well-known and important property is **Parseval's relation**, which we encountered previously in connection with trigonometric interpolation. It is an important property with some physical significance and its proof (problem 48) appeals again to orthogonality properties. Parseval's relation says that a sequence f_n and its N-point DFT F_k are related by

$$\frac{1}{N} \sum_{n=-\frac{N}{2}+1}^{\frac{N}{2}} |f_n|^2 = \sum_{k=-\frac{N}{2}+1}^{\frac{N}{2}} |F_k|^2.$$

Parseval's relation has analogous versions in terms of Fourier series and Fourier transforms. In all cases, the same physical interpretation may be given. The sum of the squares of the terms of a sequence, $\sum_n |f_n|^2$, is often associated with the **energy** of the sequence (it is also the square of the length of the vector whose components are f_n). It says that the energy of f_n may be computed using either the original sequence or the sequence of DFT coefficients. Alternatively, the energy of the input sequence equals the energy of the DFT sequence (up to the scaling factor $1/N$).

We close this section with a summary table of the basic DFT properties that have been presented up to this point. It is more than coincidence that the properties of the DFT have continuous analogs that can be expressed in terms of Fourier series and Fourier transforms. Because of this kinship, many of the most useful techniques for Fourier transforms and Fourier series can be used when the DFT is applied to discrete problems. To emphasize these remarkable similarities, Table 3.2 shows not only the DFT properties, but the corresponding continuous properties.

The Fourier series properties assume that the functions f and g are defined on the interval $[-A/2, A/2]$ and have the representations

$$f(x) = \sum_{k=-\infty}^{\infty} c_k e^{i2\pi kx/A} \quad \text{and} \quad g(x) = \sum_{k=-\infty}^{\infty} d_k e^{i2\pi kx/A},$$

where the coefficients are given by

$$c_k = \frac{1}{A} \int_{-\frac{A}{2}}^{\frac{A}{2}} f(x)e^{-i2\pi kx/A} dx \quad \text{and} \quad d_k = \frac{1}{A} \int_{-\frac{A}{2}}^{\frac{A}{2}} g(x)e^{-i2\pi kx/A} dx.$$

The Fourier transform properties assume that f and g are absolutely integrable functions on the real line ($\int_{-\infty}^{\infty} |f(x)|dx < \infty$) whose Fourier transforms are

$$\hat{f}(\omega) = \mathcal{F}\{f(x)\} = \int_{-\infty}^{\infty} f(x)e^{-i2\pi\omega x} dx$$

and

$$\hat{g}(\omega) = \mathcal{F}\{g(x)\} = \int_{-\infty}^{\infty} g(x)e^{-i2\pi\omega x} dx.$$

The table refers to the convolution and correlation properties for Fourier transforms. For completeness, we state the continuous form of the convolution theorem, which bears an expected similarity to the discrete convolution theorem. Given two functions f and g, with Fourier transforms $\hat{f}$ and $\hat{g}$, their convolution and its transform are given by

$$\mathcal{F}\{f(x) * g(x)\} = \mathcal{F}\left\{\int_{-\infty}^{\infty} f(y)g(x-y)dy\right\} = \hat{f}(\omega)\hat{g}(\omega),$$

while their correlation and its transform are

$$\mathcal{F}\{f(x) \otimes g(x)\} = \mathcal{F}\left\{\int_{-\infty}^{\infty} f(y)g(x+y)dy\right\} = \hat{f}^*(\omega)\hat{g}(\omega).$$

3.3.   Other Properties of the DFT

There are many other properties of the DFT, a few of which we will list in this section. For the most part, we will merely state the properties, with some observations about their utility. Proofs of these properties are generally relegated to the problems.

Earlier in this chapter we examined the DFT of sequences that possess certain symmetries. We now consider sequences that have no special form originally, but are then altered to produce a sequence with a special pattern. In each case the goal is to relate the DFT of the new sequence to the DFT of the unaltered sequence. It is useful to know that these altered sequences are not mentioned frivolously; many of them arise in specific applications.

TABLE 3.2
Properties of the DFT and Fourier series.

Property	Input property	DFT property	Fourier series property
Periodicity	Arbitrary	$f_{n+N} = f_n$ $F_{k+N} = F_k$	$f(x + A) = f(x)$
Linearity	Arbitrary	$\mathcal{D}\left\{\alpha f_n + \beta g_n\right\}_k$ $= \alpha F_k + \beta G_k$	FC $\left\{\alpha f(x) + \beta g(x)\right\}$ $= \alpha c_k + \beta d_k$
Shift	Arbitrary	$\mathcal{D}\left\{f_{n-j}\right\}_k = \omega_N^{-jk} F_k$	FC $\left\{f(x - y)\right\}$ $= e^{-i2\pi ky/A} c_k$
Modulation	Arbitrary	$\mathcal{D}\left\{f_n \omega_N^{nj}\right\}_k$ $= F_{k-j},\ j \in \mathbf{Z}$	FC $\left\{f(x) e^{i2\pi jx/A}\right\}$ $= c_{k-j},\ j \in \mathbf{Z}$
Hermitian symmetry	Arbitrary	$f_n = \mathcal{D}\left\{N F_k^*\right\}^*$ $F_k = \mathcal{D}^{-1}\left\{f_n^*/N\right\}^*$	– –
Reversal	Arbitrary	$\mathcal{D}\left\{f_{-n}\right\}_k = F_{-k}$	FC $\left\{f(-x)\right\} = c_{-k}$
Real	$f_n \in \mathbf{R}$ $f(x) \in \mathbf{R}$	$F_{-k} = F_k^*$	$c_{-k} = c_k^*$
Conjugate symmetric	$f_{-n} = f_n^*$ $f(-x) = f^*(x)$	$F_k^* = F_k$	$c_k^* = c_k$
Even	$f_{-n} = f_n$ $f(-x) = f(x)$	$F_{-k} = F_k$	$c_{-k} = c_k$
Odd	$f_{-n} = -f_n$ $f(-x) = -f(x)$	$F_{-k} = -F_k$	$c_{-k} = -c_k$
Real even	$f_n, f(x) \in \mathbf{R}$ $f_{-n} = f_n$ $f(-x) = f(x)$	$F_{-k} = F_k$ $F_k \in \mathbf{R}$	$c_{-k} = c_k$ $c_k \in \mathbf{R}$
Real odd	$f_n, f(x) \in \mathbf{R}$ $f_{-n} = -f_n$ $f(-x) = -f(x)$	$F_{-k} = -F_k$ $iF_k \in \mathbf{R}$	$c_{-k} = -c_k$ $ic_k \in \mathbf{R}$
Convolution	Arbitrary	$\mathcal{D}\left\{f_n * g_n\right\}_k = N F_k G_k$	–
Correlation	Arbitrary	$\mathcal{D}\left\{f_n \otimes g_n\right\}_k = N F_k^* G_k$	–

FC $\left\{f(x)\right\}$ means Fourier coefficients of f.

TABLE 3.2 (CONTINUED)
Properties of the DFT and Fourier transforms.

Property	Input property	DFT property	Fourier transform property
Periodicity	Arbitrary	$f_{n+N} = f_n$	None
Linearity	Arbitrary	$\mathcal{D}\{\alpha f_n + \beta g_n\}_k$ $= \alpha F_k + \beta G_k$	$\mathcal{F}\{\alpha f(x) + \beta g(x)\}$ $= \alpha \hat{f}(\omega) + \beta \hat{g}(\omega)$
Shift	Arbitrary	$\mathcal{D}\{f_{n-j}\}_k = \omega_N^{-jk} F_k$	$\mathcal{F}\{f(x-y)\}$ $= e^{-i2\pi \omega y} \hat{f}(\omega)$
Modulation	Arbitrary	$\mathcal{D}\left\{f_n \omega_N^{nj}\right\}_k$ $= F_{k-j},\, j \in \mathbf{Z}$	$\mathcal{F}\left\{f(x) e^{i2\pi xy}\right\}$ $= \hat{f}(\omega - y),\, y \in \mathbf{R}$
Hermitian symmetry	Arbitrary	$f_n = \mathcal{D}\left\{N F_k^*\right\}^*$ $F_k = \mathcal{D}^{-1}\{f_n^*/N\}^*$	$f(x) = \mathcal{F}\left\{\hat{f}^*(\omega)\right\}^*$ $\hat{f}(\omega) = \mathcal{F}^{-1}\{f^*(x)\}^*$
Reversal	Arbitrary	$\mathcal{D}\{f_{-n}\}_k = F_{-k}$	$\mathcal{F}\{f(-x)\} = \hat{f}(-\omega)$
Real	$f_n \in \mathbf{R}$ $f(x) \in \mathbf{R}$	$F_{-k} = F_k^*$	$\hat{f}(-\omega) = \hat{f}^*(\omega)$
Conjugate symmetric	$f_{-n} = f_n^*$ $f(-x) = f^*(x)$	$F_k^* = F_k$	$\hat{f}^*(\omega) = \hat{f}(\omega)$
Even	$f_{-n} = f_n$ $f(-x) = f(x)$	$F_{-k} = F_k$	$\hat{f}(-\omega) = \hat{f}(\omega)$
Odd	$f_{-n} = -f_n$ $f(-x) = -f(x)$	$F_{-k} = -F_k$	$\hat{f}(-\omega) = -\hat{f}(\omega)$
Real even	$f_n, f(x) \in \mathbf{R}$ $f_{-n} = f_n$ $f(-x) = f(x)$	$F_{-k} = F_k$ $F_k \in \mathbf{R}$	$\hat{f}(-\omega) = \hat{f}(\omega)$ $\hat{f}(\omega) \in \mathbf{R}$
Real odd	$f_n, f(x) \in \mathbf{R}$ $f_{-n} = -f_n$ $f(-x) = -f(x)$	$F_{-k} = -F_k$ $i F_k \in \mathbf{R}$	$\hat{f}(-\omega) = -\hat{f}(\omega)$ $i \hat{f}(\omega) \in \mathbf{R}$
Convolution	Arbitrary	$\mathcal{D}\{f_n * g_n\}_k = N F_k G_k$	$\mathcal{F}\{f(x) * g(x)\}$ $= \hat{f}(\omega) \hat{g}(\omega)$
Correlation	Arbitrary	$\mathcal{D}\{f_n \otimes g_n\}_k = N F_k^* G_k$	$\mathcal{F}\{f(x) \otimes g(x)\}$ $= \{\hat{f}(\omega)\}^* \hat{g}(\omega)$

Padded Sequences

It is a popular misconception that the FFT works only on sequences whose length is a power of two. While we shall see in Chapter 10 that this is most assuredly not the case, it is true that some FFT packages handle only limited values of N. Given a data sequence with a length that is not amenable to computation by a particular FFT package, it is not uncommon to pad the data with zeros until it has a convenient length. The padded sequence is then used as input to the DFT, often without asking about the effect of the padding. We will investigate this practice now. Let f_n be a sequence defined for $n = -M/2 + 1 : M/2$ and suppose that $N > M$, where N and M are even. Since we are working on the symmetric interval, we pad the sequence at both ends to create the augmented sequence

$$g_n = \begin{cases} 0 & \text{for} \quad n = -N/2 + 1 : -M/2, \\ f_n & \text{for} \quad n = -M/2 + 1 : M/2, \\ 0 & \text{for} \quad n = M/2 + 1 : N/2. \end{cases}$$

Letting G_k denote the N-point DFT of the padded sequence g_n, we see that

$$G_k = \mathcal{D}_N \{g_n\}_k = \frac{1}{N} \sum_{n=-\frac{N}{2}+1}^{\frac{N}{2}} g_n \omega_N^{-nk} = \frac{1}{N} \sum_{m=-M/2+1}^{M/2} f_m \omega_N^{-mk}$$

for $k = -N/2 + 1 : N/2$. Now a sleight-of-ω is needed to simplify this last sum. Using the fact that $\omega_N = \omega_M^{M/N}$, we may continue and write

$$G_k = \frac{1}{N} \sum_{m=-M/2+1}^{M/2} f_m \omega_M^{-mkM/N} = \frac{M}{N} F_{kM/N}$$

for $k = -N/2 + 1 : N/2$. We have let F_k denote the M-point DFT of the original sequence f_n. In other words, we have shown that the kth DFT coefficient of the padded sequence is equal to M/N times the (kM/N)th DFT coefficient of the original sequence, whenever kM/N is an integer! This interpretation is more meaningful if we invert it slightly; after all, the goal is to determine the coefficients F_k in terms of the computed coefficients G_k. So we might also write that

$$F_k = \frac{N}{M} G_{kN/M},$$

for $k = -M/2 + 1 : M/2$, whenever kN/M is an integer.

A specific example helps considerably in unraveling this relationship. A common situation is that in which N is chosen as an integer multiple of M, making $N = pM$, where $p > 1$ is an integer. The above relationship then appears as $F_k = pG_{pk}$, and now it is clear that the kth DFT coefficient of the original sequence is p times the (pk)th coefficient of the DFT of the padded sequence, as shown in Figure 3.8. The meaning of this property can be explained physically. By padding a sequence with zeros, we do not alter the spatial sampling rate Δx, but only increase the length of the spatial domain by increasing the number of grid points from M to N. By the reciprocity relations for each problem, we have

$$\Delta x \, \Delta \omega^{\text{old}} = \frac{1}{M} \quad \text{and} \quad \Delta x \, \Delta \omega^{\text{new}} = \frac{1}{N} = \frac{1}{pM}.$$

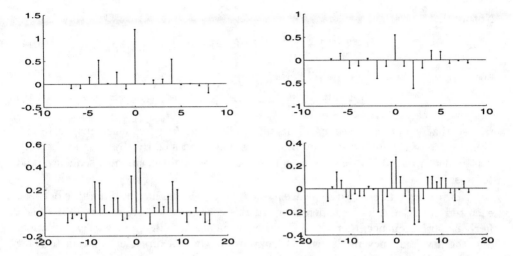

FIG. 3.8.   *The top row of figures shows the real (left) and imaginary (right) parts of the DFT of a sequence of length M = 16. The bottom row of figures shows the real (left) and imaginary (right) parts of the DFT of that same input sequence, now padded with zeros at each end to increase its length to N = 32. Observe that every other coefficient of the DFT on the bottom is half of the corresponding coefficient of the original DFT on top.*

Since $N > M$ and Δx is constant, we see that

$$\frac{\Delta\omega^{\text{old}}}{\Delta\omega^{\text{new}}} = p > 1;$$

therefore, padding with zeros decreases the grid spacing $\Delta\omega$ in the frequency domain by a factor of p. At the same time, the length of the frequency domain Ω is unchanged since $\Omega = 1/\Delta x$. Thus padding with zeros can be used as a technique for refining the frequency grid. It is important to realize that this interpretation is valid only if the padded values are correct; that is, if the function represented by f_n is indeed zero outside the interval $[-(N/2)\Delta x, (N/2)\Delta x]$. If the function is not zero over the region of zero-padding, then legitimate data is lost in the process, and errors may be introduced. An interesting effect occurs in the case when an M-periodic sequence is set to zero for $|n| \geq M/2$. The result is not increased resolution in frequency, but rather an interpolation in the frequency domain (see problem 53, as well as [21] and [36]).

Summed and Differenced Sequences

Let $g_n = f_n + f_{n+1}$ be the sequence obtained by adding each term of f_n to its right neighbor. This summing process is a sort of averaging: if the terms of f_n are constant or slowly varying, then the terms of the summed sequence will have approximately twice the magnitude of the terms of the original sequence. On the other hand, if f_n is a highly oscillatory sequence, then some cancellation will take place and the magnitude of the terms of the summed sequence will be greatly diminished. In fact, it is not difficult to show (problem 55), using the DFT definition (3.1), that

$$\mathcal{D}\{g_n\}_k = (1 + \omega_N^{-k})F_k. \tag{3.13}$$

Noting that

$$\sqrt{2} \le |1 + \omega_N^{-k}| \le 2 \quad \text{for } 0 \le |k| \le \frac{N}{4} \tag{3.14}$$

and

$$0 \le |1 + \omega_N^{-k}| < \sqrt{2} \quad \text{for } \frac{N}{4} < |k| \le \frac{N}{2}, \tag{3.15}$$

we see that summing a sequence magnifies all of the low frequency modes (3.14). While some of the high frequency modes are magnified slightly (by as much as $\sqrt{2}$), it is evident from (3.15) that the high frequency modes of the summed sequence tend to be suppressed.

Now define a differenced sequence $h_n = f_n - f_{n+1}$ by taking the difference between each term of f_n and its right neighbor. In this case the roles are reversed: if f_n is a fairly smooth sequence, then the differenced sequence h_n will experience cancellation, and the low frequency modes will be suppressed. It is a direct calculation to show (problem 55) that

$$\mathcal{D}\{h_n\}_k = (1 - \omega_N^{-k})F_k. \tag{3.16}$$

We see that now

$$0 \le |1 - \omega_N^{-k}| \le \sqrt{2} \quad \text{for } 0 \le |k| \le \frac{N}{4} \tag{3.17}$$

and

$$\sqrt{2} < |1 - \omega_N^{-k}| \le 2 \quad \text{for } \frac{N}{4} < |k| \le \frac{N}{2}. \tag{3.18}$$

In contrast to the summing operator, the difference operator amplifies high frequency coefficients (3.18) and diminishes most of the low frequency coefficients (3.17).

Folded Sum and Difference

Another interesting manipulation that may be applied to any sequence is folding. Let the folded sequence $g_n = f_n + f_{n+N/2}$ be formed by averaging each term of f_n with its partner halfway through the sequence (using the periodicity of f_n when necessary). Using the definition of the DFT, it may be shown (problem 55) that

$$\mathcal{D}\{g_n\}_k = \begin{cases} 2F_k & \text{for } k \text{ even,} \\ 0 & \text{for } k \text{ odd.} \end{cases} \tag{3.19}$$

We see that a mode with an odd frequency (for example, $f_n = \sin(2\pi n/N)$, which is an odd sequence) is exactly canceled by the folding, while a mode with an even frequency (which is an even sequence) is reinforced by the folded sum operation.

Once again, using the periodicity of f_n, let $h_n = f_n - f_{n+N/2}$ be the sequence formed by differencing each term of f_n with its partner half a period away. It then follows that

$$\mathcal{D}\{h_n\}_k = \begin{cases} 0 & \text{for } k \text{ even,} \\ 2F_k & \text{for } k \text{ odd.} \end{cases} \tag{3.20}$$

We see that the folded difference reinforces modes with an odd frequency index and annihilates modes with an even index.

3.4.   A Few Practical Considerations

In this section we will introduce three very practical matters that arise in the day-to-day use of the DFT. They are

- averaging at endpoints,

- aliasing,

- leakage.

Each of these issues will be presented in an empirical manner, as it might be observed in a typical calculation, and some observations will be made in each case. But we can promise that each of these phenomena will appear again in later chapters: averaging at endpoints is almost a campaign slogan in this book that is chanted repeatedly; and aliasing and leakage will figure in critical ways in the analysis of DFT errors in Chapter 6.

Averaging at Endpoints (AVED)

A penetrating example should suffice to make the essential points. Assume that we are approximating the Fourier coefficients of the function $f(x) = x$ on the interval $[-1/2, 1/2]$. As we can easily show, the exact values are given by

$$c_k = \int_{-\frac{1}{2}}^{\frac{1}{2}} xe^{-i2\pi kx} dx = i\frac{\cos \pi k}{2\pi k}$$

for $k = 0, \pm1, \pm2, \dots$. The Fourier coefficients are pure imaginary (since f is odd and real-valued) and they decrease as $|k|^{-1}$. In order to use the DFT to approximate the c_k's, we must sample f at the N equally spaced points $x_n = n/N$ of $[-1/2, 1/2]$. In keeping with the convention that we have established, we will take $n = -N/2+1 : N/2$. Our lesson can be communicated with a small value of N. With $N = 8$, let the samples of f be $\bar{f}_n = n/N$, as shown in Table 3.3. Also shown are the DFT coefficients $\bar{F}_k = \mathcal{D}_8 \{\bar{f}_n\}_k$.

TABLE 3.3

DFT approximations to the Fourier coefficients, with and without averaging at the endpoints; $f(x) = x$ on $[-\frac{1}{2}, \frac{1}{2}]$, with $N = 8$.

n or k	Without averaging		With averaging	
	$\bar{f}_n$	$\bar{F}_k$	f_n	F_k
−3	−3/8	−.0625 − .1509i	−3/8	−.1509i
−2	−1/4	.0625 + .0625i	−1/4	.0625i
−1	−1/8	−.0625 − .0259i	−1/8	−.0259i
0	0	.0625	0	0
1	1/8	−.0625 + .0259i	1/8	.0259i
2	1/4	.0625 − .0625i	1/4	−.0625i
3	3/8	−.0625 + .1509i	3/8	.1509i
4	1/2	.0625	0	0

A moment's glance should arouse some suspicion. Notice that the DFT coefficients $\bar{F}_k$ are not pure imaginary, as the Fourier coefficients are. An astute observer might also notice that the real part of the $\bar{F}_k$'s oscillates between $\pm.0625 = 1/16$, which means that the DFT coefficients do not decrease with $|k|$ as the Fourier coefficients do. Is there a problem?

There *is* a problem, and it can be identified in several ways. In this particular case one can argue in terms of symmetry. The function $f(x) = x$ is an odd function $(f(-x) = -f(x))$ on the interval $(-1/2, 1/2)$. If extended periodically beyond this interval, it will remain odd only if we define $f(-1/2) = f(1/2) = 0$. In sampling f for input to the DFT, the sequence of samples must also inherit this symmetry, which means $f_{-\frac{N}{2}} = f_{\frac{N}{2}} = 0$.

More generally, one must argue by analogy with the Fourier series. Given the Fourier coefficients c_k of a well-behaved function f, the function $\sum_k c_k e^{i2\pi kx}$ converges to $f(x)$ on the interval $(-1/2, 1/2)$, to the periodic extension of f outside of that interval, and to the average value at points of discontinuity. As shown in Figure 3.9, the Fourier series converges to the value

$$\frac{1}{2}\left(f\left(-\frac{1}{2}\right) + f\left(\frac{1}{2}\right)\right) = 0$$

at the points $x = \pm 1/2$. Therefore, the input to the DFT must also be defined with average values at the endpoints.

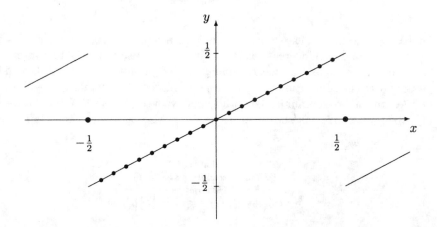

FIG. 3.9. *A function f must be sampled so that the resulting sequence takes average values at endpoints and points of discontinuity. The figure shows a linear function on $[-1/2, 1/2]$ which, if it were extended periodically beyond this interval, would be discontinuous at $x = \pm 1/2$ (solid line). Appropriate samples of the function are shown ($\bullet$) with $f_{\pm \frac{N}{2}} = 0$, which is the average value of f at $x = \pm 1/2$.*

In the example above, the sequence of samples that was fed to the DFT was not defined with an average value at the endpoints. The correct input to the DFT is shown in the third column of data in Table 3.3 as the sequence f_n, with $f_{N/2} = 0$. The resulting DFT coefficients F_k are pure imaginary and they decrease with $|k|$, as they should.

The discrepancy between the two sets of DFTs in Table 3.3 can be explained precisely. The two input sequences f_n and $\bar{f}_n$ differ only in the $n = 4$ position. In fact,

$$\bar{f}_n = f_n + \frac{1}{2}\, \delta\, (n - 4),$$

where $\delta\, (n-4)$ represents a spike with a magnitude of one at $n = 4$. Taking eight-point DFTs of both sides of this equation, and using the fact that

$$\mathcal{D}_4\, \{\delta(n - 4)\}_k = \frac{1}{8}\, \cos(\pi k) = \frac{1}{8}(-1)^k,$$

we see that

$$\bar{F}_k = F_k + \frac{1}{2} \cdot \frac{1}{8}\, (-1)^k.$$

Notice that $\bar{F}_k$ has a spurious component, $(-1)^k/16$, as observed in Table 3.3.

The lesson is now quite compelling: average values must be used for a function whose periodic extension is discontinuous at $x = \pm A/2$, or the resulting DFT will assuredly be infected with errors. The remedy is to be sure that in generating the input to the DFT, the function that is sampled on the interval $[-A/2, A/2]$ is not f, but the auxiliary function

$$g(x) = \begin{cases} f(x) & \text{for} \quad -\dfrac{A}{2} < x < \dfrac{A}{2}, \\[2mm] \dfrac{1}{2}\left(f\left(-\dfrac{A}{2}\right) + f\left(\dfrac{A}{2}\right)\right) & \text{for} \quad x = \dfrac{A}{2}. \end{cases}$$

Clearly, f and g differ only at a point, but it makes all the difference to the DFT.

The same argument applies in defining the input sequence at *any* point at which f has a discontinuity. This state of affairs leads us to coin the watchword **AVED: average values at endpoints and discontinuities**. The importance of AVED will be emphasized throughout the remaining chapters; its importance is also demonstrated in problem 58.

Aliasing

Have you ever watched a western movie (especially an old one) in which a stagecoach is traveling forward rapidly, but it looks as though the wheels are turning very slowly or even backwards? This is an example of **aliasing** (or **strobing**), a very important phenomenon to those who work in the area of signal processing. The reason for the apparent motion of the stagecoach wheel is simple. The shutter on the movie camera opens and closes at a specific rate, while the wheel on the stagecoach rotates at another rate. Suppose the wheel is turning rapidly enough that it rotates $1\frac{1}{6}$ times between consecutive openings of the camera shutter. Then in consecutive frames of the film the wheel appears to advance $\frac{1}{6}$ of a turn, when in fact it travels through $1\frac{1}{6}$ rotations. This gives the rather disconcerting illusion on film that while the stagecoach is moving at a good clip, the wheels are turning too slowly. Notice that the wheels could also have actually turned through $2\frac{1}{6}$ or $3\frac{1}{6}$ full cycles and still appeared to advance only $\frac{1}{6}$ of a turn. The same phenomenon can afflict any form of sampled data. Whenever a signal oscillates rapidly enough, a given sample rate becomes insufficient to resolve the signal. As occurred with the wagon wheel, the signal appears to be oscillating at a lower frequency than the one at which it actually oscillates.

We need to understand aliasing in order to know when it can cause difficulty with DFTs. We first introduce the important concept of a **band-limited** function. A function f is called band-limited if its Fourier transform is zero outside a finite interval $[-\Omega/2, \Omega/2]$. In other words, the "frequency content" of the function (or signal) lies below the maximum frequency $\Omega/2$. How does band-limiting affect aliasing? We have already seen in Chapter 2 that if a function is known at a finite number of sample points, then the DFT can be used to interpolate that function with a trigonometric polynomial. It makes sense that by increasing the number of samples of a function more accurate interpolating functions can be obtained. This naturally leads us to wonder: when do we have enough samples? That is, when is it possible to construct an interpolating function that is exact? Intuition suggests that there would be no sufficiently small grid spacing (or sampling rate) that would allow us to reconstruct the function exactly. Surprisingly, this isn't the case, and the statement to the contrary requires the condition of band-limiting. The celebrated theorem that gives the conditions under which a function may be reconstructed from its samples was introduced by Claude Shannon[2] [124]. It is worth stating without proof at this time.

THEOREM 3.2. SHANNON SAMPLING THEOREM. *Let f be a band-limited function whose Fourier transform is zero outside of the interval $[-\Omega/2, \Omega/2]$. If Δx is chosen so that*

$$\Delta x \leq \frac{1}{\Omega}, \tag{3.21}$$

then f may be reconstructed exactly from its samples $f_n = f(n\Delta x) = f(x_n)$ by

$$\begin{aligned} f(x) &= \sum_{n=-\infty}^{\infty} f_n \ \text{sinc}\left(\frac{\pi(x - x_n)}{\Delta x}\right) \\ &= \Delta x \sum_{n=-\infty}^{\infty} f_n \frac{\sin(\pi(x - x_n)/\Delta x)}{\pi(x - x_n)}. \end{aligned} \tag{3.22}$$

The **sinc function** is given by $\text{sinc}(x) = \sin(x)/x$, and is graphed in Figure 3.10 (see problem 51).

Rigorous proofs and discussions of this theorem abound [20], [82], [131], [158], [159], and it will appear again in Chapter 6. We will, however, make a few observations regarding this important result. The theorem tells us that, in theory, we may reconstruct a function exactly from its samples, provided that the function is band-limited *and that we sample it sufficiently often to resolve its highest frequencies*. The sampling theorem gives a prescription for selecting a sufficiently small grid spacing Δx once the band-limit Ω is known. It tells us that Δx must be chosen to satisfy (3.21). The critical sampling rate $\Delta x = 1/\Omega$ is called the **Nyquist[3] sampling rate**,

---

[2]CLAUDE SHANNON was born in 1916 and educated at the University of Michigan and MIT. His far-reaching research, done primarily at Bell Labs and MIT, either created or advanced the fields of information and communication theory, cryptography, and the theory of computation.

[3]The Swedish-born engineer HARRY NYQUIST (1889–1976) received his Ph.D. in physics from Yale University in 1917. He is best known for his contributions in telecommunications, including his discovery of the conditions necessary for maintaining stability in a feedback circuit, known as the Nyquist criterion. Nyquist held 138 patents, many of them fundamental to electronic transmission of data.

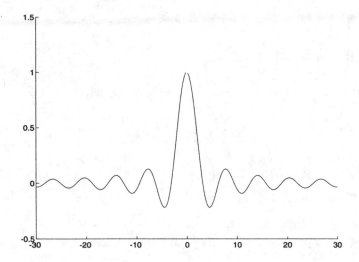

FIG. 3.10. *The* sinc *function* sinc $(x) = \sin(x)/x$, *shown in this figure, appears frequently in Fourier analysis and lies at the heart of the Shannon Sampling Theorem.*

and the frequency $\Omega/2$ is known as the **Nyquist frequency**. The Nyquist frequency is the highest frequency that can be resolved using a given sample spacing Δx. All higher frequencies will be aliased to lower frequencies. The Nyquist sampling rate is the largest grid spacing that can resolve the frequency $\Omega/2$. Observe that (3.21) also implies that in order to resolve a single wave, we must have at least two sample points per period of the wave.

As a simple example, consider Figure 3.11, which shows a single wave with a frequency of $\omega = 6$ cycles per unit length. We can use the foregoing arguments to conclude that the wave is band-limited with a Nyquist frequency of $\Omega/2 = 6$ cycles per unit length. Therefore, a sampling rate or grid spacing of $\Delta x \leq 1/12$ must be used to avoid aliasing and resolve the wave. If the wave is sampled with $\Delta x = 1/4$ as shown, the wave that is actually seen by the grid has a frequency of one cycle per unit length. The entire issue of aliasing, and all that it implies, will arise again in a fundamental way in Chapter 6 when we explore errors in the DFT.

In practice, it is rarely possible to meet the condition of band-limiting exactly. It is a fundamental fact that a function cannot be limited in both frequency (band-limited) *and* in space (or time). Since many functions (or signals) have finite duration in space or time, they cannot be strictly band-limited. However, many functions of practical use are **essentially band-limited** and have a rapid decay rate. A function f is essentially band-limiting if there exist positive constants β and μ such that

$$|\hat{f}(\omega)| \leq \beta(1 + |\omega|)^{-1-\mu},$$

which means that $|\hat{f}(\omega)|$ decays faster than $|\omega|^{-1}$ as $|\omega| \to \infty$. (A deeper and more precise statement about the rate of decay of Fourier transforms is given by the Paley–Wiener[4] Theorem [55], [110].) For such functions it is possible to choose a grid spacing Δx sufficiently small that the error in the sinc function representation

---

[4]Norbert Wiener (1894–1964) was a mathematical prodigy who spent many years at MIT as a mathematics professor. He made fundamental contributions to control theory, communication theory, analysis, and algebra, and he established the new subject of cybernetics.

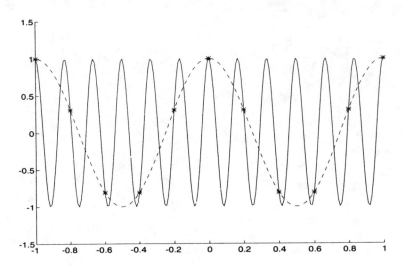

FIG. 3.11.   *A single wave with a frequency of six cycles per unit length (solid curve) is sampled (as shown by the asterisks) at a rate of $\Delta x = 1/5$, which is insufficient to resolve the wave. The wave that is actually "seen" by the sampling process has a frequency of one cycle per unit length (dashed curve).*

(3.22) is negligible. We also note that the interpolation in (3.22) requires infinitely many points. In practice, the series may be truncated, and the resulting errors can be estimated and controlled [82].

Leakage

Another phenomenon that DFT users must understand is called **leakage**. It is so named because its effect is to allow frequency components that are not present in the original waveform to "leak" into the DFT. A simple example will demonstrate the phenomenon quite vividly. We have seen that the Fourier transform of the function $f(x) = \cos(2\pi\omega_0 x)$ is given by

$$\mathcal{F}\left\{\cos(2\pi\omega_0 x)\right\}(\omega) = \frac{1}{2}\delta(\omega - \omega_0) + \frac{1}{2}\delta(\omega + \omega_0),$$

and consists of two "spikes" in the frequency domain at the points $\pm\omega_0$. Now consider the task of approximating this Fourier transform by sampling the function f on the interval $[-1/2, 1/2]$ at the N equally spaced points $x_n = n/N$. This means that $\Delta x = 1/N$ and (by the reciprocity relations) the grid spacing in the frequency domain is $\Delta\omega = 1$. How well does the DFT approximate the Fourier transform if we use a fixed value of N and choose two different values of ω_0?

The input sequence for the DFT is given by

$$f_n = \cos\left(\frac{2\pi\omega_0 n}{N}\right),$$

and the DFT can be evaluated directly from the definition (problem 59) or found in *The Table of DFTs* in the Appendix. Let's first consider the case in which $\omega_0 = 10$. The DFT is given by

$$\mathcal{D}\left\{\cos(20\pi n/N)\right\}_k = \frac{1}{2}\delta(k - 10) + \frac{1}{2}\delta(k + 10).$$

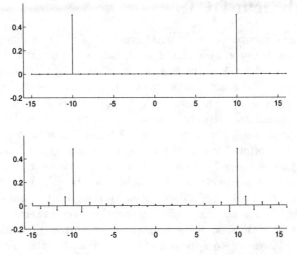

FIG. 3.12.   *The DFT of a sampled cosine wave $f(x) = \cos(2\pi\omega_0 x)$ for $N = 32$ behaves quite differently depending on the value of ω_0. In the top figure, with $\omega_0 = 10$, the DFT exactly reproduces the Fourier transform $\delta(\omega - 10)/2 + \delta(\omega + 10)/2$. In the bottom figure, with $\omega_0 = 10.13$, the DFT exhibits leakage errors in approximating the Fourier transform, because the function is sampled on a fraction of a full period.*

Noting that in this case $\omega_k = k\Delta\omega = k$, we see that the DFT approximates the Fourier transform exactly at the frequency grid points as shown in the upper graph of Figure 3.12. On the other hand, when $\omega_0 = 10.13$, the Fourier transform still takes the form of two spikes,

$$\mathcal{F}\{\cos(20.26\pi x)\} = \frac{1}{2}\delta(\omega - 10.13) + \frac{1}{2}\delta(\omega + 10.13).$$

However, the lower graph of Figure 3.12 reveals that the DFT does a very poor job of approximating the Fourier transform. In fact, for $k_0 = 2\omega_0 = 20.26$, *The Table of DFTs* in the Appendix shows that the DFT is now given by

$$\mathcal{D}\{f_n\}_k = \frac{(-1)^k \sin(10.13\pi)}{4N}\left(\frac{\sin(\frac{2\pi}{N}(k + 10.13))}{\sin^2(\frac{\pi}{N}(k + 10.13))} - \frac{\sin(\frac{2\pi}{N}(k - 10.13))}{\sin^2(\frac{\pi}{N}(k - 10.13))}\right),$$

which is hardly the representation of two spikes! In the latter case the exact Fourier transform, which should still consist of two spikes, has been contaminated by errors in the DFT that have leaked in from neighboring **sidelobes** frequencies. We will return to the matter of leakage in Chapter 6, when it arises in the analysis of DFT errors. For now it suffices to note that leakage occurs when a periodic function is truncated and sampled on an interval that is not an integer multiple of the period. In the above example, the function $f(x) = \cos(2\pi\omega_0 x)$ has $1/\omega_0$ periods on the interval $[-1/2, 1/2]$. When $\omega_0 = 10$, an integer number of complete cycles is sampled, and the truncated function is continuous when extended periodically. In the second case (when $\omega_0 = 10.13$), the truncated wave, when extended periodically, has a discontinuity. As we will soon see, when the DFT sees discontinuities in the underlying function, it reacts with large errors.

3.5.   Analytical DFTs

One of the universal distinctions that cuts across all of mathematics is one of process: it is the distinction between **analytical methods** and **numerical methods**. In a few simple words, analytical methods are "pencil and paper" techniques that often result in a formula for an exact solution to the problem at hand. For example, using the quadratic formula to solve a second-degree equation is an analytical method. On the other hand, numerical methods involve approximations (often very good ones) that are usually implemented on a computer. A numerical solution is often expressed as an algorithm and often involves the process of convergence. For example, using an iterative method (such as Newton's method) to find the roots of an equation is a numerical approach. The distinction is usually clear, but there are some gray areas. For example, finding a solution of a differential equation in the form of a Fourier series is an analytical method, since the coefficients can be given in terms of a nice formula. However, the actual evaluation of the Fourier series at specific points is usually done on a computer and involves an approximation which makes the method numerical in nature.

Despite occasional ambiguities, the distinction is quite useful, and furthermore it pervades the topic of DFTs. Given a function or a set of data samples, the evaluation of the DFT is usually done on a computer and is regarded as a numerical procedure. However, there are cases in which the DFT can be evaluated analytically, and when this is possible, the result is a single expression that gives all of the DFT coefficients in one clean sweep. Thus, there is something very satisfying about generating a DFT analytically. However, this satisfaction may not be appreciated by everyone! The thought of evaluating DFTs with a pencil and paper, instead of a computer, may seem arcane to some, rather like writing a term paper in calligraphy or reading Euclid in Greek. But for those who enjoy analytical methods (and calligraphy and Greek), this section may have some appeal. At the same time it is recommended for everyone: analytical methods for DFTs rely on the entire repertoire of properties that we have just studied, and there are many practical lessons along the way.

We will try to dispense with most preliminaries and proceed by example. However, there are two techniques that are so pervasive in evaluating DFTs analytically that they are worth stating at the start. One pattern that arises endlessly in various forms in this business is the **geometric sum**. The reader should know and be able to show (problem 64) that the sum of consecutive powers of a fixed number $a \neq 1$ (called the **ratio**) is given by

$$\sum_{k=M}^{N-1} a^k = \frac{a^M - a^N}{1 - a}.$$

More familiar forms of the geometric sum are

$$\sum_{k=0}^{N-1} a^k = \frac{1 - a^N}{1 - a} \quad \text{and} \quad \sum_{k=1}^{N-1} a^k = \frac{a - a^N}{1 - a}.$$

The second tool that is essential in more involved DFT calculations is **summation by parts**. Just as integration by parts is applied to the product of functions, summation by parts can be applied to the product of sequences. Let u_n and v_n be two sequences and let M and N be integers. To make the summation of parts rule resemble the familiar integration by parts formula ($\int u\,dv = uv - \int v\,du$), we will also let $\Delta u_n = u_{n+1} - u_n$ and $\Delta v_n = v_{n+1} - v_n$. Then the summation by parts rule says

$$\sum_{n=M+1}^{N} u_n \Delta v_n = u_N v_{N+1} - u_M v_{M+1} - \sum_{n=M+1}^{N} v_n \Delta u_{n-1}.$$

This result is most easily proved by writing out the terms of the left-hand sum, regrouping terms, and identifying the terms on the right side. A special case that arises in DFT calculations is

$$\sum_{n=-\frac{N}{2}+1}^{\frac{N}{2}} u_n \Delta v_n = u_{\frac{N}{2}} v_{\frac{N}{2}+1} - u_{-\frac{N}{2}} v_{-\frac{N}{2}+1} - \sum_{n=-\frac{N}{2}+1}^{\frac{N}{2}} v_n \Delta u_{n-1}.$$

Notice that if u_n and v_n are periodic sequences with period N, then $u_{\frac{N}{2}} = u_{-\frac{N}{2}}$ and $v_{-\frac{N}{2}+1} = v_{\frac{N}{2}+1}$, and the two "boundary terms" cancel. Thus, for periodic sequences we have

$$\sum_{n=-\frac{N}{2}+1}^{\frac{N}{2}} u_n \Delta v_n = - \sum_{n=-\frac{N}{2}+1}^{\frac{N}{2}} v_n \Delta u_{n-1}.$$

With these two handy tools at our disposal, let's do a couple of DFT calculations analytically.

Example: DFT of the square pulse. An important transform that arises both in theory and in practice is the transform of the **square pulse** (often called a **boxcar** function). We will begin with a function defined on the interval $[-A/2, A/2]$, and then sample it to generate a sequence suitable as input to the DFT. The square pulse is given by

$$f(x) = \begin{cases} 1 & \text{if } |x| < A/4, \\ 0 & \text{if } A/4 < |x| \le A/2, \\ 1/2 & \text{if } x = \pm A/4. \end{cases}$$

Notice that in anticipation of sampling this function for use in the DFT, we have defined f to have the average of its values at $x = \pm A/4$, where it is discontinuous. Notice also that if f is extended periodically beyond this basic interval, then it is continuous at $x = \pm A/2$, and we may rightfully assume that $f(\pm A/2) = 0$.

When f is sampled at the usual grid points $x_n = nA/N$ for $n = -N/2 + 1 : N/2$, the resulting sequence is

$$f_n = \begin{cases} 1 & \text{if } |n| \le N/4 - 1, \\ 0 & \text{if } -N/2 + 1 \le n \le -N/4 - 1 \text{ or } N/4 + 1 \le n \le N/2, \\ 1/2 & \text{if } n = \pm N/4. \end{cases}$$

Again notice how f_n is assigned its average value at the discontinuities.

We may now proceed with the calculation of the N-point DFT of this sequence. Using the familar definition we have that for $k = -N/2 + 1 : N/2$,

$$NF_k = \sum_{n=-\frac{N}{2}+1}^{\frac{N}{2}} f_n \omega_N^{-nk}$$

$$= \frac{1}{2}\left(\omega_N^{-\frac{Nk}{4}} + \omega_N^{\frac{Nk}{4}}\right) + \sum_{n=-\frac{N}{4}+1}^{\frac{N}{4}-1} \omega_N^{-nk}$$

$$= \cos\left(\frac{\pi k}{2}\right) + \sum_{n=0}^{\frac{N}{4}-1}(\omega_N^{-nk} + \omega_N^{nk}) - 1$$

$$= \cos\left(\frac{\pi k}{2}\right) + 2\text{Re} \sum_{n=0}^{\frac{N}{4}-1} \omega_N^{nk} - 1.$$

Let's pause and comment on a few important maneuvers that were used. The $\pm N/4$ terms were separated in the first step and gathered as a single cosine term under the authority of the Euler relations. The remaining sum was then collapsed, so it runs over the indices $n = 0 : N/4 - 1$. It is best to include the $n = 0$ term in the sum, but notice that it is "double-counted," which is why it is necessary to subtract 1 from the sum. In the last step we noted that the summand $\omega_N^{-nk} + \omega_N^{nk}$ can be written as $2\text{Re}\left\{\omega_N^{nk}\right\}$.

Now the sum can be recognized as a geometric sum with ratio ω_N^k. Evaluating the geometric sum allows the calculation to continue:

$$NF_k = \cos\left(\frac{\pi k}{2}\right) + 2\text{Re} \sum_{n=0}^{\frac{N}{4}-1} \omega_N^{nk} - 1$$

$$= \cos\left(\frac{\pi k}{2}\right) + 2\text{Re}\left[\frac{1 - \omega_N^{\frac{kN}{4}}}{1 - \omega_N^k}\right] - 1.$$

At this point the problem has been "cracked" since the DFT sum has been evaluated. However, the work is not complete. A bit of algebraic alacrity is needed to collect and simplify terms, and this is best not done in public! No matter what form of thrashing is used, the result that finally emerges is

$$F_k = \frac{\sin(\pi k/2)\sin(2\pi k/N)}{2N\sin^2(\pi k/N)} \tag{3.23}$$

for $k = -N/2 + 1 : N/2$, but $k \neq 0$. A separate calculation shows that

$$F_0 = \frac{1}{N} \sum_{n=-\frac{N}{2}+1}^{\frac{N}{2}} f_n = \frac{1}{N}\left[\frac{1}{2} + \sum_{n=-\frac{N}{4}+1}^{\frac{N}{4}-1} 1 + \frac{1}{2}\right] = \frac{1}{2}.$$

Generally, one arrives at this point without a clue as to whether the result is correct. How can we check the accuracy of our calculation? First, it does not hurt to graph the sequence F_k for selected values of N. Figure 3.13 shows the alleged DFT coefficients plotted for $N = 16$ and $N = 32$. The coefficients decay by oscillation like k^{-1}, which is apparent from the analytic result (3.23). Notice that the length of the spatial interval A does not appear in the DFT (in fact, it did not even appear in the input sequence).

There are several tests that can be used to instill some confidence in an analytical result. The first check is symmetry. In the case of the square pulse, the input sequence is real and even ($f_{-n} = f_n$). Therefore, we expect the DFT sequence to be real and even as well, and indeed it is. A second easily applied test is periodicity. The sequence F_k generated by an N-point DFT must be N-periodic. The alleged DFT (3.23) has this property.

Another test might be called the F_0 **test** or **average value test**. It is usually fairly easy to check that the DFT property

$$F_0 = \frac{1}{N} \sum_{n=-\frac{N}{2}+1}^{\frac{N}{2}} f_n$$

is satisfied. In the case of the square pulse, we have already noted that indeed $F_0 = 1/2$ is the average value of the input sequence.

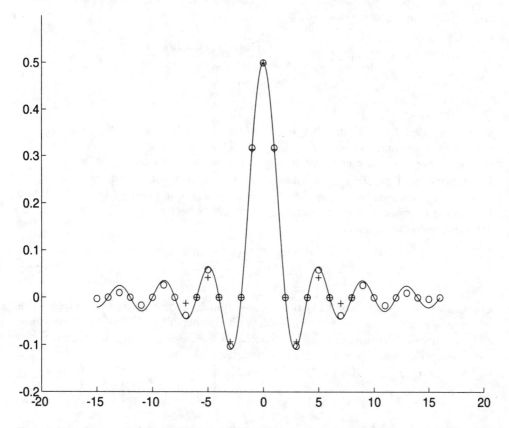

FIG. 3.13.   *The analytical DFT of the square pulse is plotted at the points $k = -N/2 : N/2$ for $N = 16$ (marked with $+$) and $N = 32$ (marked with $\circ$). The exact values of the Fourier coefficients lie on the solid curve, which is superimposed for comparison. The convergence of the DFT coefficients to the Fourier coefficients as $N \to \infty$ can be observed. Note that where coefficients were computed with both sequence lengths, those using $N = 32$ are more accurate than those using $N = 16$. Note also that within either sequence, the coefficients for small $|k|$ are more accurate than those for large $|k|$.*

These three tests are encouraging, but not conclusive. Perhaps the most convincing support one can find for an analytical DFT is the comparison with the Fourier series coefficients, c_k, and the Fourier transform, $\hat{f}$. In this case, two brief calculations demonstrate that

$$c_k = \frac{\sin(\pi k/2)}{\pi k} = \frac{1}{2}\text{sinc}\,(\pi k/2) \quad \text{and} \quad \hat{f}(\omega) = \frac{\sin(\pi \omega A/2)}{\pi \omega} = \frac{A}{2}\text{sinc}\,(\pi \omega A/2),$$

where the **sinc function** has been defined as sinc $(x) = \sin x / x$. Notice that c_k is independent of A, while $\hat{f}$ is not. As an aside, we may confirm the relationship, derived in Chapter 2, between the Fourier coefficients and the Fourier transform: evaluating $\hat{f}$ at the points $\omega_k = k/A$ of the frequency grid, we see that $\hat{f}(\omega_k) = A c_k$.

How do c_k and $\hat{f}(\omega_k)$ compare to our analytical DFT? Visually the comparison is good, as seen in Figure 3.13, in which the Fourier coefficients c_k are superimposed on the DFT coefficients. Furthermore, the agreement between the DFT and Fourier coefficients appears to improve as N increases. The comparison becomes more than visual if we let N become large in the DFT coefficients. A limit can actually be taken, and we find that (problem 65)

$$
\begin{aligned}
\lim_{N \to \infty} F_k &= \lim_{N \to \infty} \frac{\sin(\pi k / 2)\sin(2\pi k/N)}{2N \sin^2(\pi k/N)} \\
&= \frac{\sin(\pi k / 2)}{\pi k} = c_k
\end{aligned}
$$

for $k = -N/2 + 1 : N/2$. In other words, in the limit $N \to \infty$, we see that the DFT coefficients approach the Fourier coefficients. This limiting property will be explored at great lengths in Chapter 6, but it is useful to see a prior example of it.

Example: The DFT of a linear profile. We now consider an input sequence that requires another technique. Let $f(x) = x/A$ on the interval $[-A/2, A/2]$. If this function is sampled at the N points $x_n = nA/N$, where $n = -N/2 + 1 : N/2$, the resulting sequence is given by

$$
f_n = \begin{cases} n/N & \text{if } |n| \leq N/2 - 1, \\ 0 & \text{if } n = \pm N/2. \end{cases} \tag{3.24}
$$

Once again, we see (Figure 3.14) that in anticipation of the DFT, the average value of the function has been used at the points of discontinuity (AVED) which occur at the endpoints of the interval. Said differently, if the sequence f_n defined above were extended periodically, it would be an N-periodic sequence.

Now we may proceed with the DFT. Applying the DFT definition to the sequence f_n we have

$$
NF_k = \sum_{n=-\frac{N}{2}+1}^{\frac{N}{2}} \underbrace{f_n}_{u_n} \underbrace{\omega_N^{-nk}}_{\Delta v_n} = \sum_{n=-\frac{N}{2}+1}^{\frac{N}{2}} \underbrace{\frac{n}{N}}_{u_n} \underbrace{\omega_N^{-nk}}_{\Delta v_n}.
$$

Notice that the $n = N/2$ term has been kept in the sum even though $f_{N/2} = 0$. Since f_n is linear in n, we no longer have a geometric sum, and another tool is needed. In analogy with the integration of a function like $f(x) = x e^{ax}$, we consider summation by parts. As shown in the DFT sum above, the term u_n is identified with that part of the sum that is easily "differenced," whereas Δv_n is chosen so that it is (hopefully) the difference of a sequence v_n. With u_n and Δv_n selected in this manner, the summation by parts formula says that

$$
\sum_{n=-\frac{N}{2}+1}^{\frac{N}{2}} u_n \Delta v_n = u_{\frac{N}{2}} v_{\frac{N}{2}+1} - u_{-\frac{N}{2}} v_{-\frac{N}{2}+1} - \sum_{n=\frac{N}{2}+1}^{\frac{N}{2}} v_n \Delta u_{n-1}.
$$

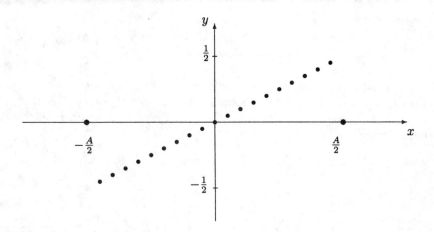

FIG. 3.14. *The linear function $f(x) = x/A$ is sampled at $N = 20$ equally spaced points of the interval $[-A/2, A/2]$ to generate the input sequence $f_n = n/N$ for $n = -N/2 + 1 : N/2$. The periodic extension of this function beyond the interval is discontinuous at $x = \pm A/2$. Therefore, the average value of the function must be used at these points (AVED), and the value $f_{\pm \frac{N}{2}} = 0$ is assigned.*

As with integration by parts, there are two subproblems that must be solved. The first is to compute Δu_{n-1}, where $u_n = f_n$ is given by

$$\{Nu_n\} = \left\{0, -\frac{N}{2} + 1, -\frac{N}{2} + 2, \ldots, -1, 0, 1, \ldots, \frac{N}{2} - 1, 0\right\}.$$

A short calculation shows that elements of the (periodic) sequence $\Delta u_{n-1} = u_n - u_{n-1}$ for $n = -N/2 + 1 : N/2$ are given by

$$\{N\Delta u_{n-1}\} = \left\{-\frac{N}{2} + 1, 1, 1, \ldots, 1, 1, -\frac{N}{2} + 1\right\}.$$

The second subproblem is to find the sequence v_n that satisfies

$$\Delta v_n = v_{n+1} - v_n = \omega_N^{-nk}.$$

Those readers who are familiar with difference equations will recognize this as a first-order constant coefficient difference equation (a creature that will be examined in greater detail in Chapter 7). It can be solved by assuming a trial solution of the form $v_n = \alpha \omega_N^{-nk}$, substituting and solving for the undetermined coefficient α. This problem is also equivalent to finding the "antidifference" of ω_N^{-nk} (analogous to antidifferentiation). By either method, the result (problem 67) is that the sequence v_n is given by

$$v_n = \frac{\omega_N^{-nk}}{\omega_N^{-k} - 1}.$$

An important fact is that v_n is periodic with period N, as is u_n.

We may now invoke the summation by parts formula. We find that

$$NF_k = \underbrace{u_{\frac{N}{2}} v_{\frac{N}{2}+1} - u_{-\frac{N}{2}} v_{-\frac{N}{2}+1}}_{0} - \sum_{n=-\frac{N}{2}+1}^{\frac{N}{2}} v_n \Delta u_{n-1}.$$

The first important simplification now occurs. Notice that because of the periodicity of both u_n and v_n, the first two "boundary terms" exactly cancel. We next recall that $\Delta u_{n-1} = 1/N$ except when $n = -N/2 + 1$ and $n = N/2$, which prompts us to split the above sum as follows:

$$
\begin{aligned}
NF_k &= -\sum_{n=-\frac{N}{2}+1}^{\frac{N}{2}} v_n \Delta u_{n-1} \\
&= -\left[\sum_{n=-\frac{N}{2}+1}^{\frac{N}{2}} \left(v_n \cdot \frac{1}{N} \right) \right] + \left(v_{\frac{N}{2}} \cdot \frac{1}{N} \right) + \left(v_{-\frac{N}{2}+1} \cdot \frac{1}{N} \right) \\
&\quad - v_{\frac{N}{2}} \underbrace{\left(-\frac{1}{2} + \frac{1}{N} \right)}_{\Delta u_{\frac{N}{2}-1}} - v_{-\frac{N}{2}+1} \underbrace{\left(-\frac{1}{2} + \frac{1}{N} \right)}_{\Delta u_{-\frac{N}{2}}}.
\end{aligned}
$$

The first two terms after the sum remove the extraneous first and last terms from the sum; the last two terms after the sum add in the correct first and last terms of the sum. The reason for splitting the sum in this awkward way is to maintain a sum over the full range of indices ($n = -N/2 + 1 : N/2$); furthermore, this sum is a geometric sum. Additional simplification leads to

$$
\begin{aligned}
NF_k &= -\frac{1}{N} \left[\sum_{n=-\frac{N}{2}+1}^{\frac{N}{2}} v_n \right] + \frac{1}{2}\left(v_{\frac{N}{2}} + v_{-\frac{N}{2}+1} \right) \\
&= -\frac{1}{N} \sum_{n=-\frac{N}{2}+1}^{\frac{N}{2}} \frac{\omega_N^{-nk}}{\omega_N^{-k} - 1} + \frac{1}{2}\left(\frac{\cos(\pi k)(1 + \omega_N^{-k})}{\omega_N^{-k} - 1} \right) \\
&= -\frac{1}{N} \frac{1}{\omega_N^{-k} - 1} \underbrace{\sum_{n=-\frac{N}{2}+1}^{\frac{N}{2}} \omega_N^{-nk}}_{N\hat{\delta}_N(k)} + \frac{1}{2}\left(\frac{\cos(\pi k)(1 + \omega_N^{-k})}{\omega_N^{-k} - 1} \right),
\end{aligned}
$$

where $k = -N/2 + 1 : N/2$. The important step is the evaluation of the geometric sum. As shown, it has a value of $N\hat{\delta}_N(k)$, which is zero unless $k = 0$.

Now only some algebraic dexterity is required to reduce this expression to a manageable form. Once it is done, the result is remarkably simple; we find that the DFT of the linear sequence is given by

$$F_k = i\frac{\cos(\pi k)\sin(2\pi k/N)}{4N \sin^2(\pi k/N)},$$

for $k = -N/2 + 1 : N/2$, with the special case that $F_0 = 0$. Most mortals emerge from such a calculation with at least a bit of skepticism about its accuracy. So let's try to verify the above DFT. First the symmetry looks good: the input sequence is real and odd; the DFT is pure imaginary and odd as it should be. Also, the periodicity test is satisfied, since F_k as defined above is an N-periodic sequence. The F_0 test is also confirmed, since

$$\sum_{n=-\frac{N}{2}+1}^{\frac{N}{2}} f_n = 0 \quad \text{and} \quad F_0 = 0.$$

Note that this test fails if the AVED warning is not heeded, that is, if the average value is not used at the point $n = N/2$.

A little more effort is needed to compare the DFT to the Fourier coefficients, but it offers further assurance. A short calculation (not surprisingly requiring integration by parts) shows that the Fourier coefficients of the function $f(x) = x/A$ on the interval $[-A/2, A/2]$ are given by

$$c_k = i\frac{\cos(\pi k)}{2\pi k}$$

for $k = \ldots, -2, -1, 1, 2, \ldots$, with $c_0 = 0$. Clearly, we already have $c_0 = F_0$. A limiting argument (problem 66) can be used to show that

$$\lim_{N\to\infty} F_k = c_k$$

for $k = -N/2 + 1 : N/2$. At this point, we have used all of the immediate tests, and they give reasonable testimony to the accuracy of the DFT calculation just presented.

Example: How one DFT begets another. Since the previous computation may have seemed arduous, it would be nice to know that additional DFTs can be squeezed from it with very little effort. The strategy relies upon the properties of the DFT and it finds frequent use. Consider the problem of approximating the Fourier coefficients of the function $g(x) = x$ on the interval $[0, A]$. If this function is sampled at the N equally spaced points $x_n = nA/N$, where $n = 0 : N - 1$, the resulting sequence is

$$g_n = \begin{cases} nA/N & \text{if } n = 1 : N-1, \\ A/2 & \text{if } n = 0. \end{cases}$$

Notice that in anticipation of using g_n as input to the DFT, the average value of the function has been used at the discontinuity at $x = 0$. To compute the DFT, one could use the alternate definition of the DFT (on the index set $n = 0 : N - 1$) and compute the coefficients directly. This exercise would resemble the previous example very closely. On the other hand, one could use the result of the previous example and save some labor. Let's follow the latter course.

Looking at Figure 3.15, we see that the input sequence g_n can be formed from the sequence $f_n = n/N$ of the previous example by (i) shifting it horizontally by $N/2$ index units (or $A/2$ physical units), (ii) shifting it vertically $A/2$ physical units, and (iii) scaling it by a factor of A. Each of these three operations can be accommodated by the DFT. In fact, we can relate the two sequences as follows:

$$g_n = A\left(f_{n-\frac{N}{2}} + \frac{1}{2}\right)$$

for $n = 0 : N - 1$. (Check, for example, that $g_0 = A\left(f_{-\frac{N}{2}} + \frac{1}{2}\right) = A/2$.) Letting $G_k = \mathcal{D}\{g_n\}_k$ we can now appeal to the relevant DFT properties to write

$$G_k = \mathcal{D}\left\{A\left(f_{n-\frac{N}{2}} + \frac{1}{2}\right)\right\}_k$$

$$= A\left(\mathcal{D}\{f_{n-\frac{N}{2}}\}_k + \mathcal{D}\left\{\frac{1}{2}\right\}_k\right) \qquad \text{(by linearity)}$$

$$= A\omega_N^{-\frac{Nk}{2}}\mathcal{D}\{f_n\}_k + \frac{A}{2}\hat{\delta}_N(k) \qquad \text{(by the shift property)}$$

$$= A\cos(\pi k)F_k + \frac{A}{2}\hat{\delta}_N(k)$$

for $k = 0 : N-1$. In order of appearance, the linearity of the DFT, the shift property, and the fact that $\mathcal{D}\{1\}_k = \hat{\delta}_N(k)$ have been used to arrive at this result.

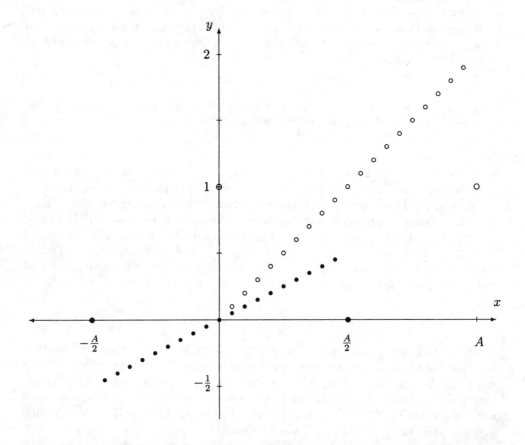

FIG. 3.15. *Having computed the DFT of one sequence, the DFT of a related sequence can often be found easily. The figure shows two linear sequences: $f_n = n/N$ (shown as •) arises from sampling the function $f(x) = x/A$ on the interval $[-A/2, A/2]$, and $g_n = nA/N$ (shown as ∘) are the samples of $g(x) = x$ on the interval $[0, A]$. The sequences differ by a scaling, a horizontal shift, and a vertical shift. All of these modifications can be accomodated by the DFT. The figure shows the case in which $A = 2$ and $N = 20$.*

Using the values of F_k found previously we have that

$$G_k = \begin{cases} A/2 & \text{if } k = 0, \\ A\cos(\pi k)F_k = i\frac{A\sin(2\pi k/N)}{4N\sin^2(\pi k/N)} & \text{if } k = 1 : N-1. \end{cases}$$

Notice that the sequence G_k is not *quite* symmetric; although it appears to be imaginary and odd, it is *not*, since $G_0 = A/2$. The sequence G_k lost its symmetry when the sequence f_n was altered to form g_n.

Examples and exercises involving analytical DFTs abound and could continue endlessly (as will be demonstrated in the problem section). However, we must conclude, and will do so with a brief look at *The Table of DFTs* which appears in the Appendix. This table is a collection of as many analytical DFTs as time, energy, and sanity allowed. A few words of explanation are needed. Each entry in the *Table of DFTs* is arranged as follows.

Discrete input name	$f_n,\ n \in \mathcal{N}$	
Graph of f_n	$F_k,\ k \in \mathcal{N}$	Graph of F_k
	$\|c_k - F_k\|,\ k \in \mathcal{N}$	
Continuum input name	$f(x),\ x \in I$	
Graph of $f(x)$	$c_k,\ k \in \mathbf{Z}$	Graph of $\|c_k - F_k\|$
	Comments	$\max \|c_k - F_k\|$

The first column has two boxes. The upper box gives the name of the input, below which are graphs of the real and imaginary parts of the discrete input sequence. The lower box contains the name of the continuum input, and the corresponding continuum input graphs. The middle column has six boxes containing, in order from top to bottom, the formula of the input sequence f_n; the analytic N-point DFT output F_k; a measure of the difference $|c_k - F_k|$; the formula of the continuum input function $f(x)$; the formula for the Fourier coefficients c_k; an entry for comments, perhaps the most important of which is the AVED warning. This means that *average values at endpoints and discontinuities* must be used if the correct DFT is to be computed. The third column consists of two boxes. The upper box displays graphically the real and imaginary parts of the DFT. The lower box gives the maximum error $\max |c_k - F_k|$, and displays graphically the error $|c_k - F_k|$ for a small (24-point) example.

Unless otherwise noted, the function is assumed to be sampled on the interval $[-A/2, A/2]$. The difference $|c_k - F_k|$ is generally in the form CN^{-p} for some constant C and some positive integer p, which should be interpreted in an asymptotic sense for $N \to \infty$; in other words, if $|F_k - c_k| = CN^{-p}$, then

$$\lim_{N \to \infty} \left| \frac{F_k - c_k}{N^p} \right| = C.$$

While this measure is different than the pointwise error that will be derived in Chapter 6, it does agree with those estimates in its dependence on N.

We close this section of esoterica with the request that readers who find challenge and joy in computing DFTs analytically submit any genuinely new DFTs that do not appear in *The Table of DFTs* to the authors.

3.6. Problems

Alternate Forms of the DFT

30. Geometry of even/noncentered DFT modes.   Consider a 12-point DFT defined on the index sets $n, k = 0 : 11$. Sketch the 12 different modes corresponding to the indices $k = 0 : 11$ and note their frequency and period. What value of k corresponds to the highest frequency mode? What is the frequency and period of that mode? Give the pairs of indices that correspond to modes with the same frequency.

31. Geometry of odd/noncentered DFT modes.   Consider an 11-point DFT defined on the index sets $n, k = 0 : 10$. Sketch the 11 different modes corresponding to the indices $k = 0 : 10$ and note their frequency and period. What values of k correspond to the highest frequency mode? What is the frequency and period of this mode? Give the pairs of indices that correspond to modes with the same frequency.

32. DFTs on an odd number of points.   Show that in general a noncentered DFT on an odd number of points (single set or double set) does not include the highest frequency mode $\cos(\pi n)$. Show that the highest frequency modes that are present correspond to indices $k = (N \pm 1)/2$ for a single set of points and $k = N, N + 1$ for a double set of points.

33. DFTs on an odd number of points.   Show that in general a centered DFT on an odd number of points (single set or double set) does not include the highest frequency mode $\cos(\pi n)$. Show that the highest frequency modes that are present correspond to indices $k = \pm N/2$ for a single set of points with N even, $k = \pm(N-1)/2$ for a single set of points with N odd, and $k = \pm N$ for a double set of points.

34. Odd/centered DFTs.   Consider the DFT defined on a centered interval with an odd number of transform points.

(a) Assuming N is even, verify the orthogonality property

$$\sum_{n=-\frac{N}{2}}^{\frac{N}{2}} \omega_{N+1}^{-n(j-k)} = (N + 1)\hat{\delta}_{N+1}(j - k).$$

(b) Given this orthogonality property and the forward DFT,

$$F_k = \frac{1}{N + 1} \sum_{n=-\frac{N}{2}}^{\frac{N}{2}} f_n \omega_{N+1}^{-nk},$$

for $k = -N/2 : N/2$, derive the corresponding inverse DFT.

(c) Verify that the sample points for this DFT, $x_n = nA/(N+1)$, do not include either endpoint of the interval $[-A/2, A/2]$. Show that $x_{\pm\frac{N}{2}}$ approach $\pm A/2$ as N increases.

35. Double/noncentered/odd DFT.   Write out the complete DFT pair for a double set of an odd number of noncentered points. Indicate the ranges of the indices clearly. Write out the grid points in the spatial and frequency domains and find the reciprocity relations that apply.

36. Double/centered/odd DFT.   Write out the complete DFT pair for a double set of an odd number of centered points. Indicate the ranges of the indices clearly. Write out the grid points in the spatial and frequency domains and find the reciprocity relations that apply.

37. Using a DFT.   Assume that you have a program that computes a double/noncentered/odd DFT. Write out the forward DFT expression. Show how it can be used to approximate the Fourier coefficients of $f(x) = x$ on the interval $[0, 2]$ using 31 points. What are the spatial grid points x_n and the frequency grid points ω_k that are used by this DFT? Give the values of the input sequence at each spatial grid point (with attention to the endpoints).

38.   Using a DFT.   Assume that you have a program that computes a double/centered/even DFT. Write out the forward DFT expression. Show how it can be used to approximate the Fourier coefficients of $f(x) = x$ on the interval $[-2, 2]$ using 32 points. In particular, what are the spatial grid points x_n and the frequency grid points ω_k that are used by this DFT? Give the values of the input sequence at each spatial grid point (with attention to the endpoints).

39. Modifying a DFT.   It is not uncommon to have a DFT program that does not fit the specifications of the problem at hand. Assume that you have a program that computes a single/centered/even DFT. Write out the forward transform and indicate the index ranges clearly. Show how it can be used to approximate the Fourier coefficients of $f(x) = x$ on the interval $[0, 2]$ using 32 points. Use the periodicity of the DFT and indicate how to define the input sequence and interpret the output.

40. Software and DFTs.   Assume that the Fourier coefficients of $f(x) = \cos(\pi x/2)$ must be approximated on the interval $[-1, 1]$. Show how the N-point input sequence must be defined for each of the software packages discussed in the text. Equally important, show how the sequence of transform coefficients should be interpreted in each case. In particular, show where the coefficients of the constant mode and the highest frequency mode(s) appear in the output list.

41. Custom DFTs.   Assume that you need a DFT program that approximates the Fourier coefficients of a given function f on the interval $[-A/4, 3A/4]$. Describe how to "hand tool" a DFT for this problem with each of the following strategies.

- Design a special DFT from scratch based on the sampling interval $[-A/4, 3A/4]$. Describe how you would define the spatial and frequency grids. How would you interpret the transform coefficients F_k; specifically, with which frequency is each F_k associated?

- Use the shift theorem for the DFT to transform this problem so that a standard DFT can be used.

- Use the shift theorem for Fourier coefficients to transform this problem so that a standard DFT can be used.

Can you think of any other strategies? Can you outline a general procedure for creating a new DFT or modifying an existing DFT so that it applies to an arbitrary sampling interval $[-pA, (1-p)A]$ of length A, where $0 < p < 1$?

Properties of the DFT

42. Hermitian symmetry.   Verify the Hermitian symmetry relations

$$f_n = \mathcal{D}\{NF_k^*\}_n^* \quad \text{and} \quad F_k = \mathcal{D}^{-1}\{f_n^*/N\}_k^*$$

using the definition of the DFT and its inverse.

43.   Shift property.   Apply the definition of the DFT (8.21) directly to the sequence $g_n = f_{n-j}$, where $n = -N/2 + 1 : N/2$ and j is a fixed integer to prove the shift property $\mathcal{D}\{f_{n-j}\}_k = \omega_N^{-jk}\,\mathcal{D}\{f_n\}_k$.

44. Even and odd input. Use the definition of the DFT to show directly that (a) the DFT of an even sequence is even, (b) the DFT of a real and even sequence is real and even, (c) the DFT of an odd sequence is odd, and (d) the DFT of a real and odd sequence is imaginary and odd.

45. Waveform decomposition. Show that an arbitrary sequence f_n can be decomposed in the form $f_n = f_n^{\text{even}} + f_n^{\text{odd}}$, where

$$f_n^{\text{even}} = \frac{f_n + f_{-n}}{2} \quad\text{and}\quad f_n^{\text{odd}} = \frac{f_n - f_{-n}}{2}.$$

Verify that f_n^{even} is an even sequence and f_n^{odd} is an odd sequence. Finally, show that

$$\mathcal{D}\{f_n^{\text{even}}\}_k = \frac{F_k + F_{-k}}{2} \quad\text{and}\quad \mathcal{D}\{f_n^{\text{odd}}\}_k = \frac{F_k - F_{-k}}{2}.$$

46. IDFT of a conjugate symmetric sequence. Show that if F_k is any conjugate symmetric sequence then $f_n = \mathcal{D}^{-1}\{F_k\}_n$ is a real-valued sequence.

47. Discrete correlation theorem. Prove that

$$\mathcal{D}\{f_n \otimes g_n\}_k = NF_k^*G_k \quad\text{and}\quad \mathcal{D}\{f_n^*g_n\}_k = F_k \otimes G_k,$$

where $f_n \otimes g_n = \sum_n f_j g_{n+j}$.

48. Parseval's relation. Prove Parseval's relation

$$\frac{1}{N}\sum_{n=-\frac{N}{2}+1}^{\frac{N}{2}} |f_n|^2 = \sum_{k=-\frac{N}{2}+1}^{\frac{N}{2}} |F_k|^2$$

using the definition of the DFT and the orthogonality property.

49. Alternate proof of the Discrete Convolution Theorem. Prove the Discrete Convolution Theorem by showing directly that $\mathcal{D}^{-1}\{NF_k g_k\}_n = f_n * g_n$. Is it a simpler proof than the one given in the text?

50. Convolution and high precision arithmetic. Note that the decimal representation of an $(n+1)$-digit integer a has the form

$$a = \sum_{k=0}^{n} a_k 10^k,$$

where $a_n \neq 0$. Using a similar representation for the $(m+1)$-digit integer b, show how convolution can be used to compute the product of these two integers exactly. In particular, show how the sets of digits a_k and b_k must be extended to perform the necessary convolution.

51. The sinc function. Consider the sinc function, $\operatorname{sinc}(x) = \sin(x)/x$, and the square pulse (or boxcar) function

$$B_a(x) = \begin{cases} 1 & \text{for } |x| < a/2, \\ 0 & \text{for } |x| > a/2, \\ 1/2 & \text{for } x = \pm a/2. \end{cases}$$

(a) Show that $B_a(bx) = B_{a/b}(x)$ for any positive real numbers a and b.

(b) Make rough sketches of sinc (x), sinc $(2x)$, and sinc $(x/2)$. Note the location of the zeros and the "width" of each function.

(c) Verify that the inverse Fourier transform of $B_a(\omega)$ is

$$\mathcal{F}^{-1}\{B_a(\omega)\} = a\ \text{sinc}\ (\pi a x).$$

(d) From part (c), it follows that

$$\mathcal{F}\{\text{sinc}\ (\pi a x)\} = \frac{1}{a}B_a(\omega).$$

Make some observations about how the shape of the sinc function and its transform vary as a is increased and decreased.

(e) From part (d) deduce that $\mathcal{F}\{\text{sinc}\ (x)\} = \pi B_{1/\pi}(\omega)$.

(f) Is the sinc function band-limited? What is the maximum frequency represented in the Fourier transform of sinc (ax)? Based on the Shannon Sampling Theorem, what is the maximum sampling rate needed to resolve the function sinc (ax)?

52. Padded sequences Let f_n be the sequence

$$f_{-5},\ f_{-4},\ \ldots f_0,\ \ldots f_5,\ f_6,$$

and suppose that its 12-point DFT is F_k for $k = -5 : 6$. Show that if g_n is formed by padding f_n with two zeros on either end, and G_k is the 16-point DFT of g_n, then

$$G_{-4} = F_{-3}, \qquad G_0 = F_0, \qquad G_4 = F_3, \qquad G_8 = F_6.$$

53. DFT interpolation Assume that f is a function that is either zero outside of the interval $[-A/2, A/2]$ or is A-periodic. The sequence f_n is formed by sampling f at N equally spaced points of $[-A/2, A/2]$. Let $F_k = \mathcal{D}_N\{f_n\}_k$ be the N-point DFT of f_n. Now extend the sequence f_n to a length of $2N$ by padding with zeros on both ends and call this new sequence g_n with a DFT $G_k = \mathcal{D}_{2N}\{g_n\}_k$. Show that the even terms of the sequence G_k match the terms of F_k and that the odd terms of G_k can be interpreted as interpolated values of the sequence F_k. Extend this observation and conclude that it is possible to interpolate at $p - 1$ points between each term of F_k by padding the original sequence with zeros to a length of pN.

54. Using the DFT to approximate integrals. Let f be continuous on $[-A/2, A/2]$. Show how the definition of the DFT

$$F_k = \frac{1}{N} \sum_{n=-\frac{N}{2}+1}^{\frac{N}{2}} f_n \omega_N^{-nk}$$

can be used to approximate the integral

$$\int_0^A f(x)e^{-i2\pi\omega x}\,dx.$$

Take care to define the input sequence correctly at points where the periodic extension of f might be discontinuous.

55. Summed and differenced sequences. Given a sequence f_n, let

$$g_n = f_n + f_{n+1}, \quad h_n = f_n - f_{n+1}, \quad p_n = f_n + f_{n+N/2}, \quad q_n = f_n - f_{n+N/2}.$$

Use the definition of the DFT along with linearity and the shift property to verify the properties (3.13), (3.16), (3.19), and (3.20):

$$\mathcal{D}\{g_n\}_k = (1 + \omega_N^{-k})F_k \quad \text{and} \quad \mathcal{D}\{h_n\}_k = (1 - \omega_N^{-k})F_k,$$

and

$$\mathcal{D}\{p_n\}_k = \begin{cases} 2F_k & \text{for } k \text{ even,} \\ 0 & \text{for } k \text{ odd,} \end{cases} \quad \text{and} \quad \mathcal{D}\{q_n\}_k = \begin{cases} 0 & \text{for } k \text{ even,} \\ 2F_k & \text{for } k \text{ odd.} \end{cases}$$

56. Average value property. Show that the average value of an input sequence f_n is given by

$$\mathcal{D}\{f_n\}_0 = F_0 = \frac{1}{N} \sum_{n=-\frac{N}{2}+1}^{\frac{N}{2}} f_n,$$

and hence, if f_n is odd, $F_0 = 0$.

57. Rarified and repeated sequences. An N-point sequence f_n can be **rarified** by preceding each of its elements by $p-1$ zeros. For example, the three-fold rarefaction of f_n is the sequence of length $3N$

$$g_n = \{0, 0, f_{-\frac{N}{2}+1}, 0, 0, \ldots, 0, 0, f_0, 0, 0, 0, f_1, \ldots, f_{\frac{N}{2}}\}.$$

Show that

$$\mathcal{D}_{pN}\{g_n\}_k = \frac{1}{p}\mathcal{D}\{f_n\}_k,$$

for $k = -N/2 + 1 : N/2$. An N-point sequence f_n can be **repeated** by preceding each of its elements by $p - 1$ copies of itself. For example, the two-fold repetition of f_n is

$$h_n = \{f_{-\frac{N}{2}+1}, f_{-\frac{N}{2}+2}, \ldots, f_0, f_0, f_1, f_1, \ldots, f_{\frac{N}{2}}, f_{\frac{N}{2}}\}.$$

Find a relationship between $\mathcal{D}_{pN}\{h_n\}$ and $\mathcal{D}\{f_n\}$.

58. Averaging at endpoints and discontinuities. The rectangular wave on the interval $[-1/2, 1/2]$ is defined by

$$f(x) = \begin{cases} -1 & \text{for } -1/2 < x < 0, \\ 1 & \text{for } 0 \le x \le 1/2. \end{cases}$$

Note that f is real and odd, and therefore its Fourier coefficients are odd and imaginary. Assume that the function is sampled at the points $x_n = n/8$, where $n = -3 : 4$, to produce the input sequence

$$\bar{f}_n = \{-1, -1, -1, 1, 1, 1, 1, 1\}.$$

Compute the eight-point DFT of $\bar{f}_n$. Is the DFT odd and imaginary? How do you explain the error? How should the input sequence be defined? Verify that when the input sequence is correctly defined, the DFT is odd and imaginary.

59. Cosine DFT. Verify that if k_0 is a fixed integer, then the N-point DFT of the sampled cosine wave is given by

$$\mathcal{D}\left\{\cos\left(\frac{2\pi n k_0}{N}\right)\right\}_k = \frac{1}{2}\left(\hat{\delta}_N(k - k_0) + \hat{\delta}_N(k + k_0)\right).$$

Why must the modular delta $\hat{\delta}_N(k \pm k_0)$ rather than the regular delta $\delta(k \pm k_0)$ be used?

60. Shift property. Let f_n be an arbitrary sequence and let n_0 be a fixed integer.

(a) Show directly using the definition of discrete cyclic convolution that

$$f_n * \delta(n - n_0) = f_{n_0}.$$

(b) Show the same result in the following, less direct way. Let F_k denote the DFT of f_n and note that

$$\mathcal{D}\left\{\delta(n - n_0)\right\}_k = e^{-i2\pi n_0 k/N}.$$

Now use the convolution theorem in the form

$$f_n * \delta(n - n_0) = \mathcal{D}^{-1}\left\{F_k e^{-i2\pi n_0 k/N}\right\}_n,$$

together with the modulation property, to conclude that $f_n * \delta(n - n_0) = f_{n_0}$.

61. Modulation by real modes. Given a real sequence f_n find the DFT of the modulated sequences $f_n \cos(2\pi n k_0/N)$ and $f_n \sin(2\pi n k_0/N)$, where k_0 is an integer.

62. Aliasing. Consider the following functions on the indicated intervals. In each case, determine the maximum grid spacing Δx and the minimum number of grid points N needed to insure that the function is sufficiently resolved to avoid aliasing.

(a) $f(x) = \sin(4\pi x)$ on $[-1/2, 1/2]$,

(b) $f(x) = \cos(20\pi x)$ on $[-1, 1]$,

(c) $f(x) = \sin(8x)$ on $[-2\pi, 2\pi]$,

(d) $f(x) = \cos^2(4x)$ on $[-\pi, \pi]$.

63. Aliasing (strobing). A spinning wheel with a light attached to the rim is photographed with a camera with a shutter speed of 16 frames per second.

(a) What is the greatest wheel speed (in revolutions per second (rps)) that can be accurately resolved by this camera without aliasing?

(b) If the wheel revolves at 20 rps, what is the apparent speed recorded by the camera?

(c) If the wheel appears to revolve at 1 rps, what are the possible true wheel speeds (in addition to 1 rps)?

(d) If the wheel appears to revolve backwards at 1 rps, what are the possible true wheel speeds?

Analytical DFTs

64. Geometric sums. Show that for a given real or complex number $a \neq 1$ and integers M and N,

$$\sum_{k=M}^{N-1} a^k = \frac{a^M - a^N}{1 - a}.$$

65. Square pulse Fourier coefficients and DFT. Compare the computed DFT of the square pulse with its Fourier coefficients by showing that

$$\lim_{N \to \infty} F_k = \lim_{N \to \infty} \frac{\sin(\pi k/2)\sin(2\pi k/N)}{2N\sin^2(\pi k/N)} = c_k = \frac{\sin(\pi k/2)}{\pi k},$$

for $k = -N/2 + 1 : N/2$.

66. Convergence test for linear sequence DFT. Consider the linear sequence $f_n = n/N$. Compare the computed DFT with its Fourier coefficients by showing that

$$\lim_{N \to \infty} F_k = \lim_{N \to \infty} i\frac{\cos(\pi k)\sin(2\pi k/N)}{4N\sin^2(\pi k/N)} = c_k = i\frac{\cos(\pi k)}{2\pi k}$$

for $k = -N/2 + 1 : N/2$ (with $F_0 = 0$).

67. An "antidifference" calculation. The summation by parts formula requires that a difference equation of the form $\Delta u_n = f_n$ be solved for the sequence u_n when f_n is given. Solve the difference equation

$$\Delta u_n = u_{n+1} - u_n = \omega_N^{-nk},$$

where $k = -N/2 + 1 : N/2$ is fixed and the solution sequence u_n is N-periodic. (Hint: Assume a solution of the form $u_n = \alpha \omega_N^{-nk}$ and determine α.)

68. New DFTs. Use *The Table of DFTs* in the Appendix to determine the DFT of the following sequences and functions on the indicated index sets or intervals. Try to minimize your efforts by using DFT properties! In each case sketch the input and be sure that average values are used when necessary.

(a) $f_n = |n|$ for $n = -N/2 + 1 : N/2$,

(b) $f_n = -n/4$ for $n = -N/2 + 1 : N/2$,

(c) $f_n = 1 - n$ for $n = 0 : N - 1$.

(d) $f(x) = 2x + 1$ on $[0, 2]$,

(e) $f(x) = e^{-2|x|}$ on $[-2, 2]$,

(f) $f(x) = xe^{-x}$ on $[0, 2]$.

69. Evaluating sums. Use the geometric series and/or summation by parts to evaluate the following sums:

(a) $\sum_{n=0}^{19} e^{-n}$, (b) $\sum_{n=0}^{9} 3^n$, (c) $\sum_{n=0}^{9} n2^{-n}$, (d) $\sum_{n=0}^{24} n^2 3^n$.

Chapter **4**

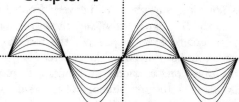

Symmetric DFTs

What immortal
hand or eye
Could frame thy
fearful symmetry?
— William Blake

117

4.1.   Introduction

The preceding chapters have been devoted to the DFT in its fullest generality, as it applies to arbitrary complex-valued input sequences. It is now time to investigate some special forms that the DFT takes when the input possesses special properties. These special properties are often called **symmetries**, and the resulting DFTs are called **symmetric DFTs**. This exploration will prove to be very fruitful for two reasons. First, in practice, the input to the DFT very often *does* possess symmetries. Second, we will soon see that when symmetries are exploited, the resulting DFTs offer savings in both computational effort and storage over the general complex DFT. These economies can be significant in large computations, particularly in more than one dimension. Symmetric DFTs are also the discrete analogs of the real, sine, and cosine (and other) forms of the Fourier series. Not surprisingly, they are indispensable in applications in which these special forms of the Fourier series arise, most notably in solving boundary value problems (as we will see in Chapter 7). Therefore, the subject of symmetric DFTs is not an idle exercise; it has a tremendous impact on issues of practical importance. In learning and writing about symmetric DFTs, we have found the subject filled with subtleties and pitfalls. There are several intertwined themes that underlie symmetric DFTs, and we will try to untangle and highlight them carefully. Here is the general four-step approach that we will follow.

Step 1: Symmetry in the input. The discusson always begins by assuming that the input sequence f_n has a particular symmetry. Three of the many possible symmetries that we will consider are:

1. **real symmetry**: f_n is real,

2. **even symmetry**: $f_n = f_{-n}$,

3. **odd symmetry**: $f_n = -f_{-n}$.

Examples of these three symmetries are shown in the simple sequences of Figure 4.1. We will see that in the presence of each of these symmetries, the DFT takes a special, simplified form. In each case, the cost (in terms of storage and computation) of evaluating the DFT of a symmetric sequence is less than in the case of a general complex sequence. We will note these savings for each symmetry.

Step 2: New symmetric transform. The next step in the development is subtle. For each symmetry, it is possible to define a *new* DFT that can be applied to an *arbitrary* real input sequence. The new DFT that arises in each case is called a **symmetric DFT**. The symmetric DFTs are generally analogs of special forms of the Fourier series. For the three cases we will consider, the symmetric DFTs and their Fourier series analogs are:

1. **real symmetry** $\rightarrow$ Real DFT (RDFT) $\Leftrightarrow$ Real Fourier Series,

2. **even symmetry** $\rightarrow$ Discrete Cosine Transform (DCT) $\Leftrightarrow$ Fourier Cosine Series,

3. **odd symmetry** $\rightarrow$ Discrete Sine Transform (DST) $\Leftrightarrow$ Fourier Sine Series.

If confusion arises at all, it is because a symmetric DFT can be applied to *any* real sequence in which the original symmetry is usually entirely absent. In other words, given a real input sequence, one could apply the RDFT, the DCT, or the DST,

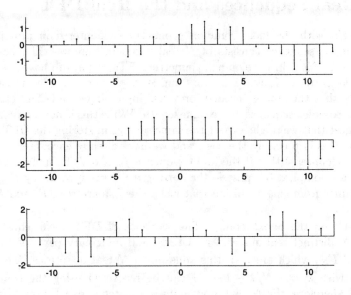

FIG. 4.1. *Periodic sequences with the three most commonly occurring symmetries are shown. The top sequence is arbitrary apart from the fact that its elements are real; the middle sequence also has the even symmetry $x_{-n} = x_n$; and the bottom sequence has the odd symmetry $x_{-n} = -x_n$.*

depending upon the problem at hand. In general, the results of the three transforms are different. In the opposite direction, given a particular problem, an appropriate solution might take the form of an inverse RDFT, an inverse DCT, or an inverse DST. This is analogous to the use of a Fourier series: given a fairly arbitrary real-valued function, one may represent it in terms of a real, cosine, or sine Fourier series.

Step 3: Pre- and postprocessing forms. With the new symmetric DFTs and their inverses defined, we will next address the question of how they can be computed efficiently. One can always use these definitions, which we will call **explicit forms** of the DFT; they are direct, but often inefficient. An improvement over the explicit forms can be found in **pre- and postprocessing** techniques. These algorithms take various forms, but the fundamental idea is always the same. The input sequence is modified in a simple preprocessing step to produce an auxiliary sequence; the auxiliary sequence is then used as input to the complex DFT (which is evaluated by a fast Fourier transform (FFT)); and the resulting sequence of coefficients is then modified in a simple postprocessing step to produce the desired DFT. In this way the complex DFT can be used to compute symmetric DFTs efficiently.

Step 4: Compact symmetric form. To complete the story and make it as currrent as possible, we will occasionally mention a more recent development that leads to very efficient symmetric DFT algorithms. This is the design of **compact symmetric FFTs** in which the pre- and postprocessing is avoided altogether. Instead the symmetry in both the input and output is built directly into the FFT. The resulting compact FFT offers savings in both computation and storage over the complex FFT. These symmetric FFTs appear to be the most efficient methods for computing symmetric DFTs.

4.2.   Real Sequences and the Real DFT

We will begin with the most common symmetry encountered in practice, that in which the input sequence consists of N real numbers that we will label f_n, where $n = -N/2 + 1 : N/2$. In developing symmetric DFTs, it pays to have a bookkeeping outlook to keep track of savings in work and storage. For example, we can conclude immediately that the storage needed for a real input sequence is half of the storage needed for a complex sequence of the same length. While this is not a profound insight, it does suggest that we might expect the same savings in storing the DFT coefficients F_k. In fact, this is precisely the case, and we have already encountered the reason. Recall from Chapter 3 that if the input sequence f_n is real, then the resulting DFT coefficients are complex, but possess the **conjugate even symmetry** $F_{-k} = F_k^*$. We can also deduce from this definition that when the f_n's are real, F_0 and $F_{\frac{N}{2}}$ are also real.

Now let's do the bookkeeping. The sequence of DFT coefficients F_k consists of exactly N distinct real quantities: the real and imaginary parts of $F_1, \ldots, F_{\frac{N}{2}-1}$ plus F_0 and $F_{\frac{N}{2}}$, which are real; this adds up to N real quantities. The remaining coefficients with $k = -N/2 + 1 : -1$ can be recovered using the conjugate even symmetry. Therefore, the N real input data are matched exactly by N distinct real values in the output. Using the fact that the input sequence is real we can rewrite the complex DFT as

$$
F_k = \frac{1}{N} \sum_{n=-\frac{N}{2}+1}^{\frac{N}{2}} f_n \omega_N^{-nk}
$$

$$
= \frac{1}{N} \sum_{n=-\frac{N}{2}+1}^{\frac{N}{2}} f_n \left[\cos\left(\frac{2\pi nk}{N} \right) - i \sin\left(\frac{2\pi nk}{N} \right) \right]. \tag{4.1}
$$

This expression is what we will call the **explicit form** of the real DFT. While it hardly looks like a simplification, there are a few important facts to be gleaned from it. First, it allows us to assess the computational cost of the real DFT and make a surprising comparison with the complex DFT. Recall that we need roughly N^2 complex multiplications and additions to compute the complex DFT from its definition. (In all that follows, the important lessons can be extracted if we make rough operation counts; it suffices to keep track of multiples of N^2.) In terms of real operations, this amounts to $4N^2$ real multiplications and $4N^2$ real additions. From the explicit form of the real DFT (4.1) a quick count shows that roughly $2N$ real multiplications and $2N$ additions are required to compute a single coefficient F_k, but because of the symmetry in the transform coefficients, only $F_0, \ldots, F_{\frac{N}{2}}$ need to be computed. Therefore, roughly N^2 real multiplications and N^2 additions are needed to compute the coefficients of the real DFT. There is a factor of four savings that arises for two reasons: the fact that f_n is real means the complex multiplications of the DFT have only half of their usual cost, and the fact that F_k is conjugate even means that only half of the coefficients need to be computed. It is worth mentioning that a bit more efficiency can be squeezed out of both the complex and real DFTs. As shown in problem 70, by "folding" the sums, the number of real multiplications in both DFTs can be reduced by another factor of two.

The first stage of the discussion of the real symmetry is complete. We have shown how the DFT responds when it is presented with a real input sequence: the explicit form is more economical in both computation and storage. The next step is to show how a new DFT can be created from this explicit form. In the case of the real symmetry it is a short step. The explicit form (4.1) can be regarded as a transformation between an arbitrary real N-vector

$$\{f_{-\frac{N}{2}+1}, \ldots, f_0, \ldots, f_{\frac{N}{2}}\}$$

and another real N-vector

$$\left\{F_0, \operatorname{Re}\{F_1\}, \operatorname{Im}\{F_1\}, \ldots, \operatorname{Re}\{F_{\frac{N}{2}-1}\}, \operatorname{Im}\{F_{\frac{N}{2}-1}\}, F_{\frac{N}{2}}\right\},$$

where we have denoted the real and imaginary parts of F_k as $\operatorname{Re}\{F_k\}$ and $\operatorname{Im}\{F_k\}$, respectively. We call this transformation the real DFT.

▶ **Real DFT (RDFT)** ◀

$$F_k = \frac{1}{N} \sum_{n=-\frac{N}{2}+1}^{\frac{N}{2}} f_n \left[\cos\left(\frac{2\pi nk}{N}\right) - i\sin\left(\frac{2\pi nk}{N}\right)\right], \qquad (4.2)$$

for $k = 0 : N/2$, or

$$\operatorname{Re}\{F_k\} = \frac{1}{N} \sum_{n=-\frac{N}{2}+1}^{\frac{N}{2}} f_n \cos\left(\frac{2\pi nk}{N}\right)$$

for $k = 0 : N/2$, and

$$\operatorname{Im}\{F_k\} = -\frac{1}{N} \sum_{n=-\frac{N}{2}+1}^{\frac{N}{2}} f_n \sin\left(\frac{2\pi nk}{N}\right)$$

for $k = 1 : N/2 - 1$. This is the first of the symmetric DFTs. From a computational point of view, the RDFT *is* a new DFT. However, it produces the same coefficients as the complex DFT; it just does it more efficiently.

For each of the symmetries that we will consider there is always the question of an inverse transform. To determine the relevant inverse DFT it is necessary to use the known symmetries in the opposite direction. The real DFT provides a good first example. We must now imagine a sequence of DFT coefficients F_k that has conjugate even symmetry; that is, $F_k = F^*_{-k}$. The goal is to recover the *real* sequence f_n that is the inverse DFT of F_k. Beginning with the complex inverse DFT and then incorporating the conjugate even symmetry of F_k, we arrive at the inverse real DFT as follows.

▶ **Inverse Real DFT** ◀

$$
\begin{aligned}
f_n &= \sum_{k=-\frac{N}{2}+1}^{\frac{N}{2}} F_k \omega_N^{nk} \\
&= F_0 + \sum_{k=1}^{\frac{N}{2}-1} (F_k \omega_N^{nk} + F_{-k} \omega_N^{-nk}) + F_{\frac{N}{2}} \cos(\pi n)
\end{aligned}
$$

$$= F_0 + \sum_{k=1}^{\frac{N}{2}-1} 2\mathrm{Re}\left\{F_k \omega_N^{nk}\right\} + F_{\frac{N}{2}} \cos(\pi n)$$

$$= F_0 + 2 \sum_{k=1}^{\frac{N}{2}-1} \left[\mathrm{Re}\left\{F_k\right\} \cos\left(\frac{2\pi nk}{N}\right) - \mathrm{Im}\left\{F_k\right\} \sin\left(\frac{2\pi nk}{N}\right) \right]$$

$$+ F_{\frac{N}{2}} \cos(\pi n), \tag{4.3}$$

where $n = -N/2 + 1 : N/2$. Notice how the conjugate even symmetry, $F_k = F_{-k}^*$, has been used to collect the complex exponentials and express them as sines and cosines. Notice also that in the end, only the independent values of F_k, for $k = 0 : N$, are used in computing the inverse. This explicit form of the inverse real DFT makes it clear that the sequence f_n is real (recall that F_0 and $F_{\frac{N}{2}}$ are real). A quick count of operations also confirms that each term of the sequence f_n requires roughly N real multiplications and additions; therefore a total of roughly N^2 real multiplications and additions are needed to evaluate the inverse real DFT; this matches the operation count for the forward real DFT.

In working with the inverse DFT many authors give the coefficients a distinct name and write

$$f_n = \frac{1}{2} a_0 + \sum_{k=1}^{\frac{N}{2}-1} \left[a_k \cos\left(\frac{2\pi nk}{N}\right) + b_k \sin\left(\frac{2\pi nk}{N}\right) \right] + \frac{1}{2} a_{\frac{N}{2}} \cos(\pi n).$$

It is then an easy matter to make the association between the complex coefficients F_k and the real coefficients a_k and b_k; it is simply

$$a_k = 2\mathrm{Re}\left\{F_k\right\} = \frac{2}{N} \sum_{n=-\frac{N}{2}+1}^{\frac{N}{2}} f_n \cos\left(\frac{2\pi nk}{N}\right)$$

for $k = 0 : N/2$, and

$$b_k = -2\mathcal{I}\{F_k\} = \frac{2}{N} \sum_{n=-\frac{N}{2}+1}^{\frac{N}{2}} f_n \sin\left(\frac{2\pi nk}{N}\right)$$

for $k = 1 : N/2 - 1$. The inverse RDFT defines a sequence f_n that is real and periodic with period N. This property suggests the analogy between the inverse RDFT and the real form of the Fourier series. Therefore, this is an appropriate form for the solution of a problem (for example, a difference equation) in which the solution must be real and periodic.

4.3. Even Sequences and the Discrete Cosine Transform

Another symmetry that appears frequently, for example in the solution of boundary value problems [79], [133] and in image processing techniques [37], is **real even symmetry**. This term describes a real sequence f_n that satisifes the condition $f_n = f_{-n}$. While a sequence *could* be complex and even, we shall assume from here onwards that an even sequence is also real. Notice that if a sequence is known to have even symmetry, then it can be stored in half of the space required for a general real sequence and one-fourth of the space required for a general complex sequence. We should anticipate a similar savings in storing the DFT coefficients, and indeed this is the case. Recall from Chapter 3 that the DFT of a real even sequence is also real and even; that is, $F_k = F_{-k}$. This means that only half of the real DFT coefficients need to be computed.

Let's determine the form of the DFT when it is applied to an even sequence. Anticipating coming developments, it is best if we begin with a real even sequence of length $2N$ and then use the definition of the complex DFT. Appealing to the symmetry of the input sequence f_n, we have that the $2N$-point DFT is

$$
\begin{aligned}
F_k &= \frac{1}{2N} \sum_{n=-N+1}^{N} f_n \omega_{2N}^{-nk} \\
&= \frac{1}{2N} \left[f_0 + \sum_{n=1}^{N-1} (f_n \omega_{2N}^{-nk} + f_{-n} \omega_{2N}^{nk}) + f_N \cos(\pi k) \right] \\
&= \frac{1}{2N} \left[f_0 + \sum_{n=1}^{N-1} f_n (\omega_{2N}^{-nk} + \omega_{2N}^{nk}) + f_N \cos(\pi k) \right] \\
&= \frac{1}{N} \left[\frac{1}{2} f_0 + \sum_{n=1}^{N-1} f_n \cos\left(\frac{\pi n k}{N}\right) + \frac{1}{2} f_N \cos(\pi k) \right],
\end{aligned}
\tag{4.4}
$$

for $k = 0 : N$. Notice how the fact that f_n is real and even allows the complex exponentials to be gathered into a single cosine term.

With this form of the DFT we see that only half of the input terms, $f_0, f_1, \ldots, f_N$, are needed to compute each F_k. We can also confirm that since the cosine is an even function of its arguments, the DFT coefficients have the property that $F_k = F_{-k}$. A quick tally shows that the cost of computing the DFT of an even sequence of length $2N$ is roughly N^2 real additions and N^2 real multiplications. Therefore, the cost of computing the DFT of an even sequence of length N is $N^2/4$ real additions and multiplications. Thus the even symmetry provides additional savings over the DFT of real sequences.

Once again, we have completed the first stage of the discussion, namely to show the form of the DFT in the presence of even symmetry. The next step is to define a genuinely new DFT. Here is the thinking that leads us there. We have seen that if the input sequence to the DFT is real and even, then its DFT can be computed using only half of its elements. Furthermore, only half of the resulting DFT coefficients are independent. One might wonder why the "other half" of the input and output sequences are even needed. In fact they are not! Expression (4.4) can be regarded as a mapping between the two arbitrary real vectors

$$
f_n = \{f_0, f_1, \ldots, f_N\} \quad \text{and} \quad F_k = \{F_0, F_1, \ldots, F_N\},
$$

both of length $N + 1$. By omitting the redundant half of each sequence, we have created a new transform between two arbitrary real vectors. With no more delay we

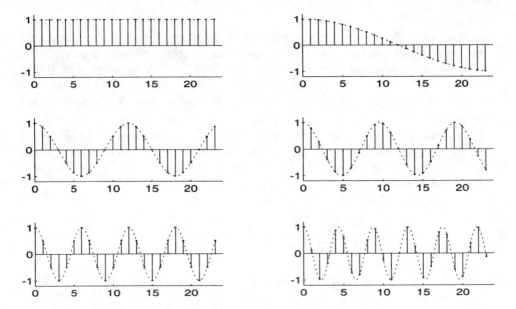

FIG. 4.2. *A few representative modes of the discrete cosine transform,* $\cos(\pi kn/N)$, *are shown on a grid with* $N = 24$ *points. Shown are modes* $k = 0, 1, 5, 11, 8,$ *and* 4 *(clockwise from upper left). Notice that the odd modes (right column) have a period of* $2N$, *while the even modes (left column) have a period of* N.

now give the explicit form of the discrete cosine transform.

▶ **Discrete Cosine Transform (DCT)** ◀

$$
\begin{aligned}
F_k &= \frac{1}{N}\left[\frac{1}{2}f_0 + \sum_{n=1}^{N-1} f_n \cos\left(\frac{\pi nk}{N}\right) + \frac{1}{2}f_N \cos(\pi k)\right] \\
&\equiv \frac{1}{N}\sum_{n=0}^{N}{}'' f_n \cos\left(\frac{\pi nk}{N}\right),
\end{aligned}
\tag{4.5}
$$

for $k = 0 : N$. The handy notation Σ'' has been introduced to denote a sum whose first and last terms are weighted by one-half. We emphasize the possibly confusing point that in using the N-point DCT, the input sequence f_n is an *arbitrary* set of $N + 1$ real numbers that carries no particular symmetry. The DCT produces another set of $N + 1$ real transform coefficients F_k that also possesses no particular symmetry. In other words, the DCT is a new and independent discrete transform!

It is worthwhile to pause for a moment and inspect the geometry of the DCT. Notice that the modes of an N-point DCT are different than the modes of the N-point real and complex DFTs. The kth DCT mode, $\cos(\pi nk/N)$, has k *half*-periods of the interval $[0, N]$. Small values of k correspond to low frequency modes with $k = 0$ being the constant mode; values of k near N correspond to high frequency modes with the $k = N$ mode oscillating between ± 1 at every grid point. An important observation is that all of the modes have a period of $2N$ grid points, but only the even modes are periodic on the interval $[0, N]$. A few representative modes of the DCT are shown in Figure 4.2.

There are a couple of ways to arrive at the inverse of the DCT. Perhaps the simplest is to begin with the inverse real DFT (4.3) of length $2N$ and assume that the coefficients F_k are real and even. This means $\text{Im}\{F_k\} = 0$, and the inverse real DFT follows immediately.

▶ **Inverse Discrete Cosine Transform (IDCT)** ◀

$$f_n = F_0 + 2 \sum_{k=1}^{N-1} F_k \cos\left(\frac{\pi nk}{N}\right) + F_N \cos(\pi n)$$

$$= 2 \sum_{k=0}^{N} {}'' F_k \cos\left(\frac{\pi nk}{N}\right) \tag{4.6}$$

for $n = 0 : N$. Since the F_k's are real we have replaced the coefficients $\text{Re}\{F_k\}$ by F_k to simplify the notation. A notable conclusion about the DCT is that it is its own inverse up to a multiplicative factor. This fact can also be demonstrated (problems 78 and 79) by using the discrete orthogonality of the cosine terms $\cos(\pi nk/N)$.

In many problems, a solution f_n is defined in terms of the inverse DCT (4.6), with the goal of finding the coefficients F_k. This is entirely analogous to assuming a solution to a continuous problem in the form of a Fourier cosine series. It is important to note the properties of a sequence defined by the IDCT. It is straightforward to check that when it is extended beyond the set $n = 0 : N$, the sequence f_n given by (4.6) is periodic with a period of $2N$. Furthermore, when extended, the sequence f_n has the property that

$$f_1 = f_{-1} \quad \text{and} \quad f_{N+1} = f_{N-1}.$$

Not surprisingly, this extended sequence is called the **even extension** of f_n. As we will see in Chapter 7, this property of the IDCT is essential in solving boundary value problems which require a solution with a zero "derivative" (or zero flux) at the boundaries.

4.4. Odd Sequences and the Discrete Sine Transform

The foregoing discussion of the discrete cosine transform is a pattern for all other symmetric DFTs. We will outline the highlights of one other important symmetry, leaving the reader to engage in the particulars in the problem section. Another common symmetry that arises in practice is that in which the real input sequence f_n has the **odd symmetry** $f_n = -f_{-n}$. (We will assume that odd sequences are real unless otherwise stated.) The task is to determine the form the DFT takes in the presence of this symmetry, and then to define the new DFT that results from it. Before diving into the computations, a few crucial observations will ease our labors.

First, note that only half of a sequence with odd symmetry needs to be stored. This suggests that there should be similar savings in storing the transform coefficients F_k, and indeed this is true. We learned in Chapter 3 that the DFT of a real odd sequence is odd and pure imaginary; thus only half of the transform coefficients need to be stored. Here is another essential observation. We will begin with a periodic input sequence of length $2N$ that has the odd symmetry $f_n = -f_{-n}$. Since $f_0 = -f_0$, it follows that $f_0 = 0$. Furthermore, $f_N = f_{-N}$ by periodicity and $f_N = -f_{-N}$ by the odd symmetry; therefore $f_N = f_{-N} = 0$. In other words, an odd periodic sequence will always have predictable zero values (see Figure 4.1). We may now proceed.

As in the case of even sequences, we will begin with a real odd sequence of length $2N$ and apply the complex DFT. We have the following simplifications due to the odd symmetry:

$$
\begin{aligned}
F_k &= \frac{1}{2N} \sum_{n=-N+1}^{N} f_n \omega_{2N}^{-nk} \\
&= \frac{1}{2N} \left[\underbrace{f_0}_{0} + \sum_{n=1}^{N-1} (f_n \omega_{2N}^{-nk} + f_{-n} \omega_{2N}^{nk}) + \underbrace{f_N \cos(\pi k)}_{0} \right] \\
&= \frac{1}{2N} \sum_{n=1}^{N-1} f_n (\omega_{2N}^{-nk} - \omega_{2N}^{nk}) \\
&= -\frac{i}{N} \sum_{n=1}^{N-1} f_n \sin\left(\frac{\pi n k}{N} \right).
\end{aligned}
\tag{4.7}
$$

This expression holds for $k = -N + 1 : N$, although we see immediately that $F_0 = F_{\pm N} = 0$. Furthermore, we may verify that $F_k = -F_{-k}$, which means that only the coefficients $F_1, \ldots, F_{N-1}$ are independent. Since only the input values $f_1, \ldots, f_{N-1}$ are required to compute the DFT of an odd sequence, the entire computation can be done with $N-1$ storage locations for the input and output. This form of the DFT also confirms that the transform coefficients are all imaginary. A quick count reveals that roughly $N^2/4$ real additions and multiplications are required to compute the independent DFT coefficients of an odd sequence of length N. As with the even symmetry, this represents a factor of four savings over the real DFT.

We now argue as we did in the case of even symmetry. Expression (4.7) can be regarded as a mapping between the two arbitrary real vectors

$$
f_n = \{f_1, \ldots, f_{N-1}\} \quad \text{and} \quad F_k = \{F_1, \ldots, F_{N-1}\},
$$

both of length $N-1$. The odd symmetry that led us to this point is nowhere to be found in either of these vectors. But we can still define this mapping and give it a name.

► **Discrete Sine Transform (DST)** ◄

$$
F_k = \frac{1}{N} \sum_{n=1}^{N-1} f_n \sin\left(\frac{\pi n k}{N} \right)
\tag{4.8}
$$

for $k = 1 : N - 1$. This is the explicit form of a new DFT defined for an arbitrary real sequence f_n. The multiplicative factor i that appears in (4.7) is not needed; without it the DST involves only real arithmetic.

If the symmetries of the DST are used in the inverse real DFT (4.3), it is not difficult to show (problem 77) that the DST is its own inverse up to a multiplicative constant. This also follows from the orthogonality of the sine modes $\sin(\pi n k/N)$

(problems 78 and 79). In either case we have the inverse of the DST.

► **Inverse Discrete Sine Transform (IDST)** ◄

$$f_n = 2 \sum_{n=1}^{N-1} F_k \sin\left(\frac{\pi n k}{N}\right). \tag{4.9}$$

for $n = 1 : N - 1$.

Notice that a sequence f_n defined by the IDST has some special properties. First, f_n is a sequence with period $2N$. Furthermore, if f_n is extended beyond the indices $n = 0 : N$, the extended sequence is odd ($f_n = -f_{-n}$), and $f_0 = f_N = f_{pN} = 0$ for any integer p. This extended sequence is called the **odd extension** of f_n. There are instances in which the solution to a problem is required to satisfy "zero boundary conditions" ($f_0 = f_N = 0$). A trial solution given by the ISDT (4.9) has this property, just as a Fourier sine series does for continuous problems.

Some conclusions of this section are summarized in Table 4.1 which shows the computational and storage costs of evaluating the DFT when the input sequence has various symmetries. These are *not* the costs of computing symmetric DFTs (which will be discussed in the next section), but rather the costs of computing the DFT of symmetric sequences. For this reason the expected savings in storage and computation can be seen as we move from the complex to the real to the even/odd DFTs.

TABLE 4.1
Cost of computing the DFT of symmetric
N-point input sequences using the explicit form.

	Complex	Real	Real even	Real odd
Storage (real locations)	$2N$	N	$N/2$	$N/2$
Real additions*	$4N^2$	N^2	$N^2/4$	$N^2/4$
Real multiplications*	$4N^{2**}$	N^{2**}	$N^2/4$	$N^2/4$

* Computation costs shown up to multiples of N^2.
** An additional factor-of-two savings can be obtained by folding
sums (see problem 70).

4.5. Computing Symmetric DFTs

We must now embark on an odyssey that will lead to more efficient methods for computing symmetric DFTs. Before doing so, a few historical thoughts will give the discussion some perspective. In the beginning the only way to compute the complex, real, cosine, and sine DFTs was by using their definitions (2.6), (4.1), (4.5), and (4.8), respectively (the explicit forms). One must now imagine the arrival of the FFT in 1965, but also realize that for (just) a few years the FFT was used to evaluate *only* the complex DFT. Given the remarkable efficiency of the FFT, it made sense to devise methods whereby the complex FFT could be used to evaluate the RDFT, DCT, and DST. These methods, which we will call **pre- and postprocessing** methods, were introduced in 1970 by Cooley, Lewis and Welch in a paper that is filled with ingenious tricks [41]. Their methods, when combined with the complex FFT, provided much faster ways to evaluate the RDFT, DCT, and DST. Other early pre- and postprocessing methods (with better stability properties) were introduced by Dollimore in 1973 [49]. Research on pre- and postprocessing algorithms for other symmetries and for advanced computer architectures has continued to the present day [33].

Soon after the pre- and postprocessing methods were established, it became evident that efficient symmetric DFTs could also be devised by building the symmetry of the input and output sequences directly into the FFT itself. The first of these methods, called **compact symmetric FFTs**, was designed around 1968. Attributed to Edson [9], it computes the RDFT and provides the template for all other compact symmetric FFTs. Another compact symmetric FFT for the DCT was given by Gentleman [65] in 1972. Further refinements on the compact symmetric FFT idea have appeared during the last 20 years [141], [19], [75], [16] for many different computer architectures. Compact symmetric FFTs have now replaced the pre- and postprocessing methods in many software packages and are generally deemed superior. Nevertheless, the pre- and postprocessing algorithms have historical significance and are still the best way to compute symmetric DFTs if only a complex FFT is available. For these reasons we will have a look at a few of the better known, easily implemented pre- and postprocessing methods for symmetric DFTs.

Let's begin with the problem of computing the DFT of real input sequences. Here is a motivating thought: the fact that the real DFT can be done with half of the storage of the complex DFT might lead an enterprising soul to surmise that it should be possible to do *either*

- two real N-point DFTs with a single complex N-point DFT *or*

- one real $2N$-point DFT with a single complex N-point DFT.

The first proposition is the easiest to demonstrate, and we will also be able to answer the second proposition in the affirmative.

Assume that we are given two real sequences g_n and h_n, both of length N. The goal is to compute their DFTs G_k and H_k using a single N-point complex DFT. In order to make full use of the complex DFT, we form the complex sequence $f_n = g_n + ih_n$ to be used for input. The complex DFT applied to the sequence f_n now takes the form

$$
\begin{aligned}
F_k &= \frac{1}{N} \sum_{n=-\frac{N}{2}+1}^{\frac{N}{2}} (g_n + ih_n)\omega_N^{-nk} \\
&= \frac{1}{N} \sum_{n=-\frac{N}{2}+1}^{\frac{N}{2}} \left[g_n \cos\left(\frac{2\pi nk}{N}\right) + h_n \sin\left(\frac{2\pi nk}{N}\right) \right] \\
&\quad - \frac{i}{N} \sum_{n=-\frac{N}{2}+1}^{\frac{N}{2}} \left[g_n \sin\left(\frac{2\pi nk}{N}\right) - h_n \cos\left(\frac{2\pi nk}{N}\right) \right].
\end{aligned}
$$

We need to compare this expression to the two desired RDFTs

$$
G_k = \frac{1}{N} \sum_{n=-\frac{N}{2}+1}^{\frac{N}{2}} g_n \left[\cos\left(\frac{2\pi nk}{N}\right) - i \sin\left(\frac{2\pi nk}{N}\right) \right]
$$

and

$$H_k = \frac{1}{N} \sum_{n=-\frac{N}{2}+1}^{\frac{N}{2}} h_n \left[\cos\left(\frac{2\pi nk}{N}\right) - i\sin\left(\frac{2\pi nk}{N}\right) \right].$$

A brief calculation (problem 80) confirms that the computed DFT F_k can be easily related to the desired DFTs G_k and H_k. The relationships will be used again in this chapter, and it is worthwhile to summarize them as the first of several procedures.

▶ **Procedure 1: Two Real DFTs by One Complex DFT** ◀

$$G_k = \frac{1}{2}(F_k + F_{-k}^*) \quad \text{and} \quad H_k = -\frac{i}{2}(F_k - F_{-k}^*), \tag{4.10}$$

for $k = -N/2 + 1 : N/2$.

This is the first example of what we call a **pre- and postprocessing algorithm**. In this case it is a scheme for computing two real DFTs for the price of a single complex DFT. The underlying idea will appear again: the given input sequences are modified (in this case combined) to form the input to the complex DFT; this is the preprocessing stage. A complex DFT is computed and then the output is modified in a simple way (4.10) to form the sequences G_k and H_k; this is the postprocessing stage.

In assessing the computational costs of pre- and postprocessing algorithms, it is customary to neglect the cost of the pre- and postprocessing. These costs are always proportional to N while the DFT step consumes a multiple of N^2 operations (or $N \log N$ operations if the FFT is used). When N is large (which is when one worries most about computation and storage costs) the pre- and postprocessing costs become negligible compared to the DFT costs.

A perceptive reader might notice that it is more efficient to compute the DFT of two real sequences by using the explicit form of the RDFT (4.2) on each sequence rather than using Procedure 1, which involves an N-point complex DFT. This is true if one uses the explicit forms for all DFTs. However, if the FFT is used to compute the complex DFT, then Procedure 1 is more efficient than two explicit RDFTs.

The second idea mentioned above, computing a real DFT of length $2N$ with an N-point complex DFT, is also worth investigating. In order to do so we must introduce the **splitting method**, which will make several more appearances in this and later chapters. We begin by splitting the input sequence f_n of length $2N$ into its even and odd subsequences,

$$g_n = f_{2n} \quad \text{and} \quad h_n = f_{2n-1}.$$

There is an unfortunate coincidence of terminology: these even and odd subsequences must not be confused with the even and odd symmetries. We may now write the complex $2N$-point DFT of f_n as

$$
\begin{aligned}
F_k &= \frac{1}{2N} \sum_{n=-N+1}^{N} f_n \omega_{2N}^{-nk} \\
&= \frac{1}{2N} \sum_{n=-\frac{N}{2}+1}^{\frac{N}{2}} \left(f_{2n}\omega_{2N}^{-2nk} + f_{2n-1}\omega_{2N}^{-(2n-1)k} \right) \\
&= \frac{1}{2N} \sum_{n=-\frac{N}{2}+1}^{\frac{N}{2}} g_n \omega_N^{-nk} + \frac{1}{2N}\omega_{2N}^{k} \sum_{n=-\frac{N}{2}+1}^{\frac{N}{2}} h_n \omega_N^{-nk}
\end{aligned}
$$

$$= \frac{1}{2}(G_k + \omega_{2N}^k H_k). \tag{4.11}$$

The fact that $\omega_{2N}^2 = \omega_N$ is the crux of the splitting method and has been used to obtain the third line of this argument.

Recall that the goal is to determine F_k from G_k and H_k. The relationship (4.11) holds for $k = -N/2 + 1 : N/2$, which determines half of the F_k's. Since the sequences G_k and H_k are N-periodic, we may replace k by $k \pm N$ in the expression (4.11) to find that

$$F_{k \pm N} = \frac{1}{2}(G_k - \omega_{2N}^k H_k),$$

which can be used to compute the remaining F_k's. We now summarize these relationships (often called **combine** or **butterfly** relations), which allow the DFT of a full-length sequence to be computed easily from the DFTs of the even and odd subsequences.

▶ **Procedure 2: Combine (Butterfly) Relations** $N \to 2N$ ◀

$$F_k = \frac{1}{2}(G_k + \omega_{2N}^k H_k),$$

$$F_{k \pm N} = \frac{1}{2}(G_k - \omega_{2N}^k H_k) \tag{4.12}$$

for $k = -N/2 + 1 : N/2$.

The butterfly relations take an ever-so-slightly different form if the alternate definition of the DFT on the sets $n, k = 0 : N - 1$ is used (problem 81).

We can now turn to the conjecture made earlier that it should be possible to compute a real DFT of length $2N$ using a single complex DFT of length N. With the two procedures already developed, we will see that it can be done. Assume that a real sequence of length $2N$ is given; call it f_n where $n = -N + 1 : N$. The goal is to compute its DFT F_k, and since the DFT is a conjugate even sequence, only $F_0, \ldots, F_N$ need to be computed. The overall strategy is this: we first compute the DFT of the even and odd subsequences of f_n simultaneously using Procedure 1, and then use Procedure 2 to combine these two DFTs into the full DFT F_k.

Here is how it is actually done. Letting $g_n = f_{2n}$ and $h_n = f_{2n-1}$, we form the N-point complex sequence $z_n = g_n + ih_n$, where $n = -N/2 + 1 : N/2$. Once its complex DFT Z_k has been computed, Procedure 1 can be used to form the transforms G_k and H_k. According to (4.10) they are given by

$$G_k = \frac{1}{2}(Z_k + Z_{-k}^*) \quad \text{and} \quad H_k = -\frac{i}{2}(Z_k - Z_{-k}^*)$$

for $k = -N/2 + 1 : N/2$, and are both conjugate even sequences. We now have the DFTs of the even and odd subsequences of the original sequence. Using Procedure 2, the sequences G_k and H_k can be combined to form the desired DFT sequence F_k. By the combine formulas (4.12) we have that

$$\begin{aligned} F_k &= \frac{1}{2}(G_k + \omega_{2N}^k H_k) \\ &= \frac{1}{4}(Z_k + Z_{-k}^*) - \frac{i}{4}\omega_{2N}^k(Z_k - Z_{-k}^*) \\ &= \frac{Z_k}{4}(1 - i\omega_{2N}^k) + \frac{Z_{-k}^*}{4}(1 + i\omega_{2N}^k) \end{aligned}$$

for $k = 0 : N/2$. The second relation of Procedure 2 can be used to compute the remaining coefficients for $k = N/2 + 1 : N$. Collecting both sets of relations, we have the following procedure.

▶ **Procedure 3: Length $2N$ RDFT from Length N Complex DFT** ◀

- Given the real sequence f_n of length $2N$, form the sequence $z_n = f_{2n} + if_{2n-1}$ and compute its N-point complex DFT Z_k where $k = -N/2 + 1 : N/2$.

- Then $F_0, \ldots, F_N$ are given by

$$F_k = \frac{Z_k}{4}(1 - i\omega_{2N}^k) + \frac{Z_{-k}^*}{4}(1 + i\omega_{2N}^k) \quad \text{for} \quad k = 0 : \frac{N}{2},$$

$$F_{k+N} = \frac{Z_k}{4}(1 + i\omega_{2N}^k) + \frac{Z_{-k}^*}{4}(1 - i\omega_{2N}^k) \quad \text{for} \quad k = -\frac{N}{2} + 1 : 0.$$

It should be verified (problem 82) that the coefficients F_k given by these relations actually have the conjugate even symmetry that we would expect of the DFT of a real input sequence, and that F_0 and F_N are real.

A quick operation count is revealing. The cost of the pre- and postprocessing method is the cost of a complex N-point DFT or roughly $4N^2$ real additions and $4N^2$ real multiplications (neglecting the cost of the pre- and postprocessing). The cost of a single real DFT of length $2N$ is $(2N)^2 = 4N^2$ real additions and multiplications, so it appears that the pre- and postprocessing method is no better than using the explicit form of the RDFT. As mentioned earlier, the pre- and postprocessing method becomes preferable when the FFT is used to compute the complex DFT. Another more efficient strategy for computing RDFTs (without FFTs) is explored in problem 86. Finally, we mention in passing that Procedure 3 can be essentially "inverted," or applied in reverse, to give an efficient method for computing the inverse RDFT of a conjugate even set of coefficients F_k [41].

We now come to the question of computing the DCT of an arbitrary real sequence. Let's start with a bad idea and refine it. Our discussion about DFTs of even sequences leads naturally to the following fact (problem 83):

> The N-point DCT of an arbitrary real sequence f_n consists of the $N + 1$ (real) coefficients $F_0, \ldots, F_N$ of the $2N$-point complex DFT of the even extension of f_n.

This says that one way to compute an N-point DCT of a sequence f_n defined for $n = 0 : N$ is to extend it evenly to form a new sequence f_n that is even over the index range $n = -N + 1 : N$. This new sequence can then be used as input to a complex DFT of length $2N$. The output F_k will be real and even, and the desired DCT coefficients will be $F_0, \ldots, F_N$. Needless to say, there is tremendous redundancy and inefficiency in computing a DCT in this manner, since the symmetries in both the input and ouput sequences have been overlooked.

We will now describe a pre- and postprocessing algorithm for the DCT. Such methods are often difficult to motivate, since they were rarely discovered in the succinct manner in which they are presented. The underlying idea is always the same: we must find an auxiliary sequence, formed in the preprocessing step, that can be fed to a complex DFT. The output of that DFT may then be modified in a postprocessing step to produce the desired DCT. Some of the procedures already developed (particularly Procedures 1 and 2) will generally enter the picture.

Given an arbitrary real sequence f_n defined for $n = 0 : N$, the goal is to compute the sequence of DCT coefficients F_k for $k = 0 : N$. We begin by forming the even extension of f_n so that it is defined for $n = -N + 1 : N$. It is important to note that the complex DFT of this even extension, which we will denote F_k, contains the desired DCT of the original sequence. The trick is to define the complex auxiliary sequence

$$z_n = \underbrace{f_{2n}}_{g_n} + i\underbrace{(f_{2n+1} - f_{2n-1})}_{h_n} \equiv g_n + ih_n$$

for $n = -N/2 + 1 : N/2$. Notice that we have defined two subsequences g_n and h_n of length N that form the real and imaginary parts of z_n. If there is any rationale for this choice of a sequence, it is that g_n is the even subsequence of z_n, and h_n can be easily related to the odd subsequence of z_n. Furthermore, z_n is a conjugate even sequence ($z_n = z_{-n}^*$), which means that its DFT Z_k is real.

Now the actual transform takes place. The sequence z_n is used as input to an N-point complex DFT to produce the coefficients Z_k. Procedure 1 can then be used to recover the DFT of the two real subsequences g_n and h_n. Recalling that Z_k is real ($Z_k = Z_k^*$), these DFTs are given by

$$G_k = \frac{1}{2}(Z_k + Z_{-k}) \quad \text{and} \quad H_k = -\frac{i}{2}(Z_k - Z_{-k})$$

for $k = -N/2 + 1 : N/2$. The DFTs of the two real subsequences g_n and h_n are now available. We are maneuvering toward an application of Procedure 2 to compute F_k in terms of G_k and H_k. But it cannot be used yet since, while g_n *is* the even subsequence of f_n, h_n is *not* its odd subsequence. Let h_n' be the odd subsequence f_{2n-1}, and let its DFT be H_k'. Then we can write

$$
\begin{aligned}
H_k &= \mathcal{D}\{h_n\}_k \\
&= \mathcal{D}\{f_{2n+1} - f_{2n-1}\}_k \\
&= \mathcal{D}\{f_{2n+1}\}_k - H_k' \\
&= \omega_N^k H_k' - H_k' \quad \text{(by the shift property)}.
\end{aligned}
$$

Notice that the shift property has been used to relate $\mathcal{D}\{f_{2n+1}\}_k$ to $\mathcal{D}\{f_{2n-1}\}_k$. It is now possible to solve for H_k', the DFT of the odd subsequence, in terms of the DFT H_k that has been computed. We have that

$$H_k' = \frac{H_k}{\omega_N^k - 1} = -\frac{i}{2}\frac{Z_k - Z_{-k}}{\omega_N^k - 1}.$$

Now we are set to use Procedure 2 since the DFTs of the even and odd subsequences of f_n, which we have labeled G_k and H_k', are known. The combine formulas of Procedure 2 give us that

$$
\begin{aligned}
F_k &= \frac{1}{2}(G_k + \omega_{2N}^k H_k') \\
&= \frac{1}{4}(Z_k + Z_{-k}) - \frac{i}{4}\omega_{2N}^k \frac{Z_k - Z_{-k}}{\omega_N^k - 1} \\
&= \frac{1}{4}\left[(Z_k + Z_{-k}) - \frac{Z_k - Z_{-k}}{2\sin(\frac{\pi k}{N})}\right]
\end{aligned}
$$

for $k = 1 : N/2$. Since Z_k is a real sequence, this expression involves only real arithmetic and provides the first half of the desired DFT F_k. A similar calculation produces the relation

$$F_{k+N} = \frac{1}{4}\left[(Z_k + Z_{-k}) + \frac{Z_k - Z_{-k}}{2\sin(\frac{\pi k}{N})}\right]$$

for $k =: -N/2 + 1 : -1$. This expression can be used to find the coefficients with $k = N/2 + 1 : N - 1$. Finally, a special case must be made for $k = 0$ and $k = N$. We can directly compute

$$H_0' = \frac{1}{N}\sum_{n=-\frac{N}{2}+1}^{\frac{N}{2}} f_{2n-1},$$

and then it follows that

$$F_0 = \frac{1}{2}(G_0 + H_0') = \frac{1}{2}(Z_0 + H_0')$$

and

$$F_N = \frac{1}{2}(G_0 - H_0') = \frac{1}{2}(Z_0 - H_0').$$

We may now collect all of these results in a single recipe for computing the DCT by pre- and postprocessing.

▶ **Procedure 4: DCT by Pre- and Postprocessing** ◀

- Given an *arbitrary* real sequence $\{f_0, \ldots, f_N\}$, extend it evenly $(f_{-n} = f_n)$ and periodically $(f_{n+2N} = f_n)$ to form the sequence

$$z_n = f_{2n} + i(f_{2n+1} - f_{2n-1}) \quad \text{for} \quad n = -N/2 + 1 : N/2.$$

- Compute the N-point complex DFT Z_k, where $k = -N/2 + 1 : N/2$.

- Then $F_0, \ldots, F_N$ are given by

$$F_k = \frac{1}{4}\left[(Z_k + Z_{-k}) - \frac{Z_k - Z_{-k}}{2\sin(\frac{\pi k}{N})}\right] \quad \text{for} \quad k = 1 : \frac{N}{2},$$

$$F_{k+N} = \frac{1}{4}\left[(Z_k + Z_{-k}) + \frac{Z_k - Z_{-k}}{2\sin(\frac{\pi k}{N})}\right] \quad \text{for} \quad k = -\frac{N}{2} + 1 : -1,$$

$$F_0 = \frac{1}{2}\left[Z_0 + \frac{1}{N}\sum_{n=-\frac{N}{2}+1}^{\frac{N}{2}} f_{2n-1}\right],$$

$$F_N = \frac{1}{2}\left[Z_0 - \frac{1}{N}\sum_{n=-\frac{N}{2}+1}^{\frac{N}{2}} f_{2n-1}\right].$$

It should be noted that the computations given in Procedure 4 can be prone to numerical instability due to the division by $\sin(\pi n k/N)$ which approaches zero for values of k near 0 and N. For small N this may not present a problem. Stable pre- and postprocessing methods have been devised essentially by inverting the relations given in Procedure 4 [49].

What about computing the DST? As with the DCT, there is a safe but inefficient way. It follows from the following assertion (problem 84):

> The N-point DST of an arbitrary real sequence f_n consists of the $N - 1$ coefficients $iF_1, \ldots, iF_{N-1}$ of the $2N$-point complex DFT of the odd extension of f_n. (Since the F_k's are imaginary, the DST coefficients are real.)

This says that one can always apply a $2N$-point complex DFT to the odd extension of a given real sequence and then find the DST coefficients in the imaginary part of the output. However, there are (much) better ways to accomplish the same end. One of the preferred methods is a pre- and postprocessing algorithm. In the interest of brevity and reader involvement, we will give only the menu for this meal; the full feast can be savored in problem 85. Here is the key idea: the pre- and postprocessing algorithm for the DCT was launched by the observation that if f_n is an even sequence then the auxiliary sequence $z_n = f_{2n} + i(f_{2n+1} - f_{2n-1})$ is conjugate even. In a similar way, if f_n is an odd sequence, then $iz_n = f_{2n-1} - f_{2n+1} + if_{2n}$ is conjugate even. Therefore, in a few words, the pre- and postprocessing algorithm for the DST can be derived by applying the arguments leading to Procedure 4 to the auxiliary sequence iz_n. The resulting Procedure 5 is given in problem 85.

Example: Numerical RDFT, DCT, and DST. For the sake of illustration, we will take $N = 12$ and analyze a real sequence f_n. The RDFT, DCT, and DST of this sequence will be computed and discussed. Before proceeding, we must make one small adjustment to facilitate a comparison between the transforms. The RDFT given in (4.2) is defined for an input sequence f_n with $n = -N/2 + 1 : N/2$, whereas the DCT and DST given in (4.5) and (4.8) are defined for $n = 0 : N$ and $n = 1 : N - 1$, respectively. To make these definitions compatible, we may use the periodicity of f_n to give an alternate (equivalent) definition of the N-point RDFT pair (problem 72).

▶ **Alternate RDFT** $(n, k = 0 : N - 1)$ ◀

$$F_k = \frac{1}{N} \sum_{n=0}^{N-1} f_n \left[\cos\left(\frac{2\pi nk}{N}\right) - i \sin\left(\frac{2\pi nk}{N}\right) \right] \tag{4.13}$$

for $k = 0 : N/2$.

▶ **Alternate Inverse RDFT** $(n, k = 0 : N - 1)$ ◀

$$f_n = F_0 + 2 \sum_{k=1}^{\frac{N}{2}-1} \left[\operatorname{Re}\{F_k\} \cos\left(\frac{2\pi nk}{N}\right) - \operatorname{Im}\{F_k\} \sin\left(\frac{2\pi nk}{N}\right) \right] + F_{\frac{N}{2}} \cos(\pi n) \tag{4.14}$$

for $n = 0 : N - 1$.

With these adjustments we can now turn to Table 4.2 and make a few pertinent comments. First look at the input column labeled f_n. Notice that values are given for $f_0, \ldots, f_{12}$, but they must be used carefully. To compute the RDFT using (4.13) with $N = 12$, we use the values of f_n with $n = 0 : 11$, with the important condition that the value used for f_0 is the average of f_0 and f_{12}. On the other hand, the DCT requires values for $n = 0 : 12$, and so an additional value for f_{12} is given for use with the DCT only. The DST requires values for $n = 1 : 11$ and assumes that $f_0 = f_{12} = 0$. With these essential observations, we can now proceed with the experiment. Table 4.2 also shows the computed values of the transform coefficients.

TABLE 4.2
RDFT, DCT, and DST of a real sequence with $N = 12$.

n, k	f_n	F_k for RDFT	F_k for DCT	F_k for DST
0	.11	$2.15(-1)$	$2.15(-1)$	$-$
1	.44	$2.87(-1) + 1.71(-2)i$	$-8.23(-2)$	$2.08(-2)$
2	.55	$-5.50(-2) + 1.03(-1)i$	$2.87(-1)$	$-1.71(-2)$
3	.22	$-5.04(-2) - 9.17(-3)i$	$-9.96(-2)$	$3.00(-1)$
4	.00	$-1.83(-2) + 5.56(-2)i$	$-5.50(-2)$	$-1.03(-1)$
5	$-.11$	$-3.04(-2) + 1.23(-3)i$	$2.73(-2)$	$1.01(-1)$
6	$-.33$	$1.38(-2)$	$-5.04(-2)$	$9.17(-3)$
7	$-.66$	$-3.04(-2) - 1.23(-3)i$	$-5.05(-2)$	$3.06(-2)$
8	$-.11$	$-1.83(-2) - 5.56(-2)i$	$-1.83(-2)$	$-5.56(-2)$
9	.33	$-5.04(-2) + 9.17(-3)i$	$1.70(-2)$	$2.45(-2)$
10	.77	$-5.50(-2) - 1.03(-1)i$	$-3.04(-2)$	$-1.23(-3)$
11	.99	$2.87(-1) - 1.71(-2)i$	$-4.49(-3)$	$-1.83(-2)$
(12)	.88	$-$	$1.38(-2)$	$-$

Note: The RDFT uses $\frac{1}{2}(f_0 + f_{12})$ for f_0. The DCT uses both f_0 and f_{12}. The DST uses f_n with $n = 1 : 11$. The parenthetical entries are exponents, e.g., $1.38(-2)$ means 1.38×10^{-2}.

Simple as this example may seem, it does offer some valuable lessons. The first is that for an arbitrary real sequence with no symmetry, the sets of RDFT, DCT, and DST coefficients are different. There are 12, 13, and 11 independent real coefficients for the RDFT, DCT, and DST, respectively, matching the number of real input entries exactly. Notice that the RDFT transform coefficients possess conjugate even symmetry since the input sequence is real. However, the DCT and DST coefficients have no particular symmetry (other than real), since the input has no additional symmetry.

A further revelation occurs if we now use the computed coefficients and evaluate the inverse transforms. Figure 4.3 shows the sequences that are generated by the inverse RDFT, DCT, and DST given by (4.14), (4.6), and (4.9). The important observation is that the values $f_1, \ldots, f_{11}$ match the original input values. However, the sequences differ at the endpoints and when they are extended beyond the interval $n = 0 : 12$. Specifically, the inverse RDFT extends to a sequence with period N, the inverse DCT extends to an even sequence with period $2N$, and the inverse DST extends to an odd sequence with period $2N$. This is in strict analogy with the real, cosine, and sine forms of the Fourier series.

We have described in detail how the explicit forms and the pre- and postprocessing methods can be used to compute three symmetric transforms. For a complete comparison, the compact symmetric FFTs mentioned earlier must also be brought into the discussion. Table 4.3 shows the cost of computing the RDFT, DCT, and DST of an arbitrary real sequence of length N by using the explicit form, a pre- and postprocessing method, and a compact symmetric FFT. It assumes that the pre- and postprocessing methods use a complex N-point FFT which has a computational cost of $3N \log N - 2N$ real additions and $2N \log N - 4N$ real multiplications. Only the two "leading order" terms in the pre- and postprocessing and compact FFT costs are

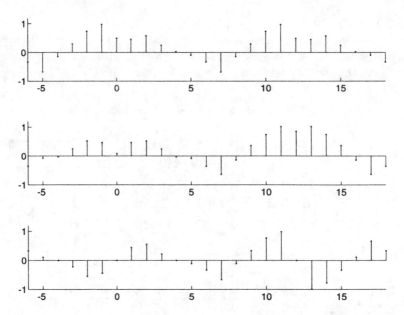

FIG. 4.3. *Having computed the RDFT, DCT, and DST of a given sequence with $N = 12$ (values given in Table 4.2), the transform coefficients are then used to reconstruct the sequence f_n according to the inverse RDFT (top figure), inverse DCT (middle figure), and inverse DST (lower figure). These sequences agree at the points $n = 1 : 11$, but differ when extended beyond this interval.*

TABLE 4.3
Cost of RDFT, DCT, and DST by explicit definition, pre- and postprocessing, and compact symmetric FFT.

	RDFT explicit	RDFT by pre- and postprocessing	RDFT by compact FFT
Real adds	$\sim N^2$	$3N \log N/2 - 2$	$3N \log N/2 - 5N/2$
Real mults	$\sim N^2$	$N \log N - 2N$	$N \log N - 3N$
	DCT explicit	DCT by pre- and postprocessing	DCT by compact FFT
Real adds	$\sim N^2$	$3N \log N/2 + N$	$3N \log N/2 - 3N$
Real mults	$\sim N^2$	$N \log N - 2N$	$N \log N - 4N$
	DST explicit	DST by pre- and postprocessing	DST by compact FFT
Real adds	$\sim N^2$	$3N \log N/2 - 3N/2$	$3N \log N/2 - 3N$
Real mults	$\sim N^2$	$N \log N - 3N$	$N \log N - 4N$

Operation counts for pre- and postprocessing and compact FFTs taken from Swarztrauber [141] modified for arbitrary sequences of length N.

included, which is sufficient for purposes of comparison. We note that costs shown are for *arbitrary* real sequences of length N (not symmetric sequences of length N).

A few comments are in order. First, note the savings that both the pre- and postprocessing and compact FFTs offer over the explicit forms. However, the pre- and postprocessing methods and the compact FFTs are identical in cost in the leading term. The differences appear in the terms that are linear in N and may be insignificant.

TABLE 4.4
Real symmetric sequences and their symmetric DFTs,
original sequence of length $2N$, DFT of length N.

Symmetry	Property	Example: $N = 3$	Boundary conditions
Even	$f_{-n} = f_n$	$\{f_2, f_1, f_0, f_1, f_2, f_3\}$	$f_1 - f_{-1} = 0$ $f_{N+1} - f_{N-1} = 0$
Odd	$f_{-n} = -f_n$	$\{-f_2, -f_1, 0, f_1, f_2, 0\}$	$f_0 = f_N = 0$
Quarter even*	$f_{-n-1} = f_n$	$\{f_1, f_0, f_0, f_1, f_2, f_2\}$	$f_0 - f_{-1} = 0$ $f_N - f_{N-1} = 0$
Quarter odd**	$f_{-n-1} = -f_n$	$\{-f_1, -f_0, f_0, f_1, f_2, -f_2\}$	$f_0 + f_{-1} = 0$ $f_N + f_{N-1} = 0$

* See problem 87.
** See problem 88.

The compact FFTs are often given the advantage because they require fewer passes through the data; for some computer architectures, this property could be far more important than operation counts. Finally, we emphasize that the DCT and DST show no computational advantage over the RDFT. This is because all three DFTs are applied to an arbitrary sequence of length N in which no symmetries are present. Of course, as claimed earlier, all three methods have roughly half the storage and computation costs of the complex DFT.

Much of the symmetric DFT territory has been covered in this section— certainly those regions that are most frequently traveled. However, there are many more symmetries that have been discovered and studied, often in rather specialized applications. We close this section with Table 4.4, which is a list (still far from exhaustive) of symmetries that arise in practice and that lead to other symmetric DFTs. Each line represents a different symmetric sequence and the associated symmetric DFT. The original sequence is assumed to be periodic with length $2N$, which leads to a new symmetric DFT of length N. Since many symmetric DFTs are used in the solution of boundary value problems, the relevant boundary condition is also shown on a computational domain defined by $n = 0 : N$.

4.6. Notes

A matrix formulation of both pre- and postprocessing algorithms and compact symmetric FFTs can be found in Van Loan [155]. The symmetries discussed in this chapter arise on a regular set of grid points. Symmetric DFTs can also be defined on staggered grids using midpoints of subintervals [144]. DFTs have also been defined on more exotic two-dimensional grids; see [5] for DFTs on hexagonal grids. The most complete collection of compact symmetric FFTs that includes versions for standard, staggered, and mixed grids (together with codes) is given by Bradford [16]. Readers interested specifically in the DCT should see Rao and Yip [116].

4.7. Problems

70. Folded RDFT. It is interesting to note that further savings can be realized in *both* the real and complex DFTs by folding the DFT sum. Show that the real N-point DFT (4.2) can be written as

$$NF_k = f_0 + \sum_{n=1}^{\frac{N}{2}-1} (f_n + f_{-n}) \cos\left(\frac{2\pi nk}{N}\right)$$

$$-i \sum_{n=1}^{\frac{N}{2}-1} (f_n - f_{-n}) \sin\left(\frac{2\pi nk}{N}\right) + f_{\frac{N}{2}} \cos(\pi k),$$

where $k = 0 : N/2$. In this form, how many real additions and multiplications are needed to compute the full set of coefficients (counting only multiples of N^2)?

71. Folded DFT. Show how the maneuver of the previous problem can also be applied to the complex DFT (2.6). What is the savings in real arithmetic in using the folded form of the DFT?

72. Alternate form of the RDFT. Show that if the real sequence f_n is defined for $n = 0 : N - 1$, then the transform sequence has the conjugate even property $F_k = F_{N-k}^*$, and is given by

$$F_k = \frac{1}{N} \sum_{n=0}^{N-1} f_n \left[\cos\left(\frac{2\pi nk}{N}\right) - i \sin\left(\frac{2\pi nk}{N}\right)\right]$$

for $k = 0 : N/2$. What value of k corresponds to the highest frequency mode? Show that the associated inverse (again using the conjugate even symmetry of the F_k's) is

$$f_n = F_0 + 2 \sum_{k=1}^{\frac{N}{2}-1} \left[\text{Re}\{F_k\} \cos\left(\frac{2\pi nk}{N}\right) - \text{Im}\{F_k\} \sin\left(\frac{2\pi nk}{N}\right)\right] + F_{\frac{N}{2}} \cos(\pi n),$$

for $n = 0 : N - 1$.

73. DCT and DST modes. Make a rough sketch of the modes that appear in the DCT and DST of length $N = 8$. In each case note the frequency and period of each mode.

74. Aliasing of DST and DCT modes. Let

$$\mathbf{w}_n^k = \cos\left(\frac{\pi nk}{N}\right)$$

be the nth component of the kth DCT mode where $k = 0 : N$. Show that the $k + 2N$ mode is aliased as the kth mode. What form does the $k + N$ mode take relative to the kth mode? Do the same conclusions apply to the DST modes?

75. Properties of the DCT. Given a sequence f_n, let $F_k = \mathcal{C}\{f_n\}_k$ denote the N-point DCT of f_n as given by (4.5). Show that the following properties are true:

    (a) Periodicity: $F_{k+2N} = F_k$ and $f_{n+2N} = f_n$.

    (b) Shift$_1$: $\mathcal{C}\{f_{n+p} + f_{n-p}\}_k = 2\cos(\frac{\pi pk}{N})\mathcal{C}\{f_n\}_k$.

(c) Shift$_2$: $F_{k+p} + F_{k-p} = 2\cos(\frac{\pi pn}{N})F_k$.

(d) Even input: If $f_{N-n} = f_n$, then $\mathcal{C}\{f_n\}_{2k+1} = F_{2k+1} = 0$.

(e) Odd input: If $f_{N-n} = -f_n$, then $\mathcal{C}\{f_n\}_{2k} = F_{2k} = 0$.

76. Properties of the DST. Given a sequence f_n, let $F_k = \mathcal{S}\{f_n\}_k$ denote the N-point DST of f_n as given by (4.8). Show that the following properties are true:

(a) Periodicity: $F_{k+2N} = F_k$ and $f_{n+2N} = f_n$.

(b) Shift$_1$: $\mathcal{S}\{f_{n+p} + f_{n-p}\}_k = 2\cos(\frac{\pi pk}{N})\mathcal{S}\{f_n\}_k$.

(c) Shift$_2$: $F_{k+p} + F_{k-p} = 2\cos(\frac{\pi pn}{N})F_k$.

(d) Even input: If $f_{N-n} = f_n$, then $\mathcal{S}\{f_n\}_{2k} = F_{2k} = 0$.

(e) Odd input: If $f_{N-n} = -f_n$, then $\mathcal{S}\{f_n\}_{2k+1} = F_{2k+1} = 0$.

77. Inverse DST. Start with a sequence of coefficients F_k of length $2N$ that is odd and pure imaginary. Apply the inverse RDFT (4.3) of length $2N$ and use the symmetry of the coefficients to derive the inverse DST as given in (4.9).

78. Orthogonality. Show that the following orthogonality properties govern the modes of the DCT and DST.

$$\sum_{n=0}^{N}{}'' \cos\left(\frac{\pi nk}{N}\right) \cos\left(\frac{\pi nj}{N}\right) = \begin{cases} N/2 & \text{if } j = k \neq 0, \\ N & \text{if } j = k = 0 \text{ or } j = k = N, \\ 0 & \text{if } j \neq k \end{cases}$$

for $k = 0 : N, j = 0 : N$, and

$$\sum_{n=1}^{N-1} \sin\left(\frac{\pi nk}{N}\right) \sin\left(\frac{\pi nj}{N}\right) = \begin{cases} N/2 & \text{if } j = k, \\ 0 & \text{if } j \neq k \end{cases}$$

for $k = 1 : N - 1, j = 1 : N - 1$.

79. Inverses from orthogonality. Derive the inverse DCT and DST directly using the above orthogonality properties. For example, in the DCT case proceed as follows. Multiply both sides of the forward DCT

$$F_k = \frac{1}{N} \sum_{j=0}^{N}{}'' f_j \cos\left(\frac{\pi jk}{N}\right)$$

by an arbitrary mode $\cos(\pi nk/N)$ and sum (using Σ'') both sides over $k = 0 : N$. Use the orthogonality to solve for f_n, where $n = 0 : N$. Carry out the same procedure to derive the inverse DST.

80. Two real DFTs by one complex DFT. Verify that if g_n and h_n are two real sequences of length N and $f_n = g_n + ih_n$, then the respective DFTs are related as given by Procedure 1:

$$G_k = \frac{1}{2}(F_k + F^*_{-k}) \quad \text{and} \quad H_k = -\frac{i}{2}(F_k - F^*_{-k}),$$

where $k = -N/2 + 1 : N/2$.

81. Alternate butterfly relations. Show that if the complex DFT is defined on the indices $n, k = 0 : N - 1$ by

$$F_k = \frac{1}{N} \sum_{n=0}^{N-1} f_n \omega_N^{-nk}$$

for $k = 0 : N - 1$, then the butterfly relations of Procedure 2 take the form

$$F_k = \frac{1}{2}(G_k + \omega_{2N}^{-k} H_k),$$

$$F_{k+\frac{N}{2}} = \frac{1}{2}(G_k - \omega_{2N}^{-k} H_k)$$

for $k = 0 : N/2 - 1$. In light of this modification, how are the relations of Procedure 3 changed?

82. 2N-point real DFT from N-point complex DFT. Verify that the DFT of a real sequence of length $2N$ as given by Procedure 3 is conjugate even. Verify that F_0 and F_N are real.

83. An inefficient DCT. Show that the N-point DCT of an arbitrary real sequence f_n consists of the $N + 1$ (real) coefficients $F_0, \ldots, F_N$ of the $2N$-point complex DFT of the even extension of f_n.

84. An inefficient DST. Show that the N-point DST of an arbitrary real sequence f_n consists of the $N - 1$ (real) coefficients $iF_1, \ldots, iF_{N-1}$ of the $2N$-point complex DFT of the odd extension of f_n.

85. Pre- and postprocessing for the DST. The following steps lead to a pre- and postprocessing method for computing the DST.

(a) Given an arbitrary real sequence f_n for $n = 1 : N$, use its odd extension to define the auxiliary sequence

$$z_n = f_{2n-1} - f_{2n+1} + if_{2n} \equiv g_n + ih_n$$

for $n = -N/2 + 1 : N/2$.

(b) Show that z_n is a conjugate even sequence and its DFT Z_k is real.

(c) Having computed Z_k, use Procedure 1 to show that

$$G_k = \frac{1}{2}(Z_k + Z_{-k}) \quad \text{and} \quad H_k = -\frac{i}{2}(Z_k - Z_{-k}).$$

(d) Letting $g'_n = f_{2n-1}$ show that

$$G'_k = \frac{G_k}{1 - \omega_N^k}.$$

(e) Having found the DFT of the even and odd subsequences of f_n, use the combine relations of Procedure 2 to write the following procedure.

▶ **Procedure 5: Pre- and postprocessing for the DST** ◀

$$F_k = \frac{1}{4}\left[(Z_k - Z_{-k}) - \frac{Z_k + Z_{-k}}{2\sin(\frac{\pi k}{N})}\right] \quad \text{for} \quad k = 1 : N/2,$$

$$F_{k+N} = \frac{1}{4}\left[(Z_k - Z_{-k}) + \frac{Z_k + Z_{-k}}{2\sin(\frac{\pi k}{N})}\right] \quad \text{for} \quad k = -\frac{N}{2} + 1 : -1.$$

86. A more efficient RDFT. The text offered three methods for computing the DFT of a real sequence of length $2N$.

- Use a $2N$-point complex DFT (with a cost of roughly $4(2N)^2 = 16N^2$ real additions and multiplications).

- Use the explicit form of the RDFT (with a cost of roughly $(2N)^2 = 4N^2$ real additions and multiplications).

- Use Procedure 3 (with a cost of roughly $4N^2$ real additions and multiplications).

Here is a more efficient method (short of calling in the FFTs, which will surpass all of these suggestions). Procedure 2 can be streamlined to combine the RDFTs of two sequences of length N to form an RDFT of length $2N$.

(a) Let f_n be the given sequence of length $2N$, where $n = -N+1 : N$. Note that its DFT F_k is conjugate even and hence only the coefficients for $k = 0 : N$ need to be computed.

(b) Let $g_n = f_{2n}$ and $h_n = f_{2n-1}$ be the even and odd subsequences of f_n, where $n = -N/2 + 1 : N/2$. Note that the N-point RDFTs G_k and H_k are also conjugate even and only the coefficients $k = 0 : N/2$ need to be stored.

(c) With these symmetries in mind, show that the combine relations of Procedure 2 can be reduced to

$$\begin{aligned} F_k &= G_k + \omega_{2N}^k H_k \quad \text{for} \quad k = 0 : N/2, \\ F_{N-k} &= G_k^* - \omega_{2N}^k H_k^* \quad \text{for} \quad k = 0 : N/2 - 1, \end{aligned}$$

and the full sequence F_k can be recovered.

(d) How many real additions and multiplications are required in this process (in multiples of N^2)?

(The recursive application of this strategy to the smaller RDFTs results in Edson's compact symmetric FFT.)

87. The quarter-wave even DFT. Assume that a real sequence f_n of length $2N$ is defined for $n = -N + 1 : N$ and satisfies the quarter-wave even (QE) symmetry $f_{-n-1} = f_n$.

(a) Show that this symmetry together with periodicity (period $2N$) implies that the sequence f_n satisfies the boundary conditions $f_0 - f_{-1} = f_N - f_{N-1} = 0$. Note that there are N independent values of f_n in this sequence of length $2N$.

(b) Beginning with the complex DFT (2.6), reverse the order of summation to show that the QE symmetry implies that $F_{-k} = F_k^*$ and $F_k = \omega_{2N}^k F_k^*$. Show that F_k can be expressed in the form $F_k = \omega_{2N}^{k/2} \tilde{F}_k$, where $\tilde{F}_k$ is real.

(c) Show that the auxiliary sequence $\tilde{F}_k$ has the properties $\tilde{F}_N = 0$, $\tilde{F}_k = \tilde{F}_{k\pm 2N}$, and $\tilde{F}_{-k} = \tilde{F}_k$.

(d) Now use the QE symmetry in the definition of the complex DFT (2.6) to derive the following symmetric DFT.

▶ **Quarter-Wave Even (QE) DFT** ◀

$$\tilde{F}_k = \frac{1}{N} \sum_{n=0}^{N-1} f_n \cos\left(\frac{\pi k}{2N}(2n+1)\right) \quad \text{for} \quad k = 0 : N-1.$$

(e) Use either the inverse complex DFT or the inverse real DFT (4.3) and the symmetry $F_k = \omega_{2N}^{k/2} \tilde{F}_k$ to find the following inverse DFT.

▶ **Inverse Quarter-Wave Even DFT** ◀

$$f_n = \tilde{F}_0 + 2 \sum_{k=1}^{N-1} \tilde{F}_k \cos\left(\frac{\pi k}{2N}(2n+1)\right) \quad \text{for} \quad n = 0 : N-1.$$

Note that by using the auxiliary sequence $\tilde{F}_k$ the transform pair can be computed entirely in terms of real quantities.

88. The quarter-wave odd DFT. Follow the pattern of the previous problem to develop the forward and inverse quarter-wave odd (QO) DFTs.

(a) Beginning with a sequence f_n of length $2N$ that possesses the QO symmetry $f_{-n-1} = -f_n$, show that f_n satisfies the boundary conditions $f_0 + f_1 = f_N + f_{N-1} = 0$.

(b) Show that with this symmetry the DFT satisfies $F_k = -\omega_{2N}^k F_k^*$; this implies that $F_k = i\omega_{2N}^{k/2} \tilde{F}_k$ where $\tilde{F}_k$ is real.

(c) Show that the auxiliary sequence $\tilde{F}_k$ satisfies $\tilde{F}_0 = 0$ and $\tilde{F}_{-k} = -\tilde{F}_k$.

(d) Derive the following symmetric transform pair.

▶ **Quarter-Wave Odd (QO) DFT** ◀

$$\tilde{F}_k = -\frac{1}{N} \sum_{n=0}^{N-1} f_n \sin\left(\frac{\pi k}{2N}(2n+1)\right) \quad \text{for} \quad k = 1 : N.$$

▶ **Inverse Quarter-Wave Odd DFT** ◀

$$f_n = -2 \sum_{k=1}^{N-1} \tilde{F}_k \sin\left(\frac{\pi k}{2N}(2n+1)\right) - \tilde{F}_N \cos(\pi n) \quad \text{for} \quad n = 0 : N-1.$$

Note that by using the auxiliary sequence $\tilde{F}_k$ the transform pair can be computed entirely in terms of real quantities.

Chapter **5**

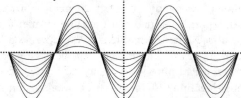

Multidimensional DFTs

*It is good to express a
matter in two ways
simultaneously so as to
give it both a right foot
and a left. Truth can
stand on one leg, to be
sure; but with two it can
walk and get about.*
– Friedrich Nietzsche

5.1.   Introduction

Up until this point every ounce of attention has been devoted to the uses and properties of the one-dimensional DFT. However, in terms of practical occurrences of the DFT and in terms of the overall computational effort spent on DFTs, the problem that arises even more frequently is that of computing **multidimensional DFTs**. We must now imagine input that consists of an **array** of data that may be arranged in a plane (for the two-dimensional case), a parallelepiped (for the three-dimensional case), or even higher-dimensional configurations. It is not difficult to appreciate how such higher-dimensional arrays of data might arise. The image on a television screen, the output of a video display device, or an aerial photograph may all be regarded as two-dimensional arrays of pixels (picture elements). A tomographic image from a medical scanning device is a two- or three-dimensional array of numbers. The solution of a differential equation in two or more spatial dimensions is generally approximated at discrete points of a multidimensional domain. A statistician doing a multivariate study of the causes of a particular disease may collect data in 20 or 30 variables and then look for correlations. All of these applications require the computation of DFTs in two or more dimensions. Fortunately, not one bit of the time spent on one-dimensional DFTs will be wasted: almost without exception, the properties of the one-dimensional DFT can be used and extended in our discussions of multidimensional DFTs. The primary task is to present the two-dimensional DFT and make sure that it is well understood. The generalization to three and more dimensions is then absolutely straightforward. It is a lovely excursion with analytical, geometrical, and computational sights in all directions. With that promise, let's begin this important and most relevant subject.

5.2.   Two-Dimensional DFTs

As with the one-dimensional DFT, multidimensional DFTs begin with input. Depending on the origins of the particular problem, the input may be given in a discrete form (perhaps sampled data) or in continuous form (a function of several variables). For the moment assume that we are given a function f defined on the rectangular region[1]

$$\left\{ (x,y) : -\frac{A}{2} \leq x \leq \frac{A}{2}, -\frac{B}{2} \leq y \leq \frac{B}{2} \right\}.$$

As in the one-dimensional case this function must be sampled in order to handle it numerically. To carry out this sampling a grid is established on the region with a uniform grid spacing of $\Delta x = A/M$ in the x-direction and $\Delta y = B/N$ in the y-direction. The resulting grid points (see Figure 5.1) are given by

$$(x_m, y_n) = (m\Delta x, n\Delta y)$$

for $n = -N/2 + 1 : N/2$ and $m = -M/2 + 1 : M/2$. The input function f can now be sampled by recording its values at the grid points to produce the sequence (or array) of numbers $f_{mn} = f(x_m, y_n)$. If the original problem had been discrete in nature, the input would already appear in this form. In anticipation of using the array f_{mn}

---

[1]Following the convention of previous chapters, we will continue to define functions on closed sets (boundary points included) with the understanding that special attention must be given to boundary points.

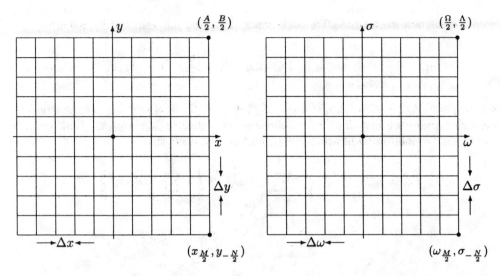

FIG. 5.1. *The two-dimensional DFT works on a spatial grid with grid spacings $\Delta x = A/M$ and $\Delta y = B/N$ in the x- and y-directions, respectively. A typical grid point is (x_m, y_n), where $m = -M/2 + 1 : M/2$ and $n = -N/2 + 1 : N/2$. Intimately related to the spatial grid is the frequency grid (to be introduced shortly) with grid spacings $\Delta \omega = \Omega/M$ and $\Delta \sigma = \Lambda/N$. A typical frequency grid point is (ω_m, σ_n), where $m = -M/2+1 : M/2$ and $n = -N/2+1 : N/2$.*

as input to a DFT, we will already begin to think of it as being **doubly periodic**, meaning that

$$f_{m\pm M,n} = f_{mn} \quad \text{and} \quad f_{m,n\pm N} = f_{mn}.$$

Furthermore, all of the familiar exhortations about endpoints and discontinuities must also be repeated. *At boundaries and discontinuities, f_{mn} must be assigned the appropriate average value* (AVED). For discontinuities at the boundaries this presents no difficulties, since the boundaries are parallel to one of the coordinate directions. For example, if $f(-A/2, y_n) \neq f(A/2, y_n)$ then $f_{\frac{M}{2},n}$ should be assigned the value $\frac{1}{2}(f(-A/2, y_n) + f(A/2, y_n))$. For internal discontinuities, there can be some subtleties in determining average values which we will not belabor here.

In order to motivate the two-dimensional DFT, we will start by considering a slightly special case that will be generalized immediately. For the moment assume that the input sequence f_{mn} is **separable**, meaning that it has the form $f_{mn} = g_m h_n$, the product of an m-dependent (or x-dependent) term and an n-dependent (or y-dependent) term. With this small assumption, we now look for a representation of the input f_{mn} in terms of the basic sine and cosine modes in each of the coordinate directions. In this separable case, we know that the individual sequences g_m and h_n have representations given by the one-dimensional inverse DFTs

$$g_m = \sum_{j=-\frac{M}{2}+1}^{\frac{M}{2}} G_j \omega_M^{mj} \quad \text{and} \quad h_n = \sum_{k=-\frac{N}{2}+1}^{\frac{N}{2}} H_k \omega_N^{nk}.$$

In these representations, $m = -M/2 + 1 : M/2$ and $n = -N/2 + 1 : N/2$. We should recognize G_j and H_k as the DFT coefficients of g_m and h_n, respectively, given by

$$G_j = \frac{1}{M} \sum_{m=-\frac{M}{2}+1}^{\frac{M}{2}} g_m \omega_M^{-mj} \quad \text{and} \quad H_k = \frac{1}{N} \sum_{n=-\frac{N}{2}+1}^{\frac{N}{2}} h_n \omega_N^{-nk},$$

where $j = -M/2 + 1 : M/2$ and $k = -N/2 + 1 : N/2$. It is now an easy matter to construct a representation for the two-dimensional sequence $f_{mn} = g_m h_n$. Multiplying the two individual representations for g_m and h_n, we have that

$$f_{mn} = g_m h_n = \left\{ \sum_{j=-\frac{M}{2}+1}^{\frac{M}{2}} G_j \omega_M^{mj} \right\} \left\{ \sum_{k=-\frac{N}{2}+1}^{\frac{N}{2}} H_k \omega_N^{nk} \right\}$$

$$= \sum_{j=-\frac{M}{2}+1}^{\frac{M}{2}} \sum_{k=-\frac{N}{2}+1}^{\frac{N}{2}} \underbrace{G_j H_k}_{F_{jk}} \omega_M^{mj} \omega_N^{nk}.$$

If we now let the product $G_j H_k$ be the new DFT coefficient F_{jk}, we have the following representation for the input sequence f_{mn}:

$$f_{mn} = \sum_{j=-\frac{M}{2}+1}^{\frac{M}{2}} \sum_{k=-\frac{N}{2}+1}^{\frac{N}{2}} F_{jk} \omega_M^{mj} \omega_N^{nk},$$

where $m = -M/2 + 1 : M/2$ and $n = -N/2 + 1 : N/2$. Furthermore, we can combine the expressions for the DFT coefficients G_j and H_k to write that

$$F_{jk} = G_j H_k = \left\{ \frac{1}{M} \sum_{m=-\frac{M}{2}+1}^{\frac{M}{2}} g_m \omega_M^{-mj} \right\} \left\{ \frac{1}{N} \sum_{n=-\frac{N}{2}+1}^{\frac{N}{2}} h_n \omega_N^{-nk} \right\}$$

$$= \frac{1}{MN} \sum_{m=-\frac{M}{2}+1}^{\frac{M}{2}} \sum_{n=-\frac{N}{2}+1}^{\frac{N}{2}} f_{mn} \omega_M^{-mj} \omega_N^{-nk},$$

which applies for $j = -M/2 + 1 : M/2$ and $k = -N/2 + 1 : N/2$. At this point we have presented little more than a conjecture, but here is the claim that needs to be verified: Given an $M \times N$ input array f_{mn}, the forward DFT is given by the following formula.

▶ **Forward Two-Dimensional DFT** ◀

$$F_{jk} = \frac{1}{MN} \sum_{m=-\frac{M}{2}+1}^{\frac{M}{2}} \sum_{n=-\frac{N}{2}+1}^{\frac{N}{2}} f_{mn} \omega_M^{-mj} \omega_N^{-nk}, \qquad (5.1)$$

for $j = -M/2+1 : M/2$ and $k = -N/2+1 : N/2$. Meanwhile the inverse DFT is given by the following formula, although the validity of the inverse is still to be verified.

▶ **Inverse Two-Dimensional DFT** ◀

$$f_{mn} = \sum_{j=-\frac{M}{2}+1}^{\frac{M}{2}} \sum_{k=-\frac{N}{2}+1}^{\frac{N}{2}} F_{jk} \omega_M^{mj} \omega_N^{nk}, \qquad (5.2)$$

for $j = -M/2+1 : M/2$ and $k = -N/2+1 : N/2$. It is this pair that we will use in all that follows. Having finally written down an alleged two-dimensional DFT, it should be said that several different paths could have been chosen to the same destination. Any of the derivations of the DFT presented in Chapter 2 (approximation of Fourier series, approximation of Fourier transform, interpolation, least squares approximation) could be used in a two-dimensional setting to arrive at the two-dimensional DFT. The derivation based on Fourier series is explored in problem 115.

Now the task is to show that (5.1) and (5.2) really *do* form a DFT pair. Before doing so, it would help to interpret these expressions for f_{mn} and F_{jk}. Just as in the one-dimensional case, the aim of the DFT is to express the two-dimensional input f_{mn} as a linear combination of **modes** that consist of sines and cosines. We can organize the various modes with the help of the two spatial indices j and k. For fixed values of j and k, these modes are functions of the two indices m and n. It is essential to understand what these modes look like. The (j, k) mode has the form

$$\omega_M^{mj} \omega_N^{nk} = e^{i2\pi\left(\frac{mj}{M} + \frac{nk}{N}\right)} = \cos\left(2\pi\left(\frac{mj}{M} + \frac{nk}{N}\right)\right) + i\sin\left(2\pi\left(\frac{mj}{M} + \frac{nk}{N}\right)\right)$$

We see that for a fixed pair (j, k), one complex mode ω_N^{jk} corresponds to one cosine mode and one sine mode.

We must now interpret the frequency indices j and k and make the first of many important geometrical observations. Consider either the sine or cosine mode given above for a fixed (j, k) pair. If we hold n fixed and sample this wave on a line of increasing m, then the resulting sequence is periodic in m with j periods (or wavelengths) every M grid points (see Figure 5.2). In other words, j determines the frequency of the mode as it varies in the m- (or x-) direction alone. Similarly, if we hold m fixed and sample the mode on a line of increasing n, we obtain a slice of the wave that has a frequency of k periods (or wavelengths) every N grid points in the n- (or y-) direction.

As in the one-dimensional case, values of j or k near zero correspond to low frequencies (long periods) in the respective directions, while if $|j|$ is near $M/2$ or $|k|$ is near $N/2$, we expect to see a high frequency (small period) in that direction. A few pictures will speak many words in showing how both high and low frequencies in either direction can be mixed in all combinations. Figure 5.3 shows several of the modes that are needed for a full representation of the two-dimensional input f_{mn}.

We will return to the fascinating geometry of the DFT modes shortly. But it's time to verify that the pair of expressions (5.1) and (5.2) has properties that we might expect of a DFT pair. It is easy to see that both the input sequence f_{mn} as given by (5.1) and the transform sequence F_{jk} as given by (5.2) are M-periodic with respect to their first index and N-periodic with respect to the second. We must also verify that the alleged forward and inverse transforms really *are* inverses of each other. As in the one-dimensional case, the inverse property relies on orthogonality. So let's first check the orthogonality of the vectors $\mathbf{w}^{jk}$ with components $\mathbf{w}_{mn}^{jk} = \omega_M^{mj} \omega_N^{nk}$. Consider the discrete inner product of two modes with frequencies (j, k) and (j_0, k_0). It looks like

$$\langle \mathbf{w}^{jk}, \mathbf{w}^{j_0 k_0} \rangle = \sum_{m=-\frac{M}{2}+1}^{\frac{M}{2}} \sum_{n=-\frac{N}{2}+1}^{\frac{N}{2}} \mathbf{w}_{mn}^{jk} (\mathbf{w}_{mn}^{j_0 k_0})^*$$

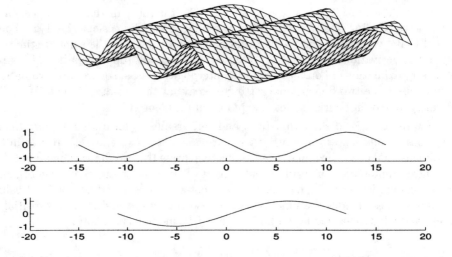

FIG. 5.2.   *Two slices reveal the anatomy of a typical two-dimensional DFT mode. These modes are fully two-dimensional waves (top figure) characterized by two frequency indices (j, k). When sliced along a line of constant n (or y), the mode is periodic in the m- or x-direction (middle figure) with a frequency of j periods every M grid points. Similarly, a slice through the mode along a line of constant m (or x) produces a wave with a frequency of k periods every N grid points (bottom figure). Here $j = 2$ and $M = 32$, while $k = 1$ and $N = 24$.*

$$= \sum_{m=-\frac{M}{2}+1}^{\frac{M}{2}} \sum_{n=-\frac{N}{2}+1}^{\frac{N}{2}} \omega_M^{-mj} \omega_N^{nk} \omega_M^{mj_0} \omega_N^{-nk_0}.$$

And now something happens that will recur many times in working with multidimensional DFTs. The problem splits into separate problems, each of which can be handled easily with one-dimensional thinking. Here is how it happens in the case of orthogonality. By the multiplicative property of the exponential $(\omega_N^a \omega_N^b = \omega_N^{a+b})$ we have that

$$\sum_{m=-\frac{M}{2}+1}^{\frac{M}{2}} \sum_{n=-\frac{N}{2}+1}^{\frac{N}{2}} \omega_M^{mj} \omega_N^{nk} \omega_M^{-mj_0} \omega_N^{-nk_0} = \sum_{m=-\frac{M}{2}+1}^{\frac{M}{2}} \omega_M^{m(j-j_0)} \sum_{n=-\frac{N}{2}+1}^{\frac{N}{2}} \omega_N^{n(k-k_0)}$$

$$= M\hat{\delta}_M(j - j_0) N\hat{\delta}_N(k - k_0).$$

We have appealed to the orthogonality of the one-dimensional vectors ω_M^{mj} and ω_N^{nk} and used the modular Kronecker delta notation

$$\hat{\delta}_N(k) = \begin{cases} 1 & \text{if } k = 0 \text{ or a multiple of } N, \\ 0 & \text{otherwise.} \end{cases}$$

Stated in words, the inner product of two modes vanishes unless the two modes are identical *or unless* the two modes *after aliasing* are identical. (We will return to the matter of aliasing in two dimensions shortly.) With this two-dimensional orthogonality property, it is a short and direct calculation (problem 89) to verify that the relations (5.1) and (5.2) really are inverses of each other.

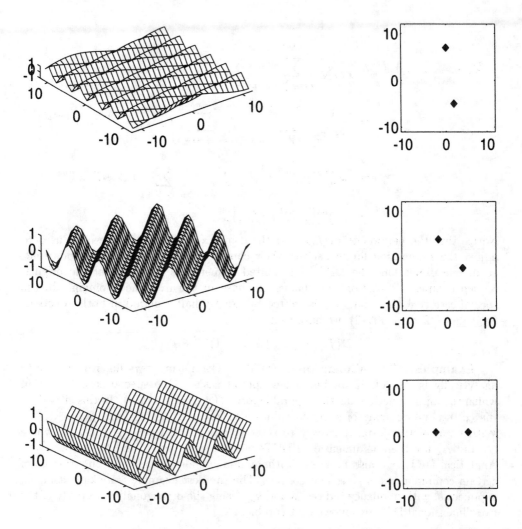

FIG. 5.3. *The two-dimensional DFT uses modes (sines and cosines) that can have different frequencies in the two coordinate directions. The figure shows three typical cosine modes on a grid with $M = N = 24$ points in both directions: (top left) $j = 1, k = 6$ (low frequency in the x-direction, high frequency in the y-direction), (middle left) $j = 3, k = 3$, and (bottom left) $j = 4, k = 0$. The three figures on the right are the corresponding frequency domain representations. Since each function consists of a single real mode, the DFTs consist of two isolated spikes. They are displayed in map view, with the diamonds at the locations of the spikes.*

There is a multitude of other one-dimensional DFT properties that carry over directly to the two-dimensional case. Most of these will be relegated to the problem section (problem 92). As a quick example, let's look at a shift property to demonstrate some common techniques. Using the operator notation, we will let $F_{jk} = \mathcal{D}\{f_{mn}\}_{jk}$, where M and N are understood. Now consider the input sequence obtained from f_{mn} by shifting it m_0 units in the positive m direction and n_0 units in the positive n direction. Its DFT is given by

$$
\begin{aligned}
\mathcal{D}\{f_{m-m_0,n-n_0}\}_{jk} &= \frac{1}{MN} \sum_{m=-\frac{M}{2}+1}^{\frac{M}{2}} \sum_{n=-\frac{N}{2}+1}^{\frac{N}{2}} f_{m-m_0,n-n_0} \omega_M^{-mj} \omega_N^{-nk} \\
&= \frac{1}{MN} \sum_{m=-\frac{M}{2}+1-m_0}^{\frac{M}{2}-m_0} \sum_{n=-\frac{N}{2}+1-n_0}^{\frac{N}{2}-n_0} f_{mn} \omega_M^{-(m+m_0)j} \omega_N^{-(n+n_0)k} \\
&= \omega_M^{-m_0 j} \omega_N^{-n_0 k} \frac{1}{MN} \sum_{m=-\frac{M}{2}+1}^{\frac{M}{2}} \sum_{n=-\frac{N}{2}+1}^{\frac{N}{2}} f_{mn} \omega_M^{-mj} \omega_N^{-nk} \\
&= \omega_M^{-m_0 j} \omega_N^{-n_0 k} \mathcal{D}\{f_{mn}\}_{jk}.
\end{aligned}
$$

Notice that the periodicity of f_{mn} and the complex exponential have been used to adjust the summation limits back to their conventional values. We conclude that, as in one dimension, the DFT of a shifted sequence is the DFT of the original sequence times a "rotation" or change of phase in the frequency domain. In the special case that the input f_{mn} is shifted by exactly half a period in both directions ($m_0 = \pm M/2, n_0 = \pm N/2$), we have that

$$
\mathcal{D}\{f_{m\pm\frac{M}{2},n\pm\frac{N}{2}}\}_{jk} = (-1)^{j+k} F_{jk}.
$$

Examples: Two-dimensional DFTs. Computing two-dimensional DFTs analytically is difficult in all but a few special cases. One special case is that in which the input function has the separable form $f(x,y) = g(x)h(y)$. In this situation, the corresponding array of sampled values also has a separable form $f_{mn} = g_m h_n$. With input of this form, it is easy to show (problem 90) that $F_{jk} = G_j H_k$, where G_j and H_k are the one-dimensional DFTs of the sequences g_m and h_n, respectively. Analytical DFTs can also be found without undue effort when the input consists of certain combinations of sines and cosines. The simplest example is a single complex exponential with frequency indices j_0 and k_0. Using the orthogonality property of the two-dimensional DFT, it follows that (problem 91)

$$
\mathcal{D}\left\{e^{i2\pi(\frac{mj_0}{M}+\frac{nk_0}{N})}\right\}_{jk} = \hat{\delta}_M(j-j_0)\hat{\delta}_N(k-k_0).
$$

We see that the DFT of a single complex mode consists of one spike located at position (j_0, k_0) in the frequency domain. Combining two complex exponential modes allows us to produce single real modes. For a fixed pair of integers j_0 and k_0, we have that

$$
\cos\left(2\pi\left(\frac{mj_0}{M}+\frac{nk_0}{N}\right)\right) = \frac{1}{2}\left(e^{i2\pi(\frac{mj_0}{M}+\frac{nk_0}{N})} + e^{-i2\pi(\frac{mj_0}{M}+\frac{nk_0}{N})}\right)
$$

and

$$
\sin\left(2\pi\left(\frac{mj_0}{M}+\frac{nk_0}{N}\right)\right) = \frac{1}{2i}\left(e^{i2\pi(\frac{mj_0}{M}+\frac{nk_0}{N})} - e^{-i2\pi(\frac{mj_0}{M}+\frac{nk_0}{N})}\right).
$$

Therefore, it follows immediately that the DFT of a single real mode consists of two spikes:

$$
\mathcal{D}\left\{\cos\left(2\pi\left(\frac{mj_0}{M}+\frac{nk_0}{N}\right)\right)\right\}_{jk} = \frac{1}{2}\hat{\delta}_M(j-j_0)\hat{\delta}_N(k-k_0)
$$

$$
+ \frac{1}{2}\hat{\delta}_M(j+j_0)\hat{\delta}_N(k+k_0)
$$

and

$$\mathcal{D}\left\{\sin\left(2\pi\left(\frac{mj_0}{M} + \frac{nk_0}{N}\right)\right)\right\}_{jk} = \frac{1}{2i}\hat{\delta}_M(j - j_0)\hat{\delta}_N(k - k_0)$$
$$- \frac{1}{2i}\hat{\delta}_M(j + j_0)\hat{\delta}_N(k + k_0).$$

The DFTs of three different single cosine modes are shown in the plots of Figure 5.3. The double spikes occur symmetrically with respect to the origin, and their position indicates the mode's frequency in both the x- and y-directions. We can also consider the product of two real modes. To change the setting a bit, assume that for a fixed pair of integers j_0 and k_0, a function of the form

$$f(x, y) = \cos\left(\frac{2\pi j_0 x}{A}\right)\sin\left(\frac{2\pi k_0 y}{B}\right)$$

is given on the domain with $-A/2 \leq x \leq A/2$ and $-B/2 \leq y \leq B/2$. We now sample this function with M points in the x-direction and N points in the y-direction and then apply the two-dimensional DFT. We can either note that the input function has the separable form mentioned above or use the identity

$$\cos A \sin B = \frac{1}{2}\left(\sin(A + B) - \sin(A - B)\right).$$

Either route leads to the same result. Letting

$$f_{mn} = \cos\left(\frac{2\pi j_0 m}{M}\right)\sin\left(\frac{2\pi k_0 n}{N}\right)$$

be the sampled form of the function f, we find that the DFT is given by

$$\mathcal{D}\{f_{mn}\}_{jk} = \frac{1}{4i}\left((\hat{\delta}_M(j - j_0) + \hat{\delta}_M(j + j_0))(\hat{\delta}_N(k - k_0) - \hat{\delta}_N(k + k_0))\right).$$

A careful inspection of this expression reveals that it amounts to four spikes (with imaginary amplitudes) located at the four pairs of frequency indices $(\pm j_0, \pm k_0)$. Other combinations of sines and cosines in product form can be handled in a similar manner. For example, the DFT of the product of two cosine modes or two sine modes will consist of four spikes with real amplitude. Problem 93 offers several more examples of analytical DFTs in two dimensions.

This would be a good place to note that we have defined the two-dimensional DFT pair on a symmetric domain with the origin $(0, 0)$ at the center. But just as in the one-dimensional case, there are often situations in which other domains are convenient. We can define the DFT pair on any domain with $M \times N$ adjacent grid points. In later chapters, we will use the two-dimensional DFT on domains with spatial indices $m = 0 : M - 1, n = 0 : N - 1$. The DFT pair for this set of indices is

$$F_{jk} = \frac{1}{MN}\sum_{m=0}^{M-1}\sum_{n=0}^{N-1}f_{mn}\omega_M^{-mj}\omega_N^{-nk}$$

for $j = 0 : M - 1, k = 0 : N - 1$, in the forward direction, and

$$f_{mn} = \sum_{j=0}^{M-1}\sum_{k=0}^{N-1}F_{jk}\omega_M^{mj}\omega_N^{nk}$$

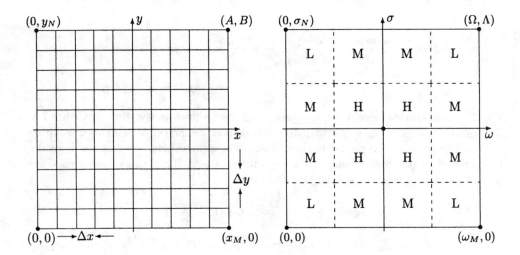

FIG. 5.4. *The two-dimensional DFT can also be defined for indices* $m, j = 0 : M - 1$ *and* $n, k = 0 : N - 1$. *With this choice, the high frequencies appear near the center (denoted "H"), the low frequencies near the corners (denoted "L"), and the frequencies near the central edges (denoted "M") are mixed frequencies, high in one direction and low in the other.*

for $m = 0 : M - 1$, $n = 0 : N - 1$ for the inverse DFT. This convention has some advantages; however, it should be noted that it does imply a rather peculiar arrangement of the frequencies. As shown in Figure 5.4, the indices for the low frequency modes are located near the corners of the frequency domain (indicated by "L"), while the high frequency (denoted "H") indices are located near the center (clustered around $j = M/2$ and $k = N/2$). Mixed modes, with a high frequency in one direction and a low frequency in the other direction, appear near the edges of the frequency domain, and are marked "M" in the figure.

5.3. Geometry of Two-Dimensional Modes

We gave only passing attention to the geometry of the two-dimensional DFT modes. There is more to be learned and it will eventually bear on the important questions of reciprocity, aliasing, and sampling. For the sake of illustration, we will work on the domain

$$\mathcal{I} = \left\{ (x, y) : -\frac{A}{2} \le x \le \frac{A}{2}, -\frac{B}{2} \le y \le \frac{B}{2} \right\},$$

and consider the continuous mode

$$w(x, y) = \cos \left(2\pi \left(\frac{jx}{A} + \frac{ky}{B} \right) \right), \tag{5.3}$$

with a fixed pair of frequency indices (j, k). The corresponding sine mode could be considered in the same way, and there is an analogous analysis for the discrete form of this mode (problem 96). Several typical modes of this form were graphed in Figure 5.3. But how would one graph such a mode in general? What is the frequency and wavelength of such a mode? These are the questions that we must investigate.

Before proceeding, let's agree on some terminolgy: since multiple-dimension DFTs are generally used in spatially dependent problems, we will use the more appropriate term *wavelength* instead of *period* throughout this discussion. This means that the units of x and y are *length*.

Remember that the frequency indices j, k and the dimensions of the domain A, B are assumed to be fixed in this discussion. We can take a first cut at analyzing this mode by using the slicing exercise that was applied earlier (Figure 5.2 still helps). Along a line of constant y, the mode given by (5.3) has a wavelength of A/j (measured in units of length) and a frequency of j/A (measured in cycles per unit length). Some additional notation is now needed: we will let μ_j and ω_j be the wavelength and frequency, respectively, of the (j, k) mode when sliced in the x-direction. The same argument in the y-direction shows that this mode, when sliced along a line of constant x, has a wavelength of B/k and a frequency of k/B that we will denote η_k and σ_k. To summarize this notation, we have that the wavelengths and frequencies in the coordinate directions are given by

$$\omega_j = \frac{1}{\mu_j} = \frac{j}{A}$$

in the x-direction and

$$\sigma_k = \frac{1}{\eta_k} = \frac{k}{B}$$

in the y-direction. However, the mode given in (5.3) is more than a slice in the coordinate directions; it is a fully two-dimensional wave. Here is the crucial insight: viewed as a function of x and y, this mode (for fixed j and k) has the property that it is constant along the parallel lines

$$\frac{jx}{A} + \frac{ky}{B} = c \tag{5.4}$$

in the (x, y) plane, where c is any constant. These lines are called the **lines of constant phase** or simply **phase lines** of this particular mode. Envisioned as a wave, the crests and troughs of a mode are parallel to its phase lines. We see by writing equation (5.4) as $y = -(Bj/Ak)x + (Bc/k)$ that all of the phase lines of the (j, k) mode have a slope of $-Bj/Ak$ (Figure 5.5); or equivalently, the phase lines have an angle of θ_{jk} with respect to the x-direction where

$$\tan \theta_{jk} = -\frac{Bj}{Ak}. \tag{5.5}$$

We can now ask questions about wavelength and frequency of the *full* wave, not of slices. For the cosine mode (5.3), with a maximum value at $x = y = 0$, there is a crest along the line

$$\ell: \quad \frac{jx}{A} + \frac{ky}{B} = 0$$

that passes through the origin. The adjacent crests lie along the phase lines for which

$$\frac{jx}{A} + \frac{ky}{B} = \pm 1,$$

as shown in Figure 5.5. Therefore, the wavelength of this particular mode is just the perpendicular distance between pairs of adjacent phase lines. A short calculation

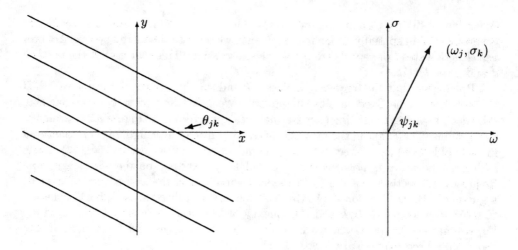

FIG. 5.5. *The modes* $\cos 2\pi(jx/A + ky/B)$ *and* $\sin 2\pi(jx/A + ky/B)$ *are constant along the phase lines* $jx/A + ky/B = c$ *(diagonal lines in the left figure). This means that the crests and troughs of these waves are aligned with the phase lines with an angle* θ_{jk} *to the horizontal. In the frequency domain the frequency vector* $(\omega_j, \sigma_k) = (j/A, k/B)$ *points in a direction orthogonal to the phase lines given by the angle* ψ_{jk}.

(problem 95) shows that the wavelength of the (j, k) mode (measured in units of length) is

$$\lambda_{jk} = \frac{1}{\sqrt{(\frac{j}{A})^2 + (\frac{k}{B})^2}} = \frac{1}{\sqrt{\mu_j^{-2} + \eta_k^{-2}}}.$$

For example, the modes with $j = 0$ (constant in the x-direction) have wavelengths of

$$\lambda_{0k} = \eta_k = \frac{B}{k},$$

as we would expect from the slicing argument. Similarly, the modes with $k = 0$ (constant in the y-direction) have wavelengths of

$$\lambda_{j0} = \mu_j = \frac{A}{j}.$$

The high frequency modes ($j = 0, k = N/2$ and $j = M/2, k = 0$) have wavelengths

$$\lambda_{0, \frac{N}{2}} = \frac{2A}{N} \quad \text{and} \quad \lambda_{\frac{M}{2}, 0} = \frac{2B}{M}.$$

But notice that the $(M/2, N/2)$ mode has an even smaller wavelength of

$$\lambda_{\frac{M}{2}, \frac{N}{2}} = \frac{2}{\sqrt{(\frac{M}{A})^2 + (\frac{N}{B})^2}}.$$

The pattern of wavelengths of two-dimensional waves is given in Table 5.1 for the case $B = 2A = 8$. The same table applies to either the sine or cosine modes. Because of symmetry, only one-quarter of the frequency domain corresponding to nonnegative indices is given.

Table 5.1
Wavelengths of two-dimensional modes with $B = 2A = 8$.

	$j = 0$	$j = 1$	$j = 2$	$j = 3$	$j = 4$
$k = 0$	∞	4	2	$4/3 = 1.33$	1
$k = 1$	8	$8/\sqrt{5} = 3.58$	$8/\sqrt{17} = 1.94$	$8/\sqrt{37} = 1.32$	$8/\sqrt{65} = .99$
$k = 2$	4	$4/\sqrt{2} = 2.82$	$4/\sqrt{5} = 1.79$	$4/\sqrt{10} = 1.26$	$4/\sqrt{17} = .97$
$k = 3$	$8/3 = 2.67$	$8/\sqrt{13} = 2.22$	$8/5 = 1.60$	$8/3\sqrt{5} = 1.19$	$8/\sqrt{73} = .94$
$k = 4$	2	$4/\sqrt{5} = 1.79$	$\sqrt{2} = 1.41$	$4/\sqrt{13} = 1.11$	$2/\sqrt{5} = .89$

Having determined the wavelength of the (j,k) mode, its frequency, measured in *cycles per unit length*, is simply the reciprocal of the wavelength. Therefore, the frequency of the (j,k) modes, denoted ν_{jk}, is given by

$$\nu_{jk} = \frac{1}{\lambda_{jk}} = \sqrt{\frac{j^2}{A^2} + \frac{k^2}{B^2}} = \sqrt{\omega_j^2 + \sigma_k^2}. \tag{5.6}$$

This expression has the valuable interpretation that ν_{jk} may be regarded as the length of a frequency *vector* with components (ω_j, σ_k). Furthermore, this frequency vector has an angle ψ_{jk} with respect to the horizontal axis of the frequency domain where

$$\tan \psi_{jk} = \frac{\sigma_k}{\omega_j} = \frac{kA}{jB}.$$

Comparing this angle ψ_{jk} to the angle of the phase lines θ_{jk} given by (5.5), it follows that θ_{jk} and ψ_{jk} differ by $\pi/2$ radians. This leads to the important observation that the phase lines of the (j,k) mode and the (j,k) frequency vector, if plotted in the same domain, are orthogonal (problem 97).

There is one final step to complete the geometrical picture of the two-dimensional DFT modes. We must now set up the actual grid in the frequency domain that is induced by the spatial grid. This will also bring us closer to the reciprocity relations. The spatial domain

$$\left\{ (x,y) : -\frac{A}{2} \le x \le \frac{A}{2}, -\frac{B}{2} \le y \le \frac{B}{2} \right\}$$

is intimately associated with a frequency domain

$$\left\{ (\omega,\sigma) : -\frac{\Omega}{2} \le \omega \le \frac{\Omega}{2}, -\frac{\Lambda}{2} \le \sigma \le \frac{\Lambda}{2} \right\},$$

where the lengths Ω and Λ will be determined momentarily. Corresponding to the $M \times N$ spatial grid, there is an $M \times N$ grid in the frequency domain with grid spacings $\Delta\omega$ and $\Delta\sigma$. From the definitions of the component frequencies $\omega_j = j/A$ and $\sigma_k = k/B$, we can anticipate that the frequency grid spacings should be given by

$$\Delta\omega = \frac{1}{A} \quad \text{and} \quad \Delta\sigma = \frac{1}{B}.$$

This makes sense physically also. It says that in the x- and y-directions separately, the smallest units of frequency ($\Delta\omega$ and $\Delta\sigma$, respectively) correspond to waves with the largest possible (finite) wavelengths: a wavelength of A in the x-direction has a frequency of $\Delta\omega = 1/A$ in that direction, and a wavelength of B in the y-direction has a frequency of $\Delta\sigma = 1/B$ in that direction. The frequencies represented by the DFT lie on this frequency grid and they are given by

$$\omega_j = j\Delta\omega = \frac{j}{A} \quad \text{and} \quad \sigma_k = k\Delta\sigma = \frac{k}{B},$$

where $j = -M/2 + 1 : M/2$ and $k = -N/2 + 1 : N/2$. It would be worthwhile to ponder Figure 5.1 to become familiar with the grids in the spatial and frequency domain, and all of the attendant notation.

With these geometrical insights about the modes of the two-dimensional DFT, we can now say a few words about two important matters: the first is the question of aliasing and the choice of grid spacings (or sampling rates), and the second is reciprocity relations. Let's begin with aliasing. Recall that if a band-limited input in one dimension is to be fully resolved, then it is necessary to sample it with at least two grid points per wavelength of every mode that appears in the signal. In one dimension, this means that if $\omega_{\max}$ is the highest frequency that appears in the input and it has a wavelength $\lambda_{\min} = 1/\omega_{\max}$, then a grid spacing Δx must be chosen such that

$$\Delta x \leq \frac{1}{2\omega_{\max}} = \frac{\lambda_{\min}}{2}.$$

A similar, but subtly different analysis can be applied in two or more dimensions. We will actually present four different arguments, each with its own virtues, to reach a final conclusion about aliasing and sampling rates.

Here is the guiding question. Consider a single mode

$$w(x, y) = \cos\left(2\pi\left(\frac{jx}{A} + \frac{ky}{B}\right)\right)$$

with frequency indices (j, k). Then, how should the grid be chosen to insure that this mode is fully resolved so as to avoid aliasing? The goal is to find the minimum values of M and N (or alternatively, the maximum values of Δx and Δy) that insure no aliasing of this mode. The same argument will also apply to the sine version of this mode. The relationship we seek is a condition on M and N in terms of A, B, j, k. Let's first review what we know about this mode.

From the foregoing discussion of the geometry of the DFT modes, we know that the mode $w(x, y)$ has crests and troughs along the phase lines

$$\frac{jx}{A} + \frac{ky}{B} = c, \quad \text{where } c \text{ is a constant;}$$

and these phase lines have an angle

$$\theta = -\tan^{-1}\left(\frac{Bj}{Ak}\right)$$

with respect to the horizontal. The wave has wavelengths of $\mu = A/j$ and $\eta = B/k$ in the x- and y-directions individually. The full wave has a wavelength of

$$\lambda = \frac{1}{\sqrt{(\frac{j}{A})^2 + (\frac{k}{B})^2}} = \frac{1}{\sqrt{\mu^{-2} + \eta^{-2}}} \tag{5.7}$$

in the direction orthogonal to the phase lines. (We will suppress the subscripts j, k to simplify the notation.) Recall that the frequency vector with components $\omega = j/A$ and $\sigma = k/B$ has a magnitude and an angle to the horizontal given by

$$\nu = \sqrt{\omega^2 + \sigma^2} \quad \text{and} \quad \psi = \tan^{-1}\left(\frac{\sigma}{\omega}\right) = \tan^{-1}\left(\frac{Ak}{Bj}\right).$$

Furthermore, the angles θ and ψ differ by $\pi/2$ radians. This geometry is shown in Figure 5.6, from which it is not difficult to show that

$$\lambda = \mu \cos \psi = \eta \sin \psi, \quad \text{which means that} \quad \mu \geq \lambda, \eta \geq \lambda.$$

In other words, the wavelength of the full wave is less than the wavelengths in each of the coordinate directions.

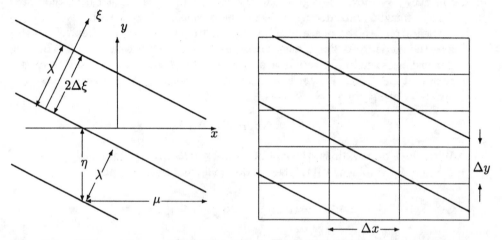

FIG. 5.6. *The sampling conditions that insure no aliasing of a two-dimensional wave can be derived from the geometry of the mode. The left figure shows the phase lines for the crests of a single mode (bold diagonal lines) with frequency indices $j = 2, k = 1$. One wavelength (λ) is the distance between two consecutive phase lines. The wavelengths in the x- and y-directions (μ and η) are also shown. A local coordinate ξ is introduced along the frequency vector perpendicular to the phase lines. The right figure shows part of a spatial grid with spacings Δx and Δy that are half of the wavelengths μ and η. The local grid spacing $\Delta \xi$ is half of the wavelength λ.*

With these preliminary remarks, we can now offer four viewpoints about aliasing and the selection of grids to avoid it.

1. **Playing it safe.** Since the wave we wish to resolve has a wavelength of λ and a frequency $\nu = 1/\lambda$, one could reason that it is best to choose the grid spacings Δx and Δy so that they individually assure no aliasing; that is,

$$\Delta x \leq \frac{1}{2\nu} = \frac{\lambda}{2} \quad \text{and} \quad \Delta y \leq \frac{1}{2\nu} = \frac{\lambda}{2}.$$

This is a conservative choice that may use more grid points than absolutely necessary. For example, if the mode is aligned strongly with the y-axis ($j >> k$), then the wave requires a small grid spacing in the x-direction, but a much coarser grid in the y-direction will suffice. Nevertheless, for an arbitrary input with frequencies of many different magnitudes and directions, this may be the safest choice.

2. **By intuition.** Arguing from a physical perspective, one might say that since the mode has wavelengths of μ and η in the x- and y-directions, then the grid spacings in those directions should satisfy the one-dimensional sampling criterion individually; to wit,

$$\Delta x \leq \frac{\mu}{2} \quad \text{and} \quad \Delta y \leq \frac{\eta}{2}.$$

If this argument is convincing, then accept it, since it gives the correct conditions! If you desire more persuasion or rigor, then consider the following two lines of thought.

3. **Geometry.** As shown in Figure 5.6, the mode $w(x, y)$ is aligned with the phase lines. Imagine introducing a local one-dimensional grid with coordinate ξ perpendicular to the phase lines (along the frequency vector). Relative to this axis the mode looks like a one-dimensional wave with wavelength λ. The one-dimensional sampling criterion says that a local grid spacing $\Delta\xi$ should be chosen such that $\Delta\xi \leq \lambda/2$. Now it is just a matter of relating $\Delta\xi$ to Δx and Δy. The geometry of Figure 5.6 shows that

$$\Delta\xi = \Delta x \cos\psi = \Delta y \sin\psi.$$

With these observations, the condition $\Delta\xi \leq \lambda/2$ can be expressed in terms of Δx and Δy (problem 101). These individual conditions become

$$\Delta x \leq \frac{\lambda}{2}\sec\psi = \frac{\mu}{2} \quad \text{and} \quad \Delta y \leq \frac{\lambda}{2}\csc\psi = \frac{\eta}{2},$$

and the intuitive argument given above is confirmed.

5. **Algebra.** The same conclusion can be reached algebraically, but let's do it with a slightly different form of the question: Assume you are told that a particular two-dimensional input on a rectangular domain with dimensions A and B has a maximum frequency of ν in the direction ψ. What are the minimum values of M and N that should be used to avoid aliasing of the input? As before, let ω and σ be the frequencies in the x- and y-directions and think of them as the *maximum* frequency components that need to be resolved by the grid. The maximum frequencies that appear on the frequency grid correspond to $j = M/2$ and $k = N/2$; they are given by $\omega = (M/2)\Delta\omega$ and $\sigma = (N/2)\Delta\sigma$. Therefore, given ψ and the facts that $\Delta\omega = 1/A$ and $\Delta\sigma = 1/B$, we have that

$$\tan\psi = \frac{\sigma}{\omega} = \frac{N\Delta\sigma}{M\Delta\omega} = \frac{NA}{MB}. \tag{5.8}$$

This relates ψ to A, B, M, N. What about ν? Recall that $\nu^2 = \omega^2 + \sigma^2$. Again thinking of $\omega = M/2A$ and $\sigma = N/2B$ as the maximum frequencies that can be resolved by the grid, we see that

$$\nu^2 = \omega^2 + \sigma^2 = \frac{1}{4}\left(\frac{M^2}{A^2} + \frac{N^2}{B^2}\right). \tag{5.9}$$

Now it is *just* a matter of algebra! In equations (5.8) and (5.9), A, B, ν, ψ are given, while the grid parameters M and N are to be determined. These two equations can be solved for M and N (problem 102) to find that

$$M = 2A\nu \cos \psi \quad \text{and} \quad N = 2B\nu \sin \psi$$

are the minimum values of M and N that avoid aliasing. To compare with the results of the earlier arguments, recall that $\Delta x = A/M$ and $\Delta y = B/N$. This gives the conditions that

$$\Delta x = \frac{\lambda \sec \psi}{2\nu} = \frac{\mu}{2} = \frac{1}{2\omega} \quad \text{and} \quad \Delta y = \frac{\lambda \csc \psi}{2\nu} = \frac{\eta}{2} = \frac{1}{2\sigma}$$

are the maximum values of Δx and Δy that insure no aliasing. We see once again that the one-dimensional sampling conditions applied to the *components* of the frequency or wavelength are the correct conditions in two dimensions.

Hopefully, at least one of the foregoing arguments demonstrates the following fact:

In order to avoid aliasing of a two-dimensional wave, grid spacings Δx and Δy must be chosen so that the one-dimensional waves obtained by slicing the full wave in the x- and y-directions are individually free of aliasing.

In applications, it is critical that the grid spacings be chosen carefully, in order to balance the need to avoid aliasing with the need to reduce the cost of complicated computations by reducing the size of the grids. Here is a simple graphical recipe for determining the maximum grid spacings that avoid aliasing of a single mode (see Figure 5.7).

▶ **Anti-Aliasing Recipe** ◀

1. Draw two adjacent phase lines ℓ_1 and ℓ_2 corresponding to the same phase of the mode (for example, two consecutive troughs or crests).

2. The maximum values of Δx and Δy that insure no aliasing are one-half the horizontal and vertical distances between ℓ_1 and ℓ_2, respectively.

It is an edifying exercise to verify (problem 103) that this recipe gives the same results as the above arguments.

We can now take the final steps to establish the reciprocity relations between the spatial and frequency domains. The fact that $\Delta \omega = 1/A$ and $\Delta \sigma = 1/B$ results in the one-dimensional relations

$$\Delta x \Delta \omega = \frac{A}{M}\frac{1}{A} = \frac{1}{M} \quad \text{and} \quad \Delta y \Delta \sigma = \frac{B}{N}\frac{1}{B} = \frac{1}{N}. \tag{5.10}$$

How are the extents of the spatial and frequency domains related? The maximum frequencies that can be represented in the x- and y-directions are given by

$$\pm\frac{\Omega}{2} = \omega_{\pm\frac{M}{2}} = \pm\frac{M}{2}\Delta\omega = \pm\frac{M}{2A} \quad \text{and} \quad \pm\frac{\Lambda}{2} = \sigma_{\pm\frac{N}{2}} = \pm\frac{N}{2}\Delta\sigma = \pm\frac{N}{2B}.$$

This results in the one-dimensional reciprocity relationships between the extents of the spatial and frequency domains:

$$A\Omega = M \quad \text{and} \quad B\Lambda = N. \tag{5.11}$$

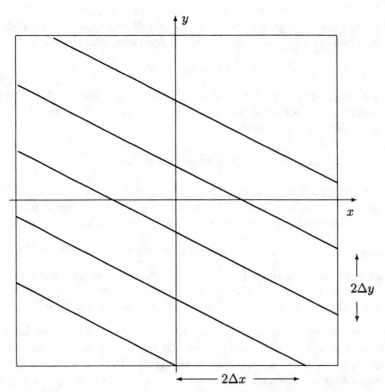

FIG. 5.7. *The figure shows consecutive phase lines for crests of a single mode (bold diagonal lines). The maximum grid spacings Δx and Δy that prevent aliasing of that mode are one-half of the horizontal and vertical distances between two consecutive phase lines.*

We see that in two dimensions the one-dimensional reciprocity relations hold independently in each direction, and they can be used in much the same way. Holding M and N fixed, an increase in either grid spacing Δx or Δy results in a corresponding increase in the extents of the spatial domain A or B, which produces a decrease in the extents Ω and Λ, which in turn leads to a decrease in the grid spacings $\Delta \omega$ and $\Delta \sigma$. Figure 5.8 is intended to be helpful in unraveling these important relationships.

These relations can also be used to form the two-dimensional reciprocity relations. Simply combining the two expressions (5.10) and (5.11), we can obtain relationships between grid *ratios*,

$$\frac{\Delta y}{\Delta x} = \frac{M}{N}\frac{\Delta \omega}{\Delta \sigma} \quad \text{and} \quad \frac{B}{A} = \frac{N}{M}\frac{\Omega}{\Lambda}, \tag{5.12}$$

or between grid *areas*,

$$\Delta x \Delta y = \frac{1}{MN\Delta \omega \Delta \sigma} \quad \text{and} \quad AB = \frac{MN}{\Omega \Lambda}. \tag{5.13}$$

The area relationships (5.13) are the true analogs of the one-dimensional reciprocity relations. Like their one-dimensional cousins, they may also be interpreted as an uncertainty principle: Holding M and N fixed, an increase in the grid resolution in the spatial domain must result in a decrease in the grid resolution in the frequency domain.

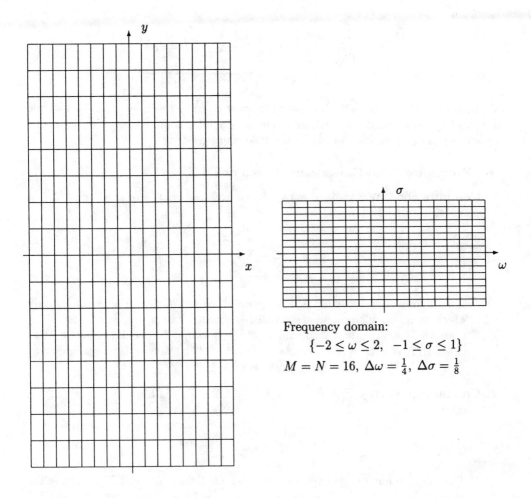

Frequency domain:
$$\{-2 \le \omega \le 2, \quad -1 \le \sigma \le 1\}$$
$$M = N = 16, \quad \Delta\omega = \tfrac{1}{4}, \quad \Delta\sigma = \tfrac{1}{8}$$

Spatial domain:
$$\{-2 \le x \le 2, \quad -4 \le y \le 4\}$$
$$M = N = 16, \quad \Delta x = \tfrac{1}{4}, \quad \Delta y = \tfrac{1}{2}$$

FIG. 5.8.   *The figure shows a spatial domain with $A = 4$ and $B = 8$ overlaid by a grid with $M = N = 16$ grid points in each direction (implying $\Delta x = 1/4$ and $\Delta y = 1/2$). The reciprocity relations determine the parameters of the $M \times N$ frequency grid. In this case, $\Omega = 4, \Lambda = 2, \Delta\omega = 1/4,$ and $\Delta\sigma = 1/8$.*

5.4. Computing Multidimensional DFTs

A few observations need to be made about the actual implementation of multidimensional DFTs. As with so much that has already transpired, we will see that it is possible to rely heavily on what we know about one-dimensional DFTs. For the purpose of computation it is useful to view the two-dimensional DFT as a two-step process. Given an $M \times N$ input array f_{mn}, we first write the forward DFT (5.1) in the form

$$F_{jk} = \frac{1}{M} \sum_{m=-\frac{M}{2}+1}^{\frac{M}{2}} \omega_M^{-mj} \underbrace{\frac{1}{N} \sum_{n=-\frac{N}{2}+1}^{\frac{N}{2}} f_{mn} \omega_N^{-nk}}_{\overline{F}_{mk}}.$$

We have used $\overline{F}_{mk}$ to label the intermediate array given by the inner sum. This simple device of splitting the double sum and defining the intermediate array $\overline{F}_{mk}$ allows us to express the computation of F_{jk} in the following two steps.

▶ **Procedure: Two-Dimensional Complex DFT** ◀

1. **Compute the intermediate array $\overline{F}_{mk}$.** For each $m = -M/2 + 1 : M/2, k = -N/2 + 1 : N/2$, compute

$$\overline{F}_{mk} = \frac{1}{N} \sum_{n=-\frac{N}{2}+1}^{\frac{N}{2}} f_{mn} \omega_N^{-nk}.$$

   This is most easily visualized if we place the input f_{mn} in a physical array in which m and j increase along rows of the array, and n and k increase along columns. Then this first step amounts to performing M one-dimensional DFTs of length N of each of the *columns* of the array. The resulting intermediate array $\overline{F}_{mk}$ can be stored over the original array.

2. **Compute the array F_{jk}.** For each $j = -M/2 + 1 : M/2, k = -N/2 + 1 : N/2$, compute

$$F_{jk} = \frac{1}{M} \sum_{m=-\frac{M}{2}+1}^{\frac{M}{2}} \overline{F}_{mk} \omega_M^{-mj}.$$

   This step amounts to doing N one-dimensional DFTs of length M of each of the *rows* of the array $\overline{F}_{mk}$. The result of this second step is just the array of DFT coefficients F_{jk}.

To summarize a fundamental theme that will be repeated often, it is possible to interpret and evaluate the two-dimensional DFT of an $M \times N$ array of data as a two-step process in which we

1. perform M DFTs of length N of the columns of the array, and then

2. perform N DFTs of length M of the rows of the array from step 1.

It should be evident that the two steps could be interchanged: the two sweeps of DFTs could be done across the rows first and columns last. The multiplication by the scaling factors $1/M$ and $1/N$ is usually deferred until the end of the computation, if it is needed at all. The power of this formulation is that it reduces the entire two-dimensional DFT to one-dimensional DFTs. All that is needed is a driver program to organize the two steps and a one-dimensional DFT subprogram that can do DFTs of fairly arbitrary length. For maximum efficiency, an FFT should be employed in the DFT subprogram. A quick operation count is revealing. If the two-dimensional DFT is done explicitly (by the definition (5.1)) then it entails the product of an $MN \times MN$

matrix and an MN-vector, or roughly M^2N^2 operations. On the other hand, if the one-dimensional DFTs are done with FFTs, the two-dimensional DFT requires N FFTs of length M and M FFTs of length N. Borrowing from Chapter 10 the fact that an N-point FFT requires on the order of $N \log N$ complex operations, we see that the two-dimensional DFT has a computational cost of roughly

$$MN(\log M + \log N) \quad \text{complex operations,}$$

which represents a significant savings over the explicit method.

The above formulation is precisely the overall strategy behind all multidimensional DFT computations. The extension to DFTs in three or more dimensions is immediate. In the three-dimensional case, three steps are needed, each requiring a set of two-dimensional DFTs along sets of parallel planes of data (problem 110). Each two-dimensional DFT can be done by the two-step procedure described above. There are many variations on this basic idea that are determined primarily by computer architecture. For single processor scalar computers, the implementation is straightforward. However, with vector computers, issues of data storage (for example, interleaving of data and stride length) determine how the arrays are stored and the order in which the two steps are performed [134]. For parallel or multiprocessing computers, additional issues of data availability and interprocessor communication arise [17], [18], [138]. Suffice it to say that when it comes to advanced computer architectures, there are still some open questions about the implementation of DFTs, and those questions will persist as long as there are new architectures.

Example: A numerical two-dimensional DFT. A brief numerical example of a two-dimensional DFT may help to solidify the ideas of this section. Consider the function

$$f(x,y) = \sin\left(3\cos\left(\frac{y}{2}\right)\sin x\right) \quad \text{on} \quad \{(x,y): -\pi \le x \le \pi, -\pi \le y \le \pi\}.$$

The function has a value of zero on the boundary of the domain and therefore has no discontinuities when extended periodically. It is sampled and the resulting array is used as input to the complex DFT implemented by the two-sweep method described above. (As we will see in the next section, the fact that f is real-valued makes the complex DFT inefficient in both computation and storage.) The resulting DFT array is shown in Figure 5.9. Notice that the DFT coefficients F_{jk} are even in the index j ($F_{-jk} = F_{jk}$), and odd in the index k ($F_{j,-k} = -F_{jk}$). These symmetries will also be explained in the next section.

5.5.   Symmetric DFTs in Two Dimensions

The preceding section dealt with the properties of the DFT applied to the most general two-dimensional input array. We have already learned that when the input possesses certain symmetries, the one-dimensional DFT can be simplified both conceptually and computationally. The same economies can be realized in two (and more) dimensions.

The simplest and most important symmetry that arises in practice is simply that of a **real** input array. So let's assume that f_{mn} is a real $M \times N$ array and then appeal to the definition of the two-dimensional complex DFT. Recall that for

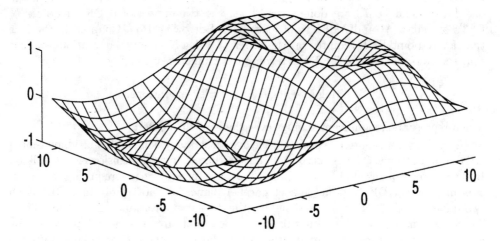

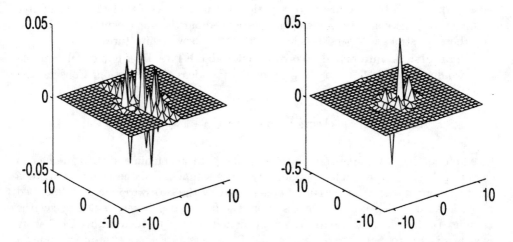

FIG. 5.9. *The function $f(x,y) = \sin(3\cos(y/2)\sin x)$ is shown (top) on the domain $[-\pi, \pi] \times [-\pi, \pi]$. For f sampled on a grid with $M = N = 24$, the two-dimensional DFT produces the arrays of transform coefficients shown below, where the real part is shown on the left and the imaginary part is on the right (with different vertical scales). Observe that as the function is primarily a sine, its DFT is dominated by the imaginary part, which has an amplitude approximately 10 times that of the real part.*

$j = -M/2 + 1 : M/2$ and $k = -N/2 + 1 : N/2$, we have

$$F_{jk} = \frac{1}{MN} \sum_{m=-\frac{M}{2}+1}^{\frac{M}{2}} \sum_{n=-\frac{N}{2}+1}^{\frac{N}{2}} f_{mn}\omega_M^{-mj}\omega_N^{-nk}.$$

The first step is to *see* what the real DFT in two dimensions actually looks like. It can be written out in an alternate form by using the real symmetry of the input array. Expanding the exponentials in the complex DFT in terms of sines and cosines, we have the following definition.

▶ **Two-Dimensional Real DFT (RDFT)** ◀

$$
\begin{aligned}
F_{jk} &= \frac{1}{MN} \sum_{m=-\frac{M}{2}+1}^{\frac{M}{2}} \sum_{n=-\frac{N}{2}+1}^{\frac{N}{2}} f_{mn} \omega_M^{-mj} \omega_N^{-nk} \\
&= \frac{1}{MN} \sum_{m=-\frac{M}{2}+1}^{\frac{M}{2}} \sum_{n=-\frac{N}{2}+1}^{\frac{N}{2}} f_{mn} \left[\cos\left(\frac{2\pi mj}{M}\right) + i \sin\left(\frac{2\pi mj}{M}\right) \right] \\
&\qquad\qquad \times \left[\cos\left(\frac{2\pi nk}{N}\right) + i \sin\left(\frac{2\pi nk}{N}\right) \right] \\
&= \frac{1}{MN} \sum_{m=-\frac{M}{2}+1}^{\frac{M}{2}} \sum_{n=-\frac{N}{2}+1}^{\frac{N}{2}} f_{mn} \cos 2\pi \left(\frac{mj}{M} + \frac{nk}{N}\right) \\
&\quad + i \frac{1}{MN} \sum_{m=-\frac{M}{2}+1}^{\frac{M}{2}} \sum_{n=-\frac{N}{2}+1}^{\frac{N}{2}} f_{mn} \sin 2\pi \left(\frac{mj}{M} + \frac{nk}{N}\right).
\end{aligned} \tag{5.14}
$$

This expression has been obtained by using the addition rules for the sine and cosine, plus the fact that the elements of f_{mn} are real.

This representation immediately exhibits the periodicity of the transform coefficients: $F_{j\pm M,k} = F_{jk}$ and $F_{j,k\pm N} = F_{jk}$. It also has the advantage that the real and imaginary parts of F_{jk} are quite evident. Notice that replacing j and k by $-j$ and $-k$, respectively, has the effect of negating the imaginary part of F_{jk} and leaving the real part unchanged. Similarly, replacing j and $-k$ by $-j$ and k, respectively, has the same effect. We can summarize this symmetry by writing

$$
\operatorname{Re}\{F_{-j,-k}\} = \operatorname{Re}\{F_{jk}\} \quad \text{and} \quad \operatorname{Im}\{F_{-j,-k}\} = -\operatorname{Im}\{F_{jk}\},
$$

or more simply $F_{-j,-k} = F_{jk}^*$.

In strict analogy with the one-dimensional case, we see that the DFT of a real sequence is **conjugate symmetric**. In practical terms, this means that only *half* of the DFT coefficients need to be computed. The remaining half is known immediately by symmetry. The particular half of the coefficients that one chooses to compute and store is somewhat arbitrary. Because of the conjugate symmetry, it is possible to work with the coefficients in any two adjacent quadrants of the frequency domain (Figure 5.10). The quadrant $j \geq 0, k \geq 0$ is familiar, and seems like a natural choice; we will also choose the quadrant $j \leq 0, k \geq 0$. To be quite specific, we must compute the F_{jk}'s for $j = -M/2 + 1 : M/2$ and $k = 0 : N/2$.

Now let's do some bookkeeping to account for all of the transform coefficients. Since there are MN *real* input data, we expect (indeed must have) exactly MN independent real quantities in the output. Referring to Figure 5.10 and using the periodicity and symmetry of F_{jk}, the following observations are critical (problem 112).

(a) $F_{00}, F_{0,\frac{N}{2}}, F_{\frac{M}{2},0}$, and $F_{\frac{M}{2},\frac{N}{2}}$ are real.

(b) $F_{j0} = F_{-j0}^*$ (which means F_{j0} must be computed only for $j = 0 : M/2$).

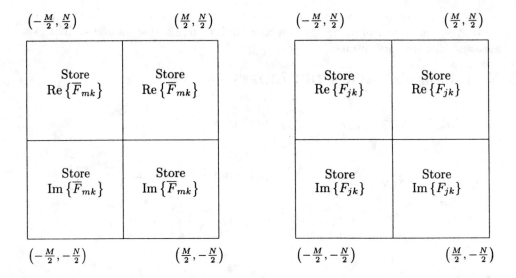

FIG. 5.10. *The DFT of a real $M \times N$ array f_{mn} can be done in two stages. The result of computing M real DFTs of length N of the columns of f_{mn} is a conjugate symmetric array $\overline{F}_{mk}$ that can be stored as shown in the left figure. The result of computing N DFTs of length M of the rows of the $\overline{F}_{mk}$ array is the conjugate symmetric DFT F_{jk}, which can be stored as shown in the right figure. The entire computation can be done in place, in the array originally occupied by the input array.*

(c) $F_{j,\frac{N}{2}} = F^{*}_{-j,\frac{N}{2}}$ (which means $F_{j,\frac{N}{2}}$ must be computed only for $j = 0$: $M/2$).

(d) The remaining F_{jk} for $j = -M/2 + 1 : M/2$ and $k = 1 : N/2 - 1$ are complex and have no symmetry (which means that their real and imaginary parts must be computed and stored).

If we now count the number of *real* quantities that must be computed and stored, the tally looks like this:

$$\underbrace{4}_{\text{(a)}} + \underbrace{(M-2)}_{\text{(b)}} + \underbrace{(M-2)}_{\text{(c)}} + \underbrace{2M\left(\frac{N}{2}\right)}_{\text{(d)}} = MN,$$

where we have indicated the source of the four terms in the list above. The match is perfect: there are MN real quantities on input and output. The practical consequence of this exercise is that since the input to the DFT consists of MN real quantities, only half of the DFT coefficients, a total of MN real quantities, needs to be computed and stored. Thus there is a factor-of-two savings in computation and storage over the complex DFT in the presence of the real symmetry in the input.

The actual computation of the real DFT can be formulated much as it was in the complex case. We start with the DFT definition (5.1)

$$F_{jk} = \frac{1}{MN} \sum_{m=-\frac{M}{2}+1}^{\frac{M}{2}} \sum_{n=-\frac{N}{2}+1}^{\frac{N}{2}} f_{mn} \omega_{M}^{-mj} \omega_{N}^{-nk},$$

for $j = -M/2 + 1 : M/2$ and $k = -N/2 + 1 : N/2$. Once again the evaluation of this double sum may be split into two stages:

$$F_{jk} = \frac{1}{M} \sum_{m=-\frac{M}{2}+1}^{\frac{M}{2}} \overline{F}_{mk} \omega_M^{-mj},$$

where

$$\overline{F}_{mk} = \frac{1}{N} \sum_{n=-\frac{N}{2}+1}^{\frac{N}{2}} f_{mn} \omega_N^{-nk}.$$

Now the symmetries of the one-dimensional real DFT must be exploited to obtain an efficient two-step method Notice that the computation of the intermediate array $\overline{F}_{mk}$ requires M *real* DFTs of length N of the columns of the array f_{mn}. Therefore, the array $\overline{F}_{mk}$ has conjugate symmetry in its second argument: $\overline{F}_{mk} = \overline{F}_{m,-k}^{*}$. This means that the complex coefficients $\overline{F}_{mk}$ need to be computed only for $k = 0 : N/2$.

We would like the second stage of the procedure, the computation of F_{jk}, to involve only *real DFTs* as well. This can be done if we split the array $\overline{F}_{mk}$ into its real and imaginary parts by writing $\overline{F}_{mk} = \text{Re}\{\overline{F}_{mk}\} + i\text{Im}\{\overline{F}_{mk}\}$. We then have that the DFT coefficients F_{jk} are given by

$$F_{jk} = \frac{1}{M} \left[\sum_{m=-\frac{M}{2}+1}^{\frac{M}{2}} \text{Re}\{\overline{F}_{mk}\} \omega_M^{-mj} + i \sum_{m=-\frac{M}{2}+1}^{\frac{M}{2}} \text{Im}\{\overline{F}_{mk}\} \omega_M^{-mj} \right].$$

In this manner the second stage of the computation consists of *real* DFTs of length M of the rows of the two arrays $\text{Re}\{\overline{F}_{mk}\}$ and $\text{Im}\{\overline{F}_{mk}\}$. But notice that these DFTs must be done only for the indices $j = -M/2 + 1 : M/2$ and $k = 0 : N/2$.

The data storage also works out exactly as one might hope provided the following observation is made: because of the conjugate symmetry of $\overline{F}_{mk}$, $\text{Im}\{\overline{F}_{m0}\} = \text{Im}\{\overline{F}_{m,\pm N/2}\} = 0$, and therefore these terms do not need to be stored. The intermediate array $\overline{F}_{mk}$ can be written over the input array by placing its real parts $\text{Re}\{\overline{F}_{mk}\}$ in the locations with $k = 0 : N/2$ and its imaginary parts in the locations with $k = -N/2 + 1 : -1$ (see Figure 5.10). The DFT coefficients F_{jk} can also be stored with their real and imaginary parts in the same locations as the real and imaginary parts of $\overline{F}_{mk}$. Therefore, the entire computation can be done in the $M \times N$ real array in which the input data arrive.

We can now summarize this two-step method for real DFTs given an $M \times N$ input array F_{mn}.

▶ **Procedure: Two-Dimensional Real DFT** ◀

1. For $m = -M/2 + 1 : M/2$, compute the M real DFTs of length N of the columns of the input array to form the arrays $\text{Re}\{\overline{F}_{mk}\}$ and $\text{Im}\{\overline{F}_{mk}\}$. For each m, these arrays are stored in rows $k = 0 : N/2$ and $k = -N/2 + 1 : -1$, respectively.

2. For $k = -N/2 + 1 : N/2$ compute the N real DFTs of length M of the rows of the arrays $\text{Re}\{\overline{F}_{mk}\}$ and $\text{Im}\{\overline{F}_{mk}\}$. The real parts of F_{jk} will appear in the rows with $k = 0 : N/2$. The imaginary parts of F_{jk} will appear in the rows with $k = -N/2 + 1 : -1$.

We see that, as in the complex case, the two-dimensional real DFT can be reduced to a sequence of one-dimensional DFTs; in this case, the one-dimensional DFTs are real DFTs. The entire calculation amounts to M real DFTs of length N and N real DFTs of length M; in practice, these DFTs should be done with efficient symmetric forms of the FFT. The two steps can be interchanged, and one order may be preferable in light of computational constraints. The multiplication by the scaling factors $1/M$ and $1/N$ is generally deferred until the end of the calculation, if it is needed at all

What about the *inverse* of the forward two-dimensional real DFT? The inverse deserves attention and illustrates some fundamental tricks that can be used in other symmetric DFTs. Imagine that we are now given an array of DFT coefficients F_{jk} with the conjugate symmetry $F_{jk} = F^*_{-j,-k}$. The goal is to reconstruct the real array f_{mn} given by

$$ f_{mn} = \sum_{j=-\frac{M}{2}+1}^{\frac{M}{2}} \sum_{k=-\frac{N}{2}+1}^{\frac{N}{2}} F_{jk} \omega_M^{mj} \omega_N^{nk} \tag{5.15} $$

for $n = -N/2 + 1 : N/2$ and $m = -M/2 + 1 : M/2$. We now look for ways to exploit the symmetry of F_{jk}, and we begin by separating the double sum of the inverse DFT as follows:

$$ f_{mn} = \sum_{j=-\frac{M}{2}+1}^{\frac{M}{2}} \omega_M^{mj} \left[F_{j0} + \sum_{k=1}^{\frac{N}{2}-1} \left(F_{jk} \omega_N^{nk} + F_{j,-k} \omega_N^{-nk} \right) + F_{j,\frac{N}{2}} \cos(\pi n) \right]. $$

Matters will quickly become overly complicated without a notational device to prevent further proliferation of ink. We will continue to use the convention that Σ'' indicates a sum in which the first and last terms are weighted by one-half and all other terms are weighted by one. With this notation we can rewrite this last line as

$$ f_{mn} = \sum_{j=-\frac{M}{2}+1}^{\frac{M}{2}} \omega_M^{mj} \sum_{k=0}^{\frac{N}{2}} {}'' \left(F_{jk} \omega_N^{nk} + F_{j,-k} \omega_N^{-nk} \right). $$

We may now split the outer sum in the same way $(j = 0, j = 1 : M/2-1, j = M/2)$. In a rather schematic form the result can be written as

$$ \begin{aligned} f_{mn} = & \left[\sum_{k=0}^{\frac{N}{2}} {}'' F_{0k} \omega_N^{nk} + F_{0,-k} \omega_N^{-nk} \right]_{j=0 \text{ terms}} \\ & + \sum_{j=1}^{\frac{M}{2}-1} \left[\omega_M^{mj} \sum_{k=0}^{\frac{N}{2}} {}'' \left(F_{jk} \omega_N^{nk} + F_{j,-k} \omega_N^{-nk} \right) \right] \\ & + \sum_{j=1}^{\frac{M}{2}-1} \left[\omega_M^{-mj} \sum_{k=0}^{\frac{N}{2}} {}'' \left(F_{-jk} \omega_N^{nk} + F_{-j,-k} \omega_N^{-nk} \right) \right] \\ & + \cos(\pi m) \left[\sum_{k=0}^{\frac{N}{2}} {}'' F_{\frac{M}{2},k} \omega_N^{nk} + F_{\frac{M}{2},-k} \omega_N^{-nk} \right]_{j=\frac{M}{2} \text{ terms}} \end{aligned} $$

These antics appear to be getting out of hand. Fortunately, it is now possible to use the Σ'' notation for the sum on j as well. With the aid of this notation (and problem 113), the previous expression for f_{mn} can miraculously be reduced to

$$f_{mn} = \sum_{j=0}^{\frac{M}{2}}{}'' \sum_{k=0}^{\frac{N}{2}}{}'' \underbrace{F_{jk}\omega_M^{mj}\omega_N^{nk} + F_{-j,-k}\omega_M^{-mj}\omega_N^{-nk}}_{\text{sum of conjugates}}$$

$$+ \sum_{j=0}^{\frac{M}{2}}{}'' \sum_{k=0}^{\frac{N}{2}}{}'' \underbrace{F_{-j,k}\omega_M^{-mj}\omega_N^{nk} + F_{j,-k}\omega_M^{mj}\omega_N^{-nk}}_{\text{sum of conjugates}}.$$

Notice how the four terms of this sum can be grouped in pairs to take advantage of the conjugate symmetry. The final step is to express f_{mn} in terms of purely real quantities. Recall that complex numbers z satisfy $z + z^* = 2\text{Re}\{z\}$. With this fact, we claim that the inverse real DFT in two dimensions is given by the following definition.

▶ **Two-Dimensional Inverse Real DFT (IRDFT)** ◀

$$f_{mn} = 2\text{Re}\left\{ \sum_{j=0}^{\frac{M}{2}}{}'' \sum_{k=0}^{\frac{N}{2}}{}'' F_{jk}\omega_M^{mj}\omega_N^{nk} + F_{-j,k}\omega_M^{-mj}\omega_N^{nk} \right\}$$

$$= \sum_{j=0}^{\frac{M}{2}}{}'' \sum_{k=0}^{\frac{N}{2}}{}'' \left[\text{Re}\{F_{jk}\} \cos\left(\frac{2\pi mj}{M} + \frac{2\pi nk}{N} \right) \right.$$

$$+ \text{Re}\{F_{-j,k}\} \cos\left(\frac{2\pi mj}{M} - \frac{2\pi nk}{N} \right)$$

$$- \text{Im}\{F_{jk}\} \sin\left(\frac{2\pi mj}{M} + \frac{2\pi nk}{N} \right)$$

$$\left. + \text{Im}\{F_{-j,k}\} \sin\left(\frac{2\pi mj}{M} - \frac{2\pi nk}{N} \right) \right]. \qquad (5.16)$$

We have called this expression the inverse *real* DFT, despite the fact that its input consists of conjugate symmetric (hence complex) data; it *does* produce real output. This expression appears rather cumbersome, but it has a few useful purposes. It demonstrates that the array f_{mn} consists of real numbers (as anticipated) that are M-periodic in the first index and N-periodic in the second index. A careful counting also confirms that the F_{jk}'s actually used in the representation (5.16) consist of exactly MN real quantities. This corroborates our previous tally, in which it was shown that there are MN real data points for the input and output. This means that the inverse real DFT can also be done with one-half of the computation and storage of the complex DFT It should be mentioned that the form given by (5.16) is never used for actual computation. Just as we did for the forward real DFT, the inverse real DFT can be reduced to a two-step procedure that requires only one-dimensional inverse real DFTs (implemented as FFTs) and MN real storage locations (problem 106).

All of the foregoing ideas can be used to further advantage when we deal with other symmetries. For example, consider an input sequence f_{mn} that is real and has the **even symmetry**:

$$f_{m,-n} = f_{mn} \quad \text{and} \quad f_{-m,n} = f_{mn}.$$

Observe that if this symmetry is known to exist in the data, it is necessary to store only one-quarter of the actual array. In order to mimic the development that we used in the one-dimensional case *and* to anticipate the most useful form of the final result, we shall assume that the input array is indexed with $m = -M + 1 : M$ and $n = -N + 1 : N$; in other words, there are $2M \times 2N$ real data points in the input array. We will now proceed as we did in the real case: first working out the symmetries that appear in the transform coefficients because of the even symmetry of the input data, then describing the expected two-step computational process. What does the two-dimensional DFT of a real even array actually look like when it is written out? We can start with the complex DFT (5.1) in the form

$$F_{jk} = \frac{1}{4MN} \sum_{m=-M+1}^{M} \omega_{2M}^{-mj} \sum_{n=-N+1}^{N} f_{mn} \omega_{2N}^{-nk},$$

and then use the symmetry of f_{mn} in each component separately. With the symmetry that $f_{m,-n} = f_{mn}$, we can write

$$F_{jk} = \frac{1}{4MN} \sum_{m=-M+1}^{M} \omega_{2M}^{-mj} \underbrace{\left[f_{m0} + 2 \sum_{n=1}^{N-1} f_{mn} \cos\left(\frac{\pi nk}{N}\right) + f_{mN} \cos(\pi k) \right]}_{2 \sum_{n=0}^{N} {}'' f_{mn} \cos\left(\frac{\pi nk}{N}\right)},$$

where we have again appealed to the Σ'' notation to denote a sum whose first and last terms are weighted by one-half. The fact that $\omega_{2N}^{nk} + \omega_{2N}^{-nk} = 2\cos(\pi nk/N)$ has also been used. The symmetry $f_{-m,n} = f_{mn}$ may now be used on the outer sum to obtain

$$F_{jk} = \frac{1}{MN} \sum_{m=0}^{M} {}'' \sum_{n=0}^{N} {}'' f_{mn} \cos\left(\frac{\pi mj}{M}\right) \cos\left(\frac{\pi nk}{N}\right). \qquad (5.17)$$

We can now unravel the symmetries in the transform F_{jk} that are induced by the symmetries of the input f_{mn}. First it is clear that all of the coefficients F_{jk} are real since the expression (5.17) involves only real quantities. Furthermore, because the cosine is even in its argument, the F_{jk}'s are even in the indices j and k; that is, $F_{-jk} = F_{jk}$ and $F_{j,-k} = F_{jk}$. Thus, from the coefficients with $j = 0 : M$ and $k = 0 : N$, it is possible to construct the entire array F_{jk}. Notice that the bookkeeping works out as it should: with the even symmetry of the arrays f_{mn} and F_{jk}, there are exactly $(M+1)(N+1)$ distinct real quantities in both arrays.

Let's stand back and see what we have done. First we noted that because of its symmetry, only one-quarter of the input array corresponding to $m = 0 : M$ and $n = 0 : N$ needs to be stored. Indeed, we see from expression (5.17) that only these input elements are required in the computation of the DFT. Furthermore, we see that because of the symmetry in the coefficients F_{jk}, only one-quarter of the transform array needs to be computed, corresponding to $j = 0 : M$ and $k = 0 : N$. We might ask why we even deal with the unused three-quarters of the data. This question prompts us to proceed just as we did in the one-dimensional case: we create a new transform that uses only the independent data points and produces only the coefficients that are actually needed. We have just derived the two-dimensional discrete cosine transform, the formula for which follows.

▶ **Two-Dimensional Discrete Cosine Transform (DCT)** ◀

$$F_{jk} = \frac{1}{MN} \sum_{m=0}^{M} {}'' \sum_{n=0}^{N} {}'' f_{mn} \cos\left(\frac{\pi mj}{M}\right) \cos\left(\frac{\pi nk}{N}\right), \qquad (5.18)$$

which is computed for $j = 0 : M$ and $k = 0 : N$. Notice that there is no longer any assumption about the symmetry of the input array f_{mn}, apart from the fact that its elements are real. And there is no particular symmetry in the transform coefficients, except that they are real also.

It is now possible (problem 108) to determine the inverse of the DCT given in (5.18). As in the one-dimensional case, the two-dimensional DCT is its own inverse up to multiplicative constants.

▶ **Two-Dimensional Inverse Discrete Cosine Transform (IDCT)** ◀

$$f_{mn} = \sum_{j=0}^{M} {}'' \sum_{k=0}^{N} {}'' F_{jk} \cos\left(\frac{\pi m j}{M}\right) \cos\left(\frac{\pi n k}{N}\right), \tag{5.19}$$

for $m = 0 : M$ and $n = 0 : N$.

A fact of importance in many applications of the DCT is that if either of the arrays f_{mn} or F_{jk} given by (5.18) or (5.19) is extended beyond the indices $m, j = 0 : M$ or $n, k = 0 : N$, then the resulting arrays are periodic with periods $2M$ and $2N$ in the first and second indices, respectively. Furthermore, both arrays are even in both directions: $f_{-mn} = f_{mn}, f_{m,-n} = f_{mn}$ and $F_{-jk} = F_{jk}, F_{j,-k} = F_{jk}$.

We can now look at the computational question and, not surprisingly, develop a two-step method for evaluating the two-dimensional DCT. Separating the two sums of the forward transform (5.18), we have that

$$F_{jk} = \frac{1}{M} \sum_{m=0}^{M} {}'' \cos\left(\frac{\pi m j}{M}\right) \underbrace{\frac{1}{N} \sum_{n=0}^{N} {}'' f_{mn} \cos\left(\frac{\pi n k}{N}\right)}_{\overline{F}_{mk}}$$

for $j = 0 : M$ and $k = 0 : N$. Once again we have defined the intermediate array $\overline{F}_{mk}$, which, for each fixed $m = 0 : M$, is simply the $(N+1)$-point DCT of the mth column of the input array f_{mn}. Thus the computation of the intermediate array $\overline{F}_{mk}$ requires $M + 1$ DCTs of length $N + 1$. In a manner that should now be familiar, this intermediate array is used for another wave of DCTs in the opposite direction. For fixed j and k, we compute

$$F_{jk} = \frac{1}{M} \sum_{m=0}^{N} {}'' \overline{F}_{mk} \cos\left(\frac{\pi m j}{M}\right),$$

which holds the one-dimensional $(M+1)$-point DCT of the kth row of the intermediate array $\overline{F}_{mk}$. The cost of doing these transforms is $N + 1$ DCTs of length $M + 1$.

Thus the two-dimensional DCT of a real input array f_{mn} can be formulated in a two-step process. Given the array f_{mn} for $m = 0 : M$ and $n = 0 : N$, we proceed directly.

▶ **Procedure: Two-Dimensional DCT** ◀

1. perform $M + 1$ one-dimensional DCTs of length $N + 1$ along the columns of f_{mn} and store (or overwrite) the results in an intermediate array $\overline{F}_{mk}$, and then

2. perform $N + 1$ one-dimensional DCTs of length $M + 1$ along the rows of the array $\overline{F}_{mk}$ and store (or overwrite) the DCT coefficients F_{jk}.

We remark as before that all of the one-dimensional DCTs should be done with symmetric FFTs, that the two steps may be done in either order, and that the normalization by the factors $1/M$ and $1/N$ can be postponed until the end of the calculation (if it is needed at all).

There is one other major symmetry that must be mentioned. We will do little more than record it and leave the derivation as a stimulating problem (problem 109). The case in which the input sequence f_{mn} is real and has the **odd symmetry**

$$f_{-m,n} = -f_{mn} \quad \text{and} \quad f_{m,-n} = -f_{mn}$$

arises frequently. It leads to the two-dimensional discrete sine transform (DST). Given an *arbitrary* real $M \times N$ input array f_{mn}, this transform pair has the following form.

▶ **Two-Dimensional Discrete Sine Transform (DST)** ◀

$$F_{jk} = -\frac{1}{MN} \sum_{m=1}^{M-1} \sum_{n=1}^{N-1} f_{mn} \sin\left(\frac{\pi m j}{M}\right) \sin\left(\frac{\pi n k}{N}\right), \tag{5.20}$$

for $j = 1 : M - 1, k = 1 : N - 1$.

▶ **Two-Dimensional Inverse Discrete Sine Transform (IDST)** ◀

$$f_{mn} = \sum_{j=1}^{M-1} \sum_{k=1}^{N-1} F_{jk} \sin\left(\frac{\pi m j}{M}\right) \sin\left(\frac{\pi n k}{N}\right), \tag{5.21}$$

for $m = 1 : M - 1, n = 1 : N - 1$. There are several important properties of this transform pair which are also stated in problem 109.

The picture may seem to be complete, and indeed it nearly is. However, we should not close without saying that the above symmetries may be combined to form hybrid transforms. For example, if an array must be even in one direction and odd in the other direction, then one could combine a DCT in one direction with a DST in the other (problem 114). There are also situations, particularly in the solution of differential equations, in which DFTs with other symmetries (for example, quarter-wave or staggered grid symmetries) must be used [134], [142], [144]. In all of these variations, in any number of dimensions, the theme is always the same: multidimensional DFTs can always be reduced to a sequence of one-dimensional DFTs, and symmetries always lead to savings in storage and computation.

5.6.   Problems

In all of the following problems assume that

$$\mathcal{I} = \{(x, y) : -A/2 \leq x \leq A/2, -B/2 \leq y \leq B/2\},$$

$j, m = -M/2 + 1 : M/2$, and $k, n = -N/2 + 1 : N/2$, unless otherwise stated.

89. Two-dimensional orthogonality and inverse property.   Use the two-dimensional orthogonality of the complex exponential,

$$\sum_{m=-\frac{M}{2}+1}^{\frac{M}{2}} \sum_{n=-\frac{N}{2}+1}^{\frac{N}{2}} \omega_M^{m(j-j_0)} \omega_M^{n(k-k_0)} = MN \hat{\delta}_M(j - j_0) \hat{\delta}_N(k - k_0),$$

to show that the forward DFT given by

$$F_{jk} = \frac{1}{MN} \sum_{m=-\frac{M}{2}+1}^{\frac{M}{2}} \sum_{n=-\frac{N}{2}+1}^{\frac{N}{2}} f_{mn} \omega_M^{-mj} \omega_N^{-nk}$$

has the inverse

$$f_{mn} = \sum_{j=-\frac{M}{2}+1}^{\frac{M}{2}} \sum_{k=-\frac{N}{2}+1}^{\frac{N}{2}} F_{jk} \omega_M^{mj} \omega_N^{nk}.$$

(Hint: Substitute the expression for f_{mn} into the expression for F_{jk} (or vice versa), choose your indices carefully, and show that an identity results.)

90. DFTs with separable input. Show that if the input to a DFT has the form $f_{mn} = g_m h_n$, then the two-dimensional DFT is given by $F_{jk} = G_j H_k$, where G_j and H_k are the one-dimensional DFTs of g_m and h_n, respectively.

91. DFTs of complex exponentials. Let j_0 and k_0 be a pair of fixed frequency indices. Show that

$$\mathcal{D}\left\{ e^{i2\pi\left(\frac{mj_0}{M} + \frac{nk_0}{N}\right)} \right\}_{jk} = \hat{\delta}_M(j - j_0)\hat{\delta}_N(k - k_0).$$

92. Two-dimensional DFT properties. Let $\mathcal{D}, \mathcal{C}$, and $\mathcal{S}$ denote the discrete two-dimensional DFT, DCT, and DST, respectively, of dimension $M \times N$. Verify the following properties, where f_{mn} and g_{mn} are input arrays and F_{jk} and G_{jk} are the corresponding arrays of transform coefficients.

(a) Periodicity of $\mathcal{D}$: $\mathcal{D}\{f_{mn}\}_{j \pm M, k \pm N} = \mathcal{D}\{f_{mn}\}_{jk}$.

(b) Periodicity of $\mathcal{S}$ or $\mathcal{C}$: $\mathcal{S}\{f_{mn}\}_{j \pm 2M, k \pm 2N} = \mathcal{S}\{f_{mn}\}_{jk}$.

(c) Linearity of $\mathcal{D}$, $\mathcal{S}$, and $\mathcal{C}$: $\mathcal{D}\{\alpha f_{mn} + \beta g_{mn}\}_{jk} = \alpha F_{jk} + \beta G_{jk}$, where α and β are constants.

(d) Shift: $\mathcal{D}\{f_{m+m_0, n+n_0}\}_{jk} = \omega_M^{-m_0 j} \omega_N^{-n_0 k} F_{jk}$.

(e) Rotation: $\mathcal{D}\{\omega_M^{-mj_0} \omega_N^{-nk_0} f_{mn}\}_{jk} = F_{j-j_0, k-k_0}$.

(f) Shift:

$$\mathcal{C}\{f_{m+p,n} + f_{m-p,n} + f_{m,n+q} + f_{m,n-q}\}_{jk} = 2\left(\cos\left(\frac{\pi p j}{M}\right) + \cos\left(\frac{\pi q k}{N}\right)\right) F_{jk},$$

with the same property holding for the DST.

(g) Convolution: $\mathcal{D}\{(f * g)_{mn}\}_{jk} = F_{jk} G_{jk}$, where

$$(f * g)_{mn} = \sum_{\ell=-\frac{M}{2}+1}^{\frac{M}{2}} \sum_{p=-\frac{N}{2}+1}^{\frac{N}{2}} f_{\ell p} g_{m-\ell, n-p}.$$

93. Analytical DFTs. Assume that the following functions f are defined on the set $\mathcal{I} = \{(x, y) : -\pi \le x \le \pi, -\pi \le y \le \pi\}$. If f is sampled with M and N points in the x- and y-directions, respectively, find the two-dimensional DFT of each function.

(a) $f(x, y) = \cos 3x$.

(b) $f(x, y) = \sin(2x + 4y)$.

(c) $f(x, y) = 3\cos(x + y) - 2\sin(x - y)$.

(d) $f(x, y) = \cos(4x - 3y)$.

(e) $f(x, y) = \cos^2(4x - 3y)$.

(f) $f(x, y) = (\cos x)(\sin 5y)$.

(g) $f(x, y) = (\sin 2x)(\sin 3y)$.

94. Two-dimensional DFT matrix.   Assume that the input f_{mn} to a two-dimensional DFT is ordered by rows (m varying most quickly) to form a vector $\mathbf{f}$ of length MN. Assume the transform array F_{jk} is ordered in the same way to form a vector $\mathbf{F}$ of length MN. Let $\mathbf{W}$ be the two-dimensional DFT matrix such that $\mathbf{F} = \mathbf{Wf}$. Write out several representative rows and columns of $\mathbf{W}$ for a grid with $M = N = 4$. What is the size of the matrix? Note the structure and symmetries of this matrix.

95. Wavelengths of continuous modes.   Show that the continuous DFT mode on $\mathcal{I}$ given by

$$w(x, y) = \cos\left(\frac{2\pi jx}{A} + \frac{2\pi ky}{B}\right)$$

has a wavelength *measured in the units of x and y* of

$$\lambda_{jk} = \frac{1}{\sqrt{\mu_j^{-2} + \eta_k^{-2}}} \quad \text{where} \quad \mu_j = \frac{A}{j} \quad \text{and} \quad \eta_k = \frac{B}{k}.$$

Furthermore, show that the phase lines of this mode have an angle of

$$\theta_{jk} = -\tan^{-1}\left(\frac{Bj}{Ak}\right)$$

with respect to the x-axis.

96. Wavelengths of discrete modes.   Verify that the discrete DFT mode

$$w_{mn} = \cos\left(\frac{2\pi mj}{M} + \frac{2\pi nk}{N}\right)$$

has a wavelength of

$$\lambda_{jk} = \frac{MN}{\sqrt{j^2 N^2 + k^2 M^2}} = \frac{1}{\sqrt{(\frac{j}{M})^2 + (\frac{k}{N})^2}}.$$

How do you interpret the units of the wavelength? Show also that the frequency of this mode is given by

$$\nu_{jk} = \sqrt{\frac{j^2}{M^2} + \frac{k^2}{N^2}},$$

and that the phase lines have an angle of

$$\theta_{jk} = -\tan^{-1}\left(\frac{Nj}{Mk}\right)$$

with respect to the direction indexed by m.

97. Phase lines and frequency vectors.   As described in the text, let θ_{jk} be the angle that the phase lines of the (j,k) mode make with the x-axis. Let ψ_{jk} be the angle between the frequency vector (ω, σ) and the w-axis. Show that θ_{jk} and ψ_{jk} differ by $\pi/2$ radians.

98. Geometry of two-dimensional modes.   Consider the following modes on the domain $\mathcal{I}$ with $A = 2\pi$ and $B = 4\pi$.

(a) $w(x,y) = \cos(2x + 3y)$,

(b) $w(x,y) = \sin(x + 2y)$,

(c) $w(x,y) = \cos(8x - y)$.

For each mode,

  i. Find the wavelength λ and the wavelengths μ and η in the x- and y-directions. Verify that $\lambda^2 = (\mu^{-2} + \eta^{-2})^{-1}$.
  ii. Find the frequency ν and its components ω and σ.
  iii. Draw the domain $\mathcal{I}$ (to scale) and the phase lines corresponding to the crests (maximum values) of the mode.
  iv. Find and indicate the angle θ that the phase lines make with the x-axis.
  v. On this same plot, indicate the direction of the frequency vector and the angle ψ that it makes with the x-axis.
  vi. Verify that θ and ψ differ by $\pi/2$ radians.

99. More geometry.   Assume that a particular DFT mode has a phase line corresponding to a crest that passes through the point $(-A/2, -B/2)$. Show that there must also be phase lines for crests passing through $(A/2, -B/2), (-A/2, B/2)$, and $(A/2, B/2)$.

100. High frequency modes.   Assume that an $M \times N$ grid is placed on a spatial domain. Draw a 3×3 block of points of this grid. Mark the points at which the mode

$$w_{mn} = \cos\left(\frac{2\pi m j}{M} + \frac{2\pi n k}{N}\right)$$

has the values $0, \pm 1$ for the modes with

(a) $j = 0, k = \dfrac{N}{2}$,   (b) $j = \dfrac{M}{2}, k = 0$,   (c) $j = \dfrac{M}{2}, k = \dfrac{N}{2}$.

Find the wavelengths and frequencies of these modes.

101. Geometry of the anti-aliasing condition.   Refer to the spatial grid shown in Figure 5.6 in which $\Delta\xi$ is the local grid spacing in the direction perpendicular to the phase lines, and Δx and Δy are the grid spacings in the x- and y-directions.

(a) Show that $\Delta\xi = \Delta x \cos\psi$ and $\Delta\xi = \Delta y \sin\psi$.

(b) Show that the local anti-aliasing condition $\Delta\xi \le \lambda/2$ implies that

$$\Delta x \le \frac{\lambda}{2}\sec\psi = \frac{\mu}{2} \quad \text{and} \quad \Delta y \le \frac{\lambda}{2}\csc\psi = \frac{\eta}{2}.$$

102. Algebra of the anti-aliasing condition. Assume that a single continuous mode is given with a frequency that has magnitude ν and an angle ψ with respect to the ω-axis. Let M and N be that minimum number of grid points in the x- and y-directions needed to resolve this mode (and avoid aliasing). Show algebraically that the conditions

$$\tan\psi = \frac{\sigma}{\omega} = \frac{N\Delta\sigma}{M\Delta\omega} = \frac{NA}{MB} \quad\text{and}\quad \nu^2 = \omega^2 + \sigma^2 = \frac{1}{4}\left(\frac{M^2}{A^2} + \frac{N^2}{B^2}\right)$$

(equations (5.8) and (5.9) of the text) imply that $M = 2A\nu\cos\psi$, $N = 2B\nu\sin\psi$, and that

$$\Delta x = \frac{1}{2\omega} = \frac{\mu}{2} \quad\text{and}\quad \Delta y = \frac{1}{2\sigma} = \frac{\eta}{2}$$

are maximum grid spacings that will resolve the mode.

103. Anti-aliasing recipe. Show that the simple graphical procedure given in the anti-aliasing recipe actually gives the maximum grid spacings Δx and Δy that insure no aliasing of a single mode.

104. Aliasing. Consider the modes

$$\text{(a) } w(x,y) = \cos\pi(8x + 12y), \qquad \text{(b) } w(x,y) = \sin\pi(16x - 20y)$$

on the domain $\mathcal{I}$ with $A = 2$ and $B = 4$. What are the minimum number of grid points M and N and the maximum grid spacings Δx and Δy required to avoid aliasing of these modes?

105. Aliasing. You are told that on a domain $\mathcal{I}$ with $A = \pi$ and $B = 4\pi$ the highest frequency that appears in the input is $\nu = 500/\pi$ (cycles per unit length). How would you design a grid (with a minimal number of grid points) to avoid aliasing of this signal? If you are also told that this maximum frequency occurs in a direction $\psi = \tan^{-1}(4/3)$, what are the minimum values of M and N that insure no aliasing?

106. Inverse real DFT algorithm. Describe a two-step algorithm analogous to that used for the forward real DFT that takes an $M \times N$ conjugate symmetric input array and produces a real output array. Use the symmetries to minimize the storage and computation requirements. Indicate clearly how the MN storage locations should be allocated.

107. Alternate index sets. Find the complex and real two-dimensional DFT pairs for the index sets $m, j = 0 : M - 1$ and $n, k = 0 : N - 1$.

108. Inverse DCT. Verify that the two-dimensional DCTs given by (5.18) and (5.19) form an inverse pair, and that the DCT is its own inverse up to a multiplicative constant.

109. Two-dimensional DST. Following the development of the two-dimensional DCT in the text, carry out the derivation and analysis of the two-dimensional DST.

    (a) Assume that the input array has dimensions $2M \times 2N$. Show that the odd symmetry $f_{m,-n} = -f_{mn}$ and $f_{-m,n} = -f_{mn}$ together with the periodicity $f_{m\pm 2M,n} = f_{mn}$ and $f_{m,n\pm 2N} = f_{mn}$ imply that $f_{mn} = 0$ whenever $m = \pm M$, $m = 0$, $n = \pm N$, or $n = 0$.

(b) Starting with a real $2M \times 2N$ input array f_{mn} with odd symmetry in both indices, use the complex DFT (5.1) to derive a transform that uses only sine modes and the input values for $m = 1 : M - 1$ and $n = 1 : N - 1$.

(c) Notice that $F_{-j,k} = -F_{jk}$ and $F_{j,-k} = -F_{jk}$, and conclude that only one-quarter of the coefficients F_{jk} need to be computed.

(d) Discarding unused coefficients, write the forward two-dimensional DST for an arbitrary real $M \times N$ input sequence.

(e) Verify that up to a multiplicative constant the DST is its own inverse.

(f) Show that the array f_{mn} given by the inverse DST (5.21) when extended is $2M$-periodic in the m-direction, $2N$-periodic in the n-direction. Furthermore, the array is odd in both directions ($f_{-mn} = -f_{mn}$ and $f_{m,-n} = -f_{mn}$).

110. Three-dimensional DFTs. Given an $M \times N \times P$ input array f_{mnp}, its three-dimensional complex DFT is defined as

$$F_{jk\ell} = \frac{1}{MNP} \sum_{m=-\frac{M}{2}+1}^{\frac{M}{2}} \sum_{n=-\frac{N}{2}+1}^{\frac{N}{2}} \sum_{p=-\frac{P}{2}+1}^{\frac{P}{2}} f_{mnp} \omega_M^{-mj} \omega_N^{-nk} \omega_P^{-p\ell}.$$

(a) Describe how you would design an efficient *three*-step method for evaluating this DFT.

(b) Carefully itemize the cost of the computation in terms of the number of one-dimensional DFTs of length M, N, and P.

(c) How much does this computation cost in terms of complex arithmetic operations if the one-dimensional DFTs are done explicitly by matrix-vector multiplication?

(d) How much does the computation cost if the one-dimensional DFTs are done with FFTs? (Assume that an N-point FFT requires on the order of $N \log N$ operations.)

111. Three-dimensional aliasing. Consider the single mode

$$w(x, y, z) = \sin \pi (16x - 20y + 24z)$$

on a parallelepiped domain with physical dimensions $8 \times 4 \times 6$. What are the minimum numbers of grid points M, N, and P that should be used in the x-, y-, and z-directions, respectively, to insure no aliasing of this mode?

112. Properties of the real DFT. Consider the real periodic transform pair given by (5.14) and (5.16).

(a) Show that the transform coefficients F_{jk} have the following symmetries.

  i. $F_{00}, F_{0,\frac{N}{2}}, F_{\frac{M}{2},0}$, and $F_{\frac{M}{2},\frac{N}{2}}$ are real,

  ii. $F_{j0} = F^*_{-j0}, F_{j,\frac{N}{2}} = F^*_{-j,\frac{N}{2}}$.

(b) Verify that there are exactly MN distinct real quantities F_{jk} defined by the forward transform (5.14) that correspond to the MN real quantities of the input array f_{mn}.

113. Σ'' notation for the inverse real DFT. Show that the Σ'' notation allows the expression (5.15) for the inverse real DFT to be reduced to the form given in (5.16).

114. Other symmetries. What symmetry do you expect in the DFT of an input array that has the symmetry $f_{-mn} = f_{mn}$ and $f_{m,-n} = -f_{mn}$? Design an efficient DFT for input with this symmetry.

115. Fourier series to the DFT. Assume that a function f is a continuous function on the rectangle $\mathcal{I}$ with $-A/2 \le x \le A/2$ and $-B/2 \le y \le B/2$. The two-dimensional Fourier series representation for f on this region is given by

$$f(x,y) = \sum_{j=-\infty}^{\infty} \sum_{k=-\infty}^{\infty} c_{jk} e^{i(\frac{2\pi jx}{A} + \frac{2\pi ky}{B})},$$

where

$$c_{jk} = \frac{1}{AB} \int_{-\frac{A}{2}}^{\frac{A}{2}} \int_{-\frac{B}{2}}^{\frac{B}{2}} f(x,y) e^{-i(\frac{2\pi jx}{A} + \frac{2\pi ky}{B})}.$$

Use this Fourier series pair to derive (or at least motivate) the two-dimensional DFT pair (5.1) and (5.2).

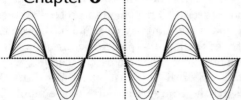

Chapter **6**

Errors in the DFT

*An approximate answer
to the right problem is
worth a good deal more
than an exact answer to
an approximate problem.*
– John Tukey

179

6.1. Introduction

Chapters 2 and 3 were devoted to introducing the DFT and discovering some of its important properties. At that time, we began to establish the relationships among the DFT, Fourier series, and Fourier transforms in rather qualitative terms. Now it is time to make these relationships more precise. Our plan is to proceed systematically and investigate a sequence of cases that will ultimately cover all of the common uses of the DFT. As we will see, the form of the input sequence dictates how the DFT is used and how its output should be interpreted. In some cases, the DFT will provide approximations to the Fourier coefficients of the input; in other cases, the DFT will provide approximations to (samples of) the Fourier transform of the input. The goal of this chapter is to understand exactly *what* the DFT approximates in each case, and then to estimate the size of the errors in those approximations. Most of the chapter will be devoted to errors in the DFT. However, it is also illuminating to explore the inverse DFT (IDFT) and to appreciate the complementarity that exists between the two transforms and their approximations. For the most part, we will be concerned with pointwise errors. For example, if the DFT component F_k is used to approximate the Fourier coefficient c_k, what is the error of that particular approximation? In the last section, however, we will look at errors in the DFT from a different perspective. When the DFT is viewed as an interpolating function for a given function, we are more interested in the error of the entire approximation, rather than the error at individual points. In this case, it makes sense to ask about the mean square error (or integrated error) in DFT approximations. Fortunately, the tools we will develop to analyze pointwise errors can also be applied to estimate the error in DFT interpolation.

This chapter walks a fine line between practical and theoretical realms. On one hand, understanding the uses of the DFT and the sources of error in the DFT is an immensely practical subject. No practitioner can use the DFT successfully without understanding what the input means, how errors arise, and how the output should be interpreted. On the other hand, it is impossible to estimate errors or even understand their sources without an occasional excursion into technical territory. This chapter presents the roughest terrain in the book. It is meant to be an exhaustive coverage of DFT errors in all of the various cases in which the DFT is used. No concessions have been made to brevity or terseness. The chapter *does* include formal theorems, and proofs will be sketched to the extent that interested readers can provide the details. Meaning and insight will be added to the theorems with detailed analytical and numerical examples. In this way a complete and balanced coverage of this important topic can hopefully be achieved.

Before embarking on this journey, we must pause and define a new semitechnical term that (to our knowledge) does not have currency in the literature. In working with DFTs, there are occasional technicalities that can arise unexpectedly. These "details" (for example, the treatment of endpoints or the issue of even/odd number of DFT points) occur often enough and in so many different forms that they deserve a name. We have coined the felicitous term **pester** to describe these various irritations. Despite their many forms, there are two properties that all pesters share: first, they cannot be ignored or they *will* cause trouble; second, once dealt with, we have never met a pester that did not ultimately hold a hidden lesson about DFTs. With that important definition, we will begin an exploration of errors in the DFT—pesters and all.

6.2. Periodic, Band-Limited Input

There is a very natural place to begin this discussion about errors in the DFT, and that is with periodic functions and sequences. As we saw in Chapter 2, if f is a piecewise smooth function on the interval $[-A/2, A/2]$ [1], then it has a representation as a Fourier series of the form

$$f(x) = \sum_{k=-\infty}^{\infty} c_k e^{i2\pi kx/A}, \tag{6.1}$$

where the complex coefficients c_k are given by the integrals

$$c_k = \frac{1}{A} \int_{-\frac{A}{2}}^{\frac{A}{2}} f(x)e^{-i2\pi kx/A}dx \quad \text{for} \quad k = 0, \pm 1, \pm 2, \ldots. \tag{6.2}$$

The series in (6.1) converges to the function f on the interval $[-A/2, A/2]$, to its *periodic extension* outside that interval, and to its average value at points of discontinuity. This series describes how the function f can be assembled as a linear combination of modes (sines and cosines), all of which have an integer number of periods on the interval $[-A/2, A/2]$. The kth mode has exactly k periods (or wavelengths) on the interval $[-A/2, A/2]$, and thus, it has a frequency of k/A cycles per unit length. The coefficient c_k is simply the amount by which the kth mode is weighted in this representation of f.

We are now in a position to investigate our first question: how well does the DFT approximate the Fourier coefficients of f? Assume that the given function f has period A; this includes the situation in which f may be defined only on the interval $[-A/2, A/2]$ and is then extended periodically. Something rather remarkable can be discovered right away with one simple calculation. Imagine that the periodic function f is sampled on the interval $[-A/2, A/2]$ at the uniformly spaced points

$$x_n = n\Delta x = \frac{nA}{N} \quad \text{for} \quad n = -\frac{N}{2} + 1 : \frac{N}{2}.$$

Denoting the sampled values $f_n = f(x_n)$, we can use the Fourier series for f to write

$$f_n = f(x_n) = \sum_{k=-\infty}^{\infty} c_k e^{i2\pi kx_n/A} = \sum_{k=-\infty}^{\infty} c_k e^{i2\pi nk/N} \quad \text{for} \quad n = -\frac{N}{2} + 1 : \frac{N}{2}.$$

Since f_n is a periodic sequence of length N, we might contemplate taking its DFT. It is not difficult to determine the outcome:

$$
\begin{aligned}
F_k &= \mathcal{D}\{f_n\}_k = \frac{1}{N} \sum_{n=-\frac{N}{2}+1}^{\frac{N}{2}} f_n e^{-i2\pi nk/N} \\[2em]
&= \frac{1}{N} \sum_{n=-\frac{N}{2}+1}^{\frac{N}{2}} \underbrace{\left\{ \sum_{j=-\infty}^{\infty} c_j e^{i2\pi nj/N} \right\}}_{f_n} e^{-i2\pi nk/N},
\end{aligned}
$$

---

[1]We repeat the convention adopted in previous chapters of using the closed interval $[-A/2, A/2]$ to denote the interval on which a function f is sampled or expanded in its Fourier series. The sampled value assigned to f at the right endpoint ($f_{N/2}$) is the average of its endpoint values.

for $k = -N/2 + 1 : N/2$. If we now rearrange the order of summation and combine the exponentials, we have that

$$F_k = \frac{1}{N} \sum_{j=-\infty}^{\infty} c_j \underbrace{\sum_{n=-\frac{N}{2}+1}^{\frac{N}{2}} e^{i2\pi n(j-k)/N}}_{N\hat{\delta}_N(j-k)} = \sum_{j=-\infty}^{\infty} c_{k+jN} \quad \text{for} \quad k = -\frac{N}{2} + 1 : \frac{N}{2}. \quad (6.3)$$

Look who appears! As indicated by the braces, we need only to use the discrete orthogonality relation for the DFT (see Chapter 2) to simplify the inner sum. The double sum in (6.3) then collapses and we are left with the following result.

▶ **Discrete Poisson Summation Formula** ◀

$$F_k = c_k + \sum_{j=1}^{\infty} (c_{k+jN} + c_{k-jN}) \quad \text{for} \quad k = -\frac{N}{2} + 1 : \frac{N}{2}. \quad (6.4)$$

We have taken some liberties with the terminology. As we will see later in the chapter, the "official" Poisson[2] Summation Formula is a relationship between a function and its Fourier *transform*. The above relationship between the DFT of a function and its Fourier coefficients is analogous to the Poisson Summation Formula; therefore, a name reflecting that similarity seems appropriate. It may not be evident right now, but this relationship between the DFT and the Fourier coefficients has a lot to tell us. To make sense of it, we need a definition, and then two quite different cases appear.

We need to review the property of band-limitedness, but now as it applies to periodic functions. If the A-periodic function f has the property that the Fourier coefficients c_k are zero for $|k| > M$, where M is some natural number, then f is said to be **band-limited**. It simply means that the "signal" f has no frequency content above the Mth frequency of M/A cycles per unit length. Expression (6.4) tells us that if f is periodic and band-limited with $M < N/2$, then $c_k = 0$ for $|k| > N/2$, and the N-point DFT exactly reproduces the N Fourier coefficients of f. By this we mean that $F_k = c_k$ for $k = -N/2 + 1 : N/2$. If f is not band-limited or is band-limited with $M > N/2$, we can expect to see errors in the DFT.

It is worth stating this important result in several different (but equivalent) ways:

- If the Fourier coefficients c_k of the function f on the interval $[-A/2, A/2]$ are zero for $k > N/2$, then the DFT of length N exactly reproduces the nonzero Fourier coefficients of f.

- Assume the Fourier coefficients on the interval $[-A/2, A/2]$ vanish for frequencies

$$\omega \geq \frac{\Omega}{2}.$$

If the highest frequency resolvable by the DFT, which is $N\Delta\omega/2 = N/(2A)$, satisfies

$$\frac{N}{2A} \geq \frac{\Omega}{2} \quad \text{or} \quad N \geq A\Omega,$$

---

[2]Encouraged by his father to be a doctor, SIMÉON DENIS POISSON (1781–1840) attracted the attention of Lagrange and Laplace when he entered the Polytechnic School at the age of 17. He made lasting contributions to the theories of elasticity, heat transfer, capillary action, electricity, and magnetism. He was also regarded as one of the finest analysts of his time.

then the N-point DFT exactly reproduces the nonzero Fourier coefficients of f. Notice how the reciprocity relation $A\Omega = N$ appears in this condition.

- A maximum frequency of $\Omega/2$ corrresponds to a minimum period (wavelength) of $\lambda = 2/\Omega$. Therefore, the above condition can also be written

$$\Delta x = \frac{A}{N} \leq \frac{1}{\Omega}.$$

This condition means that if f is sampled with at least two grid points per period, then the DFT exactly reproduces the nonzero Fourier coefficients of f.

The critical grid spacing (or sampling rate) $\Delta x = 1/\Omega$ should look familiar; we encountered it in Chapter 3 and called it the **Nyquist sampling rate**. It plays an essential role in analyzing signals and using the DFT, and we will see it again with nonperiodic band-limited functions.

The special case in which $M = N/2$ raises our first *pester*. Special care must be used with the $N/2$ mode since it is the highest frequency mode that the DFT can resolve. The Fourier series, using continuous modes, can distinguish the $-N/2$ and the $N/2$ modes. In fact, if f is real-valued, then $c_{\frac{N}{2}} = c^*_{-\frac{N}{2}}$. On the other hand, the DFT cannot distinguish between the (discrete) $\pm N/2$ modes; in fact, it combines them into one real mode of the form $\cos(\pi n)$. Combining these observations, we conclude that in the special case in which the $M = N/2$

$$F_k = \begin{cases} c_k & \text{if } |k| \leq N/2 - 1, \\ c_{-\frac{N}{2}} + c_{\frac{N}{2}} & \text{if } k = N/2 \text{ and } f \text{ is complex-valued}, \\ 2\,\text{Re}\left(c_{\frac{N}{2}}\right) & \text{if } k = N/2 \text{ and } f \text{ is real-valued}. \end{cases}$$

This result also assumes that N is even; an interesting variation arises with an odd number of DFT points (problem 126). Let's solidify these essential ideas surrounding sampling rates with a case study.

Case Study 1: Periodic, band-limited functions. Let f be the function whose nonzero Fourier coefficients are given by $c_k = 1$ for $-16 \leq k \leq 16$ and assume it is sampled at N equally spaced grid points on the interval $[-1, 1]$. The sampled values of f are given by

$$f(x_n) = \sum_{k=-16}^{16} 1 \cdot e^{i\pi k x_n} = 1 + 2\sum_{k=1}^{16} \cos(\pi k x_n) \quad \text{for} \quad n = -\frac{N}{2} + 1 : \frac{N}{2}.$$

Noting that the sample points are $x_n = 2n/N$, the input sequence to the DFT is given by

$$f_n = 1 + 2\sum_{k=1}^{16} \cos(2\pi nk/N) \quad \text{for} \quad n = -\frac{N}{2} + 1 : \frac{N}{2}.$$

We can already anticipate the minimum number of sample points that the DFT needs to reproduce the Fourier coefficients exactly. The highest frequency mode that comprises the function f has a frequency of 16 periods on the interval $[-1, 1]$ or 8 periods per unit length. Using the above notation, this means that $\Omega/2 = 8$. This highest frequency mode has a period of $1/8$ units of length (or time). Therefore, the sampling criterion that we place at least two grid points per period of every mode

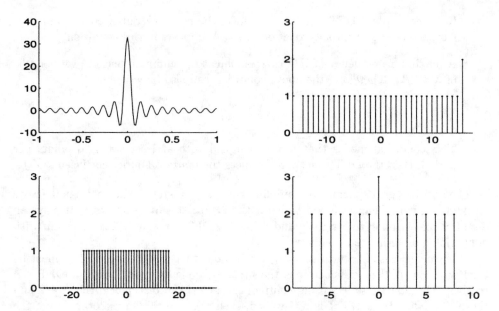

FIG. 6.1. **Case Study 1.** *A periodic function f whose nonzero Fourier coefficients are given by $c_k = 1$ for $-16 \le k \le 16$ is sampled at N equally spaced points on the interval $[-1, 1]$, and the sampled function is used as input for DFTs of various lengths. The function f itself is shown in the upper left graph, while the output of DFTs of length $N = 32$ and $N = 64$ are shown in the upper right and lower left graphs. In these two cases, all of the modes of f can be resolved exactly, since there are at least two grid points per period of every mode. The result is that all of the Fourier coefficients are to be computed exactly by the DFT. However, when a DFT of length $N = 16$ is used (lower right), there are errors in the DFT, since the input is "undersampled."*

means that we must have $\Delta x \le 1/\Omega = 1/16$. Since $\Delta x = 2/N$ we see that we must take $N \ge 32$ to resolve the Fourier coefficients exactly.

Now let's do the experiments. The input sequence f_n is fed to DFTs of various lengths, and the results are shown in Figure 6.1. The upper left graph shows the function f itself; as expected, it is real and even, since its Fourier coefficients are real and even. The upper right and lower left plots of Figure 6.1 show the DFTs of length $N = 32$ and $N = 64$, which reproduce the nonzero Fourier coefficients exactly. However, note that with $N = 32$ (the case $M = N/2$ discussed above), there is the expected doubling of the $N/2$ coefficient. To anticipate coming events, the lower right graph of Figure 6.1 shows the DFT of length $N = 16$. In this case the input no longer appears band-limited to the DFT and the input sequence is "undersampled." This simply means that there are not enough DFT points to resolve all frequencies of the input at a rate of at least two grid points per period. Therefore, the DFT coefficients are in error (see problem 122). We had a glimpse of this effect in Chapter 3, and now it will be investigated in detail.

6.3.   Periodic, Non-Band-Limited Input

We now consider a more prevalent case in which f is periodic, but not band-limited. This means that f has nonzero Fourier coefficients c_k for arbitrarily large values of k.

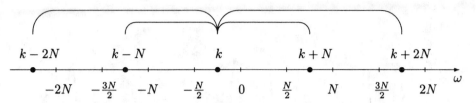

FIG. 6.2. *A periodic function with high frequency modes ($\omega > N/2A$) will undergo aliasing when sampled at N grid points on the interval $[-A/2, A/2]$. When sampled at the grid points, the kth mode is indistinguishable from the $k + mN$ mode for any integer m. The figure shows the ω-axis and the modes that are coupled through aliasing.*

The Discrete Poisson Summation Formula

$$F_k = c_k + \sum_{j=1}^{\infty}(c_{k+jN} + c_{k-jN}) \quad \text{for} \quad k = -\frac{N}{2} + 1 : \frac{N}{2}$$

is now considerably more complicated, but it has an extremely important interpretation. In this case, the DFT coefficient F_k is equal to c_k *plus* additional Fourier coefficients corresponding to higher frequencies. We see that the kth mode is linked with other modes whose index differs from k by multiples of N, as shown in Figure 6.2. Why should these modes be associated with each other? There is a good and far-reaching explanation for this effect.

For the sake of illustration assume that we are working on the interval $[-1, 1]$ with $N = 10$ grid points. Figure 6.3 shows the $k = 2$ mode with a frequency of $\omega_2 = 1$ period per unit length. Also shown are the $k + N = 12$ and $k - N = -8$ modes, which have much higher frequencies of $\omega_{12} = 6$ and $\omega_{-8} = 4$ cycles per unit length. However, all three modes take on the same values at the grid points! In other words, to someone who sees these three modes *only at the grid points*, they look identical.

This phenomenon can be verified analytically as well. On the interval $[-A/2, A/2]$ with N equally spaced grid points, the value of the kth mode, with frequency $\omega_k = k/A$, at the grid point $x_n = nA/N$ is given by

$$e^{i2\pi\omega_k x_n} = e^{i2\pi(\frac{k}{A})(\frac{nA}{N})} = e^{i2\pi nk/N}.$$

The value of the $k + mN$ mode, with frequency $\omega_{k+mN} = (k + mN)/A$, at the grid point x_n is given by

$$e^{i2\pi(\omega_{k+mN})x_n} = e^{i2\pi(\frac{k+mN}{A})(\frac{nA}{N})} = e^{i2\pi nk/N} \cdot \underbrace{e^{i2\pi mn}}_{1} = e^{i2\pi nk/N},$$

where $e^{i2\pi mn} = 1$, since m and n are integers.

In other words, the kth mode and the $k+mN$ mode (where m is any integer) agree at the grid points. This is precisely the effect called **aliasing** that was observed in Chapter 3, in which higher frequency modes masquerade as low frequency modes (see problems 116 and 117). The conclusion is that the DFT cannot distinguish a basic mode ($-N/2 + 1 \le k \le N/2$) from higher frequency modes; hence the kth coefficient computed by the DFT includes contributions not only from the basic mode, but from all of the aliased modes as well. In the case of a band-limited function, there are no

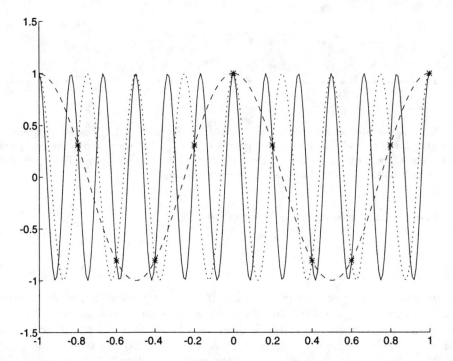

FIG. 6.3. *On a grid with $N = 10$ points, the $k = 2$ mode (dashed line) takes on the same values as the $k = 12$ mode (solid line) and the $k = -8$ mode (dotted line) at the grid points $x_n = -1, -.8, -.6, -.4, -.2, 0, .2, .4, .6, .8, 1.$*

higher frequency modes to be aliased, assuming that f is sampled at or above the critical rate.

The Discrete Poisson Summation Formula indicates how and when aliasing introduces errors to the DFT. But just how serious are these errors? This is the question that we now address. The goal is to estimate the magnitude of $|F_k - c_k|$, the error in the coefficients produced by the DFT. In order to do this we need more information about the function f, and it turns out that the most useful information about f concerns its *smoothness* or, more precisely, the number of continuous derivatives it has. With this information, a well-known theorem can be used, which is of considerable interest in its own right. Here is a central result [30], [70], [115], [158].

THEOREM 6.1. RATE OF DECAY OF FOURIER COEFFICIENTS. *Let f and its first $p - 1$ derivatives be A-periodic and continuous on $[-A/2, A/2]$ for $p \geq 1$. Let $f^{(p)}$ be bounded with at most a finite number of discontinuities on $[-A/2, A/2]$. Then the Fourier coefficients of f satisfy*

$$|c_k| \leq \frac{C}{|k|^p} \quad \text{for all} \quad k,$$

where C is a constant independent of k.

Sketch of proof: We start with the definition of the kth Fourier coefficient and integrate it by parts:

$$Ac_k = \int_{-\frac{A}{2}}^{\frac{A}{2}} f(x)e^{-i2\pi kx/A}dx = f(x)A\frac{e^{-i2\pi kx/A}}{-i2\pi k}\Big|_{-\frac{A}{2}}^{\frac{A}{2}} + \frac{A}{i2\pi k}\int_{-\frac{A}{2}}^{\frac{A}{2}} f'(x)e^{-i2\pi kx/A}dx.$$

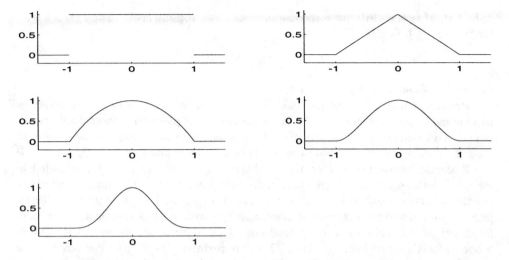

FIG. 6.4. *A piecewise monotone function on the interval* $[-A/2, A/2]$ *has the property that the interval can be subdivided into a finite number of subintervals, on which the function is either nonincreasing or nondecreasing. Several piecewise monotone functions on* $[-1, 1]$ *are shown, with smoothness corresponding to (top left)* $p = 0$, *(top right and middle left)* $p = 1$, *(middle right)* $p = 3$, *and (bottom left)* $p = 5$ *in Theorem 6.2.*

The integrated term (first term) on the right side vanishes because of the periodicity of f and the complex exponential; the remaining term is integrated again by parts. If this step is performed a total of p times and the periodicity of the derivatives is used each time, we find that

$$Ac_k = (-1)^{p+1} \left(\frac{A}{i2\pi k} \right)^p \int_{-\frac{A}{2}}^{\frac{A}{2}} f^{(p)}(x) e^{-i2\pi k x/A} dx.$$

The magnitude of this integral may be bounded by AM where $M = \sup_{[-A/2, A/2]} |f^{(p)}(x)|$, which leads to the bound

$$|c_k| \leq M \left(\frac{A}{2\pi |k|} \right)^p \equiv \frac{C}{|k|^p}. \qquad \blacksquare$$

At this point, it would be easy to use this result and deduce an error bound for the DFT. However, at the risk of raising a few technicalities, it pays to work just a bit more and obtain a stronger version of the previous theorem. This will lead to a more accurate error bound for the DFT. It turns out that the last integral in the previous proof (which has $f^{(p)}$ in the integrand) can be bounded *only* by a constant if we assume *only* that $f^{(p)}$ has a finite number of discontinuities. By placing slightly tighter conditions on $f^{(p)}$, a stronger result can be obtained. Here is the additional condition that must be imposed. A function is said to be **piecewise monotone** on an interval $[-A/2, A/2]$ if the interval can be split into a finite number of subintervals on each of which f is either nonincreasing or nondecreasing. Figure 6.4 shows examples of piecewise monotone functions with various degrees of smoothness; it suggests that this condition does not exclude most functions that arise in practice. With this definition we may now state the stronger result [30], [70], [158].

THEOREM 6.2. RATE OF DECAY OF FOURIER COEFFICIENTS. *Let* f *and its first* $p-1$ *derivatives be* A-*periodic and continuous on* $[-A/2, A/2]$ *for* $p \geq 0$. *Assume that* $f^{(p)}$ *is bounded and piecewise monotone on* $[-A/2, A/2]$. *(The case* $p = 0$ *means that* *only f itself is bounded and piecewise monotone on the interval.) Then the Fourier coefficients of f satisfy*

$$|c_k| \leq \frac{C}{|k|^{p+1}} \quad \text{for all} \quad k,$$

where C is a constant independent of k.

Proof outline: As in the previous theorem, the proof begins with p integrations by parts of the Fourier coefficients c_k. With the assumption of piecewise monotonicity and the second mean value theorem for integrals, the final integral can be bounded by a constant times k^{-1}, which provides the additional power of k in the result. ∎

It should be mentioned that the conditions of this theorem can be extended to include functions with a finite number of discontinuites with infinite jumps. All of the conditions under which this result is true are often called **Dirichlet's**[3] **conditions** [30]. As we noted, the condition of monotonicity is really not a significant restriction for functions that arise in most applications. This result may now be used to obtain a bound for the error in using the DFT to approximate the Fourier coefficients of a periodic, non-band-limited function. The result is given as follows [4], [76].

THEOREM 6.3. ERROR IN THE DFT (PERIODIC NON-BAND-LIMITED CASE). *Let f and its first $p-1$ derivatives be A-periodic and continuous on $[-A/2, A/2]$ for $p \geq 1$. Assume that $f^{(p)}$ is bounded and piecewise monotone on $[-A/2, A/2]$. Then the error in the N-point DFT as an approximation to the Fourier coefficients of f satisfies*

$$|F_k - c_k| \leq \frac{C}{N^{p+1}} \quad \text{for} \quad k = -\frac{N}{2} + 1 : \frac{N}{2},$$

where C is a constant independent of k and N.

Proof: From the Discrete Poisson Summation Formula we know that

$$F_k - c_k = \sum_{j=1}^{\infty} (c_{k+jN} + c_{k-jN}) \quad \text{for} \quad k = -\frac{N}{2} + 1 : \frac{N}{2}.$$

By Theorem 6.2, the Fourier coefficients satisfy the bound $|c_k| \leq C'|k|^{-p-1}$ for some constant C'. Hence

$$|F_k - c_k| \leq \sum_{j=1}^{\infty} \frac{C'}{|k+jN|^{p+1}} + \frac{C'}{|k-jN|^{p+1}} = \frac{C'}{N^{p+1}} \sum_{j=1}^{\infty} \frac{1}{|j+\frac{k}{N}|^{p+1}} + \frac{1}{|j-\frac{k}{N}|^{p+1}},$$

for $k = -N/2 + 1 : N/2$. Both series can be shown to converge for $p \geq 1$ [76], and we may conclude that $|F_k - c_k| \leq C/N^{p+1}$. ∎

Before turning to a case study, it would be useful to elaborate on this theorem with an eye on *The Table of DFTs* of the Appendix, which demonstrates it quite convincingly. Theorem 6.3 gives the errors in the DFT in terms of a bound that holds for all N. On the other hand, *The Table of DFTs* shows the errors in the DFTs in an asymptotic sense for large values of N. While these are not the same measure of error, it can be shown that both estimates agree in the leading power of N. Therefore, the table can be used to check the results of Theorem 6.3. We will mention several specific examples.

---

[3]Born and educated in (present day) Germany, PETER GUSTAV LEJEUNE DIRICHLET (1805–1859) interacted closely with the French mathematicians of the day. He did fundamental work in number theory (showing that Fermat's last theorem is true for $n = 5$) and in the convergence theory of Fourier series. He succeeded Gauss in Göttingen for the last four years of his life.

1. If the periodic extension of f is continuous on the interval $[-A/2, A/2]$ (implying that $f(-A/2) = f(A/2)$), but no higher derivatives are continuous, then Theorem 6.3 applies with $p = 1$. We may conclude that the error in the DFT is bounded by a multiple of N^{-2}. This situation is illustrated by cases 6a and 6b (real even harmonics) and case 8 (triangular wave), all of which are continuous, but have piecewise continuous derivatives.

2. DFTs of functions with smoothness $p > 1$ are difficult to compute analytically and do not appear in *The Table of DFTs*. Numerical examples of these cases will be shown in the next section.

3. Unfortunately, there is one case that occurs frequently in practice that Theorem 6.3 does not cover; this is the case of functions that are only piecewise continuous ($p = 0$). With a bit more work, Theorem 6.3 can be extended to this case [76], and the result is as expected: if the A-periodic extension of f is bounded and piecewise monotone, the error in the DFT is bounded by C/N, where C is a constant independent of k and N. This situation is illustrated in *The Table of DFTs* by case 5 (complex harmonic), case 6c (real odd harmonic), case 7 (linear), case 9 (rectangular wave), cases 10 and 10a (square pulse), and case 11 (exponential), all of which have discontinuities either at an interior point or at the endpoints. The typical asymptotic behavior for these cases is

$$|F_k - c_k| \sim CkN^{-2} \quad \text{for} \quad k = -\frac{N}{2} + 1 : \frac{N}{2} \quad \text{as} \quad N \to \infty.$$

There is actually a little extra meaning in this bound. For low frequency coefficients ($|k| << N/2$), the error behaves like CN^{-2}; for high frequency coefficients ($|k| \approx N/2$) the error decreases more slowly, as CN^{-1}, as predicted by the theory. This says that the errors in the low frequency coefficients are generally smaller than in the high frequency coefficients, which is often observed in computations.

Case Study 2: Periodic, non-band-limited functions. The phenomenon of aliasing can be demonstrated quite convincingly by looking at a periodic function whose period is greater than the interval from which samples are taken. Consider the 2-periodic function $f(x) = \cos(\pi x)$ on the interval $[-1/2, 1/2]$ (which is case 6b in *The Table of DFTs*). The function f is sampled at the N grid points $x_n = n/N$ for various values of N, and these sequences are used as input to the DFT. Figure 6.5 shows the function f and its periodic extension (top left). The errors $|F_k - c_k|$ in the DFTs of length $N = 16, 32, 64$ are shown in the remaining graphs. Regardless of how large N is taken, the DFT never sees a complete period of the input function, and hence the input appears non-band-limited. Therefore, the coefficients that the DFT produces are in error because the coefficients of higher frequency modes are aliased onto the corresponding low frequency modes. Indeed, as N increases, these errors decrease in magnitude in accordance with the $p = 1$ case of Theorem 6.3. Furthermore, errors in the low frequency coefficients are smaller than errors in the high frequency coefficients.

The previous case study is a specific example of a more general rule that is often overlooked in practice: when a periodic function is sampled on an interval whose length is *not* a multiple of the period, an error is introduced in the DFT (in addition to a possible aliasing error). We have already seen an example of this error in Chapter 3, and it was attributed to **leakage**. We would like to explore the phenomenon of leakage again, both from a slightly different perspective and in greater detail.

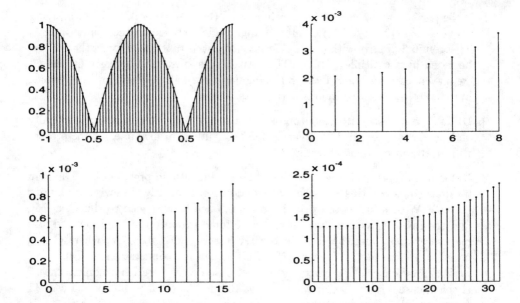

FIG. 6.5. **Case Study 2.** *The function* $f(x) = \cos(\pi x)$ *(upper left) is sampled at N uniformly spaced points on the interval* $[-1/2, 1/2]$, *and extended periodically beyond that interval. The errors in the DFTs of lengths $N = 16, 32, 64$ are shown in the upper right, lower left, and lower right plots, respectively. Because of symmetry, only the $0 \le k \le N/2$ components are shown. Note that the vertical scale is different for each of these plots. Regardless of the value of N this function, when sampled on this interval, appears non-band-limited, and there are errors in the DFT.*

Assume that the function f is A-periodic and has Fourier coefficients c_k on $[-A/2, A/2]$. We will be interested in computing the Fourier coefficients of f on a second interval $[-pA/2, pA/2]$ where $p > 1$. Let us denote the coefficients on that interval c_k'. Here is an illuminating preliminary question: what happens if the length of the second interval is a multiple of the period; that is, p is an integer? It is a significant result (problem 123) that in this case the two sets of coefficients c_k and c_k' are related by

$$c_k' = \begin{cases} c_{\frac{k}{p}} & \text{if } p \text{ divides } k, \\ 0 & \text{otherwise,} \end{cases}$$

where k is any integer. There is a clear meaning of this result: if $p > 1$ is an integer, the kth mode on the interval $[-pA/2, pA/2]$ appears as the (k/p)th mode on the interval $[-A/2, A/2]$. The remaining modes are not periodic on $[-A/2, A/2]$ and are not needed in the representation of f on $[-pA/2, pA/2]$. For example, one full period on $[-1, 1]$ looks like two full periods on $[-2, 2]$. On the other hand, one full period on $[-2, 2]$ looks like half a period on $[-1, 1]$, and this mode is not used in the representation of f on $[-2, 2]$. To make this point quite clear consider the function $f(x) = \cos 2x$. It has the following Fourier coefficients on the given intervals:

- $c_k = \frac{1}{2}(\delta(k-1) + \delta(k+1))$ on $[-\pi/2, \pi/2]$,

- $c_k' = \frac{1}{2}(\delta(k-2) + \delta(k+2))$ on $[-\pi, \pi]$, and

- $c_k'' = \frac{1}{2}(\delta(k-4) + \delta(k+4))$ on $[-2\pi, 2\pi]$.

Note that $c''_{k/4} = c'_{k/2} = c_k$ reflecting the scaling of the indices as the interval is changed.

Equally important, the DFT has the same property: if $p > 1$ is an integer and the A-periodic function f is sampled with pN points on the interval $[-pA/2, pA/2]$ the resulting coefficients F'_k are related to the original set F_k by

$$F'_k = \begin{cases} F_{\frac{k}{p}} & \text{if } p \text{ divides } k, \\ 0 & \text{otherwise,} \end{cases}$$

where $k = -pN/2 + 1 : pN/2$ (problem 124).

Those were preliminary observations. Now what happens if the A-periodic function f is sampled on the interval $[-pA/2, pA/2]$ where $p > 1$ is not an integer? In this case the interval contains at least one full period of f *plus* a fraction of a period. The crux of the issue is captured if we confine our attention to a single complex mode

$$f(x) = e^{i2\pi k_0 x/A}$$

on the interval $[-pA/2, pA/2]$, where k_0 is an integer, but p is not. In this case the length of the interval is not a multiple of the period. A short calculation (or a peek at case 5 of *The Table of DFTs*) reveals that the Fourier coefficients of f on $[-pA/2, pA/2]$ are given by

$$c'_k = \frac{\sin \pi(k - pk_0)}{\pi(k - pk_0)}, \tag{6.5}$$

where k is any integer. Since p is not an integer, $k - pk_0$ is never an integer, and none of the coefficients c'_k vanish (whether f is band-limited or not). Typically the coefficients with index closest to pk_0 will have the largest magnitude. Nearby coefficients, in what are often called the **sidelobes**, decrease in magnitude and decay to zero by oscillation like $(k - pk_0)^{-1}$. The appearance of these characteristic sidelobes is often a symptom of poor sampling of a periodic signal.

Equally important is the fact that the DFT exhibits the same effect. If a periodic function is sampled on an interval that does not contain an integer number of periods, then the resulting DFT will show discrepancies when compared to the results of sampling on a complete period. This is clearest when the same single complex mode

$$f(x) = e^{i2\pi k_0 x/A}$$

with integer frequency k_0 is sampled on the interval $[-pA/2, pA/2]$ with pN points, where p is not an integer. Assuming that $pN/2$ *is* an integer, a short calculation (problem 125 or *The Table of DFTs*) reveals that the Fourier coefficients of f on $[-pA/2, pA/2]$ in the Appendix shows that the pN-point DFT is given by

$$F'_k = \frac{\sin \pi(k - pk_0) \sin\left(\frac{2\pi(k - pk_0)}{pN}\right)}{2pN \sin^2\left(\frac{\pi(k - pk_0)}{pN}\right)}, \tag{6.6}$$

where $k = -pN/2 + 1 : pN/2$. Both the Fourier coefficients and the DFT coefficients as given by (6.5) and (6.6) have the property that *if* p is an integer, then they reduce to the expected result, namely that $c'_k = F'_k = \delta(k - pk_0)$. Formally letting $p \to 1$ in either expression (6.5) or (6.6) recovers the single spike at the central frequency (problem 125). This says that the DFT agrees with the Fourier series coefficients and

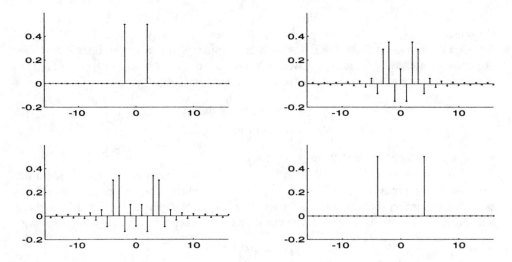

FIG. 6.6.   *The effect of sampling a periodic function on integer and noninteger multiples of its period are shown in these four graphs. The DFT coefficients of the single wave* $f(x) = \cos(2\pi x)$ *are computed on the interval* $[-p, p]$, *where* $p = 1, 1.25, 1.75, 2$ *(left to right, top to bottom). As discussed in the text, when* p *is an integer, the Fourier coefficients consist of two clean spikes (although their location must be interpreted carefully). When* p *is not an integer, leakage occurs into the sidelobes near the central frequencies. The same behavior occurs in the DFT coefficients.*

returns all zero coefficients except at the single frequency corresponding to the index k_0. On the other hand, if p is not an integer, then in general both sets c_k' and F_k' are nonzero and the sidelobes appear.

Let's interpret these expressions with the help of a few pictures. Figure 6.6 shows the DFT coefficients of the real mode $f(x) = \cos(2\pi x)$ computed on the interval $[-p, p]$ for several values of p. Observe that with $p = 1$ the set of coefficients consists of two nonzero coefficients at $k = \pm 2$. With $p = 1.25$, the wave is sampled on a fraction of a full period and nonzero coefficients appear in sidelobes around the two central frequencies. With $p = 1.75$, the wave is still sampled on a fraction of a full period and the sidelobes actually broaden. With a value of $p = 2$, the sidelobes disappear, but now the two central frequencies have moved out to $k = \pm 4$, as predicted by the above analysis. The same behavior could also be observed in the Fourier coefficients. Notice that this is essentially the same problem that was considered in Section 3.4 in our preliminary discussion of leakage. In that setting the interval $[-A/2, A/2]$ was held fixed while the frequency of the sampled function was varied.

As mentioned earlier, the effect that we have just exposed is generally called leakage, since the sidelobes drain "energy" from the central frequencies. The phenomenon also appears with other names in the literature, often resulting in confusion. Seen as an error, it is often called a **truncation error**, meaning that the periodic input function has been truncated in an ill-advised way. The effect is also called **windowing**, referring to the square "window" that is used to truncate the input. In computing Fourier coefficients, this kind of error can often be avoided with prior knowledge or estimates of the period of the input. These same ideas will recur in an unavoidable way when we consider the approximation of Fourier transforms. Let us now turn to that subject.

6.4. Replication and Poisson Summation

The next item on the agenda of DFT errors is the class of compactly supported (or spatially limited) functions. Before undertaking this mission, a small diversion is necessary. On the first pass, this excursion will seem unrelated to the discussion at hand but we will soon see that it bears heavily on everything that follows. We begin with a definition. Imagine a function f that is defined on the entire real line $(-\infty, \infty)$ and let A be any positive real number. We will now associate with the function f a new function called its **replication of period** A (or simply **replication** if A is understood). It is defined as

$$\mathcal{R}_A\{f(x)\} = \sum_{j=-\infty}^{\infty} f(x + jA).$$

The replication of period A of f is the superposition of copies of f, each displaced by multiples of A; this new function is also defined on the entire real line. The result of this operation can be seen in Figure 6.7, which shows the replication of period $A = 5$ of the function $f(x) = e^{-|x|}$. An important property is that the replication of period A *is* a periodic function with period A. Take heed: the periodic replication of a function is *not* its periodic extension (unless $f(x) = 0$ outside of an interval of length A).

There may be some solace in knowing that we have already seen the idea of replication in this chapter, although the name was not mentioned at the time. The replication operator can also be applied to a sequence in the following manner. If $\{c_n\}$ is a sequence defined for all integers n, then its replication of period N is

$$\mathcal{R}_N\{c_n\} = \sum_{k=-\infty}^{\infty} c_{n+kN}.$$

The idea is exactly the same. The replication of period N of a sequence is the superposition of copies of that sequence, each translated by multiples of N. The periodic replication is a periodic sequence as shown in Figure 6.7. The replication of a sequence appeared in the Discrete Poisson Summation Formula (6.4), which we could now write more compactly as

$$F_k = \mathcal{R}_N\{c_n\}.$$

The fact that the replication operator appears in the connection between the DFT and the Fourier series suggests that it might appear again in the relationship between the DFT and Fourier transforms. Here is how it all comes about. A short calculation will lead to the version of the Poisson Summation Formula that pertains to Fourier transforms. This result is of interest in its own right, but it also leads to greater truths.

Consider a function f that has a Fourier transform $\hat{f}$. As before, we let Δx be the grid spacing in the physical domain which determines a natural interval $[-1/(2\Delta x), 1/(2\Delta x)]$ in the frequency domain. Letting $\Omega = 1/\Delta x$, we now form the replication of period Ω of the transform $\hat{f}$. It is defined as

$$\hat{g}(\omega) \equiv \mathcal{R}_\Omega\{\hat{f}(\omega)\} = \sum_{j=-\infty}^{\infty} \hat{f}\left(\omega - \frac{j}{\Delta x}\right). \tag{6.7}$$

As discussed a moment ago, the function $\hat{g}$ is simply a superposition of copies of $\hat{f}$, each shifted by multiples of $\Omega = 1/\Delta x$. This means that $\hat{g}$ has a period of $1/\Delta x$, and

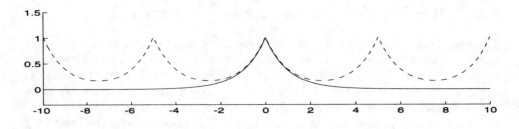

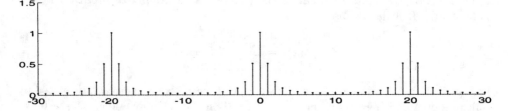

FIG. 6.7. *The replication operator may be applied to functions or sequences. The function* $f(x) = e^{-|x|}$ *(solid curve), and its replication of period 5 (denoted* $\mathcal{R}_5\{f(x)\}$*), are shown in the top figure (dashed curve). Similarly, the replication of period 20 of the sequence* $c_n = 1/(1 + |n|^{-2})$ *(denoted* $\mathcal{R}_{20}\{c_n\}$*) is shown in the bottom figure.*

as a periodic function it has a Fourier series of the form

$$\hat{g}(\omega) = \sum_{n=-\infty}^{\infty} c_n e^{i2\pi n \Delta x \omega}. \tag{6.8}$$

The coefficients in this series are given by

$$c_n = \Delta x \int_{-\frac{1}{2\Delta x}}^{\frac{1}{2\Delta x}} \hat{g}(\omega) e^{-i2\pi n \Delta x \omega} d\omega \quad \text{for} \quad n = 0, \pm 1, \pm 2, \ldots .$$

We now substitute the definition (6.7) of $\hat{g}$ into this expression for c_n and find that

$$c_n = \Delta x \int_{-\frac{1}{2\Delta x}}^{\frac{1}{2\Delta x}} \left\{ \sum_{j=-\infty}^{\infty} \hat{f}\left(\omega - \frac{j}{\Delta x}\right) \right\} e^{-i2\pi n \Delta x \omega} d\omega.$$

Now notice that the summation and the integral in this expression conspire nicely to form a single integral over the interval $-\infty < \omega < \infty$. This allows us to write

$$c_n = \Delta x \int_{-\infty}^{\infty} \hat{f}(\omega) e^{-i2\pi x_n \omega} d\omega = \Delta x f(x_{-n}),$$

where we have used $x_n = n\Delta x$ and the definition of f in terms of its Fourier transform. Using these c_n's in the Fourier series for $\hat{g}$ (6.8), we have that

$$\hat{g}(\omega) = \Delta x \sum_{n=-\infty}^{\infty} f(x_{-n}) e^{i2\pi x_n \omega} = \Delta x \sum_{n=-\infty}^{\infty} f(x_n) e^{-i2\pi x_n \omega}. \tag{6.9}$$

One more step brings us home. Now compare the two representations for $\hat{g}$ given by (6.7) and (6.9). They imply that

$$\hat{g}(\omega) = \Delta x \sum_{n=-\infty}^{\infty} f(x_n)e^{-i2\pi x_n \omega} = \sum_{j=-\infty}^{\infty} \hat{f}\left(\omega - \frac{j}{\Delta x}\right).$$

It is the second of these two equalities that is of interest. In fact, it culminates in the proof of the following theorem that will serve as the central pillar of the remainder of the chapter.

THEOREM 6.4. POISSON SUMMATION FORMULA. *Assume that f is defined on the interval $-\infty < x < \infty$, and that its Fourier transform $\hat{f}$ is defined for $-\infty < \omega < \infty$. Given a grid spacing Δx, let the sample points be given by $x_n = n\Delta x$ for integers $-\infty < n < \infty$. Then*

$$\Delta x \sum_{n=-\infty}^{\infty} f(x_n)e^{-i2\pi x_n \omega} = \sum_{j=-\infty}^{\infty} \hat{f}\left(\omega - \frac{j}{\Delta x}\right). \qquad (6.10)$$

Actually we have parted with convention; what is generally called the Poisson Summation Formula results by setting $\omega = 0$ in (6.10). This rather remarkable relationship between samples of a function and its Fourier transform is

$$\Delta x \sum_{n=-\infty}^{\infty} f(x_n) = \sum_{j=\infty}^{\infty} \hat{f}\left(\frac{j}{\Delta x}\right).$$

Quite unrelated to DFTs, this relationship says that the sum of the samples of f is a constant times the sum of the samples of $\hat{f}$ at multiples of the cut-off frequency $\Omega = 1/\Delta x$.

With this versatile and powerful result in our hands let's begin by making a few valuable observations. Some will be of immediate use; some are of interest in their own right and might help illuminate the Poisson Summation Formula.

1. Using the replication operator, the Poisson Summation Formula appears as

$$\Delta x \sum_{n=-\infty}^{\infty} f(x_n)e^{-i2\pi x_n \omega} = \mathcal{R}_\Omega\{\hat{f}(\omega)\},$$

   where $\Omega = 1/\Delta x$.

2. If f vanishes outside of some finite interval and is sampled on that interval, then the sum on the left side of the Poisson Summation Formula is finite (in fact, we will see momentarily that it is the DFT!) and the replication of the Fourier transform $\hat{f}$ can be computed exactly. This raises the question of whether a function (e.g., $\hat{f}$) can be recovered uniquely from its replication (problem 128).

3. Recall from expression (6.8) that the periodic replication of $\hat{f}$ can be written

$$\hat{g}(\omega) = \mathcal{R}_\Omega\{\hat{f}(\omega)\} = \sum_{n=-\infty}^{\infty} c_n e^{i2\pi n\Delta x \omega},$$

   where $c_n = \Delta x f(x_{-n})$. We see that the periodic replication of $\hat{f}$ can be represented as a Fourier series on the interval $[-\Omega/2, \Omega/2]$. This says that samples of f can be found by computing the Fourier series coefficients of the replication of $\hat{f}$. We will return to the implications of this observation shortly.

We now move towards a powerful connection among the Poisson Summation Formula, the DFT, and replication operators. A bit of rearranging on the left side of the Poisson Summation Formula leads to another perspective. Let the samples of f be denoted $f_n = f(x_n)$, where $x_n = n\Delta x = nA/N$. Evaluating the left side of (6.10) at $\omega = \omega_k = k/A$, we find that

$$\Delta x \sum_{n=-\infty}^{\infty} f_n e^{-i2\pi x_n \omega_k} = \Delta x \sum_{j=-\infty}^{\infty} \sum_{n=-\frac{N}{2}+1}^{\frac{N}{2}} f_{n+jN}\, e^{-i2\pi(n+jN)k/N}$$

$$= \Delta x \sum_{n=-\frac{N}{2}+1}^{\frac{N}{2}} \underbrace{\sum_{j=-\infty}^{\infty} f_{n+jN}\, e^{-i2\pi nk/N}}_{\mathcal{R}_N\{f_n\}}$$

$$= AD\{\mathcal{R}_N\{f_n\}\}_k.$$

We have used the periodicity of the complex exponential and introduced $\mathcal{R}_N\{f_n\}$. That takes care of the left side of the Poisson Summation Formula.

We now express the right side of the formula as $\mathcal{R}_\Omega\{\hat{f}(\omega)\}$ and evaluate it at $\omega = \omega_k$. An important observation is that sampling the replication of period Ω of $\hat{f}$ is the same as replicating the samples of $\hat{f}$ with period N. In other words, letting $\hat{f}_k = \hat{f}(\omega_k)$, we can write $\mathcal{R}_\Omega\{\hat{f}(\omega_k)\} = \mathcal{R}_N\{\hat{f}_k\}$. Combining the modified right and left sides of the Poisson Summation Formula leads to the following deceptively simple result.

▶ **Replication Form of the Poisson Summation Formula** ◀

$$AD\{\mathcal{R}_N\{f_n\}\}_k = \mathcal{R}_N\{\hat{f}_k\}. \tag{6.11}$$

Stand back and see what we have done! The result is the remarkable fact that (up to the constant A)

the N-point DFT of the sampled replication of f is the sampled replication of the Fourier transform of f.

The replications in the two domains have different physical periods: in the spatial domain the replication has a period of A, while in the frequency domain the replication has a period of $\Omega = N/A$. Not surprisingly, the two periods are related by the reciprocity relations. But as replications of sample sequences they both have periods of N.

Before moving ahead it might pay to indicate schematically what we have just learned. Given a function f we may denote its relationship to its Fourier transform as

$$f \xrightarrow{\mathcal{F}} \hat{f}.$$

We now see that the DFT gives the analogous relationship between the replications of the samples of f and $\hat{f}$; namely

$$A\mathcal{R}_N\{f_n\} \xrightarrow{\mathcal{D}} \mathcal{R}_N\{\hat{f}_k\}.$$

It is impossible to unravel all of the implications of these observations at once; in fact, the remainder of this chapter will be devoted to that task. Occasionally we will pause and look at a particular result or problem from the replication perspective, and very often it will provide penetrating insights.

6.5. Input with Compact Support

With the Poisson Summation Formula and the replication perspective in our quiver, we may now turn to a more general class of functions that could be used as input to the DFT. This is the class of **compactly supported** functions (also called **spatially limited functions** or **functions of finite duration**). These functions have the property that

$$f(x) = 0 \quad \text{for} \quad |x| > \frac{A}{2},$$

where $A > 0$ is some real number. The interval $[-A/2, A/2]$ is called the **interval of support**, or simply the **support**, of f. Compactly supported functions are not uncommon in applications. In most problems with spatial dependence (for example, image processing or spectroscopy), an object of finite extent is represented by a function that vanishes outside of some region. Although a compactly supported function is not periodic, we can still compute its Fourier coefficients c_k on its interval of support. The function

$$g(x) = \sum_{k=-\infty}^{\infty} c_k e^{i2\pi kx/A} \quad \text{where} \quad c_k = \frac{1}{A} \int_{-\frac{A}{2}}^{\frac{A}{2}} f(x) e^{-i2\pi kx/A} dx \qquad (6.12)$$

is an A-periodic function (called the periodic extension of f) that is identical to f on the interval $[-A/2, A/2]$ (using average values of f at the endpoints of the interval if necessary). This says that when a compactly supported function is sampled on its interval of support, and the samples are used as input for a DFT, it is as if the periodic function g had been sampled. And since the DFT "sees" samples of a periodic function, the results of the previous two sections are relevant. Notice that if the original function f is spatially limited, then it cannot also be band-limited, and hence aliasing can be expected to occur.

However, we have not dispensed with this case completely. It is not entirely analogous to the case of periodic functions. Since f is compactly supported, it also has a Fourier transform, and it is instructive to relate the Fourier transform of a compactly supported function to its Fourier coefficients. As was shown in Section 2.7, if $f(x) = 0$ when $|x| > A/2$, then its Fourier transform is given by

$$\hat{f}(\omega) = \int_{-\infty}^{\infty} f(x) e^{-i2\pi \omega x} dx = \int_{-\frac{A}{2}}^{\frac{A}{2}} f(x) e^{-i2\pi \omega x} dx. \qquad (6.13)$$

Comparing the Fourier transform (6.13) evaluated at $\omega = \omega_k = k/A$ to the Fourier coefficients of f (6.12), we discover a valuable result.

▶ $c_k \leftrightarrow \hat{f}(\omega_k)$ **Relationship** ◀

The Fourier transform of a function compactly supported on $[-A/2, A/2]$ evaluated at the frequency $\omega_k = k/A$ is a constant multiple of the corresponding Fourier coefficient:

$$c_k = \frac{1}{A} \hat{f}(\omega_k) \quad \text{for all} \quad k = 0, \pm 1, \pm 2, \ldots.$$

Hence the DFT of a compactly supported function will provide approximations to both its Fourier coefficients and its Fourier transform. The task is to estimate the errors

in these approximations. Before stating the central theorem, it might be helpful
to garner some qualitative understanding of DFT errors in the case of compactly
supported functions.

The replication form of the Poisson Summation Formula (6.11) can tell us a lot.
Recall that it says

$$AD\{\mathcal{R}_N\{f_n\}\}_k = \mathcal{R}_N\{\hat{f}_k\}.$$

In the present case, in which f has compact support on $[-A/2, A/2]$, replication of f_n
with period N does not change the sequence f_n; hence $\mathcal{R}_N\{f_n\} = f_n$. Therefore, the
Poisson Summation Formula takes the form

$$AF_k = \mathcal{R}_N\{\hat{f}_k\} = \hat{g}(\omega_k),$$

where we have used $\hat{g}(\omega_k)$ to stand for samples of the replication of $\hat{f}$. Now, the error
in the DFT, F_k, as an approximation to $\hat{f}(\omega_k)$ is $|AF_k - \hat{f}(\omega_k)|$ for $k = -N/2+1 : N/2$.
With the help of the triangle inequality, it may be written as

$$
\begin{aligned}
|AF_k - \hat{f}(\omega_k)| &= |AF_k - \hat{g}(\omega_k) + \hat{g}(\omega_k) - \hat{f}(\omega_k)| \\
&\leq \underbrace{|AF_k - \hat{g}(\omega_k)|}_{0} + \underbrace{|\hat{g}(\omega_k) - \hat{f}(\omega_k)|}_{\text{Error due to sampling}} .
\end{aligned}
$$

The first term vanishes identically by virtue of the Poisson Summation Formula.
The second term can be attributed to the sampling of the function f: sampling
in the spatial domain essentially replaces $\hat{f}$ by the replication of $\hat{f}$ which we have
called $\hat{g}$. The outcome is rather surprising; we see that, in this special case of a
compactly supported function, the error in the DFT is simply the difference between
the transform $\hat{f}$ and its replication $\hat{g}$. We need to learn more about the actual size of
that error.

We will briefly mention another perspective on DFT errors. Approximating the
Fourier transform of a compactly supported function takes place in two steps:

1. the function f must be sampled on its interval of support $[-A/2, A/2]$ (or a
   larger interval);

2. the sampled function, now a sequence of length N, is used as input to the DFT.

Each of these steps has a tangible graphical interpretation. The reader is directed
to the excellent discussion and figures of Brigham [20], [21] for the details of this
perspective. Having presented these preliminary arguments for motivation, let's now
turn to the central result that actually gives estimates of the size of errors in the DFT
approximations to the Fourier transform.

THEOREM 6.5. ERROR IN THE DFT (COMPACTLY SUPPORTED FUNCTIONS). *Let*
$f(x) = 0$ *for* $|x| \geq A/2$. *Let the A-periodic extension of f have $(p - 1)$ continuous
derivatives for $p \geq 1$ and assume that $f^{(p)}$ is bounded and piecewise monotone on*
$[-A/2, A/2]$. *If the N-point DFT is used to approximate $\hat{f}$ at the points $\omega_k = k/A$,*
then

$$|AF_k - \hat{f}(\omega_k)| \leq \frac{C}{N^{p+1}} \quad for \quad k = -\frac{N}{2} + 1 : \frac{N}{2},$$

where C is a constant independent of k and N.

Proof: The assumption that $f(x) = 0$ for $|x| \geq A/2$ reduces the Poisson Summation Formula to

$$\Delta x \sum_{n=-\frac{N}{2}+1}^{\frac{N}{2}} f(x_n) e^{-i2\pi x_n \omega_k} = \sum_{j=-\infty}^{\infty} \hat{f}\left(\omega_k - \frac{j}{\Delta x}\right),$$

where we have evaluated both sums at $\omega = \omega_k = k/A$. The sum on the left is $N\Delta x = A$ times the N-point DFT of f sampled on the interval $[-A/2, A/2]$. We can now rearrange the previous expression and apply the triangle inequality to conclude that

$$|AF_k - \hat{f}(\omega_k)| \leq \sum_{j=1}^{\infty} \left(\left|\hat{f}\left(\omega_k - \frac{j}{\Delta x}\right)\right| + \left|\hat{f}\left(\omega_k + \frac{j}{\Delta x}\right)\right|\right) = \sum_{j=1}^{\infty} \left(|\hat{f}(\omega_{k-jN})| + |\hat{f}(\omega_{k+jN})|\right).$$

The equality follows by recalling that $j/\Delta x = jN\Delta\omega = \omega_{jN}$, and that $\omega_k \pm \omega_{jN} = \omega_{k\pm jN}$. We now call in the relationship between the Fourier coefficients and the Fourier transform, $\hat{f}(\omega_k) = Ac_k$, which gives us that

$$|AF_k - \hat{f}(\omega_k)| \leq A \sum_{j=1}^{\infty} (|c_{k-jN}| + |c_{k+jN}|) \quad \text{for} \quad k = -\frac{N}{2} + 1 : \frac{N}{2}.$$

Theorem 6.2 can now be used to bound the Fourier coefficients c_k by a constant times $|k|^{-p-1}$ and then we proceed as in the proof of Theorem 6.3 to bound the series. The result is that

$$|AF_k - \hat{f}(\omega_k)| \leq \frac{C}{N^{p+1}} \quad \text{for} \quad k = -\frac{N}{2} + 1 : \frac{N}{2},$$

where C is a constant that depends on neither k nor N.   ∎

Two comments are in order. The first comment concerns another *pester* at the endpoints. The result of this theorem depends on the assumption that $f(\pm A/2) = 0$. If this requirement is relaxed to allow either $f(\frac{A}{2}^-) \neq 0$ or $f(-\frac{A}{2}^+) \neq 0$, then the periodic extension of f has a discontinuity at the endpoints, and a weaker bound results. This brings us to the second comment. If f is bounded but discontinuous (which would correspond to $p = 0$), a separate proof is required. However, the result is as one might expect and hope: if the A-periodic extension of f is bounded and piecewise monotone on $[-A/2, A/2]$ then the error in using the DFT to approximate the Fourier transform at the frequencies ω_k is bounded by C/N where C is independent of k and N [76]. Therefore, with this addendum, this theorem applies to all inputs with compact support that might be encountered in practice. We now present a case study to illustrate the conclusions of this theorem.

Case Study 3: Compactly supported functions. In this case study we consider a sequence of functions with increasing smoothness, each with compact support on the interval $[-1, 1]$. These functions are shown in Figure 6.4 and are given analytically as follows.

1. Square pulse. The function

$$f(x) = \begin{cases} 1 & \text{if} \quad |x| \leq 1, \\ 0 & \text{if} \quad |x| > 1, \end{cases}$$

has a periodic extension that is not continuous at the endpoints ± 1, and we expect that the error in the DFT should be bounded by C/N.

2. Triangular pulse. The function

$$f(x) = \begin{cases} 1 - |x| & \text{if } |x| \le 1, \\ 0 & \text{if } |x| > 1, \end{cases}$$

has a periodic extension that is continuous, but its derivative is discontinuous at $x = 0, \pm 1$. Theorem 6.5 applies with $p = 1$, and we expect the error in the DFT to be bounded by C/N^2.

3. Quadratic with cusps. The function

$$f(x) = \begin{cases} 1 - x^2 & \text{if } |x| \le 1, \\ 0 & \text{if } |x| > 1, \end{cases}$$

like the previous example, is continuous when extended periodically, but has a discontinuity in its derivative at $x = \pm 1$ ($p = 1$). The error in the DFT should decrease as C/N^2 as N increases.

4. Smooth quartic. The function

$$f(x) = \begin{cases} (1 - x^2)^2 & \text{if } |x| \le 1, \\ 0 & \text{if } |x| > 1, \end{cases}$$

has continuous derivatives of orders $p = 0, 1, 2$ when extended periodically and Theorem 6.5 applies with $p = 3$. We expect to see errors bounded by C/N^4.

5. Smoother still! The periodic extension of the function

$$f(x) = \begin{cases} (1 - x^2)^4 & \text{if } |x| \le 1, \\ 0 & \text{if } |x| > 1, \end{cases}$$

has continuous derivatives of order $p = 0, 1, 2, 3, 4$, and Theorem 6.5 applies with $p = 5$.

We define the error in the DFT approximations to the Fourier transform as

$$E_N = \max_{-\frac{N}{2}+1 \le k \le \frac{N}{2}} |F_k - \hat{f}(\omega_k)|.$$

Figure 6.8 shows how $\log E_N$ varies with $\log N$ for each of the five functions given above. The use of the logarithm identifies the rate of decrease of the errors since the graph of $\log(N^{-p})$ is a straight line with slope $-p$. We see that in all five cases the errors decrease as N^{-p-1}, as predicted by Theorem 6.5.

6.6.   General Band-Limited Functions

In our systematic survey of the DFT landscape, the next class of functions that appears consists of functions that are nonperiodic, but are **band-limited**. The notion of band-limited functions was introduced earlier in the chapter with respect to periodic functions. We now give the analogous (and conventional) definition for nonperiodic functions. A function f with a Fourier transform $\hat{f}$ is said to be band-limited, if there exists some constant Ω such that

$$\hat{f}(\omega) = 0 \quad \text{for} \quad |\omega| > \frac{\Omega}{2}.$$

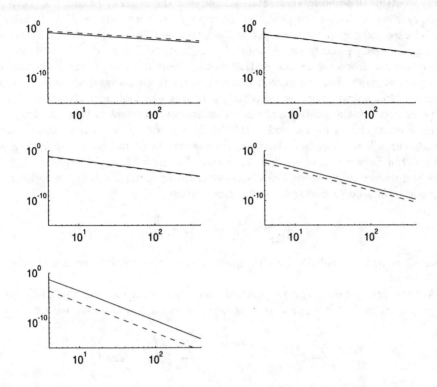

FIG. 6.8. **Case Study 3.** *The errors in DFT approximations to the Fourier transforms of the five functions of Case Study 3 are shown in this figure. The greater the degree of smoothness of f, the faster the decay rate of the errors. Each plot consists of the error curve (solid line) and the theoretical error bound $N^{-(p+1)}$ (dashed line), plotted on a log–log scale, so that a trend of $N^{-\alpha}$ appears as a line with slope $-\alpha$. The curves correspond to (top left) $p = 0$, the square pulse; (top right) $p = 1$, $f(x) = 1 - |x|$; (middle left) $p = 1$, $f(x) = 1 - x^2$; (middle right) $p = 3$, $f(x) = (1 - x^2)^2$; and (bottom left) $p = 5$, $f(x) = (1 - x^2)^4$. As predicted by the theory, the errors decrease very nearly as N^{-p-1}. It should be noted that only the slopes of the error curves are significant.*

As in the periodic case, band-limited simply means that the function f has no Fourier components with frequencies above the **cut-off frequency** $\Omega/2$. Note that the Fourier transform of a band-limited function has compact support. Two practical notes should be injected here. In most applications, it is impossible to know a priori whether a given function or "signal" is band-limited. The objective in computing the DFT is to determine the frequency structure of the function, and until it is known, a judgment about "band-limitedness" is unfounded. Perhaps more important is the fact that *truly* band-limited functions or signals are rare in practice. Fourier transforms may decrease rapidly, and at predictable rates, but functions whose Fourier transforms vanish beyond some fixed frequency are arguably idealizations. Nevertheless, with this caution now in the wind, let us proceed, because this case has several important lessons. The first path we will follow involves the Poisson Summation Formula, which has served us so well already. In contrast to the previous section in which f itself vanished outside of some interval, thus simplifying the Poisson Summation Formula, we can now see a similar simplification because $\hat{f}$ vanishes outside of some interval. First let's set some ground rules. As in the past, we will imagine that f is sampled at N equally spaced points on an interval $[-A/2, A/2]$. Notice that there is now no natural way to choose A or Δx in the spatial domain, since f has no periodicity or compact support. For this reason it makes sense to choose grid parameters in the frequency domain first; then the reciprocity relations will determine the corresponding grid parameters in the spatial domain. The first requirement is that the frequency grid must cover the entire interval $[-\Omega/2, \Omega/2]$ on which $\hat{f}$ is nonzero (the pester at the endpoint will be discussed shortly). Having chosen Ω *and* the number of sample points N, the frequency grid spacing $\Delta\omega$ is given by $\Delta\omega = \Omega/N$. Whether one appeals to the reciprocity relations or the Nyquist sampling condition (they are ultimately equivalent), the spatial grid spacing Δx must satisfy

$$\frac{1}{2\Delta x} \geq \frac{\Omega}{2} \quad \text{or} \quad \Delta x \leq \frac{1}{\Omega}$$

if aliasing is to be avoided. Finally, the length of the spatial domain is given by $A = N\Delta x$.

We are now ready to move towards an error estimate for the DFT as an approximation to $\hat{f}$. Not surprisingly, we begin with the Poisson Summation Formula

$$\Delta x \sum_{n=-\infty}^{\infty} f(x_n) e^{-i2\pi x_n \omega} = \sum_{j=-\infty}^{\infty} \hat{f}\left(\omega - \frac{j}{\Delta x}\right).$$

With the assumption that $\hat{f}(\omega) = 0$ for $|\omega| \geq \Omega/2 = 1/(2\Delta x)$, the infinite sum on the right side of the formula collapses to the single term $\hat{f}(\omega)$. The left side contains the DFT when evaluated at $\omega = \omega_k$. Therefore, evaluating both sides at $\omega = \omega_k$, the Poisson Summation Formula now takes the form

$$\Delta x \left[N F_k + \frac{1}{2}(f_{-\frac{N}{2}} + f_{\frac{N}{2}}) \cos(\pi k) + \sum_{|n| > \frac{N}{2}} f_n e^{-i2\pi x_n \omega_k} \right] = \hat{f}(\omega_k)$$

for $k = -N/2 + 1 : N/2 - 1$. We have used the familiar definition of the DFT (using average values at the endpoints) and written F_k on the left-hand side. Replacing $N\Delta x$ by A, a bound now follows by rearranging terms:

$$|A F_k - \hat{f}(\omega_k)| \leq \Delta x \left| \sum_{|n| \geq \frac{N}{2}}^{\infty} {}'' f_n \omega_N^{-nk} \right| \tag{6.14}$$

for $k = -N/2 + 1 : N/2 - 1$. The notation Σ'' has been introduced to indicate that the $\pm N/2$ terms of the sum are weighted by $1/2$. Another pester arises here concerning endpoints. Notice that if $\hat{f}(\omega_{N/2}^-) \neq 0$ or $\hat{f}(\omega_{-N/2}^+) \neq 0$ the above bound is not necessarily valid for $k = N/2$. It is a minor technicality which is avoided if Ω is chosen large enough that $\hat{f}(\omega_{\pm N/2}) = 0$.

Now the situation becomes a bit sticky, since the task is to estimate the sum on the right side of inequality (6.14). Any statement on this matter requires additional assumptions on f and its rate of decay for large $|x|$. For example, if it is known that

$|f(x)| < C|x|^{-r}$ for $|x| > A/2$ where $r \geq 1$, it is possible to approximate the sum on the right side of (6.14) by an integral which behaves asymptotically as A^{-r} for large A (problem 135). Such results may not be of great practical use, and we will not attempt to be more specific (see [36] for more results concerning the truncation of Fourier integrals). The lesson to be extracted from (6.14) is that the DFT error for band-limited functions is an error due to truncation of f in the spatial domain. It can be reduced by increasing A, which should be accomplished by increasing N with Δx fixed. Note that there is nothing to be gained by decreasing Δx (equivalently increasing Ω) since f is band-limited. We will postpone a theorem on DFT errors in the band-limited case, as it will appear as a special case of a more general theorem in the next section.

We now consider another perspective provided by the replication formulation of the Poisson Summation Formula (6.11)

$$A\mathcal{D}\{\mathcal{R}_N\{f_n\}\}_k = \mathcal{R}_N\{\hat{f}_k\}, \quad \text{where} \quad k = -\frac{N}{2} + 1 : \frac{N}{2}.$$

In the present case of a band-limited function, the replication of the samples of $\hat{f}$ becomes very simple *provided the sampling of the input is done correctly!* The replication of $\hat{f}$ occurs with a period of $1/\Delta x$. Assuming that the sampling interval Δx and the band-limit Ω satisfy the condition $\Delta x \leq 1/\Omega$, the replication of $\hat{f}$ will produce no overlapping of $\hat{f}$ with itself. This means that

$$\mathcal{R}_N\{\hat{f}_k\} = \hat{f}_k$$

for $k = -N/2 + 1 : N/2$, and no aliasing occurs in the frequency domain. Therefore, we see that if the DFT is applied to the samples of *the replicated* input sequence f_n, then the samples of the Fourier transform are produced exactly. There are a couple of interesting consequences of this observation.

First, it suggests a method for improving the DFT approximation to the Fourier transform of a band-limited function: if the function f is available over a large interval, one can compute an approximation to $\mathcal{R}_N\{f_n\}$ by taking several terms in the replication sum. This approximation can then be sampled at N points of the interval $[-A/2, A/2]$ and the samples of the replication can be used as input to the DFT. In principle, the more closely $\mathcal{R}_N\{f_n\}$ can be approximated for the input, the more closely the samples of $\hat{f}$ can be approximated. The efficacy of this strategy is examined in Case Study 7 below. The second observation has already been made, but it is quite transparent within the replication framework. The difference between the samples of f_n and the samples of $\mathcal{R}_N\{f_n\}$ can be minimized by taking the period of the replication operator as large as possible. The sampling of the input takes place over the spatial domain $[-A/2, A/2]$; therefore, increasing A decreases the overlapping of the tails of f in the replication process. If A is increased with N fixed, then Δx also increases. The reciprocity relations ($\Omega = N/A = 1/\Delta x$) tell us that the extent of the frequency grid must decrease as Δx increases *with the possibility that it will no longer cover the entire interval on which $\hat{f}$ is nonzero.* This oversight would once again cause aliasing errors in the frequency domain. On the other hand, if A is increased *and* N is increased so that Δx does not increase, then the frequency grid does not decrease in length, and the full support of $\hat{f}$ can still be sampled. As is so often the case, increasing N is a remedy for many DFT errors.

Here is a third observation that results from the replication perspective. The error that we would like to estimate is $|AF_k - \hat{f}(\omega_k)|$ for $k = -N/2 + 1 : N/2$. Using the triangle inequality we can write

$$|AF_k - \hat{f}(\omega_k)| = |AF_k - A\mathcal{D}\{\mathcal{R}_N\{f_n\}\}_k + A\mathcal{D}\{\mathcal{R}_N\{f_n\}\}_k - \hat{f}(\omega_k)|$$
$$\leq \underbrace{|AF_k - A\mathcal{D}\{\mathcal{R}_N\{f_n\}\}_k|}_{\text{Truncation error}} + \underbrace{|A\mathcal{D}\{\mathcal{R}_N\{f_n\}\}_k - \hat{f}(\omega_k)|}_{0}.$$

The source of the DFT error is the first term which can be viewed as a truncation error since it is the difference between the DFT of f_n and the DFT of the replication of f_n (which includes values of f outside the interval $[-A/2, A/2]$). The second term vanishes identically because of the replication form of the Poisson Summation Formula in this special case of a band-limited function.

We mention that there is also a graphical approach to understanding the DFT errors in this case. Let's begin by making the observation that in using the DFT to approximate the Fourier transform of a general band-limited function, three steps must be performed:

1. f is truncated to restrict it to the interval of interest $[-A/2, A/2]$,

2. f is sampled with a grid spacing Δx, which must satisfy $\Delta x \leq 1/\Omega$ if aliasing is to be avoided, and

3. the truncated, sampled version of f, now a sequence of length N, is used as input to the DFT.

Each of these steps has a known effect on f and its Fourier transform, which can be displayed graphically. Although this approach does not lead to an estimate of the error in the DFT, it does provide insight into how errors arise. As before, we cite the books of Brigham [20], [21] for the original presentation of this argument.

Finally, it would be inexcusable to discuss band-limited functions without returning to the Shannon Sampling Theorem which was first encountered in Chapter 3. Given the omnipresence of the Poisson Summation Formula in this chapter, perhaps it is not surprising that it can also be used to derive the Sampling Theorem. The following presentation is not rigorous, but it is instructive nonetheless.

Moments ago we saw that the Poisson Summation Formula for a band-limited function f is given by

$$\hat{f}(\omega) = \Delta x \sum_{n=-\infty}^{\infty} f_n e^{-i2\pi x_n \omega}.$$

To avoid endpoint pesters, assume that $\hat{f}(\omega) = 0$ for $|\omega| \geq \Omega/2$. Then the spatial grid spacing should be chosen such that $\Delta x \leq 1/\Omega$ to avoid aliasing. Since $\hat{f}$ vanishes outside of the interval $[-\Omega/2, \Omega/2]$, we can multiply both sides of the previous equation by the square pulse $\hat{p}(\omega)$, which has a value of 1 on $(-\Omega/2, \Omega/2)$ and is zero elsewhere. In other words,

$$\hat{f}(\omega) = \hat{f}(\omega)\hat{p}(\omega) = \Delta x \sum_{n=-\infty}^{\infty} f_n \left[e^{-i2\pi x_n \omega} \hat{p}(\omega) \right].$$

The goal is to extract from this equation an expression for the original band-limited function f. It appears that we are not far from this goal since $\hat{f}$ appears on the left side of this last equation. Therefore, we take an inverse Fourier transform of this equation and try to make sense of the right-hand side. Doing this, we have

$$f(x) = \Delta x \sum_{n=-\infty}^{\infty} f_n \mathcal{F}^{-1} \left[e^{-i2\pi x_n \omega} \hat{p}(\omega) \right].$$

Now recall two facts:

1. the inverse Fourier transform of the symmetric square pulse with width Ω is

$$p(x) = \mathcal{F}^{-1} \hat{p}(\omega) = \Omega \, \text{sinc} \, (\pi \Omega x) = \frac{1}{\Delta x} \, \text{sinc} \, \left(\frac{\pi x}{\Delta x} \right), \quad \text{and}$$

2. by the shift theorem for Fourier transforms

$$\mathcal{F}^{-1} \left(e^{i2\pi x_n \omega} \hat{p}(\omega) \right) = p(x - x_n) = \frac{1}{\Delta x} \, \text{sinc} \, \left(\frac{\pi(x - x_n)}{\Delta x} \right).$$

Assembling these observations, we arrive once again with the Shannon Sampling Theorem as it was presented in Chapter 3. If f is band-limited with $\hat{f}(\omega) = 0$ for $|\omega| \geq \Omega/2$, and the grid spacing Δx is chosen such that $\Delta x \leq 1/\Omega$, then

$$f(x) = \sum_{n=-\infty}^{\infty} f_n \, \text{sinc} \, \left(\frac{\pi(x - x_n)}{\Delta x} \right) = \Delta x \sum_{n=-\infty}^{\infty} f_n \frac{\sin(\pi(x - x_n)/\Delta x)}{\pi(x - x_n)}. \quad (6.15)$$

The theorem claims that if a function f is band-limited with a cut-off frequency $\Omega \leq 1/\Delta x$, then it may be reconstructed from its samples f_n. Admittedly, the reconstruction requires an infinite number of samples and the use of the sinc function (6.15) formula [131]. However, the theorem has variations and approximate versions that are of practical value. An excellent departure point for further reading on the Shannon Sampling Theorem is the tutorial review by Jerri [82], in which the author discusses the theorem and a host of related issues.

Case Study 4: Band-limited functions. As mentioned earlier, genuinely band-limited functions rarely arise in practice. In this case study, we will examine a clean but idealized band-limited function. The square pulse (or square wave) function was encountered in Chapter 3 along with its sinc function Fourier transform. We will turn this transform pair around and use the fact that

$$\mathcal{F} \left\{ \text{sinc} \, (\pi x) \right\} = \mathcal{F} \left\{ \frac{\sin(\pi x)}{\pi x} \right\} = B_1(\omega) = \begin{cases} 1 & \text{if } |\omega| < \frac{1}{2}, \\ \frac{1}{2} & \text{if } |\omega| = \frac{1}{2}, \\ 0 & \text{if } |\omega| > \frac{1}{2}. \end{cases}$$

Since B_1, the square pulse of width one, has compact support, the sinc function is band-limited. The use of the DFT to approximate the Fourier transform of the sinc function is an exquisite example of the reciprocity relations at work.

Two parameters will play leading roles in this numerical experiment. One is A, which determines the interval $[-A/2, A/2]$ from which samples of the sinc function are collected. The other parameter is the number of samples N. Once values of A and N are selected, the reciprocity relations determine the rest. The grid spacing in the frequency domain is $\Delta \omega = 1/A$, and the length of the frequency domain

TABLE 6.1
Grid parameters for Case Study 4.
DFT of a band-limited function.

Figure	N	A	$\Delta\omega = 1/A$	Ω
Upper right	32	8	$.125 = \frac{1}{8}$	4
Middle left	64	8	$.125 = \frac{1}{8}$	8
Middle right	64	16	$.0625 = \frac{1}{16}$	4
Lower left	128	32	$.03125 = \frac{1}{32}$	4
Lower right	128	64	$.015625 = \frac{1}{64}$	2

is $\Omega = N\Delta\omega = N/A$. Figure 6.9 shows just a few of the many possible DFT approximations that might be computed using different combinations of A and N. The upper left figure shows the sinc function itself. The remaining five cases and the attendant grid parameters are summarized in Table 6.1.

The numerical evidence is quite informative. The general shape of the square pulse is evident in all five DFT sequences shown and becomes more clearly defined as N increases. All of the approximations show oscillations (overshoot) near the discontinuities in the square pulse, which is due to the well-known Gibbs[4] effect. These oscillations subside with increasing N. Caused by the nonuniform convergence of the Fourier series near discontinuities, the Gibbs effect has been analyzed intensively; accounts of the Gibbs effect and methods to correct and minimize it can be found in [30], [55], and [84].

Let's see how the reciprocity relations enter the picture. With a width of one unit, there are roughly $1/\Delta\omega$ grid points under the nonzero part of the square pulse. By virtue of the reciprocity relations, the only way to increase the resolution and put more points under the pulse is to increase the length of the sampling interval A, since $\Delta\omega = 1/A$. However, if A is increased with N fixed, the effect is to decrease the length of the frequency domain Ω, as shown in moving from the middle left to the middle right figure. Note that the unit on the horizontal axes of the DFT plots is actual frequency (cycles per unit length). Therefore, if one wishes to increase the frequency resolution and maintain the same frequency domain, it is necessary to increase *both* A and N commensurately, as shown by moving from the middle right to the lower left figure. As the sequence of figures suggests, errors in the DFT decrease as both A and N are increased, and with $A = 64$ and $N = 128$ (lower right figure), the square pulse is fairly well resolved.

6.7.   General Input

We now come to the final case in which the input to the DFT does not have periodicity, compact support, or a band-limit. We will assume that these functions are defined on an interval of the real line (a, b) where a and/or b is infinite. The only other assumption is that the function f that provides the input is absolutely integrable

---

[4] JOSIAH WILLARD GIBBS (1839–1903) is generally regarded as one of the first and greatest American physicists. A professor at Yale University, he was the founder of chemical thermodynamics and modern physical chemistry.

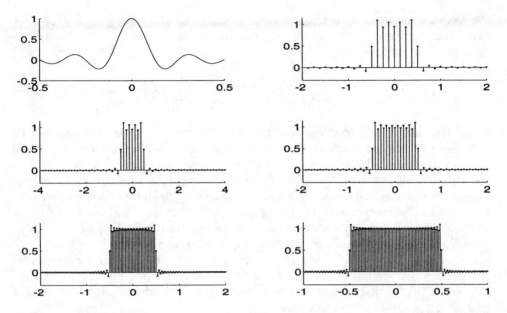

FIG. 6.9. **Case Study 4.** *The Fourier transform of the band-limited sinc function (upper left) is approximated by the DFT with various values of A and N as given in Table 6.1. The horizontal axes on the DFT plots have units of frequency (cycles per unit length).*

$(\int_a^b |f(x)|dx < \infty)$; this insures that its Fourier transform exists. Our goal is to approximate the Fourier transform $\hat{f}$ using the DFT and then to estimate the errors in that approximation. Our conclusions may be less general than in earlier cases. Nevertheless, some insight can be gained and we will be able to make some remarks that unify all of the preceding sections.

Let's return to the Poisson Summation Formula one last time, and it will tell us immediately what this case entails. Recall that it may be written

$$\Delta x \sum_{n=-\infty}^{\infty} f(x_n)e^{-i2\pi x_n \omega} = \sum_{j=-\infty}^{\infty} \hat{f}\left(\omega - \frac{j}{\Delta x}\right),$$

where Δx is the grid spacing in the spatial domain and ω is an arbitrary frequency. In the absence of compact support for f (which would make the left-hand sum finite) or band-limiting (which would make the right-hand sum finite), both of the sums are infinite in general. We can split both sums, evaluate them at $\omega = \omega_k$, and use the definition of the DFT to write

$$\Delta x \sum_{n=-\infty}^{\infty} f(x_n)e^{-i2\pi x_n \omega} = \Delta x \left[NF_k + \frac{1}{2}(f_{-\frac{N}{2}} + f_{\frac{N}{2}})\cos(\pi k) \right.$$

$$\left. + \sum_{|n|>\frac{N}{2}}^{\infty} f(x_n)e^{-i2\pi x_n \omega_k} \right]$$

$$= \hat{f}(\omega_k) + \sum_{|j|=1}^{\infty} \hat{f}\left(\omega_k - \frac{j}{\Delta x}\right).$$

We have set $\omega = \omega_k$ to anticipate the fact that F_k will approximate $\hat{f}(\omega_k)$. A slight rearrangement with $N\Delta x = A$ gives an expression for the error in AF_k as an approximation to $\hat{f}(\omega_k)$:

$$|AF_k - \hat{f}(\omega_k)| \leq \Delta x \left| \sum_{|n| \geq \frac{N}{2}}^{\infty} {}'' f(x_n) e^{-i2\pi x_n \omega_k} \right| + \left| \sum_{|j|=1}^{\infty} \hat{f}\left(\omega_k - \frac{j}{\Delta x}\right) \right|. \qquad (6.16)$$

Again the notation Σ'' indicates that the $\pm N/2$ terms in the sum are weighted by $1/2$. We can now see qualitatively how errors enter the DFT. In a very real sense this general case is a linear combination of the two previous cases. The first term on the right side of this error bound is due to the fact that f does not have compact support, and f must be truncated on a finite interval; this term was encountered in Section 6.6. The second term on the right side of this inequality arises because f is not band-limited, and hence some aliasing can be expected in the form of overlapping "tails" of $\hat{f}$. This was the same term that was handled in Section 6.5.

It turns out that a DFT error result can be stated in this case for specific classes of functions. We will offer such a result for functions with exponential decay for large $|x|$. While this hardly exhausts all functions of practical interest, it does suggest a general approach to estimating errors.

THEOREM 6.6. ERROR IN THE DFT (GENERAL FUNCTIONS WITH EXPONENTIAL DECAY). *Let the A-periodic extension of f have $(p-1)$ continuous derivatives and assume that $f^{(p)}$ is integrable and piecewise monotone on $(-\infty, \infty)$ for $p \geq 1$. Furthermore, assume that $|f(x)| < Ke^{-a|x|}$ for $|x| \geq A/2$ for some $a > 0$. If the function f is sampled on the interval $[-A/2, A/2]$ and the N-point DFT is used to approximate $\hat{f}$ at the points $\omega_k = k/A$, then*

$$|AF_k - \hat{f}(\omega_k)| \leq \frac{2Ke^{-aA/2}e^{-aA/N}}{a} + \frac{C}{N^{p+1}},$$

for $k = -N/2 + 1 : N/2$, where C is a constant independent of k and N.

The proof follows that of Theorem 6.5 with additional arguments to handle the exponential decay and the noncompact support of f. A brief sketch of the proof is in order.

Proof: As shown above (6.16), the Poisson Summation Formula for this case leads to the bound

$$|AF_k - \hat{f}(\omega_k)| \leq \Delta x \left| \sum_{|n| \geq \frac{N}{2}}^{\infty} {}'' f(x_n) e^{-i2\pi x_n \omega_k} \right| + \left| \sum_{|j|=1}^{\infty} \hat{f}\left(\omega_k - \frac{j}{\Delta x}\right) \right|.$$

The first sum can be bounded by

$$\int_{A/2-\Delta x}^{\infty} |Ke^{-ax}e^{-i2\pi\omega_k x}| dx + \int_{-\infty}^{-A/2+\Delta x} |Ke^{ax}e^{i2\pi\omega_k x}| dx = \frac{2Ke^{-a(A/2-\Delta x)}}{\sqrt{a^2 + 4\pi^2\omega_k^2}}.$$

This expression attains a maximum over k when $\omega_k = 0$, and a bound for the first sum is given by

$$\frac{2Ke^{-aA/2}e^{-aA/N}}{a}.$$

The second sum cannot be treated as it was in the proof of Theorem 6.5, since f is not compactly supported. We must first write the summand as

$$\hat{f}\left(\omega_k - \frac{j}{\Delta x}\right) = \hat{f}\left(\frac{k-jN}{A}\right) = \left(\int_{-\infty}^{-A/2} + \int_{-A/2}^{A/2} + \int_{A/2}^{\infty}\right) f(x) e^{-2\pi(k-jN)x/A} dx.$$

The second integral over $[-A/2, A/2]$ can be identified as the Fourier coefficient c_{k-jN} and treated as in the proof of Theorem 6.5. The first and third integrals must be integrated by parts p times with contributions at $x = \pm A/2$ canceling, as in the proof of Theorems 6.1 and 6.2. The outcome is that the entire second sum has a bound of the form C/N^{p+1}. Combining these two bounds we have that

$$|AF_k - \hat{f}(\omega_k)| \leq \frac{2K e^{-aA/2} e^{-aA/N}}{a} + \frac{C}{N^{p+1}}$$

for $k = -N/2 + 1 : N/2$. This proof has not been extended for the case $p = 0$, but it seems feasible. ∎

Let's make a few observations. Although the goal is to approximate the Fourier transform $\hat{f}$, the DFT is still applied on a finite interval $[-A/2, A/2]$ with a finite number of points. The error in the DFT depends on the number of sample points, the length of the interval, and the smoothness of f on that interval. Specifically, the error decreases as the length of the interval increases (approaching the interval of integration for the Fourier transform), as the number of points increases (as we have seen before), and as the smoothness of f increases (including smoothness at the endpoints $\pm A/2$). These dependencies will become clear when we present a case study.

As usual, the replication perspective also offers insight. The Poisson Summation Formula in terms of replication operators is

$$A\mathcal{D}\{\mathcal{R}_N\{f_n\}\}_k = \mathcal{R}_N\{\hat{f}_k\}.$$

In the absence of compact support or band-limiting, neither replication operation can be simplified. If the DFT could be applied to the full replication of the input, $\mathcal{R}_N\{f_n\}$, the best we could do is to produce samples of *the replication* of $\hat{f}$. Unfortunately, it is impossible to reconstruct a function or a sequence uniquely from its replication. Therefore, an error is introduced because the DFT produces samples of $\mathcal{R}_N\{\hat{f}\}$, not samples of $\hat{f}$ itself; this is the aliasing error mentioned above. But in practice the DFT cannot be applied to the exact replication of the input; therefore, a second source of error is introduced, corresponding to the truncation of f. Notice the tension that the reciprocity relations impose. The error in either of the two replication operations ($\mathcal{R}_N\{f_n\}$ or $\mathcal{R}_N\{\hat{f}_k\}$) can be reduced by increasing either A or Ω. However, unless N is increased commensurately, the reduction in one error is achieved at the expense of the other.

Case Study 5: Asymmetric exponentials. We will consider the problem of approximating the Fourier transform of functions of the form

$$f(x) = \begin{cases} e^{-ax} & \text{for} \quad x \geq 0, \\ 0 & \text{for} \quad x < 0, \end{cases}$$

where $a > 0$. Parting with the convention used in most of this book, we will consider the DFT applied to a set of N equally spaced samples taken from the asymmetric interval $[0, A]$. The way in which A and N are chosen is of critical importance, and we will investigate how the error in the DFT varies with these two parameters. We will proceed analytically and give a lustrous exhibit of the limiting relations among the DFT, the Fourier coefficients, and the Fourier transform.

Clearly, f is not compactly supported and there is no reason to suspect that it is band-limited. From a practical point of view, perhaps the first decision concerns the frequency range that needs to be resolved. Let's assume that the Fourier transform is required at frequencies in the range $[-\Omega/2, \Omega/2]$, where $\Omega/2$ is a specified maximum frequency. The reciprocity relationship now enters in a major way. In order to resolve components in this frequency range, the grid spacing Δx must satisfy $\Delta x \leq 1/\Omega$. Having chosen Δx, the choice of N, the number of sample points, determines the length of the spatial interval $A = N\Delta x$. It also determines the grid spacing in the frequency domain since $\Delta\omega = \Omega/N$. Finally, with $\Delta\omega$ specified, the actual frequencies $\omega_k = k/A$ are determined for $k = -N/2+1 : N/2$. It also follows from the reciprocity relationship that if a larger range of frequencies is desired, then Δx must be decreased. If this decrease is made with N fixed, then A decreases and $\Delta\omega$ increases; that is, resolution is lost on the frequency grid. If Δx is decreased *and* N is increased proportionally, then A and $\Delta\omega$ remain unchanged. With these qualitative remarks, let's now do some calculations.

It is possible to compute the DFT and Fourier series coefficients of f on $[-A/2, A/2]$ analytically. They are given in *The Table of DFTs* as

$$F_k = \frac{(1 - e_N^2) - i2e_N \sin\theta_k}{2N(1 - 2e_N \cos\theta_k + e_N^2)}(1 - e^{-aA}) \quad \text{and} \quad c_k = \frac{a - i2\pi\omega_k}{a^2 + 4\pi^2\omega_k^2}\frac{(1 - e^{-aA})}{A},$$

where $e_N = e^{-aA/N}$ and $\theta_k = 2\pi k/N$. A short calculation also reveals that the Fourier transform of f is given by

$$\hat{f}(\omega_k) = \frac{a - i2\pi\omega_k}{a^2 + 4\pi^2\omega_k^2},$$

where $\hat{f}$ has been evaluated at $\omega = \omega_k = k/A$. Notice that $c_k = \hat{f}(\omega_k)(1 - e^{-aA})/A$.

We can now compare F_k, c_k, and $\hat{f}(\omega_k)$. Here is the first observation. Recall that for a compactly supported function, $Ac_k = \hat{f}(\omega_k)$. For this noncompactly supported function, we see that Ac_k approaches $\hat{f}(\omega_k)$ as A becomes large (problem 132); that is,

$$\lim_{A\to\infty} (Ac_k) = \hat{f}(\omega_k).$$

To get the DFT into the picture, it will be useful to express F_k in an approximate form. Assume that A is fixed and N is large (hence Δx is small). Using a Taylor[5] series to expand $e_N, \sin\theta_k$, and $\cos\theta_k$, it can be shown (rather laboriously (problem 132)) that for $k = -N/2 + 1 : N/2$

$$F_k = \frac{a - i2\pi\omega_k}{a^2 + 4\pi^2\omega_k^2}\left(\frac{1 - e^{-aA}}{A}\right)(1 + c\Delta x) \quad \text{as} \quad N \to \infty, \ \Delta x \to 0,$$

where c is a constant that involves k, but not A or N. The term $c\Delta x$ represents the error in the Taylor series. Comparing the DFT in this form to the expressions for c_k and $\hat{f}(\omega_k)$, we can now write two most revealing relationships. We see that the DFT coefficients are related to the Fourier coefficients, for $k = -N/2 + 1 : N/2$, by

$$F_k = c_k(1 + c\Delta x). \tag{6.17}$$

Furthermore, the DFT coefficients and the Fourier transform are related according to

$$AF_k = \hat{f}(\omega_k)(1 - e^{-aA})(1 + c\Delta x). \tag{6.18}$$

---

[5]BROOK TAYLOR (1685-1731) was an English mathematician who published his famous expansion theorem in 1715. Educated at Cambridge University, he became secretary of the Royal Society at an early age, then resigned so he could write.

What does it mean? Looking at relation (6.17) first, we see that for fixed A and ω_k, the DFT coefficients approach the Fourier series coefficient as N becomes large and Δx approaches zero; that is,

$$\lim_{\substack{N \to \infty \\ \Delta x \to 0}} F_k = c_k$$

for $k = -N/2 + 1 : N/2$. Implicit in this limit (because of the reciprocity relations) is the fact that $\Delta \omega$ is fixed and $\Omega \to \infty$. In other words, letting Δx decrease allows higher frequencies to be resolved which means that the length of the frequency domain increases. However, since A remains fixed, the actual resolution in the frequency domain $\Delta \omega$ does not change.

Having let $\Delta x \to 0$ and $N \to \infty$, expression (6.18) becomes $AF_k = \hat{f}(\omega_k)(1 - e^{-aA})$. Now letting $A \to \infty$ (which also means $\Delta \omega \to 0$) we have that

$$\lim_{A \to \infty} \lim_{\substack{N \to \infty \\ \Delta x \to 0}} (AF_k) = \hat{f}(\omega_k)$$

for $k = -N/2 + 1 : N/2$, where A is held fixed in the inner limit while $\Delta x \to 0$ and $N \to \infty$.

We may also reverse the two limits above and watch the DFT approach $\hat{f}$ along another path. Imagine sampling f on larger intervals by holding Δx fixed and increasing A; this means that N increases as A increases. Notice that as A increases, $\Delta \omega$ decreases. Since we regard F_k as an approximation to $\hat{f}(\omega_k)$, it is necessary to let $\omega_k = k/A$ be fixed also. Look again at expression (6.18) and let A increase. We see that

$$\lim_{\substack{A \to \infty \\ N \to \infty}} (AF_k) = \lim_{\substack{A \to \infty \\ N \to \infty}} \hat{f}(\omega_k)(1 - e^{-aA})(1 + c\Delta x) = \hat{f}(\omega_k)(1 + c\Delta x).$$

In other words, if we hold the grid spacing Δx fixed and increase the length of the interval A by increasing N, then AF_k approaches the value of the Fourier transform at $\omega = \omega_k$ to within a relative error of $c\Delta x$. Notice that implicit in this limit, Ω is fixed and $\Delta \omega \to 0$.

If we now let Δx approach zero (meaning Ω becomes large), then AF_k approaches $\hat{f}(\omega_k)$. This two-limit process can be summarized as

$$\lim_{\Delta x \to 0} \lim_{\substack{A \to \infty \\ N \to \infty}} (AF_k) = \hat{f}(\omega_k)$$

for $k = -N/2 + 1 : N/2$, where in the inner limit Δx is fixed while $A \to \infty$. In practice, these limits are never realized computationally. However, they do indicate the sources of error in the DFT and how quickly those errors subside. In this particular case study, the error due to truncation of the interval of integration decreases exponentially with A (as predicted by Theorem 6.6). On the other hand, the error due to the lack of smoothness of f decreases only as Δx or $1/N$ (also by Theorem 6.6).

The relationships among the DFT, the Fourier series coefficients, and the Fourier transform, and the manner in which they approach each other in various limits are shown in Figure 6.10. These dependencies apply to any general (noncompactly supported, non-band-limited) input function f, although the rates of convergence in the various limits depend upon the properties of f. The previous analysis can also be carried out for the symmetric exponential function $f(x) = e^{-a|x|}$, with slightly different convergence rates (highly recommended: problem 134).

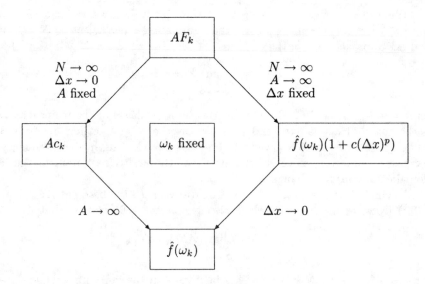

FIG. 6.10. *For a general (noncompactly supported, non-band-limited) function, the DFT F_k approximates the Fourier series coefficients c_k and the Fourier transform $\hat{f}(\omega_k)$ in various limits. With the interval $[A/2, A/2]$ fixed, F_k approaches the Fourier coefficients c_k as $N \to \infty$ and $\Delta x \to 0$ (which also implies that $\Omega \to \infty$ with $\Delta \omega$ fixed). With the sampling rate Δx fixed, F_k approaches $\hat{f}$ up to small errors proportional to $(\Delta x)^p$ as $N, A \to \infty$ (which also implies that $\Delta \omega \to 0$ with Ω fixed). Letting $A \to \infty$ in the first case or $\Delta x \to 0$ in the second case allows the AF_k to approach $\hat{f}(\omega_k)$. In order to make consistent comparisons, the frequency of interest ω_k must be held fixed in each of the limits.*

6.8. Errors in the Inverse DFT

You may agree that a significant amount of effort has been devoted to the question of errors in the *forward* DFT. And yet, in one sense, only half of the work has been done. We still have the equally important *inverse* DFT (IDFT) to consider. Be assured that the discussion of errors in the IDFT can be streamlined considerably, partly by relying on the results of the previous sections. At the same time, there are some features of the IDFT that are genuinely new, and these properties need to be pointed out carefully. First we set the stage and review a few earlier remarks.

We must now imagine starting in the frequency domain with either a sequence of coefficients c_k or a function $\hat{f}(\omega)$, either of which could be complex-valued. As the notation suggests, if we have a sequence $\{c_k\}$, it should be regarded as a set of Fourier coefficients of a function f on an interval $[-A/2, A/2]$. If we have a function $\hat{f}(\omega)$, it should be regarded as the Fourier transform of a function f. In either case, the goal is to reconstruct f, or more realistically, N samples of f at the grid points $x = x_n$ on some interval $[-A/2, A/2]$. Let's deal with these two cases separately.

Fourier Series Synthesis

First consider the case in which the input to the IDFT is a sequence $\{c_k\}$, and the task is to reconstruct the function f that has Fourier coefficients $\{c_k\}$. At this point, it is necessary to introduce some new notation. As said, we will let f represent the function with Fourier coefficients $\{c_k\}$. This means that the samples of f (let's call them $f(x_n)$) are the *exact* solution to the problem. On the other hand we will use the IDFT to compute approximations to the values of $f(x_n)$, and these approximations need a new name. We will let $\tilde{f}_n$ denote the sequence generated by the IDFT. In other words, using the definition of the IDFT,

$$\tilde{f}_n = \sum_{k=-\frac{N}{2}+1}^{\frac{N}{2}} c_k \omega_N^{nk} = \sum_{k=-\frac{N}{2}+1}^{\frac{N}{2}} c_k e^{i2\pi nk/N}$$

for $n = -N/2+1 : N/2$. Once again the reciprocity relations enter in a crucial way. If the function f is to be approximated at N equally spaced grid points on the interval $[-A/2, A/2]$, then c_k must be interpreted as the Fourier coefficient corresponding to the frequency $\omega_k = k/A$ on an interval $[-\Omega/2, \Omega/2]$, where $\Omega = A/N$. Therefore, with $x_n = nA/N$ and $\omega_k = k/A$, we have

$$\tilde{f}_n = \sum_{k=-\frac{N}{2}+1}^{\frac{N}{2}} c_k e^{i2\pi k x_n/A} \tag{6.19}$$

for $n = -N/2+1 : N/2$. Now notice that the Fourier series for $f(x_n)$ is

$$f(x_n) = \sum_{k=-\infty}^{\infty} c_k e^{i2\pi k x_n/A} \tag{6.20}$$

for $n = -N/2+1 : N/2$. If we now compare expressions (6.19) and (6.20), we can see how well the IDFT approximates the values of $f(x_n)$. Subtracting the two expressions, we have that

$$f(x_n) - \tilde{f}_n = \left(\sum_{k=-\infty}^{-\frac{N}{2}} + \sum_{k=\frac{N}{2}+1}^{\infty} \right) c_k e^{i2\pi k x_n/A}$$

for $n = -N/2+1 : N/2$. In the unlikely case that the coefficients c_k are nonzero *only* for $k = -N/2+1 : N/2$, we see that the IDFT exactly reproduces the values of $f(x_n)$; this is the case of a periodic band-limited function f. This simply says that if f has a finite number of frequency components and the IDFT has enough terms to include all of them, then f can be reconstructed exactly at the grid points.

More realistically, if the sequence c_k has nonzero values for arbitrarily large k, then the IDFT will use only N of those coefficients and the error can be bounded by

$$|\tilde{f}_n - f(x_n)| \leq \sum_{|k|=\frac{N}{2}}^{\infty} |c_k|$$

for $n = -N/2+1 : N/2$. In other words, the error in using the IDFT to approximate the Fourier series is the error in truncating the Fourier series. If additional information is available about the rate of decay of the coefficients (or, equivalently, about the smoothness of f), then this bound can be made more specific. For example, if it is known that the coefficients satisfy $c_k \leq |k|^{-p}$ for $|k| \geq N/2$, then the error can be bounded by CN^{-p}, where C is a constant independent of N (problem 136). Perhaps more important than a precise error statement are the following qualitative

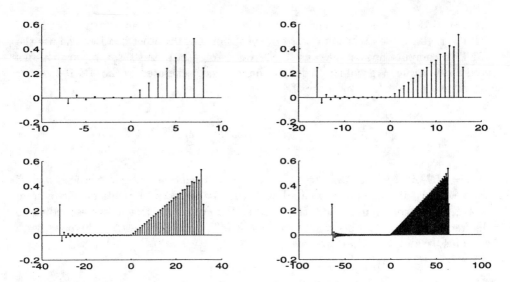

FIG. 6.11.   **Case Study 6.** *The IDFT can be used to reconstruct a function from a given set of coefficients c_k. The graphs show the reconstructions with (clockwise from upper left) $N = 16, 32, 64, 128$. The result is increasingly accurate samples of a real-valued function on an arbitrary interval.*

observations: the effect of increasing N with A fixed is to lengthen the frequency interval $[-\Omega/2, \Omega/2]$ while the grid spacing $\Delta\omega$ remains constant. The result is that higher and higher frequency components are included in the representation of f. In the limit as $N \to \infty$, the IDFT approaches the Fourier series of f at the grid points x_n of $[-A/2, A/2]$. The subtleties of using $c_{\pm\frac{N}{2}}$ in reconstructing f are examined in problem 137.

Case Study 6: Fourier series synthesis by the IDFT. This case study is a numerical demonstration of the use of the IDFT to reconstruct a function from its Fourier coefficients. We begin with the set of coefficients

$$c_0 = \frac{1}{8} \quad \text{and} \quad c_k = \frac{\cos(\pi k) - 1}{4\pi^2 k^2} + i\,\frac{\cos(\pi k)}{4\pi k}$$

for $k = -N/2 : N/2$, and use the IDFT to construct the sequence $\tilde{f}_n$ using the IDFT for various values of N. We might anticipate the outcome before we even look at the output. Since the sequence of coefficients is conjugate even ($c_k = c_{-k}^*$), we can expect that the sequence $\tilde{f}_n$ is real. As N increases, we will see more samples of that function on the *same* interval in the spatial domain. With this bit of forethought, let's look at the numerical results. Figure 6.11 shows the output of the IDFT for $N = 16, 32, 64, 128$ plotted on a fixed interval. As N increases we see the graph of a function f filled in with more resolution and more smoothness until a ramp function emerges. Notice that the DFT takes the average value of the function at the endpoint discontinuities. The oscillations that occur near the endpoints are another onset of the Gibbs effect that reflects the nonuniform convergence of the Fourier series near discontinuities. The length of the interval on which f is reconstructed is arbitrary, since the coefficient c_k simply gives the weighting of the kth mode $e^{i2\pi kx/A}$ on the interval $[-A/2, A/2]$ for any A.

We would be remiss by overlooking the replication perspective in the case of Fourier series synthesis; it has a compelling message that can be found after a brief calculation. Beginning with the Fourier series representation for $f(x_n)$ given by (6.20), we can write

$$f(x_n) = \sum_{k=-\infty}^{\infty} c_k e^{i2\pi k x_n/A}$$

$$= \sum_{j=-\infty}^{\infty} \sum_{k=-\frac{N}{2}+1}^{\frac{N}{2}} c_{k+jN} e^{i\frac{2\pi nk}{N}}$$

$$= \sum_{k=-\frac{N}{2}+1}^{\frac{N}{2}} \left(\underbrace{\sum_{j=-\infty}^{\infty} c_{k+jN}}_{\mathcal{R}_N\{c_k\}} \right) \omega_N^{nk}$$

$$= \mathcal{D}^{-1}\{\mathcal{R}_N\{c_k\}\}_n$$

for $n = -N/2 + 1 : N/2$. In a familiar maneuver, a single infinite sum has been rewritten as a double sum to introduce the replication operator. The outcome is easily explained: the N samples of the function f can be produced exactly by applying the IDFT, not to the set of Fourier coefficients, but to the *replication* of this set. This corroborates the earlier observation that if only the first N coefficients are nonzero, then the synthesis is exact. Otherwise, error is introduced because the sequence of coefficients is truncated. The only way to reduce the difference between the sequence c_k and the sequence $\mathcal{R}_N\{c_k\}$ is to increase N. However, a computational strategy is also suggested by this result. It should be possible to improve the approximation to $f(x_n)$ by using approximations to $\mathcal{R}_N\{c_k\}$ as input to the IDFT. This idea is tested in the following example.

Case Study 7: Improved Fourier synthesis. The set $c_0 = 1/2, c_k = \sin(\pi k/2)/(\pi k)$ is the set of Fourier coefficients of the square pulse of width one centered at the origin. The graphs of Figure 6.12 show various attempts to reconstruct the pulse from its set of Fourier coefficients. In all cases the number of grid points is $N = 64$. The observed improvements are due, not to increasing N, but to using approximations to $\mathcal{R}_{64}\{c_k\}$, the replication of the Fourier coefficients, as input to the IDFT. The first figure shows the output of the IDFT using only the coefficients c_k for $k = -32 : 32$. Clearly, the approximations to f_n are good near the center of the interval, but they suffer from errors near the discontinuities. The remaining figures show the results when the set c_k is replicated one, two, three times before being used as input to the IDFT. The improvement in the reconstructions after replication is significant, and comes at little additional expense. The replication is quite inexpensive and in all cases the length of the IDFT remains constant. This is a strategy which does not seem to have received much use or attention.

Inverse Fourier Transforms

We now turn to the other inversion problem, that in which a function $\hat{f}$ is given and the task is to approximate the function f that has $\hat{f}$ as a Fourier transform. Before $\hat{f}$ can be used as input for the IDFT, it must be sampled on an interval $[-\Omega/2, \Omega/2]$ with

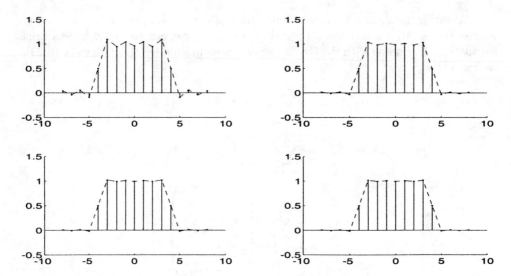

FIG. 6.12. **Case Study 7.** *The reconstruction of a square pulse from its Fourier coefficients can be improved by using a replication strategy. The IDFT, using the coefficients c_k for $k = -32 : 32$ as input, is shown in the upper left figure. If the coefficients c_k are replicated once (upper right), twice (lower left), and three times (lower right), the IDFT reconstruction shows marked improvement. In all cases the IDFT produces $N = 64$ points.*

a grid spacing of $\Delta\omega = \Omega/N$ to produce a sequence that we will denote $\hat{f}_k = \hat{f}(k\Delta\omega)$, where $k = -N/2 + 1 : N/2$. This sequence of length N can now be used as input to the IDFT, and we have that

$$\tilde{f}_n = \sum_{k=-\frac{N}{2}+1}^{\frac{N}{2}} \hat{f}_k \omega_N^{nk} = \sum_{k=-\frac{N}{2}+1}^{\frac{N}{2}} \hat{f}_k e^{i2\pi nk/N}$$

for $n = -N/2 + 1 : N/2$. (Recall that the notation $\tilde{f}_n$ has been introduced to denote the output of the IDFT.) The exponential terms in the rightmost sum will look like the integrand of the inverse Fourier transform if we write

$$\tilde{f}_n = \sum_{k=-\frac{N}{2}+1}^{\frac{N}{2}} \hat{f}_k e^{i2\pi \frac{nA}{N} \frac{k}{A}} = \sum_{k=-\frac{N}{2}+1}^{\frac{N}{2}} \hat{f}_k e^{i2\pi x_n \omega_k}$$

for $n = -N/2 + 1 : N/2$. All we have done is to abide by the reciprocity relations and let $x_n = n\Delta x$ and $\omega_k = k\Delta\omega$, where $\Delta x = A/N$ and $\Delta\omega = 1/A$.

To move onward from here and estimate the error in the IDFT as an approximation to the inverse Fourier transform, we need a version of the Poisson Summation Formula that "works in the other direction," one that will allow us to compare values of the IDFT to values of f. Fortunately the arguments of Section 6.4 leading to the Poisson Summation Formula can be carried out analogously to yield what we will call the **Inverse Poisson Summation Formula**. We will only state it and use it, leaving the derivation as a worthwhile exercise (problem 138).

▶ **Inverse Poisson Summation Formula** ◀

$$\Delta\omega \sum_{k=-\infty}^{\infty} \hat{f}(\omega_k) e^{i2\pi\omega_k x} = \sum_{j=-\infty}^{\infty} f(x - jA) \tag{6.21}$$

for $-\infty < x < \infty$.

This result, quite analogous to the (forward) Poisson Summation Formula (6.10), relates the A-periodic replication of the function f to samples of $\hat{f}$ at the points $\omega_k = k\Delta\omega = k/A$. Not surprisingly, this result is indispensable in reaching conclusions about errors in the IDFT. We will now proceed with brevity, since the arguments parallel those made in previous sections. The first case of interest is that in which f is band-limited. We will assume that the function $\hat{f}$ is sampled on the interval $[-\Omega/2, \Omega/2]$ where $\hat{f}(\omega) = 0$ for $\omega \geq \Omega/2$. The left-hand sum of (6.21) can be simplified and we have that

$$\Delta\omega \sum_{k=-\frac{N}{2}+1}^{\frac{N}{2}} \hat{f}(\omega_k) e^{i2\pi\omega_k x} = \sum_{j=-\infty}^{\infty} f(x - jA).$$

Evaluating this expression at the grid points $x = x_n$, we see that the sum on the left is the IDFT, $\tilde{f}_n$. Rearranging the terms of this expression we discover that

$$\Delta\omega\tilde{f}_n - f(x_n) = \sum_{|j|\geq 1} f(x_n - jA)$$

for $n = -N/2 + 1 : N/2$. The error in using the IDFT to reconstruct a band-limited function involves the values of f *outside* of the interval $[-A/2, A/2]$. This is a sampling error, but now in the frequency domain. In other words, the choice of a sampling interval $\Delta\omega$ induces an interval $[-A/2, A/2]$ in the spatial domain; those parts of f that do not lie in this interval are folded back onto the interval by the IDFT. It is another case of aliasing. (There is also a pester here that arose in the case of compact support for the DFT. The result requires that $\hat{f}(\omega_{-N/2}) = \hat{f}(\omega_{N/2}) = 0$; otherwise another term proportional to $\Delta\omega$ survives in the Inverse Poisson Summation Formula, and the error is bounded by $C\Delta\omega$ or C/N.)

We will not pursue this case further except to say that if additional information about f were known, for instance, the rate at which it decays for large $|x|$, then it would be possible to make more precise bounds of the error. Since f is band-limited, nothing can be gained by increasing the extent of the frequency domain $[-\Omega/2, \Omega/2]$. The way to reduce this error is to increase N with Ω fixed, which increases A, which in turn reduces the overlap in the values of $f(x - jA)$.

The other case that is easily handled by the Inverse Poisson Summation Formula is that in which f has compact support, but is necessarily not band-limited. We will assume that $f(x) = 0$ for $|x| \geq A/2$, and that $\Delta\omega$ is chosen to satisfy $\Delta\omega \leq 1/A$. Then the right-hand sum in (6.21), when evaluated at the grid points $x = x_n$, is reduced to a single term, and we have

$$\Delta\omega \sum_{k=-\infty}^{\infty} \hat{f}(\omega_k) e^{i2\pi\omega_k x_n} = f(x_n)$$

for $n = -N/2 + 1 : N/2$. Rearranging this expression to isolate the IDFT results in the error bound

$$|\Delta\omega \tilde{f}_n - f(x_n)| \leq \Delta\omega \left| \sum_{|k| \geq \frac{N}{2}}'' \hat{f}(\omega_k)\omega_N^{-nk} \right|$$

for $n = -N/2 + 1 : N/2$. The Σ'' means that the $\pm N/2$ terms of the sum are weighted by $1/2$. The expected endpoint pester appears: if $f(-A/2^+) \neq 0$ or $f(A/2^-) \neq 0$ then the result does not apply for $n = N/2$; for this single coefficient, the error is bounded by a constant times $\Delta\omega$ reflecting the discontinuity in f at the endpoints. This situation is easily avoided by taking a slightly larger value of A.

In contrast to the previous case, we see that the error in using the IDFT to approximate a compactly supported function is a truncation error: the transform $\hat{f}$ must be restricted to the finite interval $[-\Omega/2, \Omega/2]$ before it is sampled, and its values outside of that interval contribute to the error. Fortunately, there is no aliasing error in this case, provided that the sampling interval $\Delta\omega$ is chosen sufficiently small ($\Delta\omega \leq 1/A$). If additional information about $\hat{f}$ were known, it would be possible to make more specific bounds on the error. Recall that the Fourier transform of a compactly supported function is closely related to its Fourier coefficients ($Ac_k = \hat{f}(\omega_k)$). Therefore, if smoothness properties of f are known, then Theorem 6.2 can be used to describe the decay of c_k and $\hat{f}(\omega_k)$.

The qualitative lesson is most important in this case. Since the error is due to the truncation of $\hat{f}$ to the interval $[-\Omega/2, \Omega/2]$, increasing Ω (by increasing N with A and $\Delta\omega$ fixed) will decrease the error. Increasing Ω has the effect of decreasing Δx, which places more grid points in the fixed interval $[-A/2, A/2]$. There is no gain in increasing A since f is compactly supported on $[-A/2, A/2]$ and does not need to be reconstructed on a larger interval. In summary, increasing Ω by increasing N with $\Delta\omega$ fixed has the effect of increasing the resolution on the interval $[-A/2, A/2]$ on which f is represented. Thus, in this limit, we see that the IDFT approaches the Fourier series representation for f at the grid points x_n. This conclusion is mapped out in Figure 6.13, which will be discussed shortly.

There is one final case. As always, it is the most general case and perhaps the one that occurs most frequently. If f (or, equivalently, $\hat{f}$) is known to be neither compactly supported nor band-limited, then the previous two cases can be taken together both qualitatively and quantitatively. Rather than try to be more precise, we will appeal to a case study.

Case Study 8: Errors in the IDFT. In this case study we will investigate the errors in the IDFT by considering a function $\hat{f}$ whose IDFT and inverse Fourier transform can be computed explicitly. Consider the function $\hat{f}$ and its samples at the N points $\omega_k = k\Delta\omega$ on an interval $[-\Omega/2, \Omega/2]$ given by

$$\hat{f}(\omega) = e^{-a|\omega|} \quad \text{and} \quad \hat{f}_k = \hat{f}(\omega_k) = e^{-a|k|\Delta\omega}$$

for $k = -N/2 + 1 : N/2$. It is not too difficult to show that the inverse Fourier transform of $\hat{f}$ is

$$f(x) = \int_{-\infty}^{\infty} e^{-a|\omega|} e^{i2\pi\omega x} d\omega = \frac{2a}{a^2 + 4\pi^2 x^2}.$$

We will determine how the IDFT approximates and approaches f in various limits. As mentioned, it is possible to compute the IDFT $\tilde{f}_n$ of the sequence $\hat{f}_k$ explicitly. It

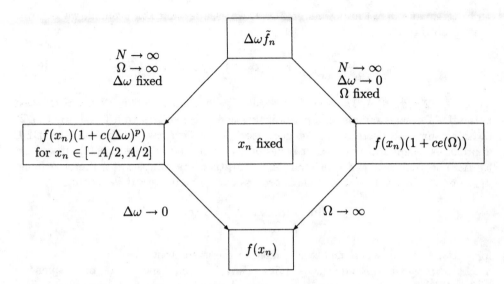

FIG. 6.13. *The IDFT $\tilde{f}_n$ approaches $f(x_n)$ in various limits. With the sampling rate $\Delta\omega$ fixed while Ω and N increase, $\Delta\omega\tilde{f}_n$ approaches f on $[-A/2, A/2]$ up to relative errors that decrease as $(\Delta\omega)^p$. (This limit also implies that A is fixed while $\Delta x \to 0$.) Alternatively, with Ω fixed while $\Delta\omega$ decreases (by letting N increase), the sequence $\Delta\omega\tilde{f}_n$ approaches $f(x_n)$ on $(-\infty, \infty)$ up to errors (denoted $e(\Omega)$) that decrease with increasing Ω. Implicit in this limit is that Δx is fixed and $A \to \infty$. Letting $\Delta\omega \to 0$ in the first case or $\Omega \to \infty$ in the second case allows $\Delta\omega\tilde{f}_n$ to approach $f(x_n)$ on $(-\infty, \infty)$.*

can be gleaned from *The Table of DFTs* in the Appendix to be

$$\tilde{f}_n = \sum_{k=-\frac{N}{2}+1}^{\frac{N}{2}} e^{-a|k|\Omega/N} e^{i2\pi nk/N} = \frac{(1 - e^{-2a\Delta\omega})(1 - e^{-a\Omega/2}\cos(\pi n))}{1 - 2e^{-a\Delta\omega}\cos(2\pi x_n\Delta\omega) + e^{-2a\Delta\omega}} \qquad (6.22)$$

for $n = -N/2 + 1 : N/2$. We have used the facts that $\Omega/N = \Delta\omega = 1/A$ and $x_n = n\Delta x$ to write $\tilde{f}_n$ in this form. At this point $\tilde{f}_n$ appears to bear little resemblance to $f(x_n)$. Recall from the Inverse Poisson Summation Formula that the quantity $\Delta\omega\tilde{f}_n$ approximates $f(x_n)$; therefore, we will look at $\Delta\omega\tilde{f}_n$ and see how it behaves in two different limits.

First, consider the effect of letting Ω and N increase while holding $\Delta\omega$ fixed (the reciprocity relations tell us that A is fixed and $\Delta x \to 0$ in this limit). This means that the range of frequencies used to reconstruct f increases, and the grid spacing in the spatial domain Δx decreases. At the same time, the interval of reconstruction in the spatial domain $[-A/2, A/2]$ as well as the grid point of interest x_n remain fixed. If we formally let $\Omega \to \infty$ in (6.22) with $\Delta\omega$ fixed, we find that

$$\lim_{\substack{\Omega\to\infty \\ N\to\infty}} (\Delta\omega\tilde{f}_n) = \Delta\omega\frac{(1 - e^{-2a\Delta\omega})}{1 - 2e^{-a\Delta\omega}\cos(2\pi x_n\Delta\omega) + e^{-2a\Delta\omega}} \equiv g(x_n)$$

for $-A/2 \leq x_n \leq A/2$. We have denoted this limit $g(x_n)$; it is a periodic function in x_n with a period of $1/\Delta\omega = A$.

We now need to relate this periodic function g to the (nonperiodic) function f which is the exact inverse Fourier transform of $\hat{f}$. This can be done if we use Taylor series to expand $g(x_n)$ in powers of $\Delta\omega$. If we assume that $\Delta\omega$ is small, we discover that (problem 133)

$$g(x_n) = \frac{2a}{a^2 + 4\pi^2 x_n^2}(1 + c(\Delta\omega)^2) = f(x_n)(1 + c(\Delta\omega)^2), \qquad (6.23)$$

where $-A/2 \leq x_n \leq A/2$ and c is a constant. In other words, letting N and Ω increase in the IDFT creates samples of a periodic function g that differs from f on $[-A/2, A/2]$ by an amount that decreases as $\Delta\omega$ decreases. Said differently, in this limit, the IDFT approaches samples of the Fourier series of f on the interval $[-A/2, A/2]$. If we now let $\Delta\omega$ approach zero (implying that $A \to \infty$), then the samples $g(x_n)$ approach the $f(x_n)$ on the interval $(-\infty, \infty)$. This process can be summarized by writing

$$\lim_{\substack{\Delta\omega \to 0}} \lim_{\substack{\Omega \to \infty \\ N \to \infty}} (\Delta\omega \tilde{f}_n) = f(x_n)$$

for $-\infty < x_n < \infty$, where $\Delta\omega$ is held fixed in the inner limit.

Now consider the alternate limit. Imagine that the range of frequencies $[-\Omega/2, \Omega/2]$ is fixed, as is the grid spacing Δx and the point of interest x_n. The effect of letting N increase is to produce a finer grid spacing $\Delta\omega$ and a larger interval of reconstruction $[-A/2, A/2]$ in the spatial domain. As before, we may start with the analytical expression (6.22) and let $\Delta\omega \to 0$ with Ω fixed. A copious use of Taylor series leads us to

$$\lim_{\substack{\Delta\omega \to 0 \\ N \to \infty}} (\Delta\omega \tilde{f}_n) = \frac{2a(1 - e^{-a\Omega/2}\cos(\pi n))}{a^2 + 4\pi^2 x_n^2} = f(x_n)(1 - \cos(\pi n)e^{-a\Omega/2}) \qquad (6.24)$$

for $-\infty < x_n < \infty$.

We see that the effect of letting $\Delta\omega$ decrease to zero while holding Ω fixed is to produce a sequence that differs from $f(x_n)$ by an amount which decreases (in this case) exponentially with Ω. Subsequently letting Ω increase, which amounts to including higher and higher frequencies in the representation of f, recovers the sampled function f. This two-limit process may be written

$$\lim_{\substack{\Omega \to \infty}} \lim_{\substack{\Delta\omega \to 0 \\ N \to \infty}} (\Delta\omega \tilde{f}_n) = f(x_n) \quad -\infty < x_n < \infty$$

for $-\infty < x_n < \infty$.

Figure 6.14 illustrates these limit paths and the manner in which $\Delta\omega \tilde{f}_n$ approaches $f(x_n)$ as A, Ω, and N change (when $a = 1/2$). First, consider the three error plots in the left column. If $A = 4$ is held fixed while N and Ω are increased (abiding by the reciprocity relation $A\Omega = N$), then the IDFT produces increasingly accurate approximations to f on the interval $[-2, 2]$. This sequence of plots follows the limit path leading to (6.23) in which, as $\Omega \to \infty$, the values of $f(n_n)$ are produced up to errors proportional to $(\Delta\omega)^2$. The errors are also proportional to $f(x_n)$ itself, which has a maximum at $x_n = 0$, explaining the maximum error at $x_n = 0$.

In the right column of Figure 6.14, we fix $\Omega = 4$ and increase N; this has the effect of decreasing $\Delta\omega$ and increasing the length of the interval $[-A/2, A/2]$ on which the samples $f(x_n)$ are generated. This sequence of plots follows the limit path leading to (6.24). Note that the errors (most noticeably in the center of the interval near $x_n = 0$) do not decrease with increasing values of N. A check of numerical values shows that

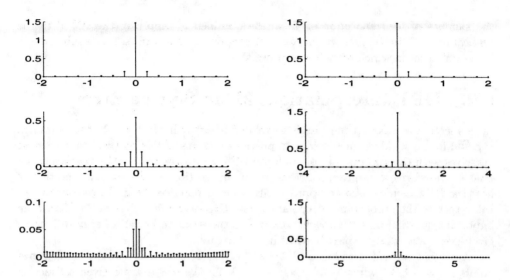

FIG. 6.14.   **Case Study 8.** *The above graphs demonstrate the limit paths of the IDFT described in Case Study 8. The goal is to approximate the inverse Fourier transform of $\hat{f}(\omega) = e^{-|\omega|/2}$ by applying the IDFT with various choices of Ω and N. In all cases, the absolute error $|f(x_n) - \Delta\omega \tilde{f}_n|$ is given at the sample points. In the left column N and Ω are increased ($N = 16, 32, 64$ and $\Omega = 4, 8, 16$) so that $A = N/\Omega$ remains constant. In the right column, N is increased ($N = 16, 32, 64$) and $\Omega = 4$ is held constant, which produces larger intervals of reconstruction in the spatial domain and decreasing grid spacings in the frequency domain ($\Delta\omega = 1/4, 1/8, 1/16$).*

the sequence $\Delta\omega \tilde{f}_n$ has reached the limit $f(x_n)(1 - \cos(\pi n)e^{-a\Omega/2})$, as predicted by (6.24); taking larger values of N cannot improve the accuracy of this approximation. Further improvements can be made only by increasing Ω. As before, since the errors are also proportional to $f(x_n)$ itself, which has a maximum at $x_n = 0$, the maximum error occurs at $x_n = 0$.

The limit paths of Case Study 8 are summarized in Figure 6.13. Like the DFT map of Figure 6.10, this figure shows the relationships between the IDFT and various forms of the function f which is to be reconstructed. If the IDFT is computed with the sampling rate $\Delta\omega$ fixed while letting N and Ω increase, the result is better approximations to the Fourier series of f on the *fixed* interval $[-A/2, A/2]$. Subsequently letting $\Delta\omega \to 0$ (and $A \to \infty$) will produce increasingly accurate approximations to $f(x_n)$ on $(-\infty, \infty)$. On the other hand, if the IDFT is computed with Ω fixed, but with $\Delta\omega$ decreasing (by letting N increase), the result is values of $f(x_n)$ up to errors that decrease as Ω increases. Subsequently letting Ω increase will reduce these errors, and the IDFT approaches $f(x_n)$ on $(-\infty, \infty)$.

We will close in a predictable manner by appealing to the replication perspective one last time. All of the discussion of this section could have centered around the replication form of the Inverse Poisson Summation Formula, and the same conclusions would have followed. We leave it as a worthy exercise (problem 139) to show that the Inverse Poisson Summation Formula (6.21) can be expressed using replication operators in the form

$$\Delta\omega \mathcal{D}^{-1}\{\mathcal{R}_N\{\hat{f}_k\}\}_n = \mathcal{R}_N\{f_n\}.$$

This says that in the absence of special cases such as band-limiting or compact support, the samples of the *replication* of f can be obtained by applying the IDFT to the *replication* of $\hat{f}$. In the presence of compact support or band-limits, this result reduces to special cases in which certain errors vanish.

6.9.   DFT Interpolation; Mean Square Error

In this section we take up one final issue related to errors in the DFT, but the problem is posed in a much different setting. In previous sections of this chapter we concerned ourselves with the question of how well the DFT approximates Fourier coefficients or Fourier transforms of a given function f. In this section we discuss the question of how the DFT can be used to approximate a given function itself. In particular, we will return to the problem of interpolation with trigonometric polynomials. This is an important question in its own right, but it also turns out to be rather easy to handle, given everything that we have learned in this chapter.

As before we will work on an interval $[-A/2, A/2]$ with N equally spaced grid points $x_n = nA/N$, where $n = -N/2 + 1 : N/2$. We will also be given a function f defined on that interval with known smoothness properties. We can now state the interpolation problem. Given the N values of the function f at the grid points x_n, find the coefficients F_k such that the trigonometric polynomial

$$\phi(x) = \sum_{k=-\frac{N}{2}+1}^{\frac{N}{2}} F_k e^{i2\pi kx/A}$$

agrees with f at the grid points; that is, $\phi(x_n) = f(x_n)$ for $n = -N/2+1 : N/2$. The term trigonometric polynomial may be confusing; it refers to the fact that ϕ consists of powers of (or is a polynomial in) $e^{i2\pi x/A}$.

The solution to this interpolation problem was actually carried out in Chapter 2 as a means of deriving the DFT. To summarize, we start with the interpolation conditions that

$$\phi(x_n) = \sum_{k=-\frac{N}{2}+1}^{\frac{N}{2}} F_k e^{i2\pi kx_n/A} = f(x_n)$$

for $n = -N/2 + 1 : N/2$. We will let $f_n = f(x_n)$ and also note that $f_{\frac{N}{2}}$ should be defined as the average of the values of f at the endpoints, $x_n = \pm A/2$ (AVED). Using the discrete orthogonality of the exponential functions allows each of the coefficients F_k to be isolated, and we find that

$$F_k = \frac{1}{N} \sum_{n=-\frac{N}{2}+1}^{\frac{N}{2}} f_n e^{-i2\pi kn/N}$$

for $k = -N/2+1 : N/2$. In other words, $F_k = \mathcal{D}\{f_n\}_k$, and we see that the coefficients of the interpolating polynomial are given by the DFT of the sequence f_n.

We now need to ask how well this function ϕ approximates f *on the entire interval* $[-A/2, A/2]$. Up until now we have discussed the error in the DFT or IDFT at particular points. When we compare two functions f and ϕ at all points of an interval, we need a new measure of error. The tools that facilitate this new measure are norms and inner products. A short review is worthwhile.

It turns out that a very convenient way to measure the difference between two functions on an interval is the **mean square error**. It is the integrated square of the difference between the two functions, and for the interval $[-A/2, A/2]$ is given by

$$||f - \phi|| = \left\{ \int_{-\frac{A}{2}}^{\frac{A}{2}} (f(x) - \phi(x))^2 dx \right\}^{\frac{1}{2}}.$$

The **norm** $||\cdot||$ that we have defined is called the **mean square norm** or often simply the L_2-**norm**. Recall that the **inner product** of two functions f and ϕ on the interval $[-A/2, A/2]$ is defined by

$$\langle f, \phi \rangle = \int_{-\frac{A}{2}}^{\frac{A}{2}} f(x)\phi^*(x)dx,$$

where ϕ^* is the complex conjugate of ϕ. Here is the important connection between inner products and mean square norms: it is easy to check that $||f||^2 = \langle f, f \rangle$. Therefore,

$$||f - \phi|| = \langle f - \phi, f - \phi \rangle = \langle f, f \rangle - \langle f, \phi \rangle - \langle \phi, f \rangle + \langle \phi, \phi \rangle.$$

There is one other fundamental property associated with inner products, and that is **orthogonality**. We say that two functions g and h are orthogonal on an interval if their inner product on that interval $\langle g, h \rangle$ vanishes. The important orthogonality property that we will need in this section concerns the trigonometric polynomials. One straightforward integral (see problem 120 and Chapter 2, problem 22) is all that is needed to show that

$$\langle e^{i2\pi kx/A}, e^{i2\pi jx/A} \rangle = \int_{-\frac{A}{2}}^{\frac{A}{2}} e^{i2\pi(k-j)x/A}dx = A\delta(j-k) = \begin{cases} A & \text{if} \quad k = j, \\ 0 & \text{if} \quad k \neq j. \end{cases}$$

This property of the complex exponentials is entirely analogous to the discrete orthogonality that lies at the heart of the DFT.

We are now ready to state and prove a result about the error in trigonometric interpolation [89].

THEOREM 6.7. ERROR IN TRIGONOMETRIC INTERPOLATION. *Let the A-periodic extension of f have $(p-1)$ continuous derivatives for $p \geq 1$ and assume that $f^{(p)}$ is bounded and piecewise monotonic on $[-A/2, A/2]$. Assume that ϕ is the trigonometric polynomial that interpolates f at the points $x_n = nA/N$, where $n = -N/2 + 1 : N/2$. Then the mean square error in ϕ as an interpolant of f satisfies*

$$||f - \phi|| \leq \frac{C}{N^{p+1}},$$

where C is a constant independent of N.

The proof of this theorem is a pleasing collection of results and ideas that have already appeared in this chapter; it merits a succinct presentation with a few details left to the reader.

Proof: First notice that we have imposed the same conditions on f that were used earlier in the chapter. This will allow us to use Theorem 6.2 to estimate the rate of decay of the Fourier coefficients of f; specifically, we know that $|c_k| \leq C'/|k|^{p+1}$ for some constant C'.

The proof is quite physical since it relies on the splitting of f into its low and high frequency parts. We will let $f = f_L + f_H$ where

$$f_L(x) = \sum_{k=-\frac{N}{2}+1}^{\frac{N}{2}} c_k e^{i2\pi kx/A} \quad \text{and} \quad f_H(x) = \left(\sum_{k=-\infty}^{-\frac{N}{2}} + \sum_{k=\frac{N}{2}+1}^{\infty} \right) c_k e^{i2\pi kx/A}.$$

Notice that f_L consists of the low frequency components that can be resolved by the DFT, whereas f_H consists of the remaining high frequency components. Since f_L and f_H are themselves functions we can form their interpolating polynomials on the same N grid points, which we will call ϕ_L and ϕ_H, respectively. This means that

$$\phi_L(x) = \sum_{k=-\frac{N}{2}+1}^{\frac{N}{2}} F_k^L e^{i2\pi kx/A} \quad \text{where} \quad F_k^L = \frac{1}{N} \sum_{n=-\frac{N}{2}+1}^{\frac{N}{2}} f_L(x_n) e^{-i2\pi nk/N}$$

and

$$\phi_H(x) = \sum_{k=-\frac{N}{2}+1}^{\frac{N}{2}} F_k^H e^{i2\pi kx/A} \quad \text{where} \quad F_k^H = \frac{1}{N} \sum_{n=-\frac{N}{2}+1}^{\frac{N}{2}} f_H(x_n) e^{-i2\pi nk/N}.$$

Four observations will now be needed to complete the proof, and they should each be verified:

1. The interpolant of f on the grid points x_n is $\phi = \phi_L + \phi_H$ (since $\phi_L(x_n) + \phi_H(x_n) = f(x_n)$).

2. $\phi_L = f_L$ (since both consist of the same N modes with the same coefficients).

3. f_H and ϕ_H are orthogonal on $[-A/2, A/2]$ since they share no common modes. This is a beautiful instance of aliasing: f_H consists entirely of high frequency modes which are aliased onto the lower frequency modes of the approximating function ϕ_H. This is a key element in the proof that follows.

4. The following Parseval relations hold:

$$\|\phi_H\|^2 = \sum_{k=-\frac{N}{2}+1}^{\frac{N}{2}} |F_k^H|^2 \quad \text{and} \quad \|f_H\|^2 = \left(\sum_{k=-\infty}^{-\frac{N}{2}} + \sum_{\frac{N}{2}+1}^{\infty} \right) |c_k|^2$$

(by the orthogonality of the sets $\{e^{i2\pi nk/N}\}$ and $\{e^{i2\pi kx/A}\}$).

We may now proceed. We start by writing the mean square difference between f and ϕ and simplifying it a bit:

$$\|f - \phi\|^2 = \|f_L + f_H - \phi_L - \phi_H\|^2 = \|f_H - \phi_H\|^2 = \|f_H\|^2 + \|\phi_H\|^2.$$

Observations 1 and 2 have been used; and the last equality follows from the orthogonality of f_H and ϕ_H (observation 3 above). The remaining work is now clear: we must find bounds on $\|f_H\|^2$ and $\|\phi_H\|^2$. Let's begin with $\|f_H\|^2$. Using observation 4 above and the fact that the Fourier coefficients c_k satisfy $|c_k| \leq C'|k|^{-p-1}$, we have that

$$\|f_H\|^2 \leq \sum_{|k|\geq \frac{N}{2}} |c_k|^2 \leq 2 \sum_{k=\frac{N}{2}}^{\infty} \left(\frac{C'}{k^{p+1}} \right)^2 \leq 2 \left(\frac{C'}{(\frac{N}{2})^{p+1}} \right)^2 \sum_{k=0}^{\infty} \frac{1}{(1 + \frac{2k}{N})^{2(p+1)}}.$$

The last series converges provided that $p > -1/2$, which for our purposes means $p \geq 0$. Therefore, we can conclude that, for some constant C_1,

$$\|f_H\|^2 \leq \frac{C_1^2}{N^{2(p+1)}}.$$

The second term $||\phi_H||^2$ is a bit more recalcitrant. The sequence F_k^H is the DFT of the function f_H, which itself has Fourier coefficients c_k that are zero for $k = -N/2 + 1 : N/2$. It follows from the Discrete Poisson Summation Formula (one last time) that

$$F_k^H = \sum_{j=1}^{\infty} (c_{k+jN} + c_{k-jN})$$

for $k = -N/2 + 1 : N/2$. We may combine this fact with observation 4 above to conclude that

$$||\phi_H||^2 = \sum_{k=-\frac{N}{2}+1}^{\frac{N}{2}} |F_k^H|^2 = \sum_{k=-\frac{N}{2}+1}^{\frac{N}{2}} \left| \sum_{j=1}^{\infty} c_{k+jN} + c_{k-jN} \right|^2.$$

Once again, we use the decay rate of the Fourier coefficients $|c_k| \le C'|k|^{-p-1}$, which leads to

$$||\phi_H||^2 \le \sum_{k=-\frac{N}{2}+1}^{\frac{N}{2}} \left| \sum_{j \neq 0} \frac{C'}{|k + jN|^{p+1}} \right|^2,$$

where the index in the inner sum runs over all integers except $j = 0$. Now, for any sum, it is true that $\sum |a_n|^2 \le (\sum |a_n|)^2$. Applying this fact to the outer sum in the previous expression, we have that

$$||\phi_H||^2 \le \left(\sum_{k=-\frac{N}{2}+1}^{\frac{N}{2}} \left| \sum_{j \neq 0}^{\infty} \frac{C'}{|k + jN|^{p+1}} \right| \right)^2.$$

Those who take a moment to write out a few terms of these two sums will discover that they can be condensed into a single sum of the form

$$||\phi_H||^2 \le \left(2 \sum_{j=\frac{N}{2}}^{\infty} \frac{C'}{|j|^{p+1}} \right)^2,$$

and now we are just about there. As we did earlier in the proof, this series can be bounded by the following maneuver:

$$||\phi_H||^2 \le \left(\frac{2C'}{(\frac{N}{2})^{p+1}} \sum_{j=0}^{\infty} \frac{1}{(1 + \frac{2j}{N})^{p+1}} \right)^2.$$

This series converges provided $p > 0$, which for our purposes means $p \ge 1$. This allows us to claim that

$$||\phi_H||^2 \le \frac{C_2^2}{N^{2(p+1)}},$$

where C_2 is independent of N. Combining the bounds on $||f_H||^2$ and $||\phi_H||^2$, together with the stonger of the two conditions on p, we have that

$$||f - \phi|| \le \frac{C}{N^{p+1}} \quad \text{for} \quad p \ge 1,$$

where C is a constant independent of N. ∎

Having worked this hard, let's try to wrest some insight from this proof. The mean square error $||f - \phi||$ consists of two parts: one from f_H and one from ϕ_H. Notice that f_H consists of high frequency components that are *not* resolved by the interpolating polynomial ϕ. Therefore, the contribution to the error from f_H represents the components of f that are lost when only N modes are used in the approximating function ϕ. This is simply a truncation error. The second contribution to the error is more interesting, but it should be no stranger. The function ϕ_H is the interpolating polynomial (which uses low frequency modes) for the high frequency modes of f. It might seem that ϕ_H cannot possibly resolve these high frequency modes of f and should be zero. Indeed, this would be the case were it not for aliasing. Because f is sampled, its high frequency modes are disguised as low frequency modes that *can* be detected by ϕ_H. These modes do not belong with the low frequency coefficients, and hence they contribute to the error. Thus, we see that there are two sources of error in trigonometric interpolation: truncation and sampling. This is precisely what we have observed throughout this chapter.

Case Study 9: Trigonometric interpolation. Graphical demonstrations of trigonometric interpolation can be made easily and convincingly. We will use a rather arbitrary polynomial

$$f(x) = (x+1)x^2(x+2)$$

that has no special symmetries on the interval $[-1, 1]$. Following the procedure outlined above, the DFT can be used to compute the coefficients of the polynomial ϕ that interpolates f at N equally spaced points on $[-1, 1]$. Figure 6.15 shows the results using $N = 4, 8, 16, 32$ points. The plots show the original function f, the interpolating polynomial ϕ, and the interpolating points x_n. First note that each interpolating polynomial does the required job: it passes through the interpolating points. Each interpolating polynomial is two-periodic, but since the periodic extension of f has a discontinuity at $x = \pm 1$, the interpolating polynomial takes the average value $\frac{1}{2}(f(-1)+f(1))$ at $x = \pm 1$. This discontinuity in f clearly degrades the accuracy of the approximations near the endpoints. In the interior of the interval, the approximations are much better, and they improve with increasing N. (Theorem 6.7 does not apply directly to this case, since it requires at least continuity of f. However, it is reasonable to suspect that mean square errors in this case of a piecewise continuous function decrease as N^{-1}.)

6.10. Notes and References

The subject of errors in the DFT is treated in a multitude of ways in a bewildering assortment of books and papers. The goal of this chapter is to collect and organize not only the results, but the frameworks in which those results are presented. Our conclusion in writing this chapter is that there are three frameworks that have emerged in the literature for analyzing DFT errors, and we have attempted to represent all three of them. In summary, they are

- Poisson Summation Formula,

- replication operators,

- graphical presentation.

This chapter relies heavily on the the Poisson Summation Formula, which is certainly the cornerstone of one framework in which DFT errors can be analyzed. The history of this remarkable result is rather difficult to trace. Most recent treatments of Fourier transforms cite the result and are consistent about its name. It appears in the

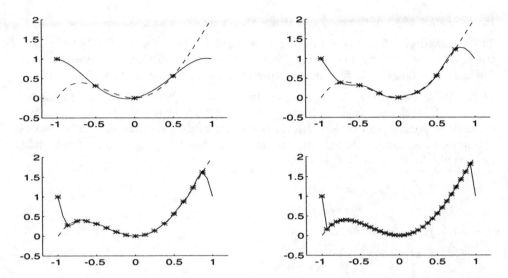

FIG. 6.15. **Case Study 9.** *The use of trigonometric interpolation is illustrated in these four figures. The function $f(x) = (x+1)x^2(x+2)$ is interpolated at N equally spaced points of the interval $[-1, 1]$ where (reading left to right, top to bottom) $N = 4, 8, 16, 32$. The original function (dashed line), the interpolation function (solid line), and the interpolating points ($*$) are shown in each figure.*

1934 book of Paley and Weiner [110] and the well-known 1924 testament of Courant and Hilbert [43]. It is the subject of two papers in 1928 by E. H. Linfoot [94] and L. J. Mordell [101], and Mordell remarks that "Poisson's formula has been ignored in the usual text-books." We were unable to trace the result back to the original work of Poisson, although it seems likely that it must appear in his treatise *Théorie mathématique de la chaleur* of 1835, in which solutions to the heat equation in many different (finite and infinite) domains and geometries are proposed.

Not entirely unrelated to the Poisson Summation Formula, but sufficiently distinct to call it a different framework, is the replication perspective. This approach has some appealing notational advantages and allows many results to be stated quite succinctly. It appears that this approach received its first expression in the literature in the work of Cooley, Lewis, and Welch in the late 1960s [40], [41]. The third approach to understanding DFT errors is the graphical approach, which lacks rigor and does not lead to concise error bounds, but certainly has acclaimed visual appeal. It seems that this approach was at least popularized, and perhaps created, by Brigham in his well-known books [20], [21].

6.11. Problems

116. Aliasing.   Consider a grid with $N = 6$ DFT points (a total of seven points including the endpoints). Find and sketch the imaginary part (the sine mode) of the $k = 2$ mode on this grid. Find and sketch the imaginary part of the $k = 8$ mode on this grid. Show that these two modes have identical values at the grid points. Find the frequency of all of the modes that are aliases for the $k = 2$ mode.

117. Aliasing.   Show that in general the modes $e^{i2\pi kx/A}$ and $e^{i2\pi(k+pN)x/A}$ have the same values at the grid points $x_n = nA/N$, where k and p are any integers. Does the same conclusion hold for the real modes $\cos(2\pi kx/A)$ and $\sin(2\pi kx/A)$?

118. Grid parameters.   Consider the following functions on the designated intervals. In each case, find the highest frequency mode that appears in the function (in units of periods per unit length), the maximum grid spacing Δx that fully resolves the function without aliasing, and the minimum number of grid points that fully resolves the function without aliasing.

(a) $f(x) = \sin(2x)$ on $[-\pi, \pi]$,

(b) $f(x) = \cos(22\pi x)$ on $[-1, 1]$,

(c) $f(x) = \sin^2(16x)$ on $[0, 4\pi]$,

(d) $f(x) = \cos^4(8\pi x)$ on $[-2, 2]$.

119. Modes on coarser grids.   Show that if the kth mode on a grid with $2N$ points is viewed on a grid with N points, then it still appears as the kth mode on this coarser grid, provided that $|k| < N/2$.

120. Inner products, orthogonality, and aliasing.   The **continuous inner product** of two functions f and g on the interval $[-A/2, A/2]$ was defined in the text as

$$\langle f, g \rangle = \int_{-\frac{A}{2}}^{\frac{A}{2}} f(x)g^*(x)dx.$$

An analogous **discrete inner product** can be defined for two sequences $f = \{f_n\}$ and $g = \{g_n\}$, where $n = -N/2 + 1 : N/2$. It is given by

$$\langle f, g \rangle_N = \sum_{n=-\frac{N}{2}+1}^{\frac{N}{2}} f_n g_n^*.$$

Two functions f and g are orthogonal on $[-A/2, A/2]$ if $\langle f, g \rangle = 0$, while two sequences f and g are orthogonal on N points if $\langle f, g \rangle_N = 0$.

(a) Verify the oft-used fact that

$$\langle e^{i2\pi kx/A}, e^{i2\pi jx/A} \rangle = A\delta(k-j) = \begin{cases} A & \text{if } k-j = 0, \\ 0 & \text{if } k-j \neq 0. \end{cases}$$

(b) Verify the equally oft-used fact that

$$\langle e^{i2\pi kn/N}, e^{i2\pi jn/N} \rangle_N = N\hat{\delta}_N(k-j) = \begin{cases} N & \text{if } k-j = mN, \ m \in \mathbf{Z}, \\ 0 & \text{if } \text{otherwise}. \end{cases}$$

(c) Argue that following claim: Aliasing can be attributed to the fact that the orthogonality of the functions $e^{i2\pi kx/A}$ is determined by $\delta(k)$ (the ordinary Kronecker delta), while orthogonality of the sequences $e^{i2\pi nk/N}$ is determined by $\hat{\delta}_N(k)$ (the modular Kronecker delta).

121. DFT approximations to Fourier coefficients. Consider each of the following functions and their periodic extension outside of the indicated interval.

(a) $f(x) = x^2$ on $[-1, 1]$,

(b) $f(x) = |x|$ on $[-1, 1]$,

(c) $f(x) = x$ on $[0, 1]$,

(d) $f(x) = \cos(3\pi x)$ on $[-1, 1]$,

(e) $f(x) = \cos(3\pi x)$ on $[-2, 2]$,

(f) $f(x) = |\sin(\pi x)|$ on $[-1, 1]$,

(g) $f(x) = x(\pi - x)$ if $0 \le x \le \pi$ and $f(x) = x(\pi + x)$ if $-\pi \le x \le 0$.

In each case carry out the following analysis:

(a) Make a sketch of f and its periodic extension.

(b) Find the sequence f_n that should be used as input to the DFT to approximate the Fourier coefficients of f (with special attention to the endpoints).

(c) Assess the smoothness of f, particularly at the endpoints.

(d) Use Theorem 6.3 to determine how the error in the DFT as an approximation to the Fourier coefficients should decrease with N.

(e) Find the DFTs, F_k, of the sequences f_n, analytically (*The Table of DFTs* in the Appendix can be used in nearly all cases).

(f) Compute the DFT for various values of N and comment on the numerical results.

122. DFT of a band-limited periodic sequence. The function considered in Case Study 1 offers a few more intriguing lessons that shed more light on the Discrete Poisson Summation Formula. Recall that the function consisted of the first 17 Fourier modes all equally weighted. When sampled at N points the resulting input sequence is

$$f_n = \sum_{k=-16}^{16} e^{i2\pi nk/N} = 1 + 2\sum_{k=1}^{16} \cos\left(\frac{2\pi nk}{N}\right)$$

for $n = -N/2 + 1 : N/2$. Evaluate the DFT coefficients of this sequence for $N = 64, 32, 16, 8$, and verify the conclusions of Case Study 1. In particular, note that if $N > 32$, then the DFT coefficients are exact ($F_k = c_k$). With $N = 32$, note that the DFT is exact except for the coefficient $F_{16} = 2$. Explain this result in light of the Discrete Poisson Summation Formula. Finally, describe the errors that arise in *all* of the DFT coefficients when $N < 32$.

123. Fourier coefficients on different intervals. Assume that f is A-periodic with Fourier coefficients c_k on the interval $[-A/2, A/2]$. If p is a positive integer, show that the coefficients c'_k of f on the interval $[-pA/2, pA/2]$ are given by

$$c'_k = \begin{cases} c_{\frac{k}{p}} & \text{if } p \text{ divides } k, \\ 0 & \text{otherwise.} \end{cases}$$

124. DFT coefficients on different intervals.   Assume that f is A-periodic and is sampled with N equally spaced points on the interval $[-A/2, A/2]$. Let the resulting DFT coefficients be F_k. If p is a positive integer, and f is sampled on the interval $[-pA/2, pA/2]$ with pN points, show that the resulting DFT coefficients are given by

$$F'_k = \begin{cases} F_{\frac{k}{p}} & \text{if } p \text{ divides } k, \\ 0 & \text{otherwise,} \end{cases}$$

for $k = -pN + 1 : pN$.

125. Sampling on nonmultiples of a period.   Consider the single mode $f(x) = e^{i2\pi k_0 x/A}$, where k_0 is an integer, and assume that it is expanded in a Fourier series on the interval $[-pA/2, pA/2]$, where $p > 1$ is *not* an integer.

  (a) Show that the Fourier coefficients are given by
  $$c'_k = \frac{\sin \pi(k - pk_0)}{\pi(k - pk_0)}.$$

  (b) Graph the coefficients for fixed k_0 and $|k| \leq 5k_0$ for several values of p with $1 \leq p \leq 2$. Notice the behavior of the coefficients as $p \to 1$ and $p \to 2$ and note the presence and size of the sidelobes (as discussed in the text).

  (c) Show analytically that as $p \to 1$, c'_k approaches the expected value $c_k = \delta(k - k_0)$.

  (d) Carry out the same analysis on the pN-point DFT of the samples of f on the interval $[-pA/2, pA/2]$. Show that
  $$F'_k = \frac{\sin \pi(k - pk_0) \sin \left(\frac{2\pi(k - pk_0)}{pN} \right)}{2pN \sin^2 \left(\frac{\pi(k - pk_0)}{pN} \right)},$$

  where $k = -pN/2 + 1 : pN/2$ (*The Table of DFTs* may be helpful).

  (e) Show that as $p \to 1$ the coefficients F'_k approach their expected values $F_k = \delta(k - k_0)$.

126. Alternate Discrete Poisson Summation Formula.   The Discrete Poisson Summation Formula takes slightly different forms depending on whether the index set is centered or whether N is even or odd.

  (a) Derive the Discrete Poisson Summation Formula for the DFT defined on the indices $n, k = 0 : N - 1$. Consider the cases in which N is even and odd.

  (b) Find the relationship (analogous to (6.4)) that relates the DFT coefficients to the Fourier coefficients. Consider the cases where f is band-limited and not band-limited. Consider N both even and odd.

127. Replication operator.   Let $c_n = 1/(|n| + 1)$ and let $f(x) = e^{-x^2}$. Make sketches of the following replications of the sequence c_n and the function f:

  (a) $\mathcal{R}_{10}\{c_n\}$,   (b) $\mathcal{R}_{50}\{c_n\}$,   (c) $\mathcal{R}_1\{f(x)\}$,   (d) $\mathcal{R}_{10}\{f(x)\}$.

128. Inverse replication.   Give an example to show a function that cannot be recovered uniquely from its replication.

129. DFT approximation to Fourier transforms.   Consider the following functions with compact support on the indicated intervals (and value zero outside of the indicated intervals).

- (a) $f(x) = 2$ on $|x| < 2$,
- (b) $f(x) = 4 - x^2$ on $|x| \leq 2$,
- (c) $f(x) = 2 - |x - 1|$ on $-1 \leq x \leq 2$,
- (d) $f(x) = \cos x$ on $|x| \leq \pi/2$,
- (e) $f(x) = 1$ on $0 < x < 1$.

In each case carry out the following analysis:

- (a) Make a sketch of f.
- (b) Find the sequence f_n that should be used as input to the DFT to approximate the Fourier transform of f (with special consideration for the endpoints).
- (c) Assess the smoothness of f (noting endpoints).
- (d) Use Theorem 6.5 to determine how the error in the DFT as an approximation to the Fourier transform should decrease with N.
- (e) At what frequencies will the DFT provide approximations to the Fourier transform?
- (f) Find the DFT F_k analytically (*The Table of DFTs* can be used in nearly all cases).
- (g) Compute the DFT for various values of N and comment on the numerical results.

130. DFT approximations to Fourier transforms.   Consider the following functions defined for $-\infty < x < \infty$.

- (a) $f(x) = (1 + x^2)^{-1}$,
- (b) $f(x) = x(x^2 + 4)^{-1}$,
- (c) $f(x) = (1 + x^4)^{-1}$,
- (d) $f(x) = e^{-x^2}$,
- (e) $f(x) = \text{sinc}\,(x/2)$ (where sinc $x = (\sin x)/x$),
- (f) $f(x) = \text{sinc}^2(x/2)$,
- (g) $f(x) = 1$ for $0 < x < 1$, $f(x) = 0$ elsewhere,
- (h) $f(x) = e^{-10|x|}$.

In each case carry out the following analysis.

- (a) Make a sketch of f.

(b) Find the sequence f_n that should be used as input to the DFT to approximate the Fourier transform of f (with special consideration for the endpoints).

(c) Assess the smoothness of f (endpoints!).

(d) Use Theorem 6.6 to determine how the error in the DFT as an approximation to the Fourier transform should decrease with N.

(e) At what frequencies will the DFT provide approximations to the Fourier transform?

(f) Find the DFT F_k analytically (*The Table of DFTs* can be used in nearly all cases).

(g) Compute the DFT for various values of N and comment on the numerical results.

131. Reciprocity relations. Consider the following functions:

(a) $f(x) = 1$ for $-1/2 < x < 1/2$ and $f(x) = 0$ elsewhere,

(b) $f(x) = \cos(\pi x/2)$.

These functions are sampled on the interval $[-A/2, A/2]$, where $A \geq 1$, and the resulting sequence is used as input for the N-point DFT. In each case, explain the effect of (i) increasing A by increasing Δx with N fixed, and (ii) increasing A by increasing N with Δx fixed. Specifically, in each case draw the corresponding grids in the frequency domain and indicate the frequencies that are represented in the DFT.

132. Details of Case Study 5. Case Study 5 explored the DFT of samples of the function $f(x) = e^{-ax}$ for $x > 0$ and $f(x) = 0$ for $x < 0$.

(a) Show that when f is sampled on the interval $[0, A]$, the DFT and Fourier coefficients are given by

$$F_k = (1 - e^{-aA}) \frac{(1 - e_N^2) - i2e_N \sin\theta_k}{2N(1 - 2e_N \cos\theta_k + e_N^2)}$$

and

$$c_k = \frac{a - i2\pi\omega_k}{a^2 + 4\pi^2\omega_k^2} \frac{1 - e^{-aA}}{A},$$

where $e_N = e^{-aA/N}$ and $\theta_k = 2\pi k/N$.

(b) Show that the Fourier transform of f is given by

$$\hat{f}(\omega_k) = \frac{a - i2\pi\omega_k}{a^2 + 4\pi^2\omega_k^2},$$

where $\hat{f}$ has been evaluated at $\omega = \omega_k = k/A$.

(c) Verify that $c_k = \hat{f}(\omega_k)(1 - e^{-aA})/A$ and that

$$\lim_{A\to\infty} (Ac_k) = \hat{f}(\omega_k).$$

(d) Use Taylor series to expand e_N, $\sin\theta_k$, and $\cos\theta_k$ for large N and small values of Δx. Show that for $k = -N/2 + 1 : N/2$

$$F_k = \frac{a - i2\pi\omega_k}{a^2 + 4\pi^2\omega_k^2} \frac{1 - e^{-aA}}{A}(1 + c\Delta x) \quad \text{as} \quad N \to \infty, \Delta x \to 0,$$

where c is a constant independent of A and N.

133. Details of Case Study 8. Assume that the function $\hat{f}(\omega) = e^{-a|\omega|}$ is sampled at the points $\omega_k = k\Delta\omega$ (where $a > 0$) to produce the sequence $\hat{f}_k = \hat{f}(\omega_k)$ for $k = -N/2 + 1 : N/2$.

(a) Show that the inverse Fourier transform of $\hat{f}$ is $f(x) = 2a/(a^2 + 4\pi^2 x^2)$.

(b) Verify that the IDFT of $\hat{f}_k$ is given by

$$
\begin{aligned}
\tilde{f}_n &= \sum_{k=-\frac{N}{2}+1}^{\frac{N}{2}} e^{-a|k|\Omega/N} e^{i2\pi nk/N} \\
&= \frac{(1 - e^{-2a\Delta\omega})(1 - e^{-a\Omega/2}\cos(\pi n))}{1 - 2e^{-a\Delta\omega}\cos(2\pi x_n\Delta\omega) + e^{-2a\Delta\omega}}
\end{aligned}
$$

for $n = -N/2 + 1 : N/2$.

(c) Show that with $\Delta\omega$ and x_n fixed

$$\lim_{\Omega\to\infty} (\Delta\omega \tilde{f}_n) = \Delta\omega \frac{(1 - e^{-2a\Delta\omega})}{1 - 2e^{-a\Delta\omega}\cos(2\pi x_n\Delta\omega) + e^{-2a\Delta\omega}} \equiv g(x_n)$$

for $-A/2 \leq x_n \leq A/2$.

(d) Verify that the function g that results from this limit is periodic in x_n with period $1/\Delta\omega = A$.

(e) Use Taylor series to expand $g(x_n)$ in powers of $\Delta\omega$ assuming $\Delta\omega \ll 1$ to show that

$$g(x_n) = \frac{2a}{a^2 + 4\pi^2 x_n^2}(1 + c\Delta\omega) = f(x_n)(1 + c\Delta\omega)$$

for $-A/2 \leq x_n \leq A/2$.

(f) Argue that as $\Delta\omega \to 0$, $g(x_n) \to f(x_n)$ on the interval $-\infty < x < \infty$.

134. DFT case studies. Carry out an analysis similar to Case Study 5 for the problem of approximating the Fourier transform of the functions

$$\text{(a) } f(x) = e^{-a|x|} \quad \text{and} \quad \text{(b) } f(x) = xe^{-a|x|}.$$

Specifically,

(a) Find the DFT and the Fourier coefficients of f when it is restricted to the interval $[-A/2, A/2]$ (or use *The Table of DFTs*).

(b) Find the Fourier transform $\hat{f}$ of f.

(c) Expand the DFT in Taylor series assuming that Δx is small and A is large.

(d) Show that both of the limit paths of Figure 6.10 lead to the sequence $\hat{f}_k$.

135. Asymptotic bounds on integrals. In estimating the error in the DFT of a noncompactly supported function that satisfies $|f(x)| \le Cx^{-r}$ for $|x| > A/2, r \ge 1$ and C a constant, the integral $\int_{A/2}^{\infty} x^{-r}e^{-i2\pi\omega x}dx$ must be bounded. Integrating by parts at least twice, show that this integral satisfies

$$\int_{A/2}^{\infty} x^{-r}e^{-i2\pi\omega x} = \frac{C}{\omega A^r} + O(A^{-r-1})$$

for $A \to \infty$, where C is a constant and $O(A^{-r-1})$ represents terms that are bounded by a constant times A^{-r-1} as $A \to \infty$.

136. IDFT error. The IDFT is to be used to reconstruct a function f from its Fourier coefficients c_k for $k = -N/2 + 1 : N/2$. Show that if the Fourier coefficients satisfy $|c_k| \le C'|k|^{-p}$ then the error in the N-point IDFT is bounded by CN^{-p} where C and C' are constants independent of k and N.

137. Fourier series synthesis. There are some subtleties in selecting the set of Fourier coefficients $\{c_k\}$ to be used for the IDFT. These requirements are the analogs of AVED (average values at endpoints and discontinuities) with respect to Fourier coefficients.

(a) Show that if $c_k = c_{-k}^*$, then the function

$$f(x) = \sum_{k=-\infty}^{\infty} c_k e^{i2\pi kx/A}$$

is real-valued, but the IDFT will produce real values of $\tilde{f}_n$ only if a symmetric set of coefficients, c_k, where $k = -N/2 : N/2$, is used as input.

(b) Show that if the c_k's are real and $c_k = c_{-k}$, then f (given above) is real and even. Furthermore, the IDFT returns a real and even sequence $\tilde{f}_n$ whether the index set $k = -N/2 : N/2$ (an odd number of points) or $k = -N/2 + 1 : N/2$ (an even number of points) is used.

138. Inverse Poisson Summation Formula. Switch the roles of the spatial and frequency domains and then mimic the derivation of the Poisson Summation Formula in the text to obtain the Inverse Poisson Summation Formula (6.21).

139. Replication form of the Inverse Poisson Summation Formula. Show that the inverse Poisson Summation Formula can be expressed using replication operators as

$$\Delta\omega \mathcal{D}^{-1}\{\mathcal{R}_N\{\hat{f}_k\}\}_n = \mathcal{R}_N\{f_n\}.$$

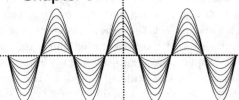

Chapter **7**

A Few Applications
of the DFT

In most cases of practice the number of given values $u_0, u_1, u_2, \ldots$ is either 12 or 24.
— E. T. Whittaker and G. Robinson
The Calculus of Observations, 1924

7.1. Difference Equations; Boundary Value Problems

Background

Difference equations have a history measured in centuries and they find frequent use today in subjects as diverse as numerical analysis, population modeling, probability, and combinatorics. And yet it is a curious fact that they are sadly neglected in the mathematics curriculum. Most students, if they encounter difference equations at all, do so *after* studying differential equations, a subject equally important, but arguably more advanced. We cannot correct this state of affairs in the confines of these few pages. However, we shall attempt to provide a qualitative survey of difference equations, and then investigate the wonderful connection between certain difference equations and the DFT. We will begin by standing back and looking at difference equations from afar, in order to supply a general map of the territory.

Like all equations, difference equations are intended to be solved, which means that they contain an unknown quantity. The unknown in a difference equation may be regarded as a sequence or a vector which we will denote either u_n or

$$\mathbf{u} = \{u_0, u_1, u_2, \ldots, u_{N-1}, u_N\}.$$

A difference equation gives a relationship between the components of this vector of the form

$$u_{n+1} = \Phi_n(u_n, u_{n-1}, \ldots, u_{n-m+1}) + f_n, \tag{7.1}$$

where $m \geq 1$ is an integer. The function Φ_n relates one component of the unknown vector to the preceding m components. The terms f_n are components of a given vector that may be viewed as input to the "system" or, in other contexts, as external forcing of the system. With this general form of the difference equation, we may now define some standard terms of classification. The **order** of the difference equation is m. If Φ is a linear function of its arguments (that is, it does not involve terms such as u_n^2 or $u_n u_{n-1}$ or $\sin(u_n)$), then the difference equation is said to be **linear**; otherwise the difference equation is **nonlinear**. If the input terms f_n are zero, the difference equation is said to be **homogeneous**; otherwise it is **nonhomogeneous**. For example, the difference equations

$$u_{n+1} = 3u_n - 2u_{n-2} + u_{n-3} + n^2 \quad \text{and} \quad u_{n+1} = 3u_n(1 - u_n)$$

are fourth-order, linear, nonhomogeneous and first-order, nonlinear, homogeneous, respectively.

We will get much more specific in just a moment, but first it is necessary to make a major distinction between two types of difference equations that arise in practice. Consider the mth-order difference equation (7.1). *If* the first m components of the solution $u_0, u_1, \ldots, u_{m-1}$ were given, then it would be possible to enumerate the remaining components by applying the difference equation explicity. In other words, given values of $u_0, u_1, \ldots, u_{m-1}$ we could then evaluate

$$
\begin{aligned}
u_m &= \phi(u_{m-1}, u_{m-2}, \ldots, u_0) + f_{m-1}, \\
u_{m+1} &= \phi(u_m, u_{m-1}, \ldots, u_1) + f_m, \\
&\vdots \quad \vdots \quad \vdots
\end{aligned}
$$

$$u_N = \phi(u_{N-1}, u_{m-2}, \ldots, u_{N-m}) + f_{N-1},$$

and the entire vector **u** would be determined. This suggests that in order to find a single solution to an mth-order difference equation, m additional conditions must be specified. The way in which these additional conditions are specified changes the character of the difference equation significantly. Here are the two cases that arise most often.

1. Given the mth-order difference equation (7.1), if the first m components are specified as

$$u_0 = a_0, \; u_1 = a_1, \; \ldots, \; u_{m-1} = a_{m-1},$$

   where the a_i's are given real numbers, then the task of solving the difference equation is called an **initial value problem (IVP)**. The easiest way to interpret this terminology is to imagine that the increasing index $n = 0, 1, 2, 3, \ldots$ represents the passing of time. The difference equation describes a particular system (for example, a bacteria culture or a bank account) as it evolves in time, and the unknown u_n represents the state of the system (the population of bacteria or the balance in the bank account) at the nth time unit. In such a **time-dependent** system, it is reasonable that the initial state of the system, as represented by the components $u_0, u_1, \ldots, u_{m-1}$, should be specified in order to determine the future state of the system. For this reason, this formulation is called an initial value problem. There are very systematic ways to solve linear difference equations/initial value problems that have counterparts in the solution of linear differential equations/initial value problems.

2. The second class of difference equations is less obvious, but equally important. Now imagine that the index $n = 0 : N$ represents spatial position within a system (for example, distance along a heat-conducting bar or position on a beam that is anchored at both ends). The unknown u_n now represents a particular property of the system at the position n when the system has reached steady state (for example, the temperature in the bar or the displacement of the beam under a load). As before, a particular solution to an mth-order difference equation in this steady state case can be determined only if m additional conditions are provided. It turns out that the conditions in a steady state problem must be specified at or near the two endpoints of the system (where the system touches the "outside world"). For this reason, a difference equation with such conditions is called a **boundary value problem (BVP)**. A second-order difference equation BVP might carry boundary conditions such as

$$u_0 = \alpha \quad \text{and} \quad u_N = \beta,$$

   which would specify, say, the temperature at the end of a rod, or the displacement at the end of a beam. There are systematic methods for solving linear difference equations that appear as BVPs. These methods usually bear little resemblance to the corresponding methods for initial value problems, reflecting the very different nature of these two types of problems.

   Hopefully this brief introduction gives a sense of the lay of the land. With this background we can now state precisely that the goal of this section is to explore difference equations that take the form of BVPs. We will limit the discussion to second-order linear difference equations, and even then make additional restrictions, with signposts to more general problems. This road (as opposed to that of initial value problems) is far less trodden; it also offers the marvelous connection with the DFT.

BVPs with Dirichlet Boundary Conditions

In this section we will consider a general family of boundary value problems and indicate how the DFT can be used to obtain solutions. We will leave it to the following section to present some specific applications in which such BVPs arise. For the moment consider the difference equation

$$au_{n+1} = -bu_n - au_{n-1} + f_n, \tag{7.2}$$

where the coefficients a and b are given real numbers, and the terms f_n are also given. According to the discussion of the previous section, this difference equation is second-order because the $(n + 1)$st term is related to the two previous terms. It is linear since the unknowns are multiplied only by the constants a and b. The equation is also nonhomogenous because of the presence of the term f_n, which can be regarded as an external input to the system. This particular equation is even more specialized since the coefficients a and b do not vary with the index n (this is the constant coefficient case), and furthermore the coefficients of $u_{n\pm1}$ are equal. Nevertheless, this is a very important special case.

This difference equation could be associated with either an initial value problem or a boundary value problem. Therefore, to complete the specification of the problem we will give the *boundary conditions*

$$u_0 = u_N = 0,$$

and quickly assert (problem 142) that the more general conditions $u_0 = \alpha, u_N = \beta$ can be handled by the same methods. Some helpful terminology can be injected at this point: a boundary condition that specifies the *value* of the solution at a boundary is called a **Dirichlet boundary condition**. There are actually several other types of admissible boundary conditions, two more of which we will consider a bit later.

The aim is to find an $(N+1)$-vector whose first and last components are zero and whose other components satisfy the difference equation (7.2). It pays to look at this problem in a couple of different ways. If the individual equations of (7.2) are listed sequentially we have the system of linear equations

$$
\begin{aligned}
bu_1 + au_2 &= f_1, \\
au_1 + bu_2 + au_3 &= f_2, \\
au_2 + bu_3 + au_4 &= f_3, \\
au_3 + bu_4 + au_5 &= f_4, \\
&\vdots \\
au_{N-2} + bu_{N-1} &= f_{N-1}.
\end{aligned}
$$

Notice that the Dirichlet boundary conditions $u_0 = u_N = 0$ have been used in the first and last equations, which results in a system of $N - 1$ equations. We could go one step further and write this system of equations in matrix form as follows:

$$
\begin{pmatrix}
b & a & \cdots & & & \cdots & 0 \\
a & b & a & 0 & & & \vdots \\
0 & a & b & a & 0 & & \\
\vdots & & \ddots & \ddots & \ddots & & \vdots \\
\vdots & & & a & b & a \\
0 & \cdots & & & 0 & a & b
\end{pmatrix}
\begin{pmatrix}
u_1 \\ u_2 \\ u_3 \\ \vdots \\ u_{N-2} \\ u_{N-1}
\end{pmatrix}
=
\begin{pmatrix}
f_1 \\ f_2 \\ f_3 \\ \vdots \\ f_{N-2} \\ f_{N-1}
\end{pmatrix}.
$$

In the language of matrices, we see that this particular boundary value problem takes the form of a symmetric **tridiagonal** system of linear equations (meaning that all of the matrix elements are zero except those on the three main diagonals). You now argue: why not use a method for solving systems of linear equations such as Gaussian elimination and be done with it? And your argument would be irrefutable! For this *one-dimensional* problem, there is no need to resort to DFTs, and indeed linear system solvers are preferable. However, for problems in two or more dimensions (coming soon), the tables are reversed and the DFT is the more economical method. So we will present the DFT solution with the assurance that the extra work will soon pay off.

There are now three ways to proceed, each instructive and each leading to the same end. We will take them in the following order:

- component perspective,

- operational perspective,

- matrix perspective.

Component Perspective

Notice that for any integer $k = 1 : N - 1$, the vector $\mathbf{u}^k$ with components

$$
\mathbf{u}_n^k = \sin\left(\frac{\pi n k}{N}\right)
$$

for $n = 0 : N$ satisfies the boundary conditions $u_0 = u_N = 0$. This observation motivates the idea of looking for a solution to the difference equation which is a linear combination of these $N - 1$ vectors. Thus, we will assume a trial solution to the difference equation that looks like

$$
u_n = 2 \sum_{k=1}^{N-1} U_k \sin\left(\frac{\pi n k}{N}\right) \tag{7.3}
$$

for $n = 0 : N$.

This representation is simply the inverse **discrete sine transform** (DST) of the vector U_k as defined in Chapter 4; the factor of 2 has been included to maintain consistency with that definition. This form of the solution contains the $N-1$ unknown coefficients U_k; so we have replaced the problem of finding the u_n's by a new problem of finding the U_k's. However, if the coefficients U_k *could* be found, then the solution

u_n could be reconstructed using the representation given in (7.3); furthermore, it can be done quickly by using the FFT.

As with any trial solution, it must be substituted into the problem at hand and then followed wherever it may lead. Substituting the expression (7.3) for u_n into the difference equation

$$au_{n+1} + bu_n + au_{n-1} = f_n,$$

we find that

$$2a \sum_{k=1}^{N-1} U_k \sin\left(\frac{\pi(n+1)k}{N}\right) + 2b \sum_{k=1}^{N-1} U_k \sin\left(\frac{\pi nk}{N}\right)$$

$$+ 2a \sum_{k=1}^{N-1} U_k \sin\left(\frac{\pi(n-1)k}{N}\right) = f_n,$$

where we understand that this relation must hold for $n = 1 : N - 1$. Now we need to combine terms with the aid of the sine addition rules. Recalling that

$$\sin(A + B) = \sin A \cos B + \cos A \sin B,$$

expanding and collecting terms simplifies this previous relationship significantly. It is now merely

$$2 \sum_{k=1}^{N-1} U_k \left(2a \cos\left(\frac{\pi k}{N}\right) + b\right) \sin\left(\frac{\pi nk}{N}\right) = f_n, \tag{7.4}$$

for $n = 1 : N - 1$. Notice how the entire left-hand side of the difference equation has been reduced to a linear combination of the terms $\sin(\pi nk/N)$ in which the solution u_n was expressed. This suggests that we ought to try to express the right-hand side, f_n, as a linear combination of the same terms. Toward this end we will give the vector f_n the representation

$$f_n = 2 \sum_{k=1}^{N-1} F_k \sin\left(\frac{\pi nk}{N}\right)$$

for $n = 1 : N - 1$. Take note that we may interpret the F_k's as the DST coefficients of the vector f_n. Since the vector f_n is known, its DST coefficients can be computed using

$$F_k = \frac{1}{N} \sum_{n=1}^{N-1} f_n \sin\left(\frac{\pi nk}{N}\right)$$

for $k = 1 : N$.

We may now return to the computation and insert this representation for f_n into the right-hand side of (7.4). If we do this and collect all terms on one side of the equation we find that the difference equation now appears as

$$\sum_{k=1}^{N-1} \left[\left(2a \cos\left(\frac{\pi k}{N}\right) + b\right) U_k - F_k\right] \sin\left(\frac{\pi nk}{N}\right) = 0 \tag{7.5}$$

for $n = 1 : N - 1$.

Let's pause to collect some thoughts before forging ahead. The original players in the difference equation were the given vector f_n and the unknown vector u_n. They are no longer in sight. They have been replaced by the DSTs of these vectors, F_k and U_k, respectively. Recall that the F_k's can be computed (since f_n is known) and that the goal is to determine U_k, from which the solution u_n can be reconstructed. With this in mind we resume the task of finding U_k in terms of F_k. The last line of computation (7.5) must be valid at each of the indices $n = 1 : N - 1$. The only way in which this can happen in general is if each term of the sum vanishes identically. Since $\sin(\pi n k/N) \neq 0$ for all $n = 1 : N - 1$, each term vanishes only if the coefficient of $\sin(\pi n k/N)$ vanishes. This observation leads to the $N - 1$ conditions

$$\left(2a \cos\left(\frac{\pi k}{N}\right) + b\right) U_k - F_k = 0$$

for $k = 1 : N - 1$. Solving for the desired coefficients U_k, we have that

$$U_k = \frac{F_k}{2a \cos(\frac{\pi k}{N}) + b}$$

for $k = 1 : N - 1$. An important note is that all of the U_k's are well defined provided that a, b, and k do not conspire to make the denominator $b + 2a \cos(\pi k/N)$ equal to zero. This can be assured if we impose a condition such as $|b| > |2a|$; this condition endows the problem with a property known as **diagonal dominance**. (Any matrix in which the magnitude of each diagonal term exceeds the sum of the magnitudes of the other terms of the same row is called diagonally dominant.)

The final step is to recover the actual solution u_n. Having determined the sequence U_k, it is now possible to find its inverse DST

$$u_n = 2 \sum_{k=1}^{N-1} U_k \sin\left(\frac{\pi n k}{N}\right)$$

for $n = 1 : N - 1$. The resulting solution satisfies the original difference equation (7.2) and the boundary conditions $u_0 = u_N = 0$. We can summarize the entire solution process in three easy steps.

1. Apply the DST to the input vector f_n to determine the coefficients F_k.

2. Solve for the coefficients U_k of the solution.

3. Apply the inverse DST to the vector U_k to recover the solution u_n.

Note that the factor of $1/N$ in the forward DST and the factor of 2 in the inverse DST can be combined in a single step as long as care is used. Let's solidify these ideas with an example that also begins to move towards one of the applications of difference equations and BVPs.

Example: Towards diffusion. If you have ever opened a door to a warm room on a winter night or spilled a drop of colored dye in a basin of calm water you have experienced the ubiquitous phenomenon of diffusion. For present purposes, it suffices to say that diffusion is characterized by the property that some "substance" (for example, heat, dye, or pollutant) spreads in such a way as to smooth out differences in concentration. A diffusing substance moves from regions of high concentration to regions of low concentration. If a door is left open long enough, the room will eventually have the same temperature as the outdoors; if left undisturbed long enough, the entire basin of water will eventually have a uniform concentration of dye. Here is an idealized model of the diffusion process. Imagine a long thin cylindrical rod

FIG. 7.1. *The process of diffusion of heat along the length of a long thin conducting rod may be idealized by a discrete model. The rod lies along the interval $[0, A]$ which is sampled at $N + 1$ equally spaced points $x_0, \ldots, x_N$. The steady state temperature (heat content) at a particular point is assumed to be the average of the temperatures at the two neighboring points plus the contribution from a possible external source.*

that has $N + 1$ uniformly spaced points along its length. As shown in Figure 7.1, the nth point has a coordinate x_n, for $n = 0 : N$. We will assume that the rod is a good conductor of heat (for example, copper) and has uniform material properties. Furthermore, we will assume that the ends of the rod ($n = 0$ and $n = N$) are held at a fixed temperature of zero; this assumption will form the boundary conditions for the problem. Finally, to make the problem interesting, we will assume that each of the points $n = 1 : N - 1$ along the interior of the rod may have a source (or sink) of heat of known intensity f_n. Our question is this: when this system consisting of a conducting rod heated externally along its length reaches equilibrium (steady state), what is the temperature at each of the points along the rod?

We will let u_n be the temperature at the nth point along the rod where $n = 0 : N$. We may argue qualitatively as follows. The fact that diffusion tries to smooth (or average) variations in temperature can be described by requiring that the steady state temperature at a point along the rod be the *average* of the temperature at the two neighboring points. Since there is also a source of heat at each point of strength f_n, we can write that the steady state temperature at the nth point is

$$u_n = \frac{u_{n+1} + u_{n-1}}{2} + f_n$$

for $n = 1 : N - 1$. By the boundary conditions we also know that $u_0 = u_N = 0$.

It should be emphasized that this is a qualitative argument since we have introduced no length scales or temperature units, and have assumed that the continuous rod can be represented as $N + 1$ sample points. Nevertheless the averaging idea does capture the essence of diffusion and does lead to a plausible BVP. Rearranging the terms of the previous relationship, we find a difference equation that looks like

$$-\frac{1}{2}u_{n-1} + u_n - \frac{1}{2}u_{n+1} = f_n$$

for $n = 1 : N - 1$, together with the boundary conditions $u_0 = u_N = 0$. This boundary value problem has exactly the form required for use of the DST. In order to produce a specific solution, let's consider a particular input vector f_n. Assume that the external heat sources along the rod have a uniform strength of $8/N^2$ units; that is, $f_n = 8/N^2$ for $n = 1 : N - 1$. Following the three-step procedure just described, we must first transform the right-hand side vector f_n, then solve for the coefficients U_k of the solution, then perform the inverse DST of U_k to recover the solution u_n.

A short calculation (problem 144) reveals that the DST of the constant vector $f_n = 8/N^2$ is given by

$$F_k = \frac{8}{N^2} \frac{\sin(\frac{\pi k}{N})(1 - \cos(\pi k))}{4N \sin^2(\frac{\pi k}{2N})}$$

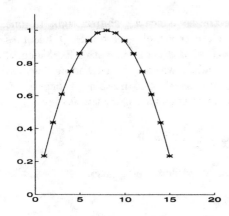
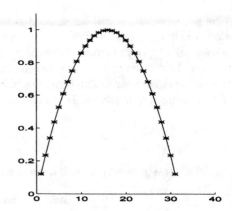

FIG. 7.2. *An idealized model of diffusion leads to a difference equation BVP that can be solved using the DST. The solution to this problem with a constant input vector $f_n = 8/N^2$ and boundary condition $u_0 = u_N = 0$ is shown for $N = 16$ (left) and $N = 32$ (right).*

for $k = 1 : N - 1$. Noting that the coefficients of the difference equation are $a = -1/2$ and $b = 1$, we find (with the help of the identity $1 - \cos 2\theta = 2 \sin \theta$) that the DST coefficients of the solution are

$$U_k = \frac{F_k}{2a \cos(\frac{\pi k}{N}) + b} = \frac{F_k}{1 - \cos(\frac{\pi k}{N})} = \frac{\sin(\frac{\pi k}{N})(1 - \cos(\pi k))}{N^3 \sin^4(\frac{\pi k}{2N})}$$

for $k = 1 : N - 1$. The final step of taking the inverse DST cannot be done analytically. However, the numerical calculation is straightforward, and the results are shown in Figure 7.2 for $N = 16$ and $N = 32$. Because of the symmetric nature of the input f_n, the solution also finds a symmetric pattern that manages to satisfy the boundary conditions $u_0 = u_n = 0$. The "hottest" point on the bar is the midpoint, while heat diffuses out of both ends of the bar in order to maintain the zero boundary conditions. The choice of scaling for f_n has the effect of making the maximum temperature independent of N.

We have now shown how the DST can be used to solve difference equations that occur as BVPs. The next move would ordinarily be to show how the discrete cosine transform and the full DFT can also be used to solve BVPs with different kinds of boundary conditions. However, before taking that step, we really should linger a bit longer with the sine transform and make two optional, but highly instructive excursions. The foregoing discussion can be presented in two other perspectives which are laden with beautiful mathematics and connections to other ideas.

Operational Perspective

It is possible to look at the solution of BVPs from an **operational** point of view. Indeed we have already used this perspective in previous chapters in discussing the DFT. We must first agree on some notation to avoid confusion. Assume that we are given a sequence u_n. Since u_n will be associated with the DST, we will assume that it has the periodicity and symmetry properties of that transform (Chapter 4), namely, $u_0 = u_N = 0$ and $u_n = u_{n+2N}$. We will denote the DST of u_n as

$$U_k = \mathcal{S}\{u_n\}_k.$$

The sequence u_{n+1} will be understood to be the sequence u_n shifted right by one unit, while u_{n-1} is u_n shifted left by one unit. (The terminology admittedly becomes somewhat slippery at times. When u_n is viewed as a list of numbers with a fixed length, it seems best to refer to it as a *vector*. However, when we imagine extending that list to use periodicity or shift properties, the term *sequence* seems more appropriate.) A short calculation (problem 145) demonstrates the shift property for the DST:

$$S\{u_{n+1} + u_{n-1}\}_k = 2\cos\left(\frac{\pi k}{N}\right)S\{u_n\}_k.$$

With this property in mind, we can turn to the difference equation

$$au_{n-1} + bu_n + au_{n+1} = f_n$$

for $n = 1 : N - 1$. We now apply the DST operator S (hence the term *operational*) to both sides of the difference equation regarding u_n and $u_{n\pm1}$ as full vectors. Regrouping terms in anticipation of the shift property we have that

$$aS\{u_{n-1} + u_{n+1}\}_k + bS\{u_n\}_k = S\{f_n\}_k.$$

Now, using the shift theorem and the notation that $S\{u_n\}_k = U_k$ and $S\{f_n\}_k = F_k$, we have that

$$2a\cos\left(\frac{\pi k}{N}\right)U_k + bU_k = F_k$$

for $k = 1 : N - 1$. As before, the terms F_k can be computed since the input vector f_n is specified. The goal is to determine the coefficients U_k, and clearly this is now possible. We find that

$$U_k = \frac{F_k}{2a\cos(\frac{\pi k}{N}) + b}$$

for $k = 1 : N - 1$, which agrees with the expression for U_k found via the component perspective. The solution to the problem can then be expressed as $u_n = S^{-1}\{U_k\}_n$.

Thus, we see that an alternative way to solve a linear difference equation with Dirichlet boundary conditions is to apply the DST as an operator to the entire equation. Other types of boundary conditions require different transforms, as we will soon see. This approach certainly has analogs in the solution of differential equations using Fourier and Laplace transforms. In all of these situations, it is necessary to know the relevant shift (or derivative) properties of the particular transform.

Matrix Perspective

It was noted earlier that the difference equation BVP

$$au_{n-1} + bu_n + au_{n+1} = f_n$$

for $n = 1 : N - 1$, with $u_0 = u_N = 0$, can be regarded as a system of linear equations. We will denote this system $\mathbf{Au} = \mathbf{f}$, where $\mathbf{u}$ and $\mathbf{f}$ are vectors of length $N - 1$ and A is an $(N-1) \times (N-1)$ symmetric tridiagonal matrix. As a system of linear equations this is not a difficult problem to solve; however, it could be even easier! We ask whether there is a change of coordinates (or change of basis) for the vectors $\mathbf{u}$ and $\mathbf{f}$ such that this system appears in an even simpler form. A change of coordinates can be represented by an invertible matrix $\mathbf{P}$ such that $\mathbf{u}$ and $\mathbf{f}$ are transformed into new vectors $\mathbf{v}$ and $\mathbf{g}$ by means of the relations $\mathbf{u} = \mathbf{Pv}$ and $\mathbf{f} = \mathbf{Pg}$. If we make these substitutions into the system of equations $\mathbf{Au} = \mathbf{f}$ we find that

$$A \underbrace{\mathbf{Pv}}_{\mathbf{u}} = \underbrace{\mathbf{Pg}}_{\mathbf{f}} \quad \text{or} \quad \underbrace{\mathbf{P^{-1}AP}}_{\mathbf{D}} \mathbf{v} = \mathbf{g}.$$

Not only are the vectors $\mathbf{u}$ and $\mathbf{f}$ transformed, the system of equations is also transformed into a new system that we will write $\mathbf{Dv} = \mathbf{g}$ where $\mathbf{D} = \mathbf{P^{-1}AP}$. Here is the question: is there a choice of the transformation matrix $\mathbf{P}$ that makes this new system particularly simple? The simplest form for a system of linear equations is a diagonal system in which the equations are decoupled and may be solved independently. So the question becomes: can $\mathbf{P}$ be chosen such that $\mathbf{D} = \mathbf{P^{-1}AP}$ is a diagonal matrix? This turns out to be one of the fundamental questions of linear algebra and it is worth looking at the solution for this particular case in which $\mathbf{A}$ is tridiagonal.

The condition $\mathbf{D} = \mathbf{P^{-1}AP}$ can be rewritten as $\mathbf{AP} = \mathbf{PD}$, where the objective is to find the elements of the matrix $\mathbf{P}$ in terms of the elements of $\mathbf{A}$. Let's denote the kth column of $\mathbf{P}$ as $\mathbf{w}^k$ and the kth diagonal element of $\mathbf{D}$ as λ_k, where $k = 1 : N - 1$. The condition $\mathbf{AP} = \mathbf{PD}$ means that the columns of $\mathbf{P}$ must each satisfy the equation

$$\mathbf{Aw}^k = \lambda_k \mathbf{w}^k \quad \text{or} \quad (\mathbf{A} - \lambda_k \mathbf{I})\mathbf{w}^k = 0$$

for $k = 1 : N - 1$, where $\mathbf{I}$ is the $(N-1) \times (N-1)$ identity matrix. For the moment, to simplify the notation a bit, let $\mathbf{w}$ represent any of the $(N-1)$ columns of the matrix $\mathbf{P}$ that we have denoted $\mathbf{w}^k$, and let λ denote any of the diagonal elements λ_k. Then the problem at hand is to find the vectors $\mathbf{w}$ and scalars λ that satisfy $(\mathbf{A} - \lambda I)\mathbf{w} = 0$. Any scalar λ that admits a nontrivial solution to this matrix equation is called an **eigenvalue** and of the matrix $\mathbf{A}$, and the corresponding nontrivial solution vector is called an **eigenvector**. Finding the eigenvalues and eigenvectors of general matrices $\mathbf{A}$ can be a challenging proposition, however we will show that for the tridiagonal matrix that corresponds to our BVP, it can be done. One approach is to note that if the homogeneous system of equations $(\mathbf{A} - \lambda I)\mathbf{w} = 0$ is to have nontrivial solutions, then the determinant $|\mathbf{A} - \lambda I|$ must vanish. This fact provides a condition (a polynomial equation) that must be satisfied by the eigenvalues.

An alternative approach is to write out the equations of the system $(\mathbf{A} - \lambda I)\mathbf{w} = 0$ in the form

$$\begin{pmatrix} b - \lambda & a & 0 & \cdots & & \cdots & 0 \\ a & b - \lambda & a & 0 & \cdots & & \vdots \\ 0 & a & b - \lambda & a & 0 & & \vdots \\ \vdots & & \ddots & \ddots & \ddots & & \vdots \\ & & & a & b - \lambda & a \\ 0 & & \cdots & 0 & a & b - \lambda \end{pmatrix} \begin{pmatrix} w_1 \\ w_2 \\ w_3 \\ \vdots \\ w_{N-2} \\ w_{N-1} \end{pmatrix} = \begin{pmatrix} 0 \\ 0 \\ 0 \\ \vdots \\ 0 \\ 0 \end{pmatrix}.$$

A typical equation in this system is

$$aw_{n-1} + (b - \lambda)w_n + aw_{n+1} = 0$$

for $n = 1 : N - 1$, where $w_0 = w_N = 0$. You should sense that we are coming back full circle, since this system of equations for the eigenvalue problem is almost identical to the original set of difference equations that sent us down this road. The only difference is that the coefficient of the "center" term w_n is $b - \lambda$ rather than b. Therefore, we can look for solutions in much the same way that we originally solved the difference equation. Solutions of the form $w_n = \sin(\pi nk/N)$ satisfy the condition $w_0 = w_N = 0$ for any $k = 1 : N - 1$. If we substitute and use the sine addition laws just as before, we find that for each $k = 1 : N - 1$

$$\left(2a \cos \left(\frac{\pi k}{N} \right) + (b - \lambda) \right) \sin \left(\frac{\pi nk}{N} \right) = 0, \tag{7.6}$$

which must hold for $n = 1 : N - 1$. Since $\sin(\pi nk/N) \neq 0$ for all $n = 1 : N - 1$, the coefficient of $\sin(\pi nk/N)$ must vanish identically, which says that the eigenvalues λ must satisfy

$$\lambda_k = b + 2a \cos \left(\frac{\pi k}{N} \right)$$

for $k = 1 : N - 1$. Corresponding to the eigenvalue λ_k we have the solution vector with components $w_n = \sin(\pi nk/N)$. Resorting to our original notation and letting $\mathbf{w}^k$ denote the solution vector associated with λ_k, we discover that the nth component of the kth eigenvector is

$$\mathbf{w}_n^k = \sin \left(\frac{\pi nk}{N} \right), \quad \text{where} \quad k, n = 1 : N - 1.$$

In other words, the (n, k) element of the matrix $\mathbf{P}$ is $\sin(\pi nk/N)$. Equivalently, the kth column of the matrix $\mathbf{P}$ is just the kth mode of the DST.

We have now accomplished the task of finding the elements of the transformation matrix $\mathbf{P}$ and the elements of the diagonal matrix $\mathbf{D}$. In principle, we could complete the solution of the difference equation. The solution of the new system $\mathbf{Dv} = \mathbf{g}$ has components $v_n = g_n/\lambda_n$. With the vector $\mathbf{v}$ determined, the solution to the original difference equation can be recovered, since $\mathbf{u} = \mathbf{Pv}$. In practice, there is usually no need to compute the eigenvectors and eigenvalues explicitly. However, this perspective does reveal the following important fact: the matrix $\mathbf{P}$ whose columns are the $N - 1$ modes of the DST is precisely the matrix that diagonalizes the tridiagonal matrix $\mathbf{A}$. In fact, the matrix perspective leads to a three-step solution method that exactly parallels the solution method described under the component and operational perspectives. The parallels are extremely revealing. Here are the three steps for solving $\mathbf{Au} = \mathbf{f}$, seen from the matrix perspective.

1. Apply $\mathbf{P}^{-1}$ to the right-hand side vector: $\mathbf{g} = \mathbf{P}^{-1}\mathbf{f}$. (This corresponds to taking the DST of the right-hand side sequence f_n.)

2. Solve the diagonal system of equations $\mathbf{Dv} = \mathbf{g}$. (This corresponds to solving for U_k from F_k by simple division.)

3. Recover the solution by computing $\mathbf{u} = \mathbf{Pv}$. (This corresponds to taking the inverse DST of U_k.)

We have now presented three different but equivalent perspectives on the use of the DST to solve a particular difference equation BVP. The component and operational perspectives provide actual methods of solution that can be carried out in practice, while the more formal matrix perspective illuminates the underlying structure of the

DST and shows why it works so effectively. We will now turn to two other commonly occurring BVPs that can be handled by the discrete cosine transform (DCT) and the DFT. In each of these cases it would be possible to use any of the three perspectives just discussed. The following discussion will be less than exhaustive, with unfinished business left for the problems.

BVPs with Neumann and Periodic Boundary Conditions

The second variety of difference equation BVP that we will consider has virtually the same form as difference equation (7.2), namely

$$au_{n-1} + bu_n + au_{n+1} = f_n \qquad (7.7)$$

for $n = 0 : N$. We consider an alternative form of the boundary conditions:

$$u_1 - u_{-1} = 0 \quad \text{and} \quad u_{N+1} - u_{N-1} = 0.$$

Whereas the Dirichlet boundary conditions of the previous section required that the solution vanish (or be specified) at the boundaries $n = 0$ and $n = N$, this new boundary condition, called a **Neumann**[1] **boundary condition**, stipulates the change in the solution at the two boundaries. (In differential equation BVPs, the Neumann condition specifies the value of the *derivative* of the solution. If the derivative at a boundary is zero, it describes the case of zero flux at that boundary.) The particular Neumann condition that we have given above requires that the solution have no change at the boundaries. Notice that with these boundary conditions u_0 and u_N are unknowns and we have $N + 1$ unknowns. If the corresponding $N + 1$ equations of (7.7) are listed sequentially we have

$$
\begin{aligned}
bu_0 + 2au_1 &= f_0, \\
au_0 + bu_1 + au_2 &= f_1, \\
au_1 + bu_2 + au_3 &= f_2, \\
au_2 + bu_3 + au_4 &= f_3, \\
&\ \ \vdots \\
2au_{N-1} + bu_N &= f_N,
\end{aligned}
$$

where the boundary conditions have been used in the first and last equations. We could go one step further and write this system of equations in matrix form as follows:

$$
\begin{pmatrix}
b & 2a & 0 & \cdots & & \cdots & 0 \\
a & b & a & 0 & \cdots & & \vdots \\
0 & a & b & a & 0 & & \vdots \\
\vdots & & \ddots & \ddots & \ddots & & \vdots \\
 & & & 0 & a & b & a \\
0 & \cdots & & & \cdots & 0 & 2a & b
\end{pmatrix}
\begin{pmatrix}
u_0 \\ u_1 \\ u_2 \\ \vdots \\ u_{N-1} \\ u_N
\end{pmatrix}
=
\begin{pmatrix}
f_0 \\ f_1 \\ f_2 \\ \vdots \\ f_{N-1} \\ f_N
\end{pmatrix}.
$$

[1] FRANZ ERNST NEUMANN (1798–1895) was professor of physics and mineralogy in Königsberg. He is credited with developing the mathematical theory of magneto-electric inductance. He also made important contributions to the subjects of spherical harmonics, hydrodynamics, and elasticity. His son Carl Neumann lived in Leipzig and is known for his work on integral equations, differential geometry, and potential theory. We believe the boundary condition was named for Carl!

The matrix associated with this BVP is tridiagonal, although not quite symmetric, and has $N + 1$ rows and columns. It would be possible to solve this system of linear equations (and hence the BVP) using Gaussian elimination. Instead, we will consider the use of another form of the DFT, since this becomes the more efficient method in higher-dimensional problems. The key observation is that a vector defined by the DCT

$$u_n = \frac{1}{2}\left[U_0 + 2\sum_{k=1}^{N-1} U_k \cos\left(\frac{\pi k}{N}\right) + U_N \cos(\pi n)\right] = 2\sum_{k=0}^{N}{}'' U_k \cos\left(\frac{\pi k}{N}\right) \qquad (7.8)$$

is $2N$-periodic and has the properties $u_1 - u_{-1} = 0$ and $u_{N+1} - u_{N-1} = 0$. (Recall that Σ'' means the first and last terms of the sum are weighted by $1/2$.) In other words, a trial solution of this form satisfies the boundary conditions of the problem. At this point we may proceed with any of the three perspectives discussed earlier. The most succinct solution is provided by the operational perspective. The component and matrix perspectives will be discussed in problems 141 and 146. As we saw earlier, the solution takes place in three steps:

1. the DCT coefficients, F_k, of the input vector, f_n, must be found,

2. the DCT coefficients, U_k, of the solution must be computed from F_k,

3. the inverse DCT must be used to recover the solution u_n.

Recall also that the operational approach requires the use of the shift properties of the relevant transform. Letting $\mathcal{C}$ represent the DCT, and letting u_{n+1} and u_{n-1} be the sequences obtained by shifting u_n right and left one unit, respectively, the critical shift property for the DCT is (problem 145)

$$\mathcal{C}\{u_{n-1} + u_{n+1}\}_k = 2\cos\left(\frac{\pi k}{N}\right)\mathcal{C}\{u_n\}_k$$

for $k = 0 : N$. We now apply the DCT to both sides of the difference equation (7.7) to obtain

$$a\mathcal{C}\{u_{n-1} + u_{n+1}\}_k + b\mathcal{C}\{u_n\}_k = \mathcal{C}\{f_n\}_k$$

for $k = 0 : N$. Using the shift property and letting $U_k = \mathcal{C}\{u_n\}_k$ and $F_k = \mathcal{C}\{f_n\}_k$, we find that

$$\left(2a\cos\left(\frac{\pi k}{N}\right) + b\right) U_k = F_k$$

for $k = 0 : N$. Finally, the coefficients F_k may be computed as the DCT of the given vector f_n; that is,

$$\begin{aligned} F_k = \mathcal{C}\{f_n\}_k &= \frac{1}{2N}\left[f_0 + 2\sum_{n=1}^{N-1} f_n \cos\left(\frac{\pi n k}{N}\right) + f_N \cos(\pi k)\right] \\ &= \frac{1}{N}\sum_{n=0}^{N}{}'' f_n \cos\left(\frac{\pi n k}{N}\right), \end{aligned}$$

for $k = 0 : N$. It is an easy matter to solve for the unknown coefficients U_k. Doing so, we find that

$$U_k = \frac{F_k}{2a\cos\left(\frac{\pi k}{N}\right) + b}$$

for $k = 0 : N$. The final step of the solution is to recover the solution u_n by applying the inverse DCT (7.8) to the vector U_k.

There is an interesting condition that arises in this BVP. Notice that as long as $|b| > |2a|$, the coefficients U_k may be determined uniquely. However, absent this condition on a and b, it might happen that the denominator $2a\cos(\pi k/N) + b$ vanishes for some value of k. If so, a solution fails to exist *unless*, for that same value of k, the numerator F_k also vanishes. In this case, U_k is arbitrary (since it satisfies the equation $0 \cdot U_k = 0$) and the solution is determined up to an arbitrary multiple of the kth mode. For example, the most common case in which this degeneracy arises is the difference equation with $b = -2a$,

$$-u_{n-1} + 2u_n - u_{n+1} = f_n,$$

together with the zero boundary conditions $u_1 - u_{-1} = u_{N+1} - u_{N-1} = 0$. The coefficient U_0 is now undefined and a solution fails to exist, unless it happens that $F_0 = 0$ also. If $F_0 = 0$, then U_0 is arbitrary and the solution is determined up to an additive constant (since the $k = 0$ mode is a constant). This makes sense, since any constant vector satisfies $-u_{n-1} + 2u_n - u_{n+1} = 0$ *and* the boundary conditions. Thus, any constant vector can be added to a solution of the BVP, and the result is another solution. It also makes sense physically since the Neumann boundary conditions $u_1 - u_{-1} = u_{N+1} - u_{N-1} = 0$ are "zero flux" conditions that prevent the diffusing substance from leaving the domain. A steady state can exist only if the net input from external sources is zero; this means that

$$F_0 = \sum_{n=0}^{N} f_n = 0.$$

We should at least mention the underlying mathematical principle at work here. This is one of many forms of the **Fredholm[2] Alternative**, which states that either the homogeneous problem has only the trivial (zero) solution or the nonhomogeneous problem has an infinity of solutions.

The third class of BVPs on our tour introduces yet another type of boundary condition to the same difference equation. We will now consider the difference equation

$$au_{n-1} + bu_n + au_{n+1} = f_n \quad \text{for} \quad n = 0 : N - 1, \tag{7.9}$$

subject to the boundary condition that $u_0 = u_N$ and $u_{-1} = u_{N-1}$. These boundary conditions stipulate that the solution should be N-periodic, and not surprisingly, they are called **periodic** boundary conditions. In this case u_0 (or u_N) becomes an unknown, and there is a total of N unknowns. If the corresponding N equations of (7.9) are listed sequentially we have the system of linear equations

$$
\begin{aligned}
bu_0 + au_1 + au_{N-1} &= f_0, \\
au_0 + bu_1 + au_2 &= f_1, \\
au_1 + bu_2 + au_3 &= f_2, \\
au_2 + bu_3 + au_4 &= f_3,
\end{aligned}
$$

---

[2]A native of Sweden, ERIC IVAR FREDHOLM worked at both the University of Stockholm and at the Imperial Bureau of Insurance during his lifetime. He is best known for his contributions to the theory of integral equations.

$$\vdots \quad \vdots \quad \vdots$$

$$au_0 + au_{N-2} + bu_{N-1} = f_{N-1}.$$

Notice that the boundary conditions have been used in the first and last equations. This set of equations can also be written in matrix form as follows:

$$
\begin{pmatrix}
b & a & 0 & \cdots & & 0 & a \\
a & b & a & 0 & \cdots & & 0 \\
0 & a & b & a & 0 & & \vdots \\
\vdots & & \ddots & \ddots & \ddots & & \vdots \\
0 & & & 0 & a & b & a \\
a & 0 & \cdots & & 0 & a & b
\end{pmatrix}
\begin{pmatrix}
u_0 \\ u_1 \\ u_3 \\ \vdots \\ u_{N-2} \\ u_{N-1}
\end{pmatrix}
=
\begin{pmatrix}
f_0 \\ f_1 \\ f_2 \\ \vdots \\ f_{N-2} \\ f_{N-1}
\end{pmatrix}.
$$

The matrix associated with this BVP is no longer tridiagonal, but **circulant** (notice the nonzero corner elements); it is symmetric and has N rows and columns.

As before, we would like to consider a DFT-based method for solving this BVP. The boundary conditions now require a (real) periodic solution vector with period N. The vector that fits this bill is given by the inverse DFT in its real form. Therefore, we will consider solutions of the form

$$u_n = \text{Re}\left\{ \sum_{k=0}^{N-1} U_k \omega_N^{nk} \right\}, \tag{7.10}$$

where $\omega_N = e^{i2\pi/N}$. It is easiest to allow the coefficients U_k to be complex, and then take the real part to assure that the solution u_n is real-valued. Note that u_n given in this form satisfies the boundary conditions. Again, there are three possible perspectives that could be adopted. For variety and brevity, we will take the component perspective.

Substituting the solution u_n as given in (7.10) and a similar representation for the input vector f_n into the difference equation (7.9), we find that

$$a \sum_{k=0}^{N-1} U_k \omega_N^{(n-1)k} + b \sum_{k=0}^{N-1} U_k \omega_N^{nk} + a \sum_{k=0}^{N-1} U_k \omega_N^{(n+1)k} = \sum_{k=0}^{N-1} F_k \omega_N^{nk},$$

where the coefficients F_k can be computed as the DFT of the input vector f_n. This equation must hold for $n = 0 : N - 1$. The shifts $n \pm 1$ that appear in the complex exponentials are easily handled (no trigonometric addition laws are needed here) and a gathering of terms gives

$$\sum_{k=0}^{N-1} \left\{ [a(\omega_N^k + \omega_N^{-k}) + b]U_k - F_k \right\} \omega_N^{nk} = 0$$

for $n = 0 : N - 1$. Once again, we argue that if this equation is to hold in general for all $n = 0 : N - 1$, then each term of the sum must vanish identically. Noting that $\omega_N^k + \omega_N^{-k} = 2\cos(2\pi k/N)$, we have that

$$\left(2a \cos\left(\frac{2\pi k}{N} \right) + b \right) U_k = F_k$$

for $k = 0 : N - 1$. Solving for the unknown coefficients U_k gives

$$U_k = \frac{F_k}{2a \cos(\frac{2\pi k}{N}) + b}$$

for $k = 0 : N - 1$. The three-step job is completed by taking the inverse DFT of the vector U_k to recover the solution u_n according to (7.10). We pause to point out the degeneracy that can occur in this case also. If we assume $|b| > |2a|$, then all of the coefficients U_k can be determined without ambiguity. Otherwise, a, b, and k may combine to make one of the U_k's indeterminate. This can lead either to the nonexistence of a solution or to an infinity of solutions. We would like to illustrate the use of a BVP with periodic conditions with a short example.

Example: A model of gossip. The process of diffusion discussed earlier has many physical applications. However, it is also used in a nonfrivolous way by sociologists to describe the spread of information known otherwise as gossip. Recall that the driving principle in diffusion is the flow of a "substance" (in this case, lurid stories) from regions of high concentration to regions of low concentration. This is certainly a plausible mechanism for the spread of gossip, since like heat, it tends to flow from "hot" (well-informed) people to "cold" people. In a steady state, large differences in the amount of gossip possessed by individuals would eventually be smoothed out, as the system tends toward equilibrium. We will idealize this process with a model that consists of N individuals (or houses or communication nodes) that have the configuration shown in Figure 7.3. Each individual can talk to two nearest neighbors and no one else. Furthermore, the group is arranged in a ring so that person $n = 0$ talks to persons $n = 1$ and $n = N - 1$. In order to capture this cyclic structure we will identify person $n = 0$ with person $n = N$. The variable (unknown) in the model will be u_n, the amount of gossip that person n has when the process has reached a steady state. (Note that, as with any diffusion process, gossip may still be transferred in the steady state, but in such a way that no individual's amount of gossip changes in time.)

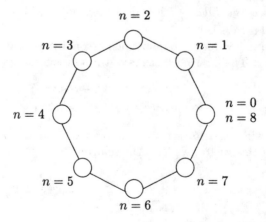

FIG. 7.3.   *A model for the steady state distribution of gossip consists of a ring of people (or communication nodes), each with two nearest neighbors. The cyclic configuration (shown here with $N = 8$) is handled by periodic boundary conditions.*

We will now model the actual diffusion process as we did before by requiring that the steady state amount of gossip possessed by person n be the average of the gossip possessed by the two neighbors. This assumption embodies the smoothing effect of diffusion. Furthermore, we will assume that there are possible sources (or sinks) of gossip, whereby new information is introduced to the system or deliberately withheld. This input to the system will be denoted f_n. Combining these two effects, we see that in a steady state the gossip content for person n satisfies

$$u_n = \frac{u_{n-1} + u_{n+1}}{2} + f_n$$

for $n = 0 : N - 1$. The cyclic geometry of the configuration makes the periodic boundary conditions $u_0 = u_N$ and $u_{-1} = u_{N-1}$ appropriate. Rearranging the difference equation we see that

$$-\frac{1}{2}u_{n-1} + u_n - \frac{1}{2}u_{n+1} = f_n$$

for $n = 0 : N - 1$. To complete the statement of the problem we must specify an input vector to describe sources and sinks of gossip. Intuition alone suggests that since this is a closed system, a steady state cannot be reached unless the net flow of gossip in and out of the system is zero. In other words, $\sum_{n=0}^{N-1} f_n = 0$, which is equivalent to $F_0 = 0$. We will see shortly that this condition also appears naturally in the course of the solution. For this particular example we will assume that there is a single source of gossip located at $n = 0$ and a single sink of gossip located at $n = N/2$, both of unit strength. This means that

$$f_n = \delta(n) - \delta\left(n - \frac{N}{2}\right)$$

for $n = 0 : N - 1$, a vector whose DFT is given by

$$F_k = \frac{1}{N}(1 - \cos(\pi k))$$

for $k = 0 : N - 1$ (see *The Table of DFTs* in the Appendix). Notice that the constraint $\sum_{n=0}^{N-1} f_n = F_0 = 0$ is satisfied.

The foregoing method for periodic boundary conditions now applies directly with $a = -1/2$ and $b = 1$. The DFT coefficients of the solution are given by

$$U_k = \frac{F_k}{1 - \cos\left(\frac{2\pi k}{N}\right)} = \frac{(1 - \cos(\pi k))}{2N \sin^2\left(\frac{\pi k}{N}\right)}$$

for $k = 1 : N - 1$, with U_0 arbitrary since it satisfies the equation $0 \cdot U_0 = 0$. Taking the inverse DFT of this vector, we have the solution

$$u_n = \frac{1}{2N} \operatorname{Re}\left\{ \sum_{k=1}^{N-1} \frac{(1 - \cos(\pi k))}{\sin^2\left(\frac{\pi k}{N}\right)} \omega_N^{nk} \right\} + U_0$$

$$= \frac{1}{2N} \sum_{k=1}^{N-1} \frac{(1 - \cos(\pi k))}{\sin^2\left(\frac{\pi k}{N}\right)} \cos\left(\frac{2\pi nk}{N}\right) + U_0$$

for $n = 0 : N - 1$.

In the absence of a reference value for gossip content, we may take $U_0 = 0$. It doesn't appear possible to simplify this expression for u_n *directly* (but see problem

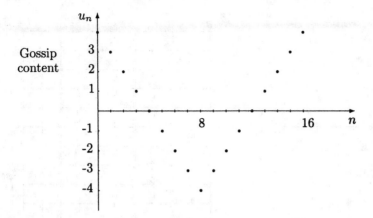

FIG. 7.4. *A diffusion model may be used to describe the steady state distribution of gossip in a cyclic configuration. The model takes the form of a difference equation with periodic boundary conditions that can be solved using the DFT. With one source and one sink of gossip, the solution with $N = 16$ participants (gossipers) consists of two linear segments as shown in the figure.*

153). However, it is easily evaluated numerically, and the results for $N = 16$ are shown in Figure 7.4. Perhaps the solution could have been anticipated from physical intuition, but certainly not from the analytical solution. We see that the solution (for any N) consists of two linear segments with a maximum value located at the source of gossip ($n = 0$) and a minimum value located at the sink ($n = N/2$). Notice that the average value of the solution is zero. The smoothing effect of the difference equation creates the linear profiles between the maximum and minimum values, which is a common feature in many steady state diffusion problems.

Fast Poisson Solvers

In this section we consider arguably the most prevalent use of DFTs and difference equations, namely the solution of ordinary and partial differential equations that take the form of boundary value problems. This is a central problem of computational mathematics and has been the scene of vigorous activity for many years [79], [133], [134], [137]. We will content ourselves with exploring a few representative problems and demonstrating the vital part that the DFT plays in the solution process. It is best to begin with a one-dimensional problem to fix ideas, then turn to higher-dimensional problems in which the DFT becomes indispensable.

Consider the ordinary differential equation (ODE)

$$- \phi''(x) + \sigma^2 \phi(x) = g(x) \quad \text{for} \quad 0 < x < A \tag{7.11}$$

subject to the Dirichlet boundary conditions $\phi(0) = \phi(A) = 0$. Notice that the additional conditions on the solution are given at the endpoints or boundaries of the domain, so this is called an ODE boundary value problem. The nonnegative coefficient σ^2 and the source function g are given. With $\sigma^2 = 0$ this equation describes steady state diffusion. (In fact, this *continuous* equation can be derived from the difference equation of the previous section by letting the spatial distance between the individual components decrease to zero, while letting the number of components N increase to infinity.)

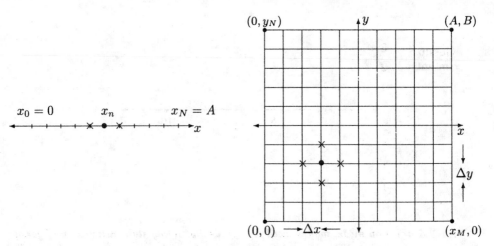

FIG. 7.5. *Numerical solutions of differential equations often take place on a computational domain or grid consisting of a finite number of points. The left figure shows the grid for the solution of a one-dimensional BVP, while the right figure shows a two-dimensional grid on a rectangular region. Second derivatives at a point (•) are approximated using function values at that point and at nearest neighbors (×).*

Here is the kind of thinking that might precede solving this BVP. One can use a Fourier series method to arrive at an analytical solution to this problem. In all likelihood, that solution will be expressed in terms of an (infinite) Fourier series that will need to be truncated and evaluated numerically. Since numerical approximation will be inevitable even with an analytical solution, one might choose to use a numerical method from the start. As we will see shortly, one numerical method is based on the DFT and mimics the analytical methods very closely. With this rationale for using a numerical approximation, let's see how it works.

The exact solution of the continuous BVP (7.11) gives the value of the solution function ϕ at every point of the interval $[0, A]$. Since we cannot compute, tabulate, or graph such an infinity of values, we must resign ourselves at the outset to dealing with a discrete problem in which the solution is approximated at a finite number of points in the domain. Toward this end, we establish a grid on the interval $[0, A]$ by defining the grid points $x_n = nA/N$ where $n = 0 : N$ (see Figure 7.5). Notice that the distance (or grid spacing) between these equally spaced points is $\Delta x = A/N$. The goal will be to approximate the solution ϕ at the grid points.

Our first task is to show how the *continuous* (differential) equation can be approximated by a *discrete* (difference) equation of the form studied in the previous section. This process is known as **discretization**, and the easiest way to do it is to approximate the derivative ϕ'' in the differential equation by a **finite difference**. A short exercise in Taylor series (problem 148) suffices to establish the following fact: if the function ϕ is at least four times differentiable, then at any of the grid points x_n

$$- \phi''(x_n) = \underbrace{\frac{-\phi(x_{n-1}) + 2\phi(x_n) - \phi(x_{n+1})}{\Delta x^2}}_{\text{finite diff. approx. to } \phi''(x_n)} + \underbrace{c_n \Delta x^2}_{\text{error}} \qquad (7.12)$$

for $n = 1 : N - 1$, where c_n is a constant. We see that the second derivative can be approximated by a simple combination of the function value at x_n and its nearest neighbors $x_{n\pm1}$. The **truncation error** in this approximation, denoted $c_n\Delta x^2$, decreases as the square of the grid spacing Δx. Thus, as N increases and the grid spacing decreases, we expect the error in our approximation to decrease.

The whole strategy behind finite difference methods is to replace derivatives by finite differences and solve the resulting discrete equations for approximations to the exact solution. We do this by using the approximation (7.12) in the original ODE (7.11), dropping the truncation error $c_n\Delta x^2$, and letting u_n denote the resulting approximation to $\phi(x_n)$. This means that the components u_n satisfy

$$\frac{-u_{n-1} + 2u_n - u_{n+1}}{\Delta x^2} + \sigma^2 u_n = g_n$$

for $n = 1 : N - 1$, where we have let $g_n = g(x_n)$. Furthermore, since u_0 and u_N are implicated in this equation, we use the boundary conditions $u_0 = u_N = 0$. A bit of rearranging will cast this equation in a familar form. Multiplying through by Δx^2 and collecting terms gives us the difference equation

$$- u_{n-1} + (2 + \sigma^2\Delta x^2)u_n - u_{n+1} = g_n\Delta x^2 \qquad (7.13)$$

for $n = 1 : N - 1$, with $u_0 = u_N = 0$. This is just the second-order BVP problem with Dirichlet boundary conditions that was discussed in the previous section, with $a = -1$, $b = 2 + \sigma^2\Delta x^2$, and $f_n = g_n\Delta x^2$. The three-step DST method can be applied directly to this difference equation, now with an eye on the question of **convergence**. The ultimate goal is to approximate the solution of the original BVP. In practice, one would solve the difference equation (7.13) for several increasing values of N until successive approximations reach a desired accuracy.

Example: Toward diffusion. Let's look at numerical solutions of the BVP

$$-\phi''(x) + 16\phi(x) = 4x\sin 4x - \cos 4x \quad \text{for} \quad 0 < x < \pi$$

with the boundary conditions $\phi(0) = \phi(\pi) = 0$. Knowing the exact solution ($\phi(x) = \frac{1}{8}x\sin 4x$), it is possible to compute the error

$$E_N = \max_{1\le n\le N-1} |\phi(x_n) - u_n|$$

for various values of N. The evidence, shown in Table 7.1, demonstrates clearly that the errors not only decrease with increasing N, but decrease almost precisely as N^{-2} or Δx^2.

TABLE 7.1
Errors in N-point DST solutions to
$\phi'' + 16\phi = 4x\sin 4x - \cos 4x$, *with* $\phi(0) = \phi(\pi) = 0$.

N	8	16	32	64
E_N	3.4(-2)	8.3(-3)	2.2(-3)	5.4(-4)

The notation $a(-n)$ means $a \times 10^{-n}$.

Before moving on to two-dimensional problems, we will just mention that two other frequently arising BVPs can be handled using DFT-based methods. The same differential equation (7.11) accompanied by the Neumann conditions $\phi'(0) = \phi'(A) = 0$

can be discretized in the same way. The result is a difference equation of the form
(7.13) that can be solved using the DCT. And finally, the same differential equation
with the periodic conditions $\phi(0) = \phi(A), \phi'(0) = \phi'(A)$ can also be solved in its
discrete form using the full DFT. Hopefully these two assertions are plausible in light
of the previous discussion. The details will be elaborated in problem 149.

Having spent considerable time on one-dimensional BVPs and having recognized
that they can be solved more efficiently using a direct (Gaussian elimination) equation
solver, we now have the groundwork to consider two-dimensional BVPs in which DFT-
based methods really *do* have an advantage over direct equation solvers. The two-
dimensional analog of the continuous BVP (7.11) is most easily posed on a rectangular
domain

$$\Omega = \{(x,y) : 0 < x < A, \ 0 < y < B\}.$$

We will denote the boundary of this domain

$$\partial\Omega = \{(x,y) : x = 0 \text{ or } x = A \text{ or } y = 0 \text{ or } y = B\}.$$

The prototype BVP is one of the classical problems of applied mathematics. It is
given by the partial differential equation (PDE)

$$-\phi_{xx}(x,y) - \phi_{yy}(x,y) + a^2\phi(x,y) = g(x,y) \quad \text{on} \quad \Omega,$$

where the nonnegative constant a^2 and the input function g are given. The PDE will
be accompanied by the boundary condition $\phi = 0$ on $\partial\Omega$. With $a^2 \neq 0$ the PDE
is called the **Helmholtz**[3] **equation**; it arises in many wave propagation problems.
With $a^2 = 0$, the equation is called the **Poisson equation**, and it governs the steady
state in diffusion processes, electrostatics, and ideal fluid flow.

We will proceed, in analogy with the one-dimensional problem, converting the
continuous PDE into a partial difference equation. As before, the first step is to
establish a grid on the domain of the problem as shown in Figure 7.5. In general,
there could be different grid spacings in the two coordinate directions; therefore, we
will let $\Delta x = A/M$ and $\Delta y = B/N$ be the grid spacings in the x- and y-directions,
respectively. The grid consists of $(M - 1)(N - 1)$ interior points, where M and
N are any positive integers. We must now use finite difference approximations to
replace ϕ_{xx} and ϕ_{yy} in the PDE. Focusing on an interior point (x_m, y_n), the simplest
approximations to the second partial derivatives are

$$-\phi_{xx}(x_m, y_n) = \frac{-\phi(x_{m-1}, y_n) + 2\phi(x_m, y_n) - \phi(x_{m+1}, y_n)}{\Delta x^2} + c_{mn}\Delta x^2$$

and

$$-\phi_{yy}(x_m, y_n) = \frac{-\phi(x_m, y_{n-1}) + 2\phi(x_m, y_n) - \phi(x_m, y_{n+1})}{\Delta y^2} + d_{mn}\Delta y^2$$

for $m = 1 : M - 1, n = 1 : N - 1$. Notice that these difference approximations
are the analogs of the difference approximation to ϕ'' used earlier. We see that the
second partial derivatives at (x_m, y_n) can be approximated by a simple combination of

---

[3]HERMANN VON HELMHOLTZ (1821–1894) was an eclectic man who lived in Berlin, Königsberg,
Bonn, and Heidelberg during his lifetime. He was trained in medicine and served as a military
physician and as a professor of physiology. As a physics professor in Berlin he did fundamental work
in electrodynamics and hydrodynamics.

function values at that point and its nearest neighbors. The truncation error, involving the constants c_{mn} and d_{mn}, decreases as the square of the grid spacing.

We proceed just as before by defining u_{mn} as the approximation to $\phi(x_m, y_n)$. Partial derivatives in the PDE are replaced by finite differences, the truncation errors are neglected, and the approximate solution u_{mn} is found to satisfy the difference equation

$$\frac{-u_{m-1,n} + 2u_{mn} - u_{m+1,n}}{\Delta x^2} + \frac{-u_{m,n-1} + 2u_{mn} - u_{m,n+1}}{\Delta y^2} + a^2 u_{mn} = g_{mn} \quad (7.14)$$

for $m = 1 : M - 1, n = 1 : N - 1$. We have also used g_{mn} to denote $g(x_m, y_n)$. The boundary conditions for ϕ carry over to the discrete approximation in the form

$$u_{0n} = u_{Mn} = u_{m0} = u_{mN} = 0$$

for $m = 0 : M$ and $n = 0 : N$.

There are several ways to proceed from here. One could write out the individual equations of (7.14) in matrix form and discover that the difference equation corresponds to a square matrix with $(M - 1)(N - 1)$ rows and columns. The matrix consists mostly of zeros (called a **sparse matrix**) and has a block tridiagonal structure. The fact that the matrix is sparse makes a direct use of Gaussian elimination very impractical. Taking a lesson from the one-dimensional case, we will use a DFT-based method. The fact that the BVP has Dirichlet boundary conditions in both the x- and y-directions suggests that a discrete sine transform is appropriate. In fact, if we assume a solution of the form

$$u_{mn} = \sum_{j=1}^{M-1} \sum_{k=1}^{N-1} U_{jk} \sin\left(\frac{\pi m j}{M}\right) \sin\left(\frac{\pi n k}{N}\right) \quad (7.15)$$

for $m = 0 : M$ and $n = 0 : N$, it is easy to see that the boundary conditions are satisfied immediately. This representation is the two-dimensional inverse DST of the vector U_{jk} as discussed in Chapter 5.

It helps to rearrange equation (7.14) into the form

$$- u_{m-1,n} - u_{m+1,n} - \gamma^2(u_{m,n-1} + u_{m,n+1}) + (2 + 2\gamma^2 + a^2\gamma^2)u_{mn} = f_{mn} \quad (7.16)$$

before proceeding. We have used γ to denote the ratio $\Delta x / \Delta y$ and let $f_{mn} = \gamma^2 g_{mn}$. Taking the component approach of the previous section we assume a representation similar to (7.15) for the vector f_{mn} with coefficients F_{jk}, and then substitute into the difference equation (7.16). Working carefully with the multitude of indices, we have that

$$\sum_{j=1}^{M-1} \sum_{k=1}^{N-1} \left\{ U_{jk} \left[-\sin\left(\frac{\pi(m-1)j}{M}\right) - \sin\left(\frac{\pi(m+1)j}{M}\right) \right] \sin\left(\frac{\pi n k}{N}\right) \right.$$
$$- \gamma^2 \left[\sin\left(\frac{\pi(n-1)k}{N}\right) + \sin\left(\frac{\pi(n+1)k}{N}\right) \right] \sin\left(\frac{\pi m j}{M}\right)$$
$$\left. + (2 + 2\gamma^2 + \gamma^2 a^2)\sin\left(\frac{\pi m j}{M}\right)\sin\left(\frac{\pi n k}{N}\right) \right\}$$
$$= \sum_{j=1}^{M-1} \sum_{k=1}^{N-1} F_{jk} \sin\left(\frac{\pi m j}{M}\right) \sin\left(\frac{\pi n k}{N}\right),$$

which holds for $m = 1 : M - 1$ and $n = 1 : N - 1$. Perhaps it is a testimony to the special power of the DST that it can render this monstrous equation benign. The coefficients F_{jk} are known (or computable) as the two-dimensional DST

$$F_{jk} = -\frac{1}{MN} \sum_{m=1}^{M-1} \sum_{n=1}^{N-1} f_{mn} \sin\left(\frac{\pi m j}{M}\right) \sin\left(\frac{\pi n k}{N}\right)$$

for $j = 1 : M - 1$, $k = 1 : N - 1$, while the coefficients U_{jk} are the objects of the quest. (The minus sign appears in the definition of the two-dimensional DST given in Chapter 5.) Using the sine addition laws as before and collecting terms, we find that

$$\sum_{j=1}^{M-1} \sum_{k=1}^{N-1} \left[\left(-2\cos\left(\frac{\pi j}{M}\right) - 2\gamma^2 \cos\left(\frac{\pi k}{N}\right) + (2 + 2\gamma^2 + \gamma^2 a^2)\right) U_{jk} - F_{jk}\right]$$

$$\times \sin\left(\frac{\pi m j}{M}\right) \sin\left(\frac{\pi n k}{N}\right) = 0$$

for $m = 1 : M - 1$ and $n = 1 : N - 1$. We argue in a familiar fashion that this equation can hold in general for all relevant values of m and n only if the coefficient of each of the modes vanishes identically. This leads to the condition

$$U_{jk} = \frac{F_{jk}}{-2(\cos(\frac{\pi j}{M}) + \gamma^2 \cos(\frac{\pi k}{N})) + 2 + 2\gamma^2 + \gamma^2 a^2}$$

$$= \frac{F_{jk}}{4\sin^2(\frac{\pi j}{2M}) + 4\gamma^2 \sin^2(\frac{\pi k}{2N}) + \gamma^2 a^2},$$

for all of the points $m = 1 : M - 1, n = 1 : N - 1$. Having determined the coefficients U_{jk} of the solution, it is a straightforward task to perform the inverse DST according to the representation (7.15) to recover the solution u_{mn}. We see that in two dimensions the same basic three-step procedure applies.

1. The two-dimensional DST must be applied to the input vector f_{mn} to find the coefficients F_{jk}.

2. A simple algebraic step gives the coefficients U_{jk}.

3. The two-dimensional inverse DST of the vector U_{jk} must be done to produce the solution u_{mn}.

We hasten to point out that in practice all of the DSTs in this procedure are done using specialized versions of the FFT, tailored for both two dimensions and for the sine symmetries.

This approach generalizes in several directions. First, the same PDE with Neumann conditions ($\phi_n = 0$ on $\partial\Omega$, where ϕ_n is the derivative in the normal (orthogonal) direction to the boundary) can be discretized and solved using the two-dimensional DCT. Similarly, the same PDE with periodic boundary conditions in both directions is amenable to the full DFT. A little contemplation shows that the two coordinate directions are independent in this solution method. Therefore, with a different type of boundary condition on each pair of parallel sides of the domain, the relevant DFT can be applied in each coordinate direction. One step further, if different types of boundary conditions are specified on a single pair of parallel sides (for example, a Dirichlet condition for $x = 0$ and a Neumann condition for $x = A$), then there are additional **quarter-wave** DFTs to handle these cases. Finally, with just a bit more industry, the same ideas can be applied to the three-dimensional version of the the BVP posed on cubical or parallelepiped domains. In this case, the three-dimensional versions of the relevant DFTs are applied in the three coordinate directions. Perhaps most important of all, the same basic three-step method still applies in all of these extended cases. There are limitations also. DFT-based methods do not work directly for problems posed on irregular (nonrectangular) domains. The symmetry of the matrix associated with the discrete problem is also crucial for the success of these methods; therefore, they do not apply to variable coefficient problems. However, other boundary value problems may be amenable to solution using other discrete transforms (see Chapter 8).

In closing, we must make an attempt to assess the performance of these DFT-based methods. The discrete problem that arises from the BVP (7.11) has roughly MN unknowns (actually $(M-1)(N-1)$ unknowns for the pure Dirichlet problem, MN for the pure periodic problem, and $(M+1)(N+1)$ for the pure Neumann problem). This means that the matrix of coefficients associated with the problem has approximately MN rows and columns. If Gaussian elimination is used mindlessly to solve this system of equations (neglecting the fact that most of the matrix coefficients are zero), roughly $(MN)^3$ arithmetic operations are needed. If the regular zero-nonzero structure of the coefficient matrix *is* taken into account, there are methods that require roughly $(MN)^{3/2}$ arithmetic operations. What is the cost of the DFT-based methods? We need to anticipate the fact (confirmed in Chapter 10) that an N-point DFT (or DST or DCT) can be evaluated in roughly $N \log N$ arithmetic operations when implemented with the fast Fourier transform (FFT). With this in mind, the operation count for the three-step DFT method can be tallied as follows.

1. The forward DFT of f_{mn} consists of M DFTs of length N plus N DFTs of length M for a total cost of roughly $MN \log N + NM \log M$ arithmetic operations.

2. The computation of U_{jk} costs roughly MN operations.

3. The inverse DFT of U_{jk} consists of M DFTs of length N plus N DFTs of length M for a total cost of roughly $MN \log N + NM \log N$ arithmetic operations.

Adding these individual costs we see that the DFT method (using the FFT) requires on the order of

$$2(MN \log N + NM \log M) + MN \quad \text{arithmetic operations.}$$

The point is easiest to make if we consider a grid with $M = N$. Then the cost of finding the N^2 unknowns is on the order of $N^2 \log N$. An operation count of N^2 would be considered optimal since *some* work must be done to determine each of the N^2 unknowns. Therefore, we conclude that the DFT-based methods are very nearly optimal. For this reason these methods are often collectively called **fast Poisson solvers**. It is worth noting that an improvement can be made over the scheme just noted by combining DFTs and tridiagonal solvers. This hybrid method is generally used in software packages, and it is worth investigating, but that will have to be the subject of problem 155.

Notes

The preceding discussion of fast Poisson solvers is just a teaspoon from the sea of research that has been done on the subject. Those interested in the history and evolution of the field should begin with the landmark papers by Hockney [79] and Buzbee, Golub, and Nielson [29], then graduate to the papers by Swarztrauber [133], [134], [137]. Comprehensive software packages have been written for fast Poisson solvers; the most notable in supercomputer environments are *FISHPACK* [135] and *CRAY-FISHPACK* [143].

7.2. Digital Filtering of Signals

It is probably no exaggeration to say that digital signal processing occupies more computers, more of the time, than any other application. Furthermore, much of this computational effort is devoted to DFTs in the form of FFTs. We cannot hope to treat the DFT in signal processing in the few pages allotted here. Instead, we will be content to look at only one application, digital filtering, and even this is much too broad a topic to be treated fully in a short discussion. This introduction will give the reader a taste of how the DFT is used in signal processing, and those finding the topic interesting may be motivated to pursue other sources. Certainly, there is no shortage of literature on the topic; a visit to the signal processing section in any library will convince you of that.

We already encountered the basic concept behind digital filtering in the development of the circular convolution property of the DFT. It is worth the time to review that development here. For our purposes a **digital signal** is a sequence of numbers occurring at regular intervals, often obtained by recording some fluctuating voltage or current in an electronic device that measures some physical quantity. Such data generally represent values of a continuous function that is sampled regularly in time, although this need not be the case.

As a simple motivating example, consider the signal shown in the upper graph of Figure 7.6. The signal consists of $N = 24$ samples and is constructed from three oscillatory components with frequencies of one, five, and six periods on the sampling interval. In fact, the signal is given by

$$f_n = \cos(2\pi n\Delta x) + \frac{1}{2}\cos(10\pi n\Delta x) + \frac{1}{3}\cos(12\pi n\Delta x)$$

for $n = -11 : 12$, where $\Delta x = 1/24$. Suppose that a new sequence g_n is formed by computing a five-point weighted running average of f_n. Specifically, assume that f_n is periodic ($f_{n\pm24} = f_n$) and let

$$g_n = \frac{1}{8}f_{n-2} + \frac{1}{4}f_{n-1} + \frac{1}{4}f_n + \frac{1}{4}f_{n+1} + \frac{1}{8}f_{n+2}$$

for $n = -11 : 12$. The sequence g_n is plotted in the lower graph of Figure 7.6. It is much smoother than the original, and the running average appears to have "killed" most of the high frequency components, leaving a signal dominated by the low frequency component. We have, in fact, filtered the signal f_n by a **low-pass filter**, an operator that eliminates high frequency components of a signal while allowing low frequency components to survive (pass) essentially intact.

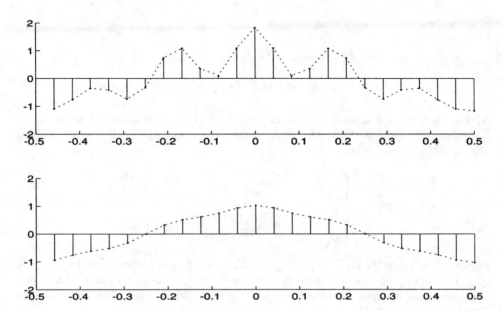

FIG. 7.6.   *A simple example of filtering is given, in which an input signal (top) consisting of three frequency components is filtered (bottom) by constructing a running weighted average of every five consecutive values. Observe that this has the effect of eliminating the higher frequency components of the signal while leaving the low frequency essentially unaltered.*

Although this is a simple example, it is instructive to examine it more closely, in order to see precisely how the filtering was accomplished. Note first that the process of forming the five-point weighted running average may be expressed as a cyclic convolution

$$g_n = f_n * h_n \tag{7.17}$$

$$= \sum_{j=-11}^{12} f_j h_{n-j}, \tag{7.18}$$

where h_n is the sequence

$$\{h_{-11}, h_{-10}, \ldots, h_{-1}, h_0, h_1, \ldots, h_{10}, h_{11}, h_{12}\}$$

whose entries are

$$\left\{0, 0, 0, 0, 0, 0, 0, 0, 0, \frac{1}{8}, \frac{1}{4}, \frac{1}{4}, \frac{1}{4}, \frac{1}{8}, 0, 0, 0, 0, 0, 0, 0, 0, 0, 0\right\}.$$

Recall the cyclic convolution property of the DFT, which holds that the N-point DFT of the cyclic convolution of two sequences equals N times the pointwise product of the DFTs of the two sequences:

$$\mathcal{D}\{f_n * h_n\}_k = NF_k H_k.$$

In this example both f_n and h_n are real even sequences, which implies that both F_k and H_k are also real even sequences. From *The Table of DFTs* of the Appendix, it may be seen that

$$F_k = \frac{1}{6}\hat{\delta}_N(k-6) + \frac{1}{4}\hat{\delta}_N(k-5) + \frac{1}{2}\hat{\delta}_N(k-1)$$
$$+ \frac{1}{2}\hat{\delta}_N(k+1) + \frac{1}{4}\hat{\delta}_N(k+5) + \frac{1}{6}\hat{\delta}_N(k+6).$$

Noting that h_n is a square pulse with amplitude 1/4 (with averaged values at its endpoints, satisfying AVED), *The Table of DFTs* may again be used (see also problem 158) to determine that

$$H_k = \begin{cases} \dfrac{1}{24} & \text{if } k = 0, \\[2ex] \left(\dfrac{1}{4}\right)\dfrac{\sin(k\pi/6)\sin(k\pi/12)}{48\sin^2(k\pi/24)} & \text{if } k \neq 0. \end{cases}$$

Figure 7.7 displays the DFT sequences F_k, H_k, and DFT of the convolution, NF_kH_k. It is now apparent how the filtering process works. The amplitudes of the high frequency components have been diminished greatly, while the amplitudes of the low frequency components have been left essentially unchanged.

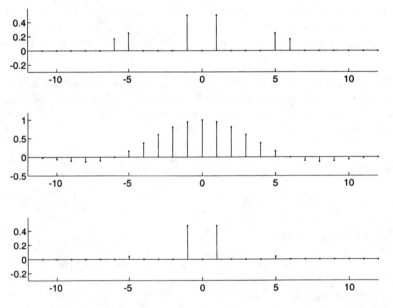

FIG. 7.7. *The frequency domain representation of a low-pass filter shows clearly how the filtering process works. The DFT F_k of the input signal (top) shows that the signal is composed of the ± 1, ± 5, and ± 6 modes. It may be observed that the DFT of the filter H_k (middle) consists of low frequency modes with high amplitude, and higher frequency modes with decreasing amplitude. The product NF_kH_k (bottom) gives the DFT of the filtered output. Comparison of the spectra of the input (top) and output (bottom) signals shows that the high frequency modes have been attenuated. Note that both input and output sequences are real and even, hence only the real parts of the DFTs are nonzero.*

Based on this example, it seems reasonable to define filtering as the process of forming the convolution of a signal with a second sequence; the second sequence is called a **filter**. To filter a digital signal we can either perform the cyclic convolution of the signal with the filter in the time or spatial domain or compute the DFTs of both the filter and the signal, multiply them together pointwise, and compute the IDFT of the result. A simple observation leads to an understanding of the popularity of the latter approach. Computing the filtered output by convolution methods is an $O(N^2)$ operation, since two sequences of length N are multiplied together pointwise for each element of the filtered output, of which there are N. Using FFTs, however, the two DFTs and the IDFT are computed in $O(N \log N)$ operations each, while the pointwise product of the DFTs F_k and H_k entails another N operations. Hence, the cost of the filtering operation using FFTs is proportional to $N \log N$ while the convolution method has a cost proportional to N^2. (To be fair, we point out that time domain filtering can be made competitive with frequency domain methods if filters are selected that are much shorter than the data stream. Details can be obtained in any signal processing text, e.g., [70] and [108].)

We will now examine several types of filters that are commonly used, and how they are designed. It is convenient to discuss filters in the setting of functions, rather than discrete sequences, so that the basic concepts may be developed without the necessity of considering sampling, aliasing, and leakage. We may proceed in this manner because of the convolution property for functions (see Chapter 3), namely that if

$$g(t) = \int_{-\infty}^{\infty} f(\tau)h(t-\tau)d\tau = f * h, \tag{7.19}$$

then

$$g(t) = \mathcal{F}^{-1}\left\{\hat{f}(\omega)\,\hat{h}(\omega)\right\}. \tag{7.20}$$

The function $h(t)$ is called the **filter**, while its transform $\hat{h}(\omega)$ is often called the **transfer function**. If the input function is an **impulse**, that is, if $f(t) = \delta(t)$, then the output $g(t) = h(t)$, by the properties of the delta function. For this reason, the filter $h(t)$ is also called the **impulse response** of the system.

It is useful to recall that since $\hat{f}(\omega)$, $\hat{h}(\omega)$, and $\hat{g}(\omega)$ are, in general, complex-valued functions, we may write them in the amplitude-phase form

$$\hat{f}(\omega) = |\hat{f}(\omega)|e^{i\phi_f(\omega)}, \cdot \quad \hat{h}(\omega) = |\hat{h}(\omega)|e^{i\phi_h(\omega)}, \quad \text{and} \quad \hat{g}(\omega) = |\hat{g}(\omega)|e^{i\phi_g(\omega)},$$

where the amplitude of $\hat{f}$ (for example) is given by

$$|\hat{f}(\omega)| = \sqrt{\text{Re}\left\{\hat{f}(\omega)\right\}^2 + \text{Im}\left\{\hat{f}(\omega)\right\}^2}$$

and the phase of $\hat{f}$ is given by

$$\phi_f(\omega) = \tan^{-1}\left(\frac{\text{Im}\left\{\hat{f}(\omega)\right\}}{\text{Re}\left\{\hat{f}(\omega)\right\}}\right).$$

With this notation, we may observe that the frequency domain representation of the filtered output

$$\hat{g}(\omega) = \hat{f}(\omega)\hat{h}(\omega) \tag{7.21}$$

may be written in amplitude-phase form as

$$|\hat{g}(\omega)|\,e^{i\phi_g(\omega)} = |\hat{f}(\omega)|\,|\hat{h}(\omega)|\,e^{i[\phi_f(\omega)+\phi_h(\omega)]}.$$

The action of the filtering operation can be summed up by observing that the **amplitude spectrum** of the output is the product of the amplitudes of the input and the filter,

$$|\hat{g}(\omega)| = |\hat{f}(\omega)||\hat{h}(\omega)|, \tag{7.22}$$

while the **phase spectrum** of the output is the pointwise sum of the phases of the input and the filter,

$$\phi_{\hat{g}}(\omega) = \phi_{\hat{f}}(\omega) + \phi_{\hat{h}}(\omega). \tag{7.23}$$

Of course, these properties have discrete analogs. If f_n is an input signal whose DFT is F_k, then we can write the DFT in amplitude-phase form as

$$F_k = |F_k| e^{i\phi_k},$$

where

$$|F_k| = \sqrt{\mathrm{Re}\{F_k\}^2 + \mathrm{Im}\{F_k\}^2} \quad \text{and} \quad \phi_k = \tan^{-1}\left(\frac{\mathrm{Im}\{F_k\}}{\mathrm{Re}\{F_k\}}\right).$$

The same amplitude and phase properties of the output spectrum can be deduced (problem 159).

Filter Design

There are many different types of filters and we employ properties (7.22) and (7.23) to develop examples of some of the most common filter types and to give insight into the process of filter design. A few words regarding strategy of filter design are in order. As we have seen repeatedly, much of the power and beauty of Fourier transforms and DFTs arise because they allow us to work in either the time domain *or* the frequency domain. This means that we may consider either the action of the filter on the signal directly (time domain design) or the action of the filter on the spectrum of the signal (frequency domain design). An example of the former approach was used to open this section, where the filter operator was viewed as a moving average acting on the input sequence.

Another example of time domain design is that of **distortionless filters**, filters that do not distort the shape of the input, but are purely **time-shifting filters**. For an input signal $f(t)$, let's assume that the desired output is $f(t - t_0)$. Thus,

$$g(t) = f(t - t_0),$$

from which the time-shifting property of the Fourier transform gives

$$\hat{g}(\omega) = \hat{f}(\omega)e^{-i2\pi\omega t_0}.$$

Comparison with (7.21) shows that the transfer function of the filter is

$$\hat{h}(\omega) = e^{-i2\pi\omega t_0},$$

which in turn implies that

$$|\hat{h}(\omega)| = 1 \quad \text{and} \quad \phi_h(\omega) = -2\pi\omega t_0.$$

The amplitude spectrum of the output is unchanged from that of the input, corresponding to $|\hat{h}(\omega)| = 1$, while the phase spectrum has undergone a linear phase shift by the amount $-2\pi\omega t_0$ (see problem 160). Recalling that $\mathcal{F}^{-1}\left\{e^{-i2\pi\omega t_0}\right\} = \delta(t - t_0)$, we see that convolving a function f with a shifted delta function yields

$$\int_{-\infty}^{\infty} f(\tau)\delta(t - t_0 - \tau)d\tau = f(t - t_0).$$

This confirms that the filter operator will indeed produce a shift of the input sequence.

The discrete form of this filter may be developed in an analogous fashion. Suppose the desired output is $g_n = f_{n-n_0}$, where the input is f_n and n_0 is an integer. Again, applying the shift theorem, we find that

$$G_k = F_k e^{-i2\pi k n_0/N},$$

so that the DFT of the desired filter (the transfer function) is

$$H_k = e^{-i2\pi k n_0/N}.$$

The amplitude and phase spectra of this filter are given by

$$|H_k| = |e^{-i2\pi k n_0/N}| = 1$$

and

$$\phi_k = \tan^{-1}\left(\frac{-\sin(2\pi k n_0/N)}{\cos(2\pi k n_0/N)}\right) = -\frac{2\pi k n_0}{N}.$$

An example of this filter is shown in Figure 7.8. The figure shows the amplitude and phase spectra for the filter, as well as the input and output signals. One feature should be noticed, as it can have disturbing consequences if no provision is made for it. This is the **wrap around** effect caused by the shift operator acting on signals that are assumed to be N-periodic. Portions of the signal originally occurring near the end of the finite-duration signal are shifted beyond the end of the signal, but the implicit periodicity of the DFT causes these portions to appear at the beginning of the signal. In the next section, we will see a problem in seismic migration in which the wrap around effect is a commonly encountered difficulty.

Perhaps the most common type of filter is the **amplitude distortion** filter, which may be characterized by the fact that it affects only the amplitude spectrum of the signal, and does not alter the phase spectrum. Since the effect of convolution is to add the phase spectra of the signal and the filter, this implies that amplitude distortion filters must have a phase spectrum that vanishes identically. Such filters are called **zero phase** filters. Since the phase spectrum of the filter h_n is

$$\phi_k = \frac{\text{Im}\{H_k\}}{\text{Re}\{H_k\}},$$

it follows that $\text{Im}\{H_k\} = 0$, and we conclude that amplitude distortion filters (zero phase filters) are real-valued sequences.

The most common of these filters are **band-pass** filters. As the name implies, a band-pass filter passes frequencies within a specified band, while distorting or eliminating all other frequencies of the signal. From this description it seems natural to perform the filter design in the frequency domain, and indeed, this is the approach we shall take.

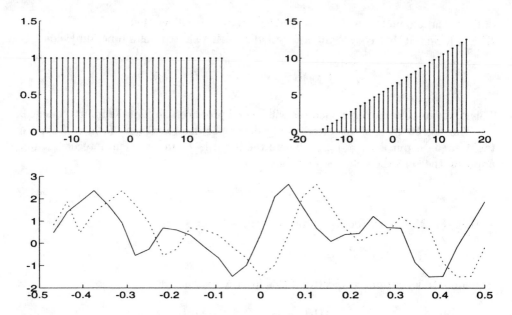

FIG. 7.8. *The amplitude spectrum of the discrete time-shifting filter is shown in the upper left graph, and the phase spectrum is shown in the upper right graph. The unfiltered input f_n is shown by the solid curve in the graph at the bottom of the figure, while the filtered (time-shifted) output g_n is shown by the dotted curve in the lower graph.*

Low-Pass Filter

Let us begin by considering an idealized **low-pass** filter, that is, a filter which passes low frequency components of the signal while eliminating the high frequency components. Such a filter has a frequency domain representation given by

$$\hat{h}(\omega) = \begin{cases} 1 & \text{if} & |\omega| \leq \omega_c, \\ 0 & \text{if} & |\omega| > \omega_c, \end{cases} \tag{7.24}$$

where ω_c is some specified *cut-off frequency*. The filter has an amplitude spectrum of unity for $|\omega| < \omega_c$ and an amplitude spectrum of zero for higher frequencies.

The time domain representation of this filter is

$$h(t) = \mathcal{F}^{-1}\left\{\hat{h}(\omega)\right\} = \int_{-\omega_c}^{\omega_c} e^{i2\pi\omega t} d\omega \tag{7.25}$$

$$= \frac{\sin(2\pi\omega_c t)}{\pi t} = 2\omega_c \ \text{sinc}(2\pi\omega_c t). \tag{7.26}$$

To apply the filter to an input signal, we have a choice of two methods: either the input signal may be convolved directly with $2\omega_c\text{sinc}(2\pi\omega_c t)$ or (using the convolution theorem) the Fourier transform of the input signal may be multiplied by $\hat{h}(\omega)$ followed by the inverse Fourier transform of that product.

As with the shift filter, the discrete form of the low-pass filter may be obtained in

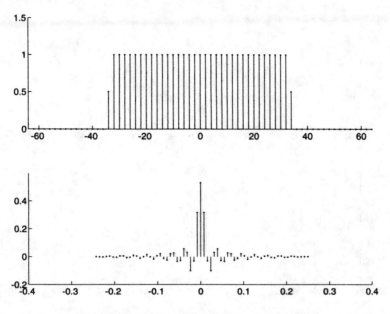

FIG. 7.9. *The amplitude of the idealized low-pass filter in the frequency domain (top) and the time domain (bottom) are shown in this figure. Observe that in the frequency domain the filter values at the cut-off frequencies are 1/2, satisfying the AVED condition.*

a straightforward fashion. It is given by

$$H_k = \begin{cases} 1 & \text{if} & |k| < k_c, \\ \frac{1}{2} & \text{if} & |k| = k_c, \\ 0 & \text{if} & k_c < |k| \leq \frac{N}{2}, \end{cases}$$

where k_c is the index associated with the desired cut-off frequency. Note that the values of $H_{\pm k_c} = \frac{1}{2}$ must be used since they are the average values at the discontinuity (AVED).

It is a direct calculation (problem 156 or a fact from *The Table of DFTs*) to show that the time domain representation of the discrete filter is given by

$$h_n = \frac{\sin\left(\frac{\pi n k_c}{N}\right)\sin\left(\frac{2\pi n}{N}\right)}{2\sin^2\left(\frac{\pi n}{N}\right)}.$$

Figure 7.9 displays a low-pass filter in both the time and frequency domains. The filter is generated using $N = 64$ and a time sample rate $\Delta t = 1/128$, so that the total length of the filter is $T = N\Delta t = 0.5$ seconds and the frequency sample rate (by the reciprocity relations) is $\Delta\omega = 2$ hertz (cycles per second). The low-pass filter passes all frequencies $|\omega| \leq \omega_c = 34$ hertz. A perfect all-pass filter has a frequency response of unity for all frequencies, and thus has a time domain representation consisting of a single spike at $t = 0$ with amplitude N. Hence, in this and the following figures, we divide all time domain representations by N, so that the amplitudes of the filters may be compared.

Examination of Figure 7.9 reveals that in the time domain the idealized low-pass filter is characterized by many small amplitude oscillations near the central lobe. These oscillations, known as **sidelobes**, are caused by the sinusoidal nature of the

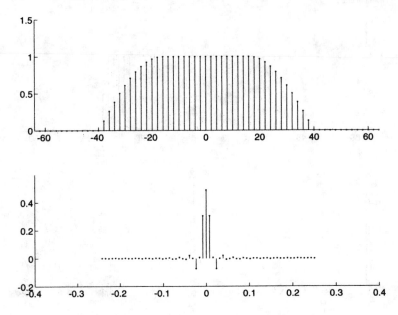

Fig. 7.10. *The amplitude of the modified low-pass filter is displayed in the frequency domain (top) and the time domain (bottom).*

numerator of a sinc function. When applied to a signal, such filters often produce many spurious oscillations near large "impulses" in the output, a phenomenon known as **ringing**. This is caused largely by the abrupt cut-off of the filter in the frequency domain. One way to counteract ringing is to smooth the sharp cut-off of the low-pass filter edge in the frequency domain. There are many ways to accomplish this; for example, a modified filter of the form

$$H_k = \begin{cases} 1 & \text{for} \quad |k| \le k_c - m, \\ \sin\left(\dfrac{\pi(k_c - k)}{2m}\right) & \text{for} \quad k_c - m < |k| \le k_c, \\ 0 & \text{for} \quad k_c < |k| \end{cases}$$

can be used, where k_c and m are positive integers. The frequency and time domain representations of this filter are shown in Figure 7.10. The parameters are $N = 64$, $\Delta t = 1/128$, $T = 0.5$, and $\Delta\omega = 2$, as in Figure 7.9. The parameters for the frequency attenuation are $k_c = 20$ and $m = 12$, so that the frequencies are attenuated for $16 \le |\omega| \le 40$ hertz. While the ranges of frequencies passed by the filters in Figures 7.9 and 7.10 are substantially the same, it may be seen that the sidelobes of the latter filter are significantly diminished.

Band-Pass Filters

By judicious application of the linearity and shifting of the continuous and discrete Fourier transforms many other filters can be created. Among the more common filters is the **band-pass** filter, which, as its name implies, attenuates all frequencies except those within a prespecified band in the spectrum. Taking $\hat{h}(\omega)$ to be the idealized

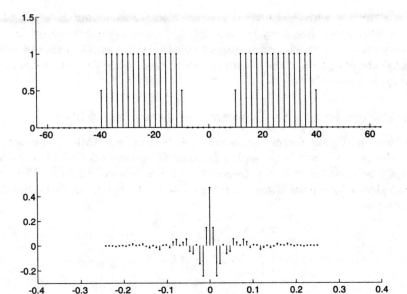

FIG. 7.11. *A typical idealized band-pass filter is shown in the frequency domain (top) and time domain (bottom). It is a symmetric filter centered at $\omega_0 = \pm 25$ hertz with half-bandwidth $\omega_c = 15$ hertz. Note that the AVED condition is applied at the edges of the band-pass and that the sharp cut-offs on both ends of the band results in a filter with very large sidelobes.*

low-pass filter (7.24), we may construct an idealized band-pass filter $\hat{b}(\omega)$ centered at $\omega = \pm \omega_0$ of width ω_c by

$$
\hat{b}(\omega) = \begin{cases} 1 & \text{for} \quad \omega_0 - \omega_c \le |\omega| \le \omega_0 + \omega_c, \\ 0 & \text{for} \quad |\omega| < \omega_0 - \omega_c \ \text{or} \ |\omega| > \omega_0 + \omega_c, \end{cases} \tag{7.27}
$$

$$
= \hat{h}(\omega - \omega_0) + \hat{h}(\omega + \omega_0).
$$

Figure 7.11 shows a typical idealized band-pass filter in the frequency and time domains. As in Figures 7.9 and 7.10, the sampling rates in time and frequency and the extent of the time domain are given by the parameters $N = 64$, $\Delta t = 1/128$, $\Delta \omega = 2$, and $T = 0.5$. The band of unattenuated frequencies is centered at $\omega = \pm 25$ hertz and the half-bandwidth is given by $\omega_c = 15$ hertz.

Letting $h(t)$ be the time domain representation of the idealized low-pass filter (7.26), the time domain representation of the band-pass filter can be obtained with the help of the frequency shift property, which implies that

$$
\mathcal{F}^{-1} \left\{ \hat{h}(\omega - \omega_0) \right\} = h(t) e^{i 2\pi \omega_0 t} \quad \text{and} \quad \mathcal{F}^{-1} \left\{ \hat{h}(\omega + \omega_0) \right\} = h(t) e^{-i 2\pi \omega_0 t}.
$$

By linearity we obtain the desired time domain representation of the idealized band-pass filter,

$$
\begin{aligned}
b(t) &= \mathcal{F}^{-1} \left\{ \hat{h}(\omega - \omega_0) \right\} + \mathcal{F}^{-1} \left\{ \hat{h}(\omega + \omega_0) \right\} \\
&= h(t) \left(e^{i 2\pi \omega_0 t} + e^{-i 2\pi \omega_0 t} \right) \\
&= 2 h(t) \cos(2 \pi \omega_0 t).
\end{aligned}
$$

As was the case with the idealized low-pass filter, constructing the idealized band-pass filter with sharp frequency cut-offs produces a filter with large sidelobes, and therefore leads to a filtered output characterized by ringing. The same technique of "softening" the sharpness of the frequency cut-off used to alleviate the problem before may be applied again here.

Filter Design from the Generalized Low-Pass Filter

Rather than developing filters from scratch, it is beneficial to have a general filter design technique. We will begin with an idealized low-pass filter which is "softened" by applying a linear taper in the frequency domain to avoid ringing. The result is the **generalized low-pass filter**, or **ramp filter**, which has the frequency domain representation

$$
\hat{h}_L(\omega; \omega_1, \omega_2) = \begin{cases} 1 & \text{for} \quad |\omega| \leq \omega_1, \\ \dfrac{\omega_2 - \omega}{\omega_2 - \omega_1} & \text{for} \quad \omega_1 \leq |\omega| \leq \omega_2, \\ 0 & \text{for} \quad \omega_2 \leq |\omega|. \end{cases}
$$

The notation $\hat{h}_L(\omega; \omega_1, \omega_2)$ is used to remind us that the filter, a function of ω, is dependent on the two parameters ω_1 and $\omega_2 > \omega_1$. It is not difficult (problem 157 is recommended) to find that the time domain representation of the filter is given by

$$
h_L(t; \omega_1, \omega_2) = \begin{cases} \omega_2 + \omega_1 & \text{for} \quad t = 0, \\ \dfrac{\cos(2\pi\omega_1 t) - \cos(2\pi\omega_2 t)}{2\pi^2 t^2 (\omega_2 - \omega_1)} & \text{for} \quad t \neq 0. \end{cases}
$$

The generalized low-pass filter is illustrated in Figure 7.12, where both frequency and time domain representations are displayed for two different choices of the frequency parameters ω_1 and ω_2. Observe that the longer taper produces a filter with a narrower main lobe of greater amplitude and (importantly) greatly reduced sidelobes.

Armed with the generalized low-pass filter, it is easy to produce several common types of filters, such as band-pass, high-pass, multiple-band, and notch (band-reject) filters. We will illustrate only three general filters, and those we design only as frequency domain representations in the continuous case. The process of converting the continuous frequency domain representation to a discrete filter should be familiar and will be exercised in problem 161.

1. To create a **generalized band-pass filter** tapering linearly from zero at $\pm\omega_1$ to unity at $\pm\omega_2$ and tapering linearly back from unity at $\pm\omega_3$ to zero at $\pm\omega_4$, we simply form the combination

$$
\hat{h}_B(\omega; \omega_1, \omega_2, \omega_3, \omega_4) = \hat{h}_L(\omega; \omega_3, \omega_4) - \hat{h}_L(\omega; \omega_1, \omega_2),
$$

   where $0 < \omega_1 < \omega_2 < \omega_3 < \omega_4$. This filter is illustrated in Figure 7.13.

2. To create a **high-pass** filter that rejects all frequencies below ω_1 and passes all frequencies above ω_2 with a linear taper between these frequencies, we subtract the appropriate band-pass filter from the **all-pass** filter $\hat{h}_A(\omega) = 1$:

$$
\hat{h}_H(\omega; \omega_1, \omega_2) = 1 - \hat{h}_L(\omega; \omega_1, \omega_2).
$$

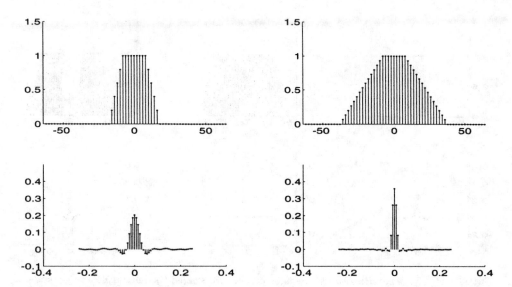

FIG. 7.12.   *The frequency domain response of the generalized low-pass filter is unity for* $|\omega| \leq \omega_1$ *and tapers linearly to zero at* $\omega = \pm\omega_2$. *The graph on the top left illustrates the choice of parameters* $\omega_1 = 8$ *hertz and* $\omega_2 = 20$ *hertz, while in the graph on the top right the choices are* $\omega_1 = 8$ *hertz and* $\omega_2 = 40$ *hertz. The corresponding time domain representations are shown in the two lower graphs.*

3. To create a **notch filter** or **band-reject filter** that passes all frequencies except those within a specified band (with tapers) we form

$$\hat{h}_R(\omega, \omega_1, \omega_2, \omega_3, \omega_4) = 1 - \hat{h}_B(\omega; \omega_1, \omega_2, \omega_3, \omega_4).$$

These last two filters (and their component filters) are illustrated in Figure 7.14.

Notes and References

We have illustrated only the simplest of filters, and have barely scratched the surface of this topic. Modifications (and improvements) to the filters presented here may be developed immediately, by replacing the linear tapers and their sharp corners with filters whose spectral cut-offs are "softer," and whose spectra are thus smoother.

   A closely related topic is the application of windows to the input signal. A signal $f(t)$ of finite duration may be viewed as the result of multiplying an infinitely long signal $g(t)$ by a function $w_c(t)$ that has the value of unity if $|t| \leq t_c$, and the value of zero otherwise. The function $w_c(t)$ is a simple example of a **window**. This idealized window has difficulties that can be alleviated by using windows with a smooth transition from unity (in the center) to zero at the edges. Such windows often bear the names of the individuals who developed them, and a list reads like a Who's Who of the pioneers of digital signal processing. Those interested in further exploration of digital filters should see problem 162 and references such as [21], [70], and [108].

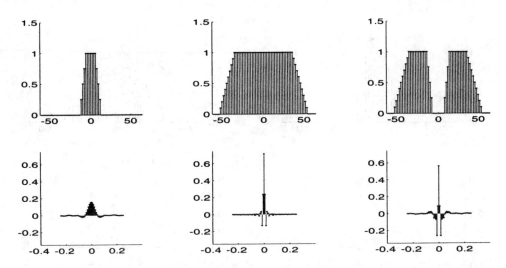

FIG. 7.13. *The generalized band-pass filter can be constructed by subtracting one generalized low-pass filter from another. The upper and lower graphs at left show the frequency domain and time domain representations of a generalized low-pass filter whose frequency response tapers from unity at $\omega_1 = 6$ hertz to zero at $\omega_2 = 16$ hertz. The upper and lower graphs in the center are for a low-pass filter tapering from unity at $\omega_3 = 36$ hertz to zero at $\omega_4 = 58$ hertz. The generalized band-pass filter results from the subtraction of the first filter from the second, and is displayed in the graphs at the right. The frequency domain response tapers linearly from 0 to unity between $\pm\omega_1 = 6$ and $\pm\omega_2 = 16$ hertz, is unity between $\pm\omega_2 = 16$ and $\pm\omega_3 = 36$ hertz, and tapers linearly to zero at $\pm\omega_4 = 58$ hertz.*

7.3. FK Migration of Seismic Data

A Crash Course in Seismic Exploration

The science of reflection seismology is used extensively for geological studies and in the exploration for oil and gas. The basic principle is quite simple. An incident sound wave is generated (by an explosion or by vibrations) at the surface of the earth, which propagates down through the rock layers. At the discontinuities between layers some fraction of the sound wave energy is reflected back toward the surface as an echo, while the remainder is transmitted further down through the rock layers as an incident wave. The echos are recorded by sensitive **geophones** as they arrive back at the surface and the resulting data are used to analyze both the depth and composition of the subsurface layers.

We know that the earth is generally formed of layers of rock piled one on top of another. This is evident in every road-cut along the highway. While we see these rock layers violently folded and broken in the mountains, throughout most of the world they have suffered only minor deformation. The individual layers of rock are relatively homogeneous, and their properties may be assumed to be reasonably consistent throughout a given layer. Hence the discontinuities between layers, such as the change from sandstone to limestone, or an abrupt change in density, should be mappable. Indeed, the reflection seismograph is the primary tool for mapping the nature of the subsurface layering.

There are two important principles that underlie the propagation of seismic waves.

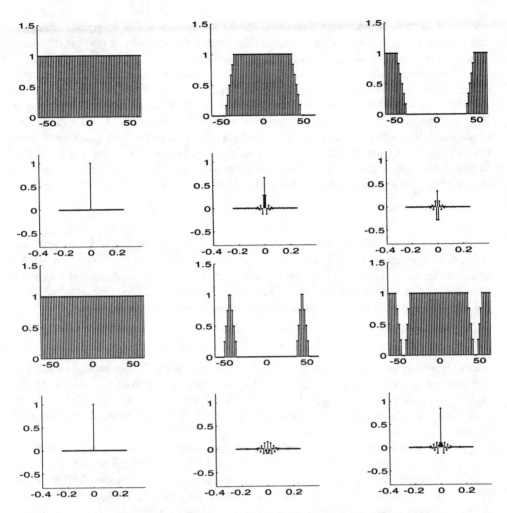

FIG. 7.14. *The figure shows the frequency domain construction (top row) and time domain representation (second row) of a high-pass filter, with zero frequency response below the frequency $\omega_1 = \pm 36$ hertz and a linear taper to unity at $\omega_2 = \pm 50$ hertz. The third and fourth rows show the frequency and time domain representations of a notch filter, with a taper to zero between $\omega_1 = \pm 36$ hertz and $\omega_2 = \pm 46$ hertz, and a taper back up to unity between $\omega_3 = \pm 46$ hertz and $\omega_4 = \pm 56$ hertz.*

The first, known as **Huygens'**[4] **principle**, states that every point on a wavefront can be regarded as a new source of waves. Given the location of the wavefront at a time t_0, the position of the wavefront at time $t_0 + \Delta t$ can be determined by drawing arcs of radius $c\Delta t$ from many points on the wavefront, where c is the **velocity** of the wavefront in the medium (in this case, the **seismic velocity**; that is, the speed of the pressure wave or sound, which may vary from point to point in the medium). Provided that sufficiently many arcs are drawn, the envelope of the arcs gives the position of the

---

[4]Born in the Hague, Holland in 1629, Christian Huygens is regarded as one of the greatest scientists of the seventeenth century. He was a respected friend of Sir Isaac Newton and a member of the French Academy of Sciences. Huygens proposed the wave theory of light, wrote the first treatise on probability, and made fundamental contributions to geometry. He died in Holland in 1695.

wavefront at time $t_0 + \Delta t$ (see Figure 7.15). As a consequence of Huygens' principle, the wavefront may be accurately described by the use of **raypaths**, lines that originate at the source and are always normal to the wavefronts.

The second fundamental principle is **Snell's**[5] **law**. When a wavefront impinges on an interface between layers, part of the energy is reflected, remaining in the same medium as the incoming wave, while the remainder is refracted into the neighboring layer with an abrupt change in direction of propagation. Snell's law dictates the angles of reflection and refraction, and states that the **angle of incidence** θ_1, measured from the normal to the interface, equals the **angle of reflection** θ_{rfl} (see Figure 7.15). Furthermore, the **angle of refraction** θ_2, measured from the normal on the opposite side of the interface, is related to the angle of incidence by

$$\frac{\sin \theta_1}{c_1} = \frac{\sin \theta_2}{c_2},$$

where c_1 and c_2 are the velocities of sound in the media through which incident and refracted waves travel (see Figure 7.15). The effect of the interface on the amplitudes of the reflected and refracted waves is complicated. In the case of normal incidence (when $\theta_1 = 0$ radians) the ratio of amplitudes of the reflected and incident waves is given by a quantity known as the **reflection coefficient**, which is

$$R = \frac{A_r}{A_i} = \frac{\rho_2 c_2 - \rho_1 c_1}{\rho_2 c_2 + \rho_1 c_1},$$

where ρ_1 and c_1 are the density and velocity of sound in the layer through which the incident wave travels, while ρ_2 and c_2 are the density and velocity of sound of the layer through which the refracted wave travels. Perhaps it is not startling to learn that these two fundamental laws are related in that Huygens' principle can be used to derive Snell's law (problem 163).

Ideally, the earth is composed of layers that are homogeneous and flat. If we place a source and a geophone at the same point on the surface, we would record reflections only from points directly below the source. This is because the angle of incidence is zero, so that the reflected wave returns to the surface along the same raypath taken by the incident wave, and the refracted wave continues downward without a change in direction. In this case the depth of the interface could be determined simply by multiplying one-half of the **travel time** (the elapsed time between activating the source and recording the echo) by the estimated velocity of sound in the subsurface layer. We do not generally acquire data in this fashion, however, for a number of reasons (not the least of which is the effect of exploding dynamite near sensitive recording equipment!). Instead, the data set is acquired using the **common midpoint** (**CMP**) method: the source and geophone are located equidistant from a common midpoint on the surface. In the ideal earth described above, the reflection received at the geophone originates from the same point on the subsurface reflector as though the source and geophone were located at the same point. Since the earth is not composed of flat, uniform layers this model is not exact, but it is sufficiently accurate for most situations (see Figure 7.16). For this reason, we may imagine that the arriving reflections were generated *at points on the subsurface interface where the raypath from the CMP is normal to the interface.* Such points are called the **normal incidence points (NIP)**.

---

[5] Born in 1591, WILLEBROD SNELL (actually Snel) was a Dutch astronomer, lawyer, and mathematician. He is best known for the triangulation method in geodesy and for Snell's Law of Optics.

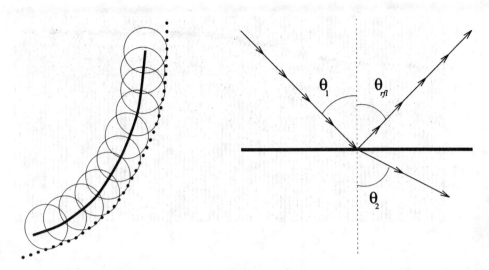

FIG. 7.15. *The left figure illustrates how Huygens' principle can be used to locate wavefronts. The wavefront at time $t + \Delta t$ may be found by drawing arcs of radius $c\Delta t$ around many points along the wavefront at time t. The envelope of all such arcs (dotted curve) gives the new position of wavefront. Snell's law (illustrated on the right) governs the reflection and refraction of waves in two different media. The angle of incidence θ_1 equals the angle of reflection θ_{rfl}, while the relationship between the angle of incidence and the angle of refraction θ_2 is $c_2 \sin \theta_1 = c_1 \sin \theta_2$, where c_1 and c_2 are the seismic velocities in the two media.*

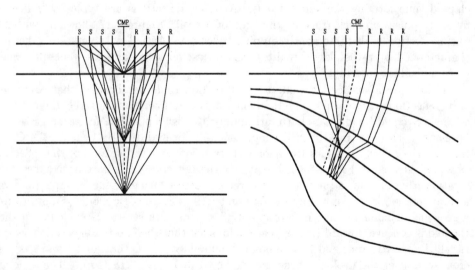

FIG. 7.16. *The common midpoint (CMP) method assumes that the source and receiver are located so that the reflection is generated at the normal incidence point (NIP), the point where a raypath from the common midpoint is normal to the reflector. For flat layers (left) this is a good model, while for rocks that have been deformed by geological forces (right) the model is less accurate.*

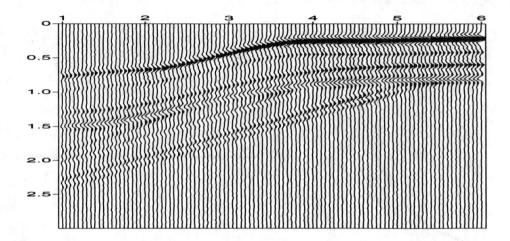

FIG. 7.17.   *A synthetic seismic section consists of many traces or curves representing*
movement of the geophones as a function of time. Time $t = 0$ corresponds to the initiation
of the seismic disturbance, and is at the top of the figure. Increasing time is downward.
Reflections are shown as large amplitude "wiggles" on the traces.

Typically, a seismic data set is presented in the form of a **seismic section** or
record section. Recordings, called **traces**, are made that correspond to many CMPs
that have been laid out along a line on the surface of the earth. Each trace, after
suitable data processing, is plotted as a curve below each CMP surface location, with
elapsed time forming the vertical axis. Arriving reflections are typically indicated
by short pulses on each trace at the time of arrival. Figure 7.17 shows a synthetic
seismic section with three gently dipping reflectors. Actual data sets are generally
characterized by rather low signal-to-noise ratios, however, and are generally much
more difficult to interpret.

If, at a given surface location, a record section *could* be obtained that contained
only reflections originating directly beneath the source, some of the difficulty of
seismology could be alleviated. Unfortunately, this isn't possible, for several reasons.
Foremost is the fact that the NIP is not generally located directly beneath the CMP.
Instead, the reflection comes from somewhere below and to one side of the reflector,
as was illustrated in Figure 7.16. Frequently the reflections originate from subsurface
locations that are out of the plane of the record section. An extreme example (and yet
a common one) is shown in Figures 7.18 and 7.19. The geologic model is a **syncline**,
a geologic phenomenon in which originally flat-lying rock strata are folded such that
the fold is concave upward (the opposite fold, such that the fold is concave downward,
is called an **anticline**, and is a common accumulation structure for oil and gas). A
cross-section of the geologic structure is shown in Figure 7.18 (top). The syncline
acts as a hemispherical "mirror," focusing the seismic raypaths on a focal point far
below the surface (Figure 7.18, bottom). As a result, the reflections at arriving at a
given surface location are *not* generated directly below the surface, but rather, from
points far to one side. Note that several CMP locations detect reflections generated
at points on both sides of the structure. The resulting record section is shown in
Figure 7.19. The focusing effect of the stream channel has the effect of inverting

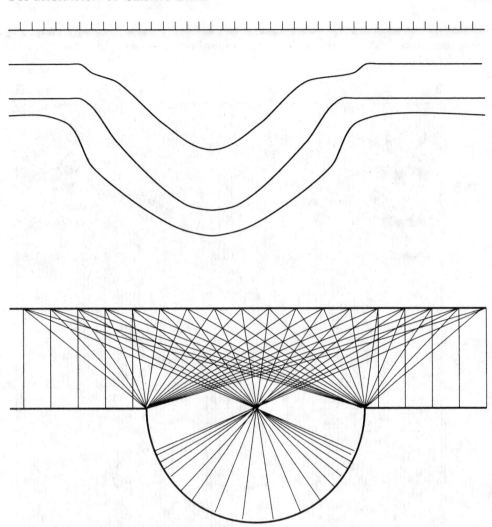

Fig. 7.18. *A simple geological model of a syncline (top) is used to illustrate the difficulty of interpreting seismic data. The syncline has the effect of focusing the raypaths like a mirror, as may be seen from the idealized sketch below, where one reflector and some of the reflection and diffraction raypaths are shown. The resulting record section is shown in Figure 7.19.*

the apparent structure, so that a geologic object which is concave-up appears on the record section as being concave-down, like a dome! This "criss-crossed" pattern of reflections is a characteristic seismic response for this geology, and is referred to as a **bowtie**. The record section is further complicated by the fact that the sharp corners of the stream channel give rise to **diffractions**, which appear on the record section as secondary, smaller "domes." The bowtie phenomenon is especially pernicious because underground domes are favorable locations for the accumulation of oil and gas. It is difficult to imagine the amount of money that has been spent drilling for oil in bowties!

We can now state the problem that **seismic migration** is designed to alleviate. The goal of seismic migration is to move the reflections on the seismic section so that they are located accurately in the subsurface. Properly corrected, a reflection will appear on the seismic trace corresponding to the CMP located vertically above the reflecting point, and at a travel time that would be correct were the travel path in fact vertical.

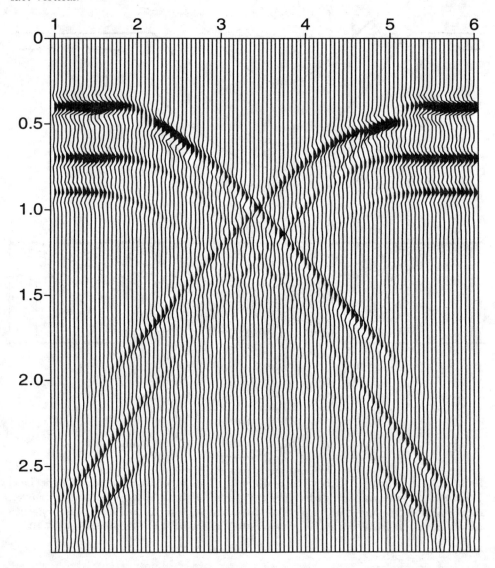

Fig. 7.19.   *The record section resulting from the geologic model in Figure 7.18 is shown. Because of the focusing effect and the diffractions, the record section is characterized by artificial ramp- and dome-like structures, which are highly misleading to the interpreter.*

Frequency-Wavenumber (FK) Migration

In actual practice, a record section is produced by setting off many sources, in many locations, at many different times, and recording the results on many separate receivers, each at a different surface location. To form the record section, the data are subjected to a sequence of data processing steps entailing the application of filters, corrections, and adjustments. The details are too complicated (and numerous) to discuss here, but the net result is that after processing, the data set appears as though all sources were fired simultaneously while all receivers were active.

A mathematical model for seismic migration can be developed through the following thought experiment. Suppose that the receivers are located along the surface, but that the sources, instead of being at the surface, are located along each of the reflectors in the subsurface, with a charge proportional to the reflection coefficient of the interface. At time $t = 0$ all charges are fired simultaneously, and the resulting waves propagate toward the surface. Furthermore, suppose that in this model all the seismic velocities are precisely half of the actual seismic velocities. This is called the **exploding reflector model**, and the record section produced in this idealized way would be essentially the same as the section actually acquired in the field. Thus, we may use a mathematical model based on this thought experiment to design a method for migrating the reflections on the record section to their correct spatial locations.

Using this conceptual model, we will let $u(x, z, t)$ denote the pressure wavefield, where x is the horizontal spatial variable, z is the vertical spatial variable, and t is the time variable. The earth is assumed to be flat, with $z = 0$ at the surface and increasing in the downward direction. (A correction for surface topography is one of many adjustments made during data processing, so the flat earth assumption is valid here.) The time $t = 0$ represents the instant at which the "exploding reflectors" are fired. Note that there is no dependence on y, the third spatial dimension. For conventional seismic data the earth is assumed to vary only with x and z; that is, reflections are assumed to originate in the plane of the section. This assumption is fairly good where the geological deformation is mild. Treatment of fully three-dimensional seismology is a much more complicated problem and will not be discussed here.

Under the assumptions of the thought experiment, it is possible to derive a governing equation (excellent references are the books by Telford et al. [146] and Dobrin [48]) which is a partial differential equation called the **wave equation**; it has the form

$$\frac{\partial^2 u}{\partial t^2} = c^2 \left(\frac{\partial^2 u}{\partial x^2} + \frac{\partial^2 u}{\partial z^2} \right), \qquad \begin{array}{l} -\infty < x < \infty, \\ 0 < z < \infty, \\ 0 < t < \infty, \end{array} \qquad (7.28)$$

where c is the seismic velocity in the subsurface. In actuality, c varies throughout the subsurface, so that $c = c(x, z)$, but we will assume a constant velocity. Framed in this way the record section, consisting of measured data, is $u(x, 0, t)$; it may be viewed as a boundary condition. The desired output is the **reflector section** or **depth section**, the distribution of the reflectors in the subsurface, and is given by $u(x, z, 0)$.

We now proceed in a straightforward manner. Let $\widehat{U}(k_x, k_z, t)$ be the Fourier transform of $u(x, z, t)$ with respect to x and z, where the frequency variables, or **wavenumbers**, in the spatial directions are denoted by k_x and k_z. Thus,

$$\widehat{U}(k_x, k_z, t) = \int_{-\infty}^{\infty} \int_{-\infty}^{\infty} u(x, z, t) \, e^{-i2\pi(k_x x + k_z z)} dx \, dz. \qquad (7.29)$$

Upon taking the two-dimensional Fourier transform with respect to the spatial variables of both sides of the wave equation (7.28), applying the derivative property (see problem 165), and interchanging the order of the integration and the differentiation with respect to t, we find that (7.28) becomes the ordinary differential equation

$$\frac{d^2 \widehat{U}}{dt^2} + 4\pi^2 c^2 (k_x^2 + k_z^2) \, \widehat{U}(k_x, k_z, t) = 0. \qquad (7.30)$$

At this point we could solve this ordinary differential equation directly. However, it proves to be advantageous to apply yet another Fourier transform, with respect to the variable t. Denoting the frequency variable associated with t by ω, we let $\hat{u}(k_x, k_z, \omega)$ be the Fourier transform of the function $\widehat{U}(k_x, k_z, t)$:

$$\hat{u}(k_x, k_z, \omega) = \int_{-\infty}^{\infty} \widehat{U}(k_x, k_z, t) e^{-i2\pi\omega t} dt.$$

Notice that $\hat{u}(k_x, k_z, \omega)$ is the full *three*-dimensional Fourier transform of $u(x, z, t)$. We may now take the Fourier transform of the ordinary differential equation (7.30) with respect to t and use the derivative property once again to obtain

$$-4\pi^2\omega^2 \, \hat{u}(k_x, k_z, \omega) + 4\pi^2 c^2 (k_x^2 + k_z^2) \, \hat{u}(k_x, k_z, \omega) = 0.$$

Since we are not interested in a solution of the form $u(x, z, t) = 0$, the factor $\hat{u}(k_x, k_z, \omega)$ may be canceled from both sides to yield

$$c^2(k_x^2 + k_z^2) = \omega^2. \tag{7.31}$$

This important relationship between the variables in the transform domain (k_x, k_z, ω) is known as the **dispersion relation**. Using it in (7.30) allows us to simplify that equation and write

$$\frac{d^2\widehat{U}}{dt^2} + 4\pi^2\omega^2 \, \widehat{U} = 0. \tag{7.32}$$

This differential equation has two linearly independent solutions,

$$e^{i2\pi\omega t} \qquad \text{and} \qquad e^{-i2\pi\omega t}.$$

Thus, the general solution to (7.32) is given by

$$\widehat{U}(k_x, k_z, t) = P(k_x, k_z) \, e^{i2\pi\omega t} + Q(k_x, k_z) \, e^{-i2\pi\omega t},$$

where $P(k_x, k_z)$ and $Q(k_x, k_z)$ are arbitrary functions of the wavenumbers (independent of t) that must be chosen to satisfy initial or boundary conditions.

If we could find $P(k_x, k_z)$ and $Q(k_x, k_z)$, we could also find $\widehat{U}(k_x, k_z, t)$, and the solution to the wave equation could be constructed by taking the two-dimensional inverse Fourier transform of $\widehat{U}(k_x, k_z, t)$. Formally, the general solution to the wave equation would look like

$$u(x, z, t) = \int_{-\infty}^{\infty} \int_{-\infty}^{\infty} P(k_x, k_z) \, e^{i2\pi\omega t} \, e^{i2\pi(k_x x + k_z z)} dk_x dk_z$$

$$+ \int_{-\infty}^{\infty} \int_{-\infty}^{\infty} Q(k_x, k_z) \, e^{-i2\pi\omega t} \, e^{i2\pi(k_x x + k_z z)} dk_x dk_z.$$

Without knowing the solution in detail, we can conclude from this general solution that it will consist of a linear combination of the functions

$$e^{i2\pi(k_x x + k_z z + \omega t)} \qquad \text{and} \qquad e^{-i2\pi(-k_x x - k_z z + \omega t)}.$$

The first of these functions is constant along the plane in (x, z, t) space given by

$$k_x x + k_z z + \omega t = \text{constant}.$$

Thus (with the help of problem 166 and the geometry of two-dimensional modes of Chapter 5), we see that $e^{i2\pi(k_x x + k_z z + \omega t)}$ represents a **plane wave** propagating in the xz-plane in the direction of the vector $(-k_x, -k_z)$. Using the dispersion relation, the plane wave travels in the direction of the *unit* vector

$$\left(-\frac{ck_x}{\omega}, -\frac{ck_z}{\omega} \right). \tag{7.33}$$

It is now critical to determine which way (up or down) this wave propagates. The dispersion relation (7.31) may be written as

$$\omega = \pm ck_z \sqrt{\frac{k_x^2}{k_z^2} + 1}. \tag{7.34}$$

By selecting the solution with the plus sign, we can ensure that ω has the same sign as k_z. In that case, the second component of the direction vector (7.33) is negative, and since we have defined z to be positive in the downward direction, this tells us that $e^{i2\pi(k_x x + k_z z + \omega t)}$ represents an *upgoing* plane wave. A similar argument reveals that $e^{-i2\pi(-k_x x - k_z z + \omega t)}$ is a downgoing plane wave. Since we are interested only in the upgoing (reflected) waves, we will choose $Q(k_x, k_z) = 0$, and seek a solution to (7.32) of the form

$$\widehat{U}(k_x, k_z, t) = P(k_x, k_z)\, e^{i2\pi\omega t}. \tag{7.35}$$

The initial condition for the ordinary differential equation (7.30) may now be used to advantage. By setting $t = 0$ in the previous line (7.35), it is clear that

$$\widehat{U}(k_x, k_z, 0) = P(k_x, k_z).$$

On the other hand, setting $t = 0$ in (7.29) tells us that

$$\widehat{U}(k_x, k_z, 0) = \int_{-\infty}^{\infty} \int_{-\infty}^{\infty} u(x, z, 0) e^{-i2\pi(k_x x + k_z z)}\, dx\, dy.$$

Here is the key to the whole calculation: we see that $\widehat{U}(k_x, k_z, 0) = P(k_x, k_z)$ is the Fourier transform of $u(x, z, 0)$, which is the reflector section that we seek. Therefore, the entire migration problem can be reduced to finding the function $P(k_x, k_z)$ and taking its two-dimensional inverse Fourier transform.

We have now worked the *unknown* reflector section into the discussion. We must now incorporate the *known* record section. Here is how it is done. Recall that the function $\widehat{U}(k_x, k_z, t) = P(k_x, k_z)e^{i2\pi\omega t}$ is the Fourier transform, with respect to the two spatial variables, of the wavefield $u(x, z, t)$. Therefore, its inverse Fourier transform is

$$
\begin{aligned}
u(x, z, t) &= \int_{-\infty}^{\infty} \int_{-\infty}^{\infty} \widehat{U}(k_x, k_z, t) e^{i2\pi(k_x x + k_z z)}\, dk_x\, dk_z \\
&= \int_{-\infty}^{\infty} \int_{-\infty}^{\infty} P(k_x, k_z)\, e^{i2\pi(k_x x + k_z z + \omega t)}\, dk_x\, dk_z.
\end{aligned}
\tag{7.36}
$$

Letting $z = 0$ in this expression (evaluating it at the earth's surface) yields a representation for the record section $u(x, 0, t)$. It is

$$u(x, 0, t) = \int_{-\infty}^{\infty} \int_{-\infty}^{\infty} P(k_x, k_z)\, e^{i2\pi(k_x x + \omega t)}\, dk_x\, dk_z. \tag{7.37}$$

Another representation for the record section $u(x, 0, t)$ can be obtained by writing it in terms of its Fourier transform with respect to x and t, which we shall denote

$$H(k_x, \omega) = \int_{-\infty}^{\infty} \int_{-\infty}^{\infty} u(x, 0, t) e^{-i2\pi(k_x x + \omega t)} dx dt.$$

Since the record section $u(x, 0, t)$ is known, $H(k_x, \omega)$ can be found by taking the Fourier transform of the record section with respect to x and t. We can also write $u(x, 0, t)$ as the inverse Fourier transform of $H(k_x, \omega)$, which looks like

$$u(x, 0, t) = \int_{-\infty}^{\infty} \int_{-\infty}^{\infty} H(k_x, \omega) e^{i2\pi(k_x x + \omega t)} dk_x d\omega. \tag{7.38}$$

Now we can make the crucial comparison. From expressions (7.37) and (7.38) we see that

$$P(k_x, k_z) dk_z = H(k_x, \omega) d\omega.$$

Therefore, $P(k_x, k_z)$ can be obtained from the Fourier transform (with respect to x and t) of the record section by

$$P(k_x, k_z) = H(k_x, \omega) \frac{d\omega}{dk_z}. \tag{7.39}$$

The dispersion relation (7.34) now enters in a fundamental way and produces the effect of migration. Differentiating the dispersion relation (problem 164) yields

$$\frac{d\omega}{dk_z} = \frac{c}{\sqrt{k_x^2/k_z^2 + 1}}.$$

Substituting for both ω and $d\omega/dk_z$ in the previous relation (7.39) we find

$$P(k_x, k_z) = H(k_x, \omega) \frac{d\omega}{dk_z} = \frac{c}{\sqrt{k_x^2/k_z^2 + 1}} H\left(k_x, \underbrace{ck_z \sqrt{\frac{k_x^2}{k_z^2} + 1}}_{\omega}\right). \tag{7.40}$$

With the function $P(k_x, k_z) = \widehat{U}(k_x, k_z, 0)$ in hand, the reflector section, $u(x, z, 0)$, may be recovered by taking one last inverse Fourier transform:

$$u(x, z, 0) = \int_{-\infty}^{\infty} \int_{-\infty}^{\infty} P(k_x, k_z) e^{i2\pi(k_x x + k_z z)} dk_x dk_z. \tag{7.41}$$

The path to this conclusion may have seemed somewhat convoluted. But it is now possible to stand back and summarize it rather succinctly. The process of converting the record section $u(x, 0, t)$ to the reflector section $u(x, z, 0)$ requires three basic steps.

1. Compute the function $H(k_x, \omega)$ by finding the two-dimensional Fourier transform, with respect to x and t, of the record section $u(x, 0, t)$.

2. Compute the function $P(k_x, k_z)$ from $H(k_x, \omega)$ by equation (7.40).

3. Compute the reflector section $u(x, z, 0)$ by finding the two-dimensional inverse Fourier transform, with respect to k_x and k_z, of $P(k_x, k_z)$.

Frequency-Wavenumber Migration with the DFT

How can we implement the method described above? Let us suppose that we have M source and receiver locations at the surface, laid out along a straight line at regular intervals, with a spacing of Δx. We are at liberty to place the origin anywhere along the line, and to stay with our symmetric interval convention we select our origin so that the record section is nonzero over the intervals $-A/2 \leq x \leq A/2$, with traces recorded at $j\Delta x$ for $j = -M/2 + 1 : M/2$. The total width of the interval in the x-direction is denoted A, so that $A = M\Delta x$. For each surface location we have a recording of the seismic trace over the interval $0 \leq t \leq T$ and there are N samples of the trace, with a sample interval of Δt, so that $T = N\Delta t$.

To perform the first step of the algorithm we must compute $H(k_x, \omega)$, which is given by

$$H(k_x, \omega) = \int_{-\frac{A}{2}}^{\frac{A}{2}} \int_0^T u(x, 0, t) e^{-i2\pi(k_x x + \omega t)} \, dt \, dx. \tag{7.42}$$

Both integrals may be approximated with a trapezoidal rule (with an appropriate adjustment to satisfy the AVED requirement). Therefore, we define a sequence $\{g_{jn}\}$ for $j = -M/2 + 1 : M/2$ and $n = 0 : N - 1$, made up of samples of u given by

$$g_{jn} = \begin{cases} u(j\Delta x, 0, n\Delta t) & \text{for } j = -\frac{M}{2} + 1 : \frac{M}{2} - 1, \ n = 1 : N - 1, \\[2mm] \frac{1}{2}\left(u(-\frac{A}{2}, 0, n\Delta t) + u(\frac{A}{2}, 0, n\Delta t)\right) & \text{for } j = \frac{M}{2}, \ n = 1 : N - 1, \\[2mm] \frac{1}{2}\left(u(j\Delta x, 0, 0) + u(j\Delta x, 0, T)\right) & \text{for } j = -\frac{M}{2} + 1 : \frac{M}{2} - 1, \ n = 0, \\[2mm] \frac{1}{4}\left(u(-\frac{A}{2}, 0, 0) + u(\frac{A}{2}, 0, 0) \right. \\ \left. \quad + u(-\frac{A}{2}, 0, T) + u(\frac{A}{2}, 0, T)\right) & \text{for } j = \frac{M}{2}, \ n = 0. \end{cases}$$

Using the sample intervals $\Delta x = A/M$ and $\Delta t = T/N$ we find by the reciprocity relations that we may take $\Delta k_x = 1/A$ and $\Delta \omega = 1/T$. Under these conditions, a trapezoidal rule approximation to the integral in (7.42) is given by

$$H(l\Delta k_x, m\Delta \omega) \approx \frac{AT}{MN} \sum_{j=-\frac{M}{2}+1}^{\frac{M}{2}} \sum_{n=0}^{N-1} g_{jn} e^{-i2\pi(l\Delta k_x)(j\Delta x)} e^{-i2\pi(m\Delta \omega)(n\Delta t)}$$

$$= \frac{AT}{MN} \sum_{j=-\frac{M}{2}+1}^{\frac{M}{2}} \sum_{n=0}^{N-1} g_{jn} \, \omega_M^{-jl} \omega_N^{-nm},$$

where $l = -M/2 + 1 : M/2$ and $m = 0 : N - 1$. We will denote the approximate values of $H(l\Delta k_x, m\Delta \omega)$ by H_{lm} and then make the enterprising observation that H_{lm} is just AT times the two-dimensional DFT of g_{jn}. It is easily written as

$$H_{lm} = AT \mathcal{D} \{g_{jn}\}_{lm}.$$

To perform the second step of the process, we must compute the function $P(k_x, k_z)$ from $H(k_x, \omega)$ by equation (7.40). Before doing this step it is necessary to determine the lengths of the domains and the sample rates for the depth variables z and the wavenumber k_z. We are at liberty to select any depth interval, since $u(x, z, t)$ clearly exists to great depth. Since our model assumes that c, the velocity of sound in the earth, is constant, it seems reasonable to choose the depth interval to be $[0, cT/2]$. The reflection from any reflector deeper than $D = cT/2$ will arrive at the surface after the recorders have been shut off.

While in theory we could change the sampling so that the number of samples used for k_z in $P(k_x, k_z)$ is *not* the same as the number used for ω in $H(k_x, \omega)$, it will certainly be simpler to choose the same number of samples, namely N. This in turn implies that $\Delta z = D/N$, from which the reciprocity relations give us $\Delta k_z = 1/D$. Now the migration step must be done in a discrete manner. Using the relationship (7.40),

$$P(k_x, k_z) = H(k_x, \omega)\frac{d\omega}{dk_z} = \frac{c}{\sqrt{k_x^2/k_z^2 + 1}}H(k_x, \omega),$$

we must form array $P_{lm} = P(l\Delta k_x, m\Delta k_z)$ from the array $H(l\Delta k_x, m\Delta\omega) = H_{lm}$ that was computed in the first step. The identification of these two arrays must be done for $l = -M/2 + 1 : M/2$ and $m = 0 : N - 1$, but there are some subtleties involved. Ordinarily some interpolation will be needed to obtain the samples $P(l\Delta k_x, m\Delta k_z)$ on a regular grid with respect to the variable k_z. Many interpolation schemes are available, but a choice must be made with care. One danger arises from the fact that the sample intervals Δx and Δz generally differ greatly. As a result, the phenomenon of **dip aliasing** may occur because the dip of the migrated reflections cannot be resolved using the horizontal sample interval Δx. The reflections will have an apparent dip that is less than the true dip. We will not digress into the arcana of dip aliasing.

To complete the migration we must perform the final step in the algorithm, the calculation of the reflector section from the function $P(k_x, k_y)$. It will be no surprise to learn that an IDFT is used to perform this step. In fact, it should be predictable that, letting u_{jn} represent the samples of $u(x, z, 0)$ and P_{lm} represent the samples of $P(k_x, \omega)$, the inverse transform is given by

$$u_{jn} = \sum_{l=-\frac{M}{2}+1}^{\frac{M}{2}} \sum_{m=0}^{N-1} P_{lm}\omega_M^{jl}\omega_N^{nm} = \mathcal{D}^{-1}\{P_{lm}\}_{jn}, \qquad (7.43)$$

where $j = -M/2 + 1 : M/2$ and $n = 0 : N - 1$. It should be verified (problem 167) that with the given domains and grid spacings, this IDFT does, in fact, provide approximations to the reflector section $u(x, z, 0)$ at the appropriate spatial grid points.

Example: FK migration. A simple example of the frequency-wavenumber migration process is shown in Figure 7.20. The synthetic seismic data from Figure 7.19 has been migrated using the algorithm described above. Recall that the geological model consisted of layers that had been folded into a syncline. The seismic section generated by this model is characterized by the bowtie pattern. Note that after migration, the data appears much more like the geological model, since the reflections shown on the original synthetic section have been moved so that they now appear on the trace overlying the reflecting points that generated them.

Of course, this example is highly idealistic. Real data, unlike the synthetic data used here, is generally characterized by a rather low signal-to-noise ratio, rendering the input data far more difficult to interpret. In particular, the record section is used to estimate the velocity to be used in the migration process, which was *known* in generating the example. Hence, reality generally falls short of the success portrayed here.

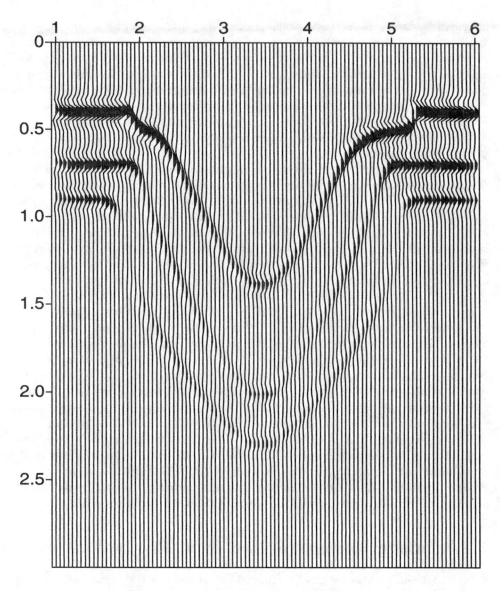

FIG. 7.20. *The application of frequency-wavenumber migration to the synthetic seismic data in Figure 7.19 yields dramatic improvement in the data, as may be seen here. The migration process has moved the reflections on the section so that they appear on the trace overlying the correct physical location of the reflectors.*

Notes and References

In reality a host of complications arise with the above procedure, and many highly sophisticated data processing routines have been developed to address them. We will mention two of these issues very briefly. The first, and most obvious drawback to the scheme concerns the use of a constant seismic velocity c. Obviously, the velocity distribution of the subsurface is unknown. Additional information often exists that can be used to make good estimates of the subsurface velocity. Even without such information, one could perform the migration several times, using several different values of c, and choose the best image of the subsurface. A more serious concern is that the velocity of the subsurface is not constant. There are many local variations and even the regional average velocity varies systematically, generally increasing with depth. Without the constant velocity assumption Fourier-based migration cannot be applied, since c cannot be moved outside the Fourier transform integrals. Approximate Fourier methods have been developed for variable velocity cases, but are much more difficult to apply. Current research addresses ways to implement migration schemes using variable velocity.

The second difficulty is the appearance of **ghost reflections** due to the use of the DFT. Ghost reflections are reflections that should occur at or near the bottom of the reflector section, but actually appear near the top after migration (and vice versa). Similarly, reflectors that should appear at the extreme right or left sides may show up at the opposite sides of the migrated section. The cause of this problem is simple, as is one method of curing it. Upon close examination of the effect of migration in the frequency domain (7.40), it will be observed that both the real and imaginary parts of $H(k_x, \omega)$ are moved the same amount for each ω. The *amount* that the data are moved differs for each ω, however. Hence the change is a phase shift, as well as an amplitude change. For this reason FK migration is often referred to as **phase shift migration**. We have seen in Section 7.2 that the effect of a linear phase shift in the frequency domain is a constant shift in the time domain, and that the periodicity of the DFT then produces the wrap around effect. While the phase shift given by (7.40) is certainly nonlinear, over the range of frequencies common to reflection seismology the effect is often a nearly linear phase shift, and the ghost reflections are indeed caused by the wrap around effect. A simple cure for the wrap around effect is to pad the data with zeros prior to the migration process. Padding is generally done at the bottom of the record section, and along one or both sides. A sufficient number of zeros must be used to insure that only zeros are "wrapped around" into the data. Of course, the introduction of these zeros affects the accuracy of the DFT as an approximation to the Fourier transform. The seismic data processor must select the migration parameters carefully, to balance the benefits of zero padding with the degradation in accuracy.

7.4.   Image Reconstruction from Projections

Suppose an x-ray is passed along a straight line through a homogeneous object. If the length of the travel path is x and the x-ray has an initial intensity I_0, then the intensity, I, of the x-ray that emerges from the object satisfies

$$I = I_0 e^{-ux},$$

where u, the **linear attenuation coefficient**, is dependent on the material making up the object. If the x-ray passes through one material along a path of length x_1, and then through another material along a path of length x_2, and so on through a number of layers, the emerging x-ray will be attenuated according to

$$I = I_0 e^{-\left(\sum_i u_i x_i\right)},$$

where u_i is the attenuation coefficient of the ith material. We see that passing an x-ray through a nonhomogeneous object may be modeled by letting the number of materials increase while the length of the travel path through each material decreases. Upon passage to the limit, the decay of the x-ray behaves according to

$$I = I_0 e^{-\left(\int_L u(x)dx\right)},$$

where $u(x)$ is the linear attenuation function, and L designates the line along which the x-ray passes through the object.

Consider passing many x-rays through an object along lines that lie in the same plane. Suppose that the linear attenuation function in the plane of the x-rays can be described by a function of two variables $u(x, y)$. The attenuation function of an object may depend on many factors, but in most cases a principal factor is the density of the material [78]. Therefore, we will refer to $u(x, y)$ as the **density** function, or simply the **image**. For each x-ray the travel path can be parameterized by the distance, s, traveled along the path, L. The attenuation in the x-ray intensity for the path L is

$$I = I_0 e^{-\left(\int_L u(x,y)ds\right)};$$

and taking logs of both sides we have

$$-\log\left(\frac{I}{I_0}\right) = \int_L u(x, y)ds.$$

The basic problem of Computer Aided Tomography (CAT) is to reconstruct the density function $u(x, y)$ from measurements of the x-ray attenuation $\log(I/I_0)$ along many paths through the object.

The Radon Transform and Its Properties

We will now let $u(x, y)$ be an arbitrary function of two spatial variables that is nonzero over some region D in the xy-plane. The xy-plane is often called the **image** or **object plane**. We will regard u as the density of a planar object that occupies the region D. The **Radon transform** of u is defined as the set of all line integrals of u,

$$\mathcal{R}\left\{u(x, y)\right\} = \int_L u(x, y)ds,$$

where L is any line passing through D. The transform is named for Johann Radon[6] [114], who studied the transform and discovered inversion formulae by which the function $u(x, y)$ may be determined from $\mathcal{R}u$.

The Radon transform $\mathcal{R}\{u\}$ is also a function of two variables that must be interpreted carefully. The two variables in the transform domain can be any two parameters that uniquely specify a line in the plane. For example, let ρ be a real number and ϕ be an angle measured from the positive x-axis. Then the condition

$$\rho = x \cos\phi + y \sin\phi \tag{7.44}$$

determines a line L in the xy-plane, normal to the unit vector $\vec{\xi} = (\cos\phi, \sin\phi)^T$, and a distance ρ from the origin, measured along $\vec{\xi}$ (see Figure 7.21 and problem 168).

---

[6]JOHANN RADON (1887–1956) was born in Bohemia (the former Czechoslovakia). He held professorships throughout Europe including the Universities of Hamburg and Vienna. He made lasting contributions to the calculus of variations, differential geometry, and integration theory.

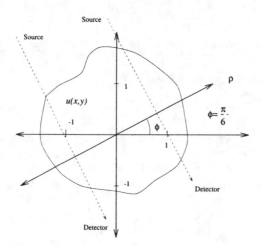

FIG. 7.21.   *The figure shows the geometry of the Radon transform $[\mathcal{R}u](\rho, \phi)$. For given values of ρ and ϕ, the transform gives the value of the line integral of u along the line orthogonal to the vector $(\cos\phi, \sin\phi)^T$ whose signed orthogonal distance from the origin is ρ. The figure shows the two lines corresponding to $\rho^2 = \cos^2(\pi/6) = 0.75$ and $\phi = \pi/6$, with equations $x\cos(\pi/6) + y\sin(\pi/6) = \pm\cos(\pi/6)$.*

Therefore, the coordinates (ρ, ϕ) determine a line uniquely and can be used as the variables of the Radon transform in the following way. Recall the sifting (or testing) property of the Dirac δ distribution,

$$\int_{-\infty}^{\infty} f(x)\delta(x)dx = f(0)$$

in one dimension, and

$$\int_{-\infty}^{\infty}\int_{-\infty}^{\infty} f(x,y)\delta(x,y)dxdy = f(0,0)$$

in two dimensions. Then the Radon transform can be specified by

$$[\mathcal{R}u](\rho, \phi) = \int_{-\infty}^{\infty}\int_{-\infty}^{\infty} u(x,y)\delta(\rho - x\cos\phi - y\sin\phi)dxdy \ , \qquad (7.45)$$

where $0 \leq \phi \leq \pi$ and $-\infty < \rho < \infty$.

Figure 7.21 shows the geometry of the Radon transform, in which ϕ gives the angle of the ray that is normal to the line L and is measured clockwise from the positive x-axis. Then ρ is the signed distance from the origin to the point where the line L meets the ray. A word of caution is needed: the variables ρ and ϕ should not be confused with polar coordinates. A function of polar coordinates must be single-valued at the origin, but there is no reason to suppose that the line integrals of u passing through the origin have the same value for all ϕ.

The Radon transform may be thought of as a projection operator. Specifically, if the transform $[\mathcal{R}u]$ is considered as a function of ρ with a parameter ϕ, then the profile given by the set of all line integrals, for fixed ϕ, as ρ varies, is a projection of $u(x,y)$ onto the one-dimensional subspace in which ρ varies between $-\infty$ and ∞. Henceforth we shall use the term **projection** to indicate the values of the Radon transform corresponding to a fixed value of ϕ. Taken over all angles, the transform is referred to as the **set of projections**. This terminology reflects medical tomography techniques in which many parallel x-rays are passed through an object and collected on the other side, all at a fixed angle ϕ. This single pass constitutes a projection. The entire apparatus is then rotated to a new angle ϕ and another projection is made.

Example: The characteristic function of a disk. Consider the density function consisting of the characteristic function of a disk of radius R, centered at the origin; that is,

$$u(x,y) = \begin{cases} 1 & \text{for } x^2 + y^2 \le R^2, \\ 0 & \text{otherwise.} \end{cases}$$

Since this object is symmetric with respect to the projection angle ϕ, its Radon transform is independent of ϕ and a single projection will suffice to determine the Radon transform. Consider the projection angle $\phi = 0$, so that the ρ-axis of the projection corresponds to the x-axis of the image space. The line integrals are then taken along lines parallel to the y-axis, at a distance ρ from the origin. For all values of $|\rho| > R$, the lines of integration do not intersect $u(x,y)$, and the projection has zero value. For $|\rho| \le R$ the transform is

$$[\mathcal{R}u](\rho, 0) = \int_{-\sqrt{R^2-\rho^2}}^{\sqrt{R^2-\rho^2}} dy \ = 2\sqrt{R^2 - \rho^2} \ .$$

Invoking the symmetry of $u(x,y)$, the Radon transform is

$$[\mathcal{R}u](\rho, \phi) = \begin{cases} 2\sqrt{R^2 - \rho^2} & \text{for } |\rho| \le R, \\ 0 & \text{otherwise.} \end{cases} \tag{7.46}$$

This example is shown graphically in Figure 7.22. Notice that both the function $u(x,y)$ and the transform $[\mathcal{R}u](\rho, \phi)$ are displayed on Cartesian grids. For the transform $[\mathcal{R}u](\rho, \phi)$ the grid uses ρ and ϕ axes, while values of the transform are displayed as height above the $\rho\phi$-plane. Another simple image whose Radon transform can be computed analytically is presented in problem 169.

The Radon transform has been well studied, largely because it has proven useful in many diverse areas. Rigorous treatments and exhaustive bibliographies can be found in numerous works [47], [62], [78], [104]. Certain properties of the Radon transform are essential in developing the DFT-based inversion methods of this study, so they are discussed briefly here. For a detailed treatment, see the books by Deans [47] or Natterer [104].

1. **Existence and smoothness.** The line specified by $\rho = x\cos\phi + y\sin\phi$ is, of course, of infinite extent. For the transform to exist at all, it is necessary that $u(x,y)$ be integrable along L. It is common to require that the function u belong to some reasonable space of functions. The following discussion is motivated by the medical imaging problem, so we are concerned primarily with functions possessing the following characteristics. First, we assume compact support, specified by $u(x,y) = 0$ for $|x| \ge A$, $|y| \ge A$, or in polar coordinates by $u(r,\theta) = 0$ for $r \ge A$. Second, we impose no smoothness requirements. We do not, for example, insist that the density function $u(x,y)$ be continuous. It is easy to imagine objects imbedded within objects (for example, bone in flesh) with sharp discontinuities. The density function may or may not be differentiable, although it often will be differentiable almost everywhere. We will assume that the density function is bounded, and that in turn the projections are bounded.

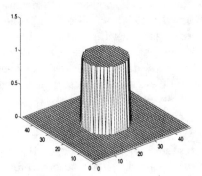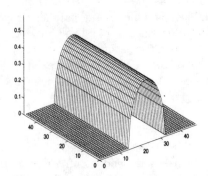

FIG. 7.22. *The characteristic function of a disk is shown on the left. The x-axis is parallel to the lower right edge, while the y-axis parallels the lower left edge. The Radon transform of the characteristic function of a disk is shown on the right. The ρ-axis is parallel to the lower right edge, while the ϕ-axis parallels the lower left edge.*

2. **Linearity.** The Radon transform of a linear combination of functions may be expressed as

$$\mathcal{R}\{\alpha u(x,y) + \beta v(x,y)\} = \int_{-\infty}^{\infty} \int_{-\infty}^{\infty} (\alpha u + \beta v)\delta(\rho - x\cos\phi - y\sin\phi)dxdy$$

$$= \alpha[\mathcal{R}u](\rho,\phi) + \beta[\mathcal{R}v](\rho,\phi).$$

Therefore, the Radon transform is a linear operator.

3. **Shift property.** Given a function $u(x,y)$, consider the effect of transforming $u(x-a, y-b)$. We see that

$$[\mathcal{R}u(x-a, y-b)](\rho,\phi) = \int_{-\infty}^{\infty} \int_{-\infty}^{\infty} u(x-a, y-b)\delta(\rho - x\cos\phi - y\sin\phi)dxdy .$$

Letting $v = x - a$ and $w = y - b$ yields

$$[\mathcal{R}u(x-a, y-b)](\rho,\phi)$$

$$= \int_{-\infty}^{\infty} \int_{-\infty}^{\infty} u(v,w)\delta(\rho - a\cos\phi - b\sin\phi - v\cos\phi - w\sin\phi)dvdw$$

$$= [\mathcal{R}u](\rho - a\cos\phi - b\sin\phi, \phi). \tag{7.47}$$

Thus, the effect of shifting the function $u(x,y)$ is to shift each projection a distance $a\cos\phi + b\sin\phi$ along the ρ axis.

4. **Evenness.** The Radon transform is an even function of (ρ,ϕ), in that

$$[\mathcal{R}u](-\rho, \phi+\pi) = \int_{-\infty}^{\infty} \int_{-\infty}^{\infty} u(x,y)\delta(-\rho - x(-\cos\phi) - y(-\sin\phi))dxdy$$

$$= \int_{-\infty}^{\infty} \int_{-\infty}^{\infty} u(x,y)\delta(x\cos\phi + y\sin\phi - \rho)dxdy$$

$$= [\mathcal{R}u](\rho,\phi).$$

There are several more properties of the Radon transform that are useful for numerous applications, but the foregoing serve as an adequate introduction for our purposes. The interested reader is referred to problem 170 for further properties.

The Central Slice Theorem

For our purposes, the most important property of the Radon transform is given by the Central Slice Theorem. This fundamental theorem relates the Fourier transform of a function to the Fourier transform of its Radon transform, and in so doing, provides the basis for several methods for inverting the Radon transform.

THEOREM 7.1. CENTRAL SLICE. *Let the image $u(x, y)$ have a two-dimensional Fourier transform, $\hat{u}(\omega_x, \omega_y)$, and a Radon transform, $[\mathcal{R}u](\rho, \phi)$. If $\widehat{\mathcal{R}u}(\omega, \phi)$ is the one-dimensional Fourier transform, with respect to ρ, of the projection $[\mathcal{R}u](\rho, \phi)$, then*

$$\hat{u}(\omega_x, \omega_y) = \widehat{\mathcal{R}u}(\omega, \phi),$$

where $\omega^2 = \omega_x^2 + \omega_y^2$ and $\phi = \tan^{-1}(\omega_y/\omega_x)$. That is, the Fourier transform of the projection of u perpendicular to the unit vector $(\cos \phi, \sin \phi)^T$ is exactly a slice through the two-dimensional Fourier transform of $u(x, y)$ in the direction of the unit vector.

Proof: Consider the Fourier transform of a projection. For fixed ϕ, the one-dimensional Fourier transform of $[\mathcal{R}u](\rho, \phi)$, with respect to ρ, is

$$
\begin{aligned}
\widehat{\mathcal{R}u}(\omega, \phi) &= \int_{-\infty}^{\infty} [\mathcal{R}u](\rho, \phi) e^{-i2\pi\omega\rho} d\rho \\
&= \int_{-\infty}^{\infty} \left(\int_{-\infty}^{\infty} \int_{-\infty}^{\infty} u(x, y)\delta(\rho - x\cos\phi - y\sin\phi)dxdy \right) e^{-i2\pi\omega\rho} d\rho \\
&= \int_{-\infty}^{\infty} \int_{-\infty}^{\infty} u(x, y) \left(\int_{-\infty}^{\infty} \delta(\rho - x\cos\phi - y\sin\phi) e^{-i2\pi\omega\rho} d\rho \right) dxdy \\
&= \int_{-\infty}^{\infty} \int_{-\infty}^{\infty} u(x, y) e^{-i2\pi(x\omega\cos\phi + y\omega\sin\phi)} dxdy.
\end{aligned}
$$

Defining the frequency variables in the component directions by

$$
\begin{aligned}
\omega_x &= \omega\cos\phi, \\
\omega_y &= \omega\sin\phi,
\end{aligned}
$$

we have $\omega^2 = \omega_x^2 + \omega_y^2$ and $\phi = \tan^{-1}(\omega_y/\omega_x)$. Note that (ω, ϕ) *are* genuine polar coordinates. The variables (ω_x, ω_y) cover all of $\mathbf{R}^2$ as ϕ varies over the interval $[0, \pi)$ and ω varies over $(-\infty, \infty)$. But then the expression for $\widehat{\mathcal{R}u}(\omega, \phi)$, the one-dimensional Fourier transform of the Radon transform, becomes

$$\widehat{\mathcal{R}u}(\omega, \phi) = \int_{-\infty}^{\infty} \int_{-\infty}^{\infty} u(x, y) e^{-i2\pi(x\omega_x + y\omega_y)} dxdy,$$

which is precisely the two-dimensional Fourier transform of $u(x, y)$.   ∎

The power of this theorem is that it leads, in a very straightforward manner, to a simple, elegant method for inverting the Radon transform, and recovering an image from its projections.

The Fourier Transform Method of Image Reconstruction

Let us suppose that the Radon transform of an unknown image $u(x, y)$ is the function $g(\rho, \phi)$; that is,

$$[\mathcal{R}u](\rho, \phi) = g(\rho, \phi).$$

Image reconstruction is an inverse problem: given the collected projection data $g(\rho, \phi)$, we seek the function $u(x, y)$. If we can form the Fourier transforms of $g(\rho, \phi)$, for all values of ϕ, then the assemblage of the Fourier transforms, $\hat{g}(\omega, \phi)$, defines a two-dimensional function (in polar coordinates) which, by the Central Slice Theorem, must equal $\hat{u}(\omega_x, \omega_y)$. In principal, the image $u(x, y)$ can be recovered from $\hat{g}(\omega_x, \omega_y)$ by a two-dimensional inverse Fourier transform

$$u(x, y) = \int_{-\infty}^{\infty} \int_{-\infty}^{\infty} \hat{g}(\omega_x, \omega_y) e^{i(x\omega_x + y\omega_y)} d\omega_x d\omega_y. \tag{7.48}$$

Letting $\mathcal{F}_n$ and $\mathcal{F}_n^{-1}$ represent the n-dimensional Fourier transform and n-dimensional inverse Fourier transform operators, respectively, this procedure can be written more compactly as

$$u = \mathcal{F}_2^{-1} \mathcal{F}_1 g, \tag{7.49}$$

as long as we keep track of the appropriate transform variables. This formula can be used to develop a practical inversion method. The algorithm proceeds by reading (7.49) right to left. Let g_ϕ be the Radon transform data for a fixed angle ϕ. Then:

1. For each ϕ, compute $\hat{g}_\phi = \mathcal{F}_1 g_\phi$, the one-dimensional Fourier transform of each projection in the data set. Assemble the transforms $\hat{g}_\phi$ into a two-dimensional function $\hat{g}(\omega, \phi)$, which is the two-dimensional Fourier transform $\hat{u}(\omega_x, \omega_y)$ where $\omega_x = \omega \cos \phi$ and $\omega_y = \omega \sin \phi$.

2. From $\hat{u}(\omega_x, \omega_y)$, find the unknown image $u(x, y)$ by way of a two-dimensional inverse Fourier transform operator.

The DFT-Based Image Reconstruction Method

In practice, of course, there is only a finite number of projections and only a finite number of samples along each projection. Therefore, in order to apply (7.49), the problem must be discretized. Assume that there are M equally spaced projection angles $\phi_j = j\pi/M$, where $j = 0 : M - 1$. This means that the direction vector specifying a given ray is $(\cos \phi_j, \sin \phi_j)^T$. It will also be assumed that each projection is evenly sampled at the points $\rho_n = n\Delta\rho = 2n/N$ on the interval $[-1, 1]$, for $n = -N/2 + 1 : N/2$. We will denote the set of projection data by

$$g_{nj} = g(\rho_n, \phi_j) = g\left(\frac{2n}{N}, \frac{j\pi}{M}\right)$$

for $n = -N/2 + 1 : N/2$ and $j = 0 : M - 1$. If we assume that the problem has been scaled so that $u(x, y) = 0$ for $\sqrt{x^2 + y^2} \geq 1$, then $g(\rho, \phi) = 0$ for $|\rho| \geq 1$. Therefore, $g_{N/2,j} = 0$. For convenience, we will assume N to be an even integer. The algorithm can easily be modified to accommodate odd N as well.

The first step of the algorithm is to calculate $\mathcal{F}_1 g_j$, where $g(\rho, \phi_j) = 0$ for $|\rho| \geq 1$. The integral

$$\hat{g}(\omega, \phi_j) = \int_{-1}^{1} g(\rho, \phi_j) e^{-i2\pi\omega\rho} d\rho$$

must be approximated for $j = 0, 1, \ldots, M - 1$. The trapezoid rule can be called in for this approximation using the grid points $\rho_n = n\Delta\rho = 2n/N$. By the reciprocity relation $\Delta\rho\Delta\omega = 1/N$, we see that $\Delta\omega = 1/2$, and therefore the appropriate grid points in the frequency domain are $\omega_k = k\Delta\omega = k/2$. Combining these facts the trapezoid rule approximation takes the form

$$\hat{g}(\omega_k, \phi_j) \approx \frac{2}{N} \sum_{n=-\frac{N}{2}+1}^{\frac{N}{2}} g_{nj} \, e^{-i2\pi\rho_n\omega_k} = \frac{2}{N} \sum_{n=-\frac{N}{2}+1}^{\frac{N}{2}} g_{nj} \, e^{-i2\pi nk/N}. \tag{7.50}$$

Note that the requirement that $g(-1, \phi) = g(1, \phi) = 0$ allows us to set $g_{\pm N/2} = 0$ for $(g_{-N/2} + g_{N/2})/2$ in the trapezoid rule. It also satisfies the AVED requirement.

We recognize the sum in (7.50) as the DFT of the sequence $\{g_{nj}\}_{n=-N/2+1}^{N/2}$, so that, for each fixed j,

$$\hat{g}(\omega_k, \phi_j) \approx 2G_{kj} \; = \; 2\mathcal{D}\{g_{nj}\}_k \, .$$

This gives frequency samples G_{kj} at $k\Delta\omega = k/2$ for $k = -N/2+1 : N/2$. Making this choice for approximating $\mathcal{F}_1$ is equivalent to approximating the Fourier coefficients c_k of the two-periodic function $g(\rho, \phi_j)$ on $[-1, 1]$. Equally important, computing the DFT for each of the M projections ϕ_j, for $m = 0 : M - 1$, produces data on a polar coordinate grid in the frequency domain (see Figure 7.23).

The next step in the algorithm is to compute the inverse transform (7.48) by applying $\mathcal{F}_2^{-1}$ to the transformed projections $\hat{g}(\omega_k, \phi_j)$. It is possible to discretize (7.48) directly, using the data in polar coordinates. For example, we might write

$$u(x, y) \approx \sum_{j=0}^{M-1} \sum_{k=-\frac{N}{2}+1}^{\frac{N}{2}} \hat{g}(k\Delta\omega, j\Delta\phi) e^{ik\Delta\omega(x \cos \phi_j + y \sin \phi_j)} |\omega| \Delta\omega\Delta\phi, \tag{7.51}$$

where $\Delta\omega = 1/2$ and $\Delta\phi = \pi/M$, which uses the data produced in step 1. We are at liberty to reconstruct at any convenient sampling of (x, y) within the region $\{(x, y) | x^2 + y^2 < 1\}$. It seems, however, to be almost universal to select $\Delta x = \Delta y = 2/N$. This is done so that the final reconstructed image has the same sampling along the x- and y-axes as $\Delta\rho$ in the projection set. It also means that image values for N^2 points must be computed.

Equation (7.51) is not a particularly good discretization for (7.48), because the summation requires $O(MN)$ operations per point. Since the reconstruction must be done for N^2 image points, this portion of the inversion would require a prohibitive $O(MN^3)$ operations. The difficulty stems from the fact that $\hat{g}_{kj}$ is a data set on a polar coordinate grid in the frequency domain, for which an FFT algorithm is not available.

In order to reconstruct an image of N^2 pixels on $[-1, 1] \times [-1, 1]$, and do so *efficiently* (with an FFT), it is essential to have the transform data, $\hat{g}(\omega, \phi)$, on a Cartesian frequency grid, as shown in Figure 7.23. Such a grid gives transform data $G_{mp} = \hat{g}(m\Delta\omega_x, p\Delta\omega_y)$ for $m, p = -N/2 + 1 : N/2$. If the grid spacing on the image

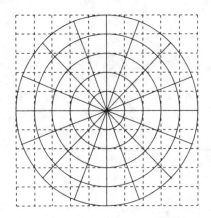 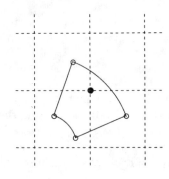

FIG. 7.23. *According to the Central Slice Theorem, the Fourier transform of the image can be computed on a polar grid in the frequency domain (solid lines). Before peforming the inverse transform, the transform must be interpolated to a Cartesian grid (dashed lines). The left figure shows the two grids, while the relationship between the grids is detailed on the right. A simple interpolation uses a weighted average of the transform values at the four nearest polar grid points (o) to produce a value at the enclosed Cartesian grid point (•).*

is to conform to the grid spacing of the projection data, then $\Delta x = \Delta y = 2/N$; this implies that $\Delta \omega_x = \Delta \omega_y = 1/2$ by the reciprocity relation.

With the transform data on a Cartesian grid, the operator $\mathcal{F}_2^{-1}$ can be approximated using the two-dimensional IDFT

$$u(q\Delta x, s\Delta y) \approx \frac{1}{4} \sum_{m=-\frac{N}{2}+1}^{\frac{N}{2}} \sum_{p=-\frac{N}{2}+1}^{\frac{N}{2}} G_{mp}\, e^{i2\pi(mq+ps)/N} \qquad (7.52)$$

for $q, s = -N/2+1 : N/2$, where the factor of $1/4$ arises as the product $\Delta \omega_x \Delta \omega_y$. This double sum *can* be computed efficiently using a two-dimensional FFT, in $O(N^2 \log_2 N)$ operations. In order to obtain $G_{mp} \approx \hat{g}(m\Delta \omega_x, p\Delta \omega_y)$ on a Cartesian grid it is necessary to interpolate the values G_{kj} from the polar grid. This is the central issue of the Fourier reconstruction method. With this in mind, the symbolic description of the algorithm, (7.49), is modified to read

$$u = \mathcal{F}_2^{-1}\, \mathcal{I}_{\rho\theta}^{xy}\, \mathcal{F}_1 g\,, \qquad (7.53)$$

where $\mathcal{I}_{\rho\theta}^{xy}$ is an interpolation operator that maps the polar representation of $\hat{g}(\omega, \phi)$ to the Cartesian representation $\hat{g}(\omega_x, \omega_y)$. The details of the interpolation problem need not concern us here, as they have no bearing on the use of the DFT in the image reconstruction problem. The interested reader will find more than enough discussion of the problem in the references. For the sake of brevity and simplicity, we will assume that the interpolation is carried out by a simple process in which each value on the Cartesian grid is a weighted average of the four nearest values on the polar grid. The geometry of this interpolation is illustrated in Figure 7.23.

The inversion algorithm may be summarized in the following steps, where we assume the discretized projection data $g_{nj} = g(\rho_n, \phi_j)$ are given.

1. For each $j = 0 : M - 1$, find $G_{kj} \approx \mathcal{F}_1 g(\rho, \phi_j)$ by applying forward DFTs:

$$G_{kj} = \frac{2}{N} \sum_{n=-\frac{N}{2}+1}^{\frac{N}{2}} g_{nj} e^{-i2\pi nk/N}$$

for $k = -N/2 + 1 : N/2$, using M FFTs of length N.

2. Interpolate the polar frequency data G_{kj} onto a Cartesian grid. A simple scheme for this has the form

$$G_{mp} = w_a G_a + w_b G_b + w_c G_c + w_d G_d,$$

where G_a, G_b, G_c, and G_d are the four points on the polar grid nearest to the target G_{mp}. The weighting coefficients (w_a, w_b, w_c, w_d) are unspecified here, but are selected according to the type of interpolation desired.

3. Approximate the image $u(x, y) = \mathcal{F}_2^{-1} \hat{u}(\omega_x, \omega_y)$ with a two-dimensional IDFT

$$u(q\Delta x, s\Delta y) = \frac{1}{4} \sum_{m=-\frac{N}{2}+1}^{\frac{N}{2}} \sum_{p=-\frac{N}{2}+1}^{\frac{N}{2}} G_{mp} e^{i2\pi(mq+sp)/N}$$

for $q, s = -N/2 + 1 : N/2$, using an N^2-point two-dimensional FFT for speed.

The entire algorithm is summarized in the map of Figure 7.24, which shows the relationships between the four computational arrays.

It is not difficult to estimate the computational cost of the algorithm, which consists of the cost of approximating the forward transforms, the cost of the interpolation, and the cost of the two-dimensional inverse transform. To simplify the argument we will assume that $M = N$ in the method just presented. As we shall see in Chapter 10, with the use of the FFT, the N forward transforms entail $O(N \log N)$ operations each. The operation count can be reduced by one-half with specialized FFTs for real data. For most interpolation schemes, $O(1)$ operations are required for each of the N^2 points to be interpolated, hence we will use $O(N^2)$ as the cost of the interpolation. (It should be noted, however, that some of the extremely accurate interpolations based on the Shannon Sampling Theorem incur costs approaching $O(N^3)$.) The cost of computing the two-dimensional IDFT is the cost of N one-dimensional IDFTs, each of length N. We will assume that an inverse fast Fourier transform (IFFT) is employed to perform this computation. Since the IFFT and the FFT have the same operation count, this means that the inverse transform step of the algorithm has a cost of $O(N^2 \log_2 N)$. Putting the three phases of the algorithm together, the total cost of the DFT reconstruction method is

$$C_1 N^2 \log N + C_2 N^2 + C_3 N^2 \log N, \tag{7.54}$$

where the constants C_1 and C_3 depend on the specific FFT algorithms used and the constant C_2 depends on the method of interpolation.

Example: A simple model of a brain. A simple example of image reconstruction from projections is shown in Figure 7.25. The desired image is shown on the left. It consists of a thin, hollow, high density ellipse filled with low density material. Within the low density material are several elliptical regions of various

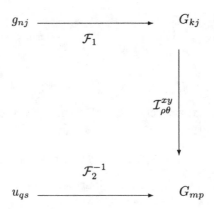

FIG. 7.24.   A map showing the relationships of the four computational arrays of the image reconstruction problem. The input data is the set of projections, g_{nj}, shown in the upper left corner. The reconstructed image is u_{qs}, in the lower left. We arrive at the reconstruction by way of the frequency domain. First, the one-dimensional DFT is taken along each projection, resulting (according to the Central Slice Theorem) in the polar-grid representation in frequency, G_{kj} (upper right). This is interpolated to a Cartesian grid, giving G_{mp}, the two-dimensional DFT of the desired image (lower right). Finally, a two-dimensional IDFT leads back from the frequency domain, and yields the reconstructed image u_{qs}.

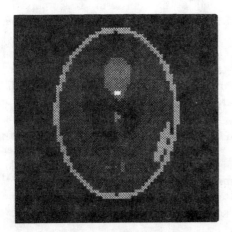
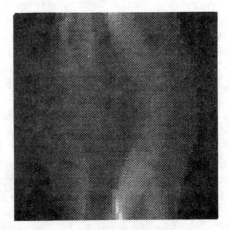

FIG. 7.25.   A simple example of the process of image reconstruction from projections is shown. The desired image is shown on the left. The "data," or projections, are shown on the right. Here the data are presented in image format, where the value of the Radon transform is represented by the shading, with white representing large values and black representing small values. The bottom edge of the projection data plot parallels the ρ axis, while the left-hand edge parallels the ϕ axis. The reconstructed image is shown in Figure 7.26.

densities. This simple model, modified from [125], might represent the problem of imaging a human brain. The "data," or projections, are shown on the right. There are 64 projections, each sampled at 64 points in $[-1:1]$. They were obtained by sampling the analytic Radon transform of the model, which is easy to compute (see problem 171 for a similar, although simpler, model). Finally, the reconstructed image is shown (using a 64×64 grid) in Figure 7.26. The reconstruction has good overall quality, and all the features of the "exact" image have been essentially resolved. Evident in the reconstructed image are reconstruction artifacts, features that appear as negative-image reflections and curves. These phenomena are characteristic of reconstructions by the Fourier method, and are caused by wrap around (leakage) and aliasing. They may be minimized by padding the data with zeros prior to processing, similar to the techniques applied in the FK migration procedure [76], [104].

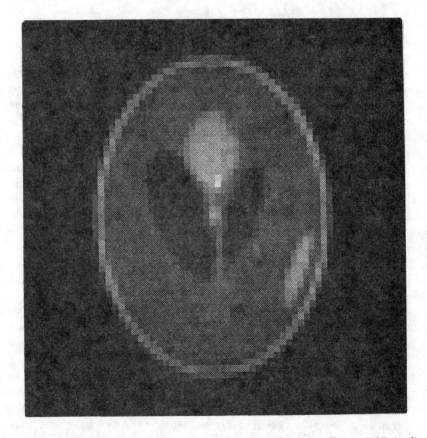

FIG. 7.26.   *The reconstructed image from the data shown in Figure 7.25 is displayed here, on a 64×64 grid. The reconstruction is generally good, and all the features of the "exact" image have been essentially resolved. Reconstruction "artifacts," features that appear as negative-image reflections and curves, are present in the reconstruction, although they are not prominent. These phenomena are characteristic of reconstructions by the Fourier method, and are caused by wrap around (leakage) and aliasing.*

Notes and References

The Radon transform literature is confusing when it comes to the topic of error analysis. Much of the error analysis that is done is heuristic in nature, and different workers hold differing views about the types and importance of the errors in reconstruction algorithms. In the Fourier reconstruction literature, the most commonly cited sources of error are [92], [129], [130]:

1. undersampling of the projections, $g(\rho, \phi)$;

2. error in approximating the Fourier transforms of the projections, $\hat{g}(\omega, \phi)$;

3. truncation of the frequency domain;

4. interpolation error in $\mathcal{I}_{\rho\theta}^{xy}$;

5. undersampling of the frequency data, $\hat{u}(\omega_x, \omega_y)$;

6. error in approximating u by $\mathcal{F}_2^{-1}$.

   The general consensus among workers in this field is that the most significant source of error in the algorithm is the operator $\mathcal{I}_{\rho\theta}^{xy}$, which maps the frequency domain data from a polar grid to a Cartesian grid. Many interpolation schemes have been proposed in the literature, including nearest-neighbor interpolation [78], [104], a simple bilinear scheme [78], bi-Lagrangian interpolation [76], and schemes based on the Shannon Sampling Theorem [102], [103], [104], [129], [130]. The discussions tend to be qualitative, although careful error analysis is included in [76], [102], and [104]. Typically, the image reconstructions based on the Fourier transform are characterized by several types of reconstruction errors called **artifacts**. The more sophisticated interpolations based on the sampling theorem [131] generally produce much better images than those made using simpler interpolation methods, however, they also tend to be quite expensive to compute.

   An alternate approach to the interpolation problem can be taken by devising a reconstruction algorithm based on discretizing (7.48) directly—that is, by performing the inverse Fourier transforms in polar coordinates. This can be done efficiently by treating the polar grid as an "irregular Cartesian grid," and employing methods for computing FFTs on irregular grids. Such methods are fairly recent additions to the FFT family. Irregular-grid FFTs were developed for the Radon transform problem in [76] and [77], while a more general and exhaustive treatment may be found in [54].

   The sources of error listed above are generally treated individually in the literature. In [76], however, it is shown that a bound on the error of the reconstruction consists of the sum of three terms, one arising from each of the three steps in the algorithm: forward transform, interpolation, and inverse transform. As we have seen, the error in the transforms in either direction is bounded by N^{-p} for some $p > 0$, depending on the smoothness of the data. We may expect, then, that the errors in these two steps can be controlled by acquiring more data, that is, by increasing N. Of more concern, however, is the error of interpolation. For simple polynomial interpolation [76], the error is $O(\Delta\omega^r)$, where r is some positive number related to the order of the interpolation (e.g., linear, cubic, etc.). This error is also dependent on $\Delta\phi$; however, the transforms are typically quite smooth in the radial direction, and there tends to be little error generated by the angular interpolation. What makes the radial interpolation error so troublesome is that, unlike the transform error, we cannot expect that simply using more data will help. In fact, from the reciprocity relation

$$\frac{\Delta\rho}{2}\Delta\omega = \frac{1}{2N},$$

we observe that doubling the number of samples on each projection by cutting the spatial sample rate does not alter the frequency sample rate at all. The quality of the radial interpolation cannot improve by this technique, as *the same samples* are used to interpolate frequencies even with smaller values of $\Delta\rho$. There are ways to improve the resolution in the frequency domain, but they involve using and understanding the reciprocity relations which appear yet again in a fundamental way.

7.5. Problems

Boundary Value Problems

140. Some difference equation BVPs.   Consider the difference equation

$$-u_{n-1} + 4u_n - u_{n+1} = f_n.$$

Solve this equation with

$$\text{(a) } f_n = 1, \qquad \text{(b) } f_n = \delta(n) - \delta\left(n - \frac{N}{2}\right),$$

subject to the boundary conditions

   (a) $u_0 = u_N = 0$,
   (b) $u_1 - u_{-1} = 0, u_{N+1} - u_{N-1} = 0$,
   (c) $u_0 = u_N, u_{-1} = u_{N-1}$

(six BVPs in all). In each case, use the appropriate form of the DFT with either the component perspective or the operational perspective. If necessary, express the solution in terms of the coefficients U_k.

141. Matrix perspective.   Consider the three matrices $\mathbf{A}$ given by

$$\begin{pmatrix} 2 & -1 & 0 \\ -1 & 2 & -1 \\ 0 & -1 & 2 \end{pmatrix}, \quad \begin{pmatrix} 2 & -1 & 0 & -1 \\ -1 & 2 & -1 & 0 \\ 0 & -1 & 2 & -1 \\ -1 & 0 & -1 & 2 \end{pmatrix}, \quad \begin{pmatrix} 2 & -2 & 0 & 0 & 0 \\ -1 & 2 & -1 & 0 & 0 \\ 0 & -1 & 2 & -1 & 0 \\ 0 & 0 & -1 & 2 & -1 \\ 0 & 0 & 0 & -2 & 2 \end{pmatrix}$$

that correspond to the difference equation $-u_{n-1} + 2u_n - u_{n+1} = f_n$, with $N = 4$, with Dirichlet, periodic, and Neumann boundary conditions, respectively. In each case find the matrix $\mathbf{P}$ that diagonalizes the given matrix and the eigenvalues that lie on the diagonal of the matrix $\mathbf{D} = \mathbf{P}^{-1}\mathbf{AP}$.

142. The trick for nonhomogeneous boundary conditions.   Consider the BVP

$$au_{n-1} + bu_n + au_{n+1} = f_n \quad \text{for} \quad n = 1 : N - 1,$$

subject to the boundary conditions $u_0 = \alpha, u_N = \beta$, where α and β are given real numbers. Show that the vector

$$v_n = u_n - \alpha - \frac{\beta - \alpha}{N} n$$

satisfies the same BVP with a different input vector f_n and homogeneous boundary conditions ($v_0 = v_N = 0$). How would you apply this idea to a BVP with nonhomogeneous Neumann boundary conditions $u_1 - u_{-1} = \alpha, u_{N+1} - u_{N-1} = \beta$?

143. Solving BVPs with nonhomogeneous boundary conditions. Use the techniques of the previous problem to solve the difference equation

$$-u_{n-1} + 2u_n - u_{n+1} = 0,$$

subject to the two sets of boundary conditions

(a) $u_0 = 0, u_N = 1$,   (b) $u_1 - u_{-1} = 0, u_{N+1} - u_{N-1} = 1$.

144. A DST. Verify that the DST of the constant vector $f_n = 1$ is

$$F_k = \frac{\sin(\frac{\pi k}{N})(1 - \cos(\pi k))}{4N \sin^2(\frac{\pi k}{2N})}$$

for $k = 1 : N - 1$, as claimed in this chapter.

145. Shift properties. Given the sequence u_n, let u_{n+1} and u_{n-1} denote the sequences obtained by shifting u_n right and left one unit, respectively. Show that if $\mathcal{S}, \mathcal{C},$ and $\mathcal{D}$ represent the DST, the DCT, and the DFT, respectively, on N points, then

$$\mathcal{S}\{u_{n+1} + u_{n-1}\}_k = 2\cos\left(\frac{\pi k}{N}\right) \mathcal{S}\{u_n\}_k, \tag{7.55}$$

$$\mathcal{C}\{u_{n+1} + u_{n-1}\}_k = 2\cos\left(\frac{\pi k}{N}\right) \mathcal{C}\{u_n\}_k, \tag{7.56}$$

$$\mathcal{D}\{u_{n+1} + u_{n-1}\}_k = 2\cos\left(\frac{2\pi k}{N}\right) \mathcal{D}\{u_n\}_k. \tag{7.57}$$

146. Component perspective for the DCT. Use the component perspective described in the chapter to derive the method of solution for the BVP

$$au_{n-1} + bu_n + au_{n+1} = f_n$$

for $n = 0 : N$, subject to the homogeneous Neumann boundary condition $u_1 - u_{-1} = 0, u_{N+1} - u_{N-1} = 0$.

147. Operational perspective for the DFT. Use the operational perspective described in this chapter to derive the method of solution for the BVP

$$au_{n-1} + bu_n + au_{n+1} = f_n$$

for $n = 0 : N - 1$, subject to periodic boundary conditions $u_0 = u_N, u_{-1} = u_{N-1}$.

148. Finite difference approximations. Assume that ϕ is at least four times differentiable in the interval $(x - \Delta x, x + \Delta x)$. Show that $\phi'(x)$ and $\phi''(x)$ may be approximated by the following finite difference approximations with the indicated truncation errors:

$$\phi'(x) = \frac{\phi(x + \Delta x) - \phi(x - \Delta x)}{2\Delta x} + c\Delta x^2,$$

$$\phi''(x) = \frac{\phi(x - \Delta x) - 2\phi(x) + \phi(x + \Delta x)}{\Delta x^2} + d\Delta x^2,$$

where c and d are constants independent of Δx. (Hint: Expand $\phi(x \pm \Delta x)$ in Taylor series.)

149. Neumann and periodic boundary conditions. Following the development of the chapter, describe the DCT and DFT methods for approximating solutions of the BVP

$$-\phi''(x) + a^2\phi(x) = f(x) \quad \text{for} \quad 0 < x < A,$$

subject to the Neumann and periodic boundary conditions

(a) $\phi'(0) = \phi'(A) = 0,$ (b) $\phi(0) = \phi(A), \; \phi'(0) = \phi'(A).$

(Hint: Use the approximation from problem 148, $\phi'(0) \approx (u_1 - u_{-1})/(2\Delta x)$, with a similar expression for $\phi'(A)$.)

150. Two-dimensional shift properties. The solution of two-dimensional BVPs is usually easier using the operational perspective, which requires the relevant shift properties. Verify the following shift properties for the two-dimensional DST ($\mathcal{S}$), DCT ($\mathcal{C}$), and DFT ($\mathcal{D}$).

$$\mathcal{S}\{u_{m+1,n} + u_{m-1,n}\}_{jk} = 2\cos(\tfrac{\pi j}{M})\mathcal{S}\{u_{mn}\}_{jk},$$

$$\mathcal{S}\{u_{m,n+1} + u_{m,n-1}\}_{jk} = 2\cos(\tfrac{\pi k}{N})\mathcal{S}\{u_{mn}\}_{jk},$$

$$\mathcal{C}\{u_{m+1,n} + u_{m-1,n}\}_{jk} = 2\cos(\tfrac{\pi j}{M})\mathcal{C}\{u_{mn}\}_{jk},$$

$$\mathcal{C}\{u_{m,n+1} + u_{m,n-1}\}_{jk} = 2\cos(\tfrac{\pi k}{N})\mathcal{C}\{u_{mn}\}_{jk},$$

$$\mathcal{D}\{u_{m+1,n} + u_{m-1,n}\}_{jk} = 2\cos(\tfrac{2\pi j}{M})\mathcal{D}\{u_{mn}\}_{jk},$$

$$\mathcal{D}\{u_{m,n+1} + u_{m,n-1}\}_{jk} = 2\cos(\tfrac{2\pi k}{N})\mathcal{D}\{u_{mn}\}_{jk}.$$

151. A two-dimensional BVP with periodic boundary conditions. Consider the two-dimensional BVP

$$a(u_{m-1,n} + u_{m+1,n}) + b(u_{m,n-1} + u_{m,n+1}) + cu_{mn} = f_{mn},$$

subject to the periodic boundary conditions

$$u_{0,n} = u_{M,n}, \; u_{-1,n} = u_{M-1,n}, \; u_{m,0} = u_{m,N}, \; u_{m,-1} = u_{m,N-1}$$

for $m = 0 : M - 1, n = 0 : N - 1$.

(a) Show that a solution of the form

$$u_{mn} = \text{Re}\left\{\sum_{j=0}^{M-1}\sum_{k=0}^{N-1} U_{jk}\omega_M^{mj}\omega_N^{nk}\right\}$$

satisfies the boundary conditions.

(b) Show that if f_{mn} is given a similar representation with coefficients F_{jk}, then the coefficients of the solution are given by

$$U_{jk} = \frac{F_{jk}}{2a \sin^2(\frac{\pi m}{M}) + 2b \sin^2(\frac{\pi k}{N}) + c}$$

for $j = 0 : M - 1, k = 0 : N - 1$.

152. A fourth-order BVP. Consider the fourth-order BVP

$$a u_{n-2} + b u_{n-1} + c u_n + b u_{n+1} + a u_{n+2} = f_n$$

for $n = 1 : N - 1$, subject to the boundary conditions

$$u_0 = u_N = 0 \quad \text{and} \quad u_{-1} + u_1 = u_{N-1} + u_{N+1} = 0$$

(corresponding to $\phi(0) = \phi(A) = \phi''(0) = \phi''(A) = 0$ for an ODE). Find the appropriate discrete transform that satisfies the boundary conditions and use it to find the coefficients of the solution u_n in terms of the coefficients of the input f_n.

153. Solution to the gossip problem Consider the solution

$$u_n = \frac{1}{2N} \sum_{k=1}^{N-1} \frac{(1 - \cos(\pi k))}{\sin^2(\frac{\pi k}{N})} \cos\left(\frac{2\pi n k}{N}\right) + U_0$$

for $n = 0 : N - 1$, given to the gossip problem in the text. How would you analyze and simplify this solution analytically?

(a) Verify that the solution is N-periodic: $u_n = u_{n+N}$.

(b) Show that the solution is even on the interval $[0, N]$: $u_n = u_{N-n}$.

(c) Use the analytical DFT of the triangular wave given in *The Table of DFTs* of the Appendix to show that the solution u_n consists of two linear segments and may be expressed as $u_n = |n - N/2| - N/4$. Plot the solution for various values of N.

154. A model of altruism. The following model suggests how diffusion might be used (or perhaps should not be used) to describe the distribution of wealth within a multicomponent system. Imagine a collection of $N+1$ economic units (people, families, towns, cartels) that may have supplies and demands of income. The underlying law of altruism is that each unit must at all times attempt to distribute its wealth equally among its two nearest neighbors (longer range interactions could also be considered), so that in a steady state the wealth of each unit is the average of the wealth of the two neighboring units.

(a) Argue that such a system would satisfy the following difference equation in a steady state:

$$-\frac{1}{2} u_{n-1} + u_n - \frac{1}{2} u_{n+1} = f_n \quad \text{for} \quad n = 0 : N,$$

where u_n is the steady state wealth of the nth unit, and f_n represents external sources or sinks of wealth for the nth unit.

(b) Interpret the boundary conditions $u_1 = u_{-1}, u_{N+1} = u_{N-1}$.

(c) Argue mathematically and physically that a solution can exist only if $\sum_{n=0}^{N} f_n = 0$.

(d) Find a solution for this problem for an arbitrary input f_n and for the specific input $f_n = \sin(2\pi n/N)$.

155. Improvement to DFT fast Poisson solvers. Consider the partial difference equation (7.16)

$$-u_{m-1,n} - u_{m+1,n} - \gamma^2(u_{m,n-1} + u_{m,n+1}) + (2 + 2\gamma^2 + a^2\gamma^2)u_{mn} = f_{mn},$$

subject to the boundary conditions

$$u_{0n} = u_{Mn} = u_{m0} = u_{mN} = 0,$$

where $m = 1 : M - 1$ and $n = 1 : N - 1$. Rather than applying *two* sweeps of DFTs (one in each direction), an improvement can be realized if the DFT is applied in one direction only and tridiagonal systems of equations are solved in the remaining direction. Here is how it works.

(a) Assume a solution to the BVP of the form

$$u_{mn} = 2 \sum_{j=1}^{M-1} U_{jn} \sin\left(\frac{\pi m j}{M}\right)$$

for $m = 1 : M - 1$ and $n = 1 : N - 1$. Notice that this amounts to applying a DST in the m- (or x-) direction only. The right-hand sequence f_{mn} is given a similar representation with coefficients

$$F_{jn} = \frac{1}{M} \sum_{m=1}^{M-1} f_{mn} \sin\left(\frac{\pi m j}{M}\right)$$

for $j = 1 : M - 1$ and $n = 1 : N - 1$. Now substitute these representations for u_{mn} and f_{mn} into the BVP. After simplifying and collecting terms, show that the BVP takes the form

$$\sum_{j=1}^{M-1} \left[\alpha_j U_{jn} - \gamma^2(U_{j,n-1} + U_{j,n+1}) - F_{jn}\right] \sin\left(\frac{\pi m j}{M}\right) = 0,$$

where

$$\alpha_j = \left(2 + 2\gamma^2 + a^2\gamma^2 - 2\cos\left(\frac{\pi j}{M}\right)\right)$$

and the coefficients U_{jn} are the new unknowns.

(b) This equation must hold for all $m = 1 : M - 1$ and $n = 1 : N - 1$, which means that each term of the sum must vanish independently. Therefore, for each fixed $j = 1 : M - 1$ the U_{jn}'s must satisfy the linear equations

$$\alpha_j U_{jn} - \gamma^2(U_{j,n-1} + U_{j,n+1}) = F_{jn}$$

for $n = 1 : N - 1$. Notice that the work entailed in this step is the solution of $M - 1$ tridiagonal systems each of which has the $(N - 1)$ unknowns $U_{j1}, \ldots, U_{j,N-1}$.

(c) Once the coefficients U_{jn} have been determined the solution can be recovered by performing inverse DSTs of the form

$$u_{mn} = 2 \sum_{j=1}^{M-1} U_{jn} \sin \left(\frac{\pi m j}{M} \right)$$

for $m = 1 : M - 1$ and $n = 1 : N - 1$. (The factors of 2 and $1/M$ in the forward and inverse transforms can be carefully combined.)

(d) Using the fact that the cost of solving an $N \times N$ symmetric system of tridiagonal equations is roughly $3N$ operations and that an N-point FFT costs approximately $N \log N$ operations, find the cost of this modified method when applied to a partial difference equation with MN unknowns. In particular, show that the modified method offers savings over the DFT method proposed in the text. Show that the method can also be formulated by doing the DSTs in the n direction and solving tridiagonal systems in the m direction. If $M > N$ which strategy is more efficient?

Digital Filters

156. Discrete low-pass filter. Show that the discrete time domain representation of the low-pass filter

$$H_k = \begin{cases} 1 & \text{if} & |k| \leq k_c, \\ \frac{1}{2} & \text{if} & |k| = k_c, \\ 0 & \text{if} & |k| > k_c, \end{cases}$$

is given by

$$h_n = \frac{\sin \left(\frac{\pi n k_c}{N} \right) \sin \left(\frac{2\pi n}{N} \right)}{2 \sin^2 \left(\frac{\pi n}{N} \right)}.$$

157. Discrete low-pass filter. Show that the generalized low-pass filter

$$\hat{h}(\omega; \omega_1, \omega_2) = \begin{cases} 1 & \text{for} & |\omega| \leq \omega_1, \\ \dfrac{\omega_2 - \omega}{\omega_2 - \omega_1} & \text{for} & \omega_1 \leq |\omega| \leq \omega_2, \\ 0 & \text{for} & \omega_2 \leq |\omega|, \end{cases}$$

has a time domain representation given by

$$h(t; \omega_1, \omega_2) = \begin{cases} \omega_2 + \omega_1 & \text{for} & t = 0, \\ \dfrac{\cos(2\pi\omega_1 t) - \cos(2\pi\omega_2 t)}{2\pi^2 t^2 (\omega_2 - \omega_1)} & \text{for} & t \neq 0. \end{cases}$$

(Hint: Use symmetry to convert the Fourier integral to a cosine integral and integrate by parts.)

158. Square pulse. Verify that the 24-point DFT of the square pulse filter h_n with components

$$\left\{ 0, 0, 0, 0, 0, 0, 0, 0, 0, \frac{1}{8}, \frac{1}{4}, \frac{1}{4}, \frac{1}{4}, \frac{1}{8}, 0, 0, 0, 0, 0, 0, 0, 0, 0, 0 \right\}$$

is given by

$$
H_k = \begin{cases} \dfrac{1}{24} & \text{if } k = 0, \\[2ex] \left(\dfrac{1}{4}\right) \dfrac{\sin(k\pi/6)\sin(k\pi/12)}{48\sin^2(k\pi/24)} & \text{if } k \neq 0. \end{cases}
$$

159. Amplitude-phase relations. Assume that a discrete filter h_n is applied to a signal f_n in the form of a convolution to produce the filtered signal $g_n = f_n * h_n$. Letting F_k, G_k, and H_k represent corresponding DFTs of these signals, show that

$$
|G_k| = |F_k||H_k| \qquad \text{and} \qquad \phi_G = \phi_F + \phi_H,
$$

where ϕ_F, ϕ_G, and ϕ_H are the phases of F_k, G_k, and H_k, respectively.

160. Time-shifting filters. Show that the time-shifting filter $h(t)$ that takes the input $f(t)$ into the output $g(t) = f(t - t_0)$ has the properties $|\hat{h}(\omega)| = 1$ and $\phi_h(\omega) = -2\pi\omega t_0$.

161. Continuous to discrete filters. Given the parameters $\omega_1, \omega_2, \omega_3$, and ω_4, devise the discrete versions of the generalized band-pass filter, $\hat{h}_B(\omega; \omega_1, \omega_2, \omega_3, \omega_4)$, the high-pass filter, $\hat{h}_H(\omega; \omega_1, \omega_2)$, and the notch filter, $\hat{h}_R(\omega; \omega_1, \omega_2, \omega_3, \omega_4)$, given in the text. Is it possible to find the time domain representations of these filters?

162. Window functions. The process of filtering a signal is closely related to the process of truncating a signal. As shown in the text, the use of a square pulse (rectangular window) to truncate a signal has some unpleasant side-effects. Consider the following window functions on the interval $[-A/2, A/2]$.

(a) Bartlett (triangular) window:

$$
h_B(x) = \begin{cases} 1 - \dfrac{2|x|}{A} & \text{for } |x| \leq A/2, \\ 0 & \text{otherwise.} \end{cases}
$$

(b) Hanning (cosine) window:

$$
h_H(x) = \begin{cases} \cos^2\left(\dfrac{\pi x}{A}\right) & \text{for } |x| \leq A/2, \\ 0 & \text{otherwise.} \end{cases}
$$

(c) Parzen window:

$$
h_P(x) = \begin{cases} 1 - 24\left(\dfrac{x}{A}\right)^2 + 48\left|\dfrac{x}{A}\right|^3 & \text{for } |x| \leq A/4, \\ 2\left(1 - \dfrac{2|x|}{A}\right)^3 & \text{for } A/4 < |x| < A/2, \\ 0 & \text{otherwise.} \end{cases}
$$

In each case plot the window function in the time domain. Then, either analytically, numerically, or with the help of *The DFT Table* in the Appendix, find the frequency domain representation of each window function. How do the properties of the frequency domain representations (such as the width of the central lobe and the amplitude of the side lobes) compare with each other and with the rectangular window?

FK Migration

163. Snell's law. Suppose an incident plane wave arrives at a flat interface between two media having densities ρ_1 and ρ_2 and sound velocities c_1 and c_2, respectively. Assume that the wavefront arrives with angle of incidence $0 < \theta_1 < \pi/2$. Using a straightedge, a compass, and a bit of trigonometry, use Huygens' principle to derive Snell's law, and to show that the angle of reflection equals the angle of refraction.

164. Dispersion relation. Given the dispersion relation $\omega^2 = c^2(k_x^2 + k_z^2)$, verify that

$$\frac{d\omega}{dk_z} = \frac{c}{\sqrt{k_x^2/k_z^2 + 1}}.$$

165. Derivative properties of the Fourier transform. Using the definition of the Fourier transform given in the text verify (using integration by parts) the following Fourier transform derivative properties.

(a) $\mathcal{F}\left\{\dfrac{df}{dx}\right\}(\omega) = i2\pi\omega \hat{f}(\omega)$,

(b) $\mathcal{F}\left\{\dfrac{d^2 f}{dx^2}\right\}(\omega) = -4\pi^2\omega^2 \hat{f}(\omega)$,

(c) $\mathcal{F}\left\{\dfrac{\partial^k f}{\partial x_j^k}(x_1, x_2, \ldots)\right\}(\omega_{x_1}, \omega_{x_2}, \ldots) = (i2\pi\omega_{x_j})^k \hat{f}(\omega_{x_1}, \omega_{x_2}, \ldots)$.

166. Plane waves in space-time. Consider a function of the form

$$u(x, z, t) = e^{i2\pi(k_x x + k_z z + \omega t)}$$

(or consider the real and imaginary parts of this function). Using the dispersion relation $\omega^2 = c^2(k_x^2 + k_z^2)$ show that this function is constant along the lines in the xz-plane

$$k_x x + k_z z = \text{constant},$$

and may be interpreted as a wave propagating in the direction given by the unit vector

$$\left(-\frac{ck_x}{\omega}, -\frac{ck_z}{\omega}\right).$$

167. Inverse DFT. Verify that with the grid parameters $\Delta x = A/M, \Delta k_x = 1/A, \Delta z = D/N$, and $\Delta k_z = 1/D$, the IDFT given in (7.43)

$$u_{jn} = \sum_{l=-M/2+1}^{M/2} \sum_{m=0}^{N-1} P_{lm}\omega_M^{jl}\omega_N^{nm} = \mathcal{D}^{-1}\{P_{lm}\}_{jn}$$

gives an approximation to the function

$$h(x, z) = \begin{cases} u(x, z, 0) & \text{for} \quad 0 < z < D, \quad -\dfrac{A}{2} \le x \le \dfrac{A}{2}, \\[2ex] \dfrac{1}{2}u(x, 0, 0) + \dfrac{1}{2}u(x, D, 0) & \text{for} \quad z = 0, z = D, \quad -\dfrac{A}{2} \le x \le \dfrac{A}{2}, \end{cases}$$

at the spatial grid points (x_j, z_n).

Image Reconstruction

168. X-ray paths. The Radon transform consists of line integrals in the xy-plane along lines given by

$$x \cos \phi + y \sin \phi = \rho,$$

where ρ is a real number and $0 \leq \phi < \pi$ is an angle measured counterclockwise from the positive x-axis. For each of the following choices of ρ and ϕ, sketch the corresponding line and interpret ρ and ϕ.

(a) $\rho = 2, \phi = \dfrac{\pi}{4}$,   (b) $\rho = -2, \phi = \dfrac{\pi}{3}$,   (c) $\rho = 2, \phi = \dfrac{3\pi}{4}$.

169. Analytical Radon transforms. Let $u(x, y) = e^{-(x^2 + y^2)}$. Show that the Radon transform of $u(x, y)$ is given by

$$[\mathcal{R}u](\rho, \phi) = \sqrt{\pi} e^{-\rho^2}.$$

(Hint: Use the definition of the Radon transform given in equation (7.45), and the change of variables

$$
\begin{aligned}
u &= x \cos \phi + y \sin \phi, \\
v &= -x \sin \phi + y \cos \phi,
\end{aligned}
$$

which is simply a rotation of axes. Then perform the integration, invoking the property of the delta function.)

170. Properties of the Radon transform. Let $\vec{\xi} = (\cos \phi, \sin \phi)^T$ be the unit vector specifying the ρ-axis for a given angle ϕ. Then the line specified by $\rho = x \cos \phi + y \sin \phi$ is also specified by $\rho = \vec{x} \cdot \vec{\xi}$, where $\vec{x} = (x, y)^T$. The Radon transform may be written as

$$[\mathcal{R}u](\rho, \vec{\xi}) = \int_{\mathbf{R}^2} u(\vec{x}) \delta(\rho - \vec{x} \cdot \vec{\xi}) d\vec{x}.$$

(a) Show that the Radon transform has a **scaling** property, that is, show that

$$[\mathcal{R}u](\alpha\rho, \alpha\vec{\xi}) = \frac{1}{|\alpha|}[\mathcal{R}u](\rho, \vec{\xi}),$$

where α is a scalar. Show that the evenness property can be obtained from the scaling property by setting $\alpha = -1$.

(b) Show that the Radon transform has a linear transformation property. Let

$$A = \begin{pmatrix} a_{11} & a_{12} \\ a_{21} & a_{22} \end{pmatrix}$$

be any nonsingular matrix and let $B = A^{-1}$. Show that the Radon transform of a function $u(A\vec{x})$ is related to the transform of $u(\vec{x})$ by

$$[\mathcal{R}u(A\vec{x})](\rho, \vec{\xi}) = |\det(B)| [\mathcal{R}u(\vec{x})](\rho, B^T\vec{\xi}).$$

(c) Let $w(\vec{x})$ be the characteristic function of an ellipse, given by

$$w(\vec{x}) = \begin{cases} 1 & \text{for } \left(\frac{x}{a}\right)^2 + \left(\frac{y}{b}\right)^2 \le R^2, \\ 0 & \text{otherwise.} \end{cases}$$

The ellipse can be generated from a circle by a linear change of variables, that is,

$$w(\vec{x}) = u(A\vec{x}),$$

where $u(\vec{x})$ is the characteristic function of a disk given by (7.46), and

$$A = \begin{pmatrix} \dfrac{1}{a} & 0 \\ 0 & \dfrac{1}{b} \end{pmatrix}.$$

Use the scaling and linear transformation properties to show that the Radon transform of the characteristic function of an ellipse is given by

$$[\mathcal{R}w](\rho, \phi) = \begin{cases} \dfrac{2ab}{|\vec{\zeta}|} \sqrt{R^2 - \left(\dfrac{\rho}{|\vec{\zeta}|}\right)^2} & \text{for } \rho^2 \le |\vec{\zeta}|^2 R^2, \\ 0 & \text{otherwise ,} \end{cases}$$

where $\vec{\zeta} = \left(A^{-1}\right)^T \vec{\xi}$.

171. Model problem.   Using linearity and the results of problem 170, determine the Radon transform of a "skull model" consisting of a thin, high density elliptical shell containing lower density material. That is, if $w(x, y)$ is the characteristic function of an ellipse and $z(x, y)$ is the characteristic function of a slightly smaller confocal ellipse, the Radon transform of the skull model may be computed by $[\mathcal{R}w](\rho, \phi) - .75[\mathcal{R}z](\rho, \phi)$. Sketch profiles of the resulting Radon transform corresponding to several different values of ϕ.

Chapter **8**

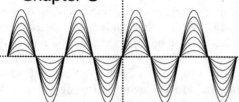

Related Transforms

*Training is everything.
The peach was once a
bitter almond;
cauliflower is nothing but
cabbage with a college
education.*
– Mark Twain

8.1.   Introduction

Like the Zen image of a finger pointing at the moon, this chapter is intended to point the way toward other abundant lands that are related in some way to the DFT. But, as in the Zen lesson, woe to those who mistake this brief discussion for a complete treatment of the vast subject of related transforms. The most we can possibly do is open a few doors, suggest some guiding references, and offer a glimpse of what lies beyond. The transforms included in this section were chosen for one of two reasons. The Laplace and Chebyshev transforms appear because they can be reduced, in their discrete form, to the DFT (and hence the FFT). On the other hand, Legendre and other orthogonal polynomial transforms, while not reducible to the DFT, share many analogous properties and applications, and are also worthy of recognition at this time. So with apologies for brevity, but with hopes of providing a valuable reconnaissance of the subject, we shall begin.

8.2.   The Laplace Transform

Used to solve initial value problems that arise in mechanics, electrodynamics, fluid dynamics, and engineering systems analysis, the **Laplace transform** is one of the fundamental tools of applied mathematics. It is a prototype for many transform methods and is often the first transform that students encounter. Laplace[1] enunciated the transform that bears his name in his 1820 treatise *Théorie analytique des probabilités* and used it to solve difference equations. With a few notational changes, here is the transform that Laplace devised. Given an integrable function h on the interval $(0, \infty)$, its Laplace transform is

$$\mathcal{L}\{h(t)\} = H(s) = \int_0^\infty h(t)e^{-st}dt. \tag{8.1}$$

The choice of t as the independent variable reflects the most common situation in which h is a time-dependent or causal function. The transform variable s is complex, and if the input f decays such that $\lim_{t\to\infty} h(t)e^{-bt} = 0$ for some real number b, then H is defined for all complex values of s with $\text{Re}\{s\} > b$. The forward transform (8.1) may be evaluated analytically for many commonly occurring functions; it has also been tabulated extensively [3], and it may be accurately approximated using numerical methods [119].

The inversion of the Laplace transform is a more challenging problem. Given a function $H(s)$ of a complex variable, the inverse Laplace transform, $h(t) = \mathcal{L}^{-1}\{H(s)\}$, is defined in terms of a contour integral in the complex plane. In this section we will present a method for inverting the Laplace transform that appeals to the Fourier transform. Not surprisingly, the actual implementation of the method implicates the DFT (and hence the FFT). The method, attributed to Dubner and Abate [50], is analyzed in the valuable paper of Cooley, Lewis, and Welch [41].

The key to most methods for the numerical inversion of the Laplace transform is to express the complex transform variable in the form $s = c + i2\pi\omega$, where c and ω

---

[1]Born in poverty, PIERRE SIMON LAPLACE (1749–1827) spent most of his life in Parisian prosperity once d'Alembert recognized his talents and assured him a faculty position at *Ecole Militaire* at a young age. His scientific output was prodigious; his most notable contributions were to the subjects of astronomy, celestial mechanics, and probability. Laplace's equation, which governs the steady state potential of many physical fields, was proposed in 1785.

are real. In so doing, e^{-st} is periodic in ω with period one. The forward transform (8.1) then becomes

$$H(c + i2\pi\omega) = \int_0^\infty h(t)e^{-ct}e^{-i2\pi\omega t}\,dt.$$

For a fixed value of c, this expression defines a Fourier transform relation between the functions $F(\omega) = H(c + i2\pi\omega)$ and

$$f(t) = \begin{cases} h(t)e^{-ct} & \text{if} \quad t > 0, \\ 0 & \text{if} \quad t < 0. \end{cases}$$

To compute the inverse Laplace transform of a given function H, one could, in principle, proceed by

- choosing a value of c and sampling F at selected values of ω,

- applying the inverse DFT to the samples of F to approximate f at selected grid points, and

- computing the values of h at the sample points from $h(t) = f(t)e^{ct}$.

However, there are some subtleties, one of which is the choice of the parameter c that has suddenly appeared. Therefore, a few more remarks are still in order.

Since the goal is to construct the function h on an interval $[0, A]$ in the time domain, the first choice is the value of A. Once a value of A and a number of grid points N are selected, then the reciprocity relation $A\Omega = N$ determines the extent of the frequency interval Ω. We will assume for the moment that an appropriate value of the parameter c can be chosen. Then the function $F(\omega) = H(c+i2\pi\omega)$ is sampled at N equally spaced points of the interval $[-\Omega/2, \Omega/2]$ to produce the samples $F_k = F(\omega_k)$. Notice that average values of F must be used at endpoints and discontinuities (AVED). In particular, this means that

$$F_{\frac{N}{2}} = \frac{1}{2}\left(F(-\omega_{\frac{N}{2}}) + F(\omega_{\frac{N}{2}})\right) = \text{Re}\left\{F(\omega_{\frac{N}{2}})\right\},$$

where we have used the fact that if h is a real-valued function (which is often the case), then F is a conjugate symmetric function with $F(-\omega) = F^*(\omega)$.

With N samples of F properly generated, the IDFT can be used to produce approximations f_n to $f(t_n)$. The grid points are $t_n = nA/N = n/\Omega$, where $n = 0 : N - 1$. Most likely some rearrangement of the DFT output (or input) will be needed to reconcile the two index sets $k = -N/2 + 1 : N/2$ and $n = 0 : N - 1$. As we have seen, the periodicity of the input and output sequences allows either sequence to be shifted. The final step is to calculate the approximations to $h(t_n)$ from the relationship $h(t_n) = f(t_n)e^{ct_n}$. We will denote these approximations $h_n = f_n e^{ct_n}$.

We can now comment qualitatively on the role of the parameter c. In theory, c can be taken as any real number greater than the real part of the largest pole of H. But in practice, there are some numerical considerations. A large value of $c > 0$ has the beneficial effect of making the output sequence $f_n = h(t_n)e^{-ct_n}$ decay rapidly, thus reducing the overlapping in the replication of f that inevitably takes place in the time domain. As we observed earlier, this error can also be reduced by decreasing $\Delta\omega$, which increases the period of the replication of f. On the other hand, if c is large, errors in the computed sequence f_n will be magnified when multiplied by e^{ct_n} to produce the final sequence h_n. A nice analysis of this optimization problem with respect to the parameter c can be found in the Cooley, Lewis, and Welch paper [41]. Rather than try to reproduce it here, we will resort to a numerical demonstration to show the effect of different choices of c.

Example: Numerical inversion of the Laplace transform. A family of convenient test problems for the Laplace transform inversion is given by $h(t) = t^{n-1}e^{at}/(n-1)!$, which has the Laplace transform $H(s) = (s-a)^{-n}$, where a is any real number and $n \geq 1$ is an integer (we agree that $0! = 1$). We will exercise the method outlined above on this problem with $n = 2$ and $a = -2$. We have computed approximations to the inverse Laplace transform h on two different intervals $[0,2]$ and $[0,4]$ using several different values of c and N. In each case an error was determined using the exact values of h and the error measure

$$\text{Error} = \max_n |h(t_n) - h_n|.$$

The graphs of Figure 8.1 offer a concise summary of several pages of numerical output from this experiment. The four curves in this figure show how the errors in the approximations vary with respect to the parameter c in the four cases $A = 2, 4$ and $N = 128, 256$. Several observations should be made. First note the sensitive dependence of the errors on the choice of c *for fixed values* of A and N. In each case there is a narrow interval of optimal values of c, and straying from this interval degrades the approximations significantly. Furthermore, the optimal value of c varies considerably with the particular choice of A and N. This relationship $c = c(A, N)$ is of great practical interest, however it appears that in general it must be determined experimentally. It should be said that once an optimal value of c is determined, good approximations to the inverse Laplace transform can be found with a predictable decrease in errors as N increases. In this particular case, with a discontinuity in the function F, errors decrease approximately as N^{-1}, assuming that optimal values of c are chosen for each N.

In closing we mention another now-classical method for the numerical inversion of the Laplace transform that is also described in Cooley, Lewis, and Welch [41], [111], [161], [163]. It assumes that the function $h(t)$ can be expanded in a series of Laguerre[2] polynomials (suggested by the exponential kernel e^{-st} of the Laplace transform). Perhaps surprisingly, the method ultimately leads to a Fourier series whose coefficients must be approximated. Hence the DFT makes another appearance in this method as well. There are also more recent methods for the inversion of the Laplace transform. The fractional Fourier transform can be applied to this computation [8] and wavelet (multipole) methods have also been proposed [10].

8.3.   The z-Transform

At this point, it would be tempting to move ahead into new waters; but we are at an exquisitely critical juncture in which several pieces of a large picture are about to come together. Throughout this book we have explored the relationship between

---

[2]EDMOND NICOLA LAGUERRE (1834–1886) published over 140 papers in his lifetime, over half of which were in geometry. After serving in the army for ten years, he was a tutor at the Ecole Polytechnique and a professor in the Collège de France. He is best known for the family of orthogonal polynomials named in his honor.

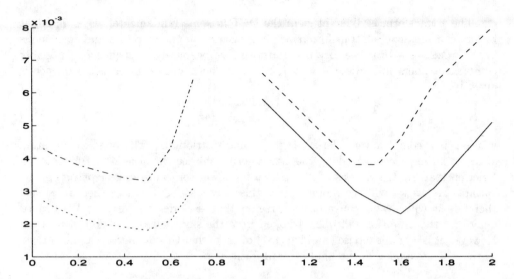

FIG. 8.1. *The four curves show how the error (on the vertical axis) in approximations to the inverse Laplace transform vary with the parameter c (on the horizontal axis) in four cases: $A = 2, N = 128$ (dashed line), $A = 2, N = 256$ (solid line), $A = 4, N = 128$ (dash-dot), and $A = 4, N = 256$ (dotted line). The test problem is the approximation of $h(t) = te^{-2t}$ from its Laplace transform $H(s) = (s + 2)^{-2}$. Note how the errors depend quite sensitively on the choice of c for fixed values of A and N.*

the DFT and the Fourier transform in some detail. We have just introduced the Laplace transform, illustrated its relationship to the Fourier transform, and found the (perhaps surprising) manner in which the DFT may be used to approximate the Laplace transform. If we were to stand back and look at the DFT, the Fourier transform, and the Laplace transform from a distance, we might see an arrangement something like the following diagram:

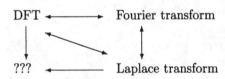

The three double-headed arrows indicate pathways that we have already traveled, and it might be tempting to conclude that the picture is complete. However, the connection between the DFT and the Laplace transform that was presented in the previous section was made via the Fourier transform. One might wonder whether the Laplace transform has its own discrete transform, and, if it does, whether it can be related to the DFT? The answer to both questions is affirmative, and the missing piece (denoted ???) of the above diagram is called the z-**transform**. Besides the aesthetics of completing the above picture, there are many reasons to study the z-transform. One reason is its extreme utility in signal processing and systems analysis applications [108]. Equally important is the fact that it provides yet another link to the DFT.

The z-transform is different from the DFT in some fundamental ways. It might be called a *semidiscrete* transform since it is discrete in one direction and continuous in the other. To define the forward transform, we begin with a sequence of possibly complex numbers u_n, where $n = 0, 1, 2, 3, \ldots$. The z-transform of this sequence is given by

$$\mathcal{Z}\{u_n\} = U(z) = \sum_{n=0}^{\infty} u_n z^{-n}, \tag{8.2}$$

where z is a complex variable, subject to some restrictions. The operator notation $\mathcal{Z}$ denotes the process of taking the z-transform. We see that the z-transform is the function that has the u_n's as coefficients in its power series of *negative* powers of the variable z. As we will see momentarily, this power series is not defined at $z = 0$; therefore it will always converge in a region that excludes the origin. We cannot deny that the definition (8.2) has dropped from the sky with very little justification. However, it is a brief exercise (problem 181) to show that this definition can be derived as the discrete analog of the Laplace transform

$$\tilde{U}(s) = \int_0^{\infty} u(t)e^{-st}dt$$

after the change of variable $z = e^s$ is made. For the moment, let's take this definition as given and use it to compute some z-transforms.

Example: z-transform of a step sequence. Consider the sequence

$$u_n = \begin{cases} 0 & \text{if } n < 0, \\ 1 & \text{if } n \geq 0. \end{cases}$$

Applying the definition of the z-transform we have that

$$\mathcal{Z}\{u_n\} = U(z) = \sum_{n=0}^{\infty} 1 \cdot z^{-n} = \sum_{n=0}^{\infty} \left(\frac{1}{z}\right)^n. \tag{8.3}$$

At this point we appeal to a hopefully familiar result that gets plenty of use in the z-transform business: the geometric series. The general form of the geometric series that we will need repeatedly is

$$\sum_{n=0}^{\infty} (g(z))^n = \frac{1}{1 - g(z)} \quad \text{for all } z \text{ for which } |g(z)| < 1.$$

Using this result in (8.3) gives us our first z-transform:

$$\mathcal{Z}\{u_n\} = U(z) = \frac{1}{1 - 1/z} = \frac{z}{z - 1}.$$

This transform is valid provided that the convergence condition of the geometric series is met. Therefore, we must require $|1/z| < 1$ or $|z| > 1$, which means that this transform is valid outside of the unit circle centered at the origin in the complex plane.

Example: Geometric sequence. In this case consider the sequence $u_n = a^n$, where a is any complex constant and $n = 0, 1, 2, 3, \ldots$. Applying the definition of the z-transform, we have that

$$\mathcal{Z}\{u_n\} = U(z) = \sum_{n=0}^{\infty} a^n z^{-n} = \sum_{n=0}^{\infty} \left(\frac{a}{z}\right)^n.$$

Appealing to the geometric series again, we have that

$$\mathcal{Z}\{u_n\} = U(z) = \sum_{n=0}^{\infty} a^n z^{-n} = \frac{1}{1 - a/z} = \frac{z}{z - a}$$

for $|z| > |a|$. Notice that as in the previous example, this z-transform is valid only in a certain region of the complex plane: in this case, for all points *outside* of a circle of radius $|a|$ centered at the origin.

Example: Sine and cosine sequences. The previous z-transform can be applied in the special case $a = e^{i\theta}$, where θ is a fixed real angle. We can then deduce that

$$
\begin{aligned}
\mathcal{Z}\{\cos(n\theta)\} &= \mathcal{Z}\left\{\frac{1}{2}\left(e^{in\theta} + e^{-in\theta}\right)\right\} \\
&= \frac{1}{2}\left(\frac{z}{z - e^{i\theta}} + \frac{z}{z - e^{-i\theta}}\right) \\
&= \frac{z - \cos\theta}{z - 2\cos\theta + z^{-1}}.
\end{aligned}
$$

This z-transform is valid provided $|z| > 1$, since $a = e^{i\theta}$ and $|a| = 1$.

Similarly (problem 175),

$$\mathcal{Z}\{\sin(n\theta)\} = \frac{\sin\theta}{z - 2\cos\theta + z^{-1}},$$

provided $|z| > 1$.

So far we have not mentioned anything about an inverse z-transform, which is the process of recovering the sequence u_n from a given function $U(z)$. We will see shortly that it exists and what it looks like. However, there are several indirect approaches to finding the inverse z-transform. We will introduce the operator notation $\mathcal{Z}^{-1}$ to indicate the inverse z-transform. Therefore, if we are given a function $U(z)$, then its inverse z-transform is the sequence

$$u_n = \mathcal{Z}^{-1}\{U(z)\}_n$$

for $n = 0, 1, 2, 3, \ldots$. Here is an example to demonstrate one approach.

Example: An inverse z-transform. Not surprisingly, the geometric series can also be used to discover inverse z-transforms. Let's consider the function

$$U(z) = \frac{1}{z - a}$$

where a is a complex constant. The idea is to rewrite this function so that it can be identified as the sum of a geometric series. Notice that the z-transform involves *negative* powers of the variable z, that is, $(1/z)^n$. Looking for powers of $1/z$, a few steps of algebra lead us to

$$
\begin{aligned}
U(z) &= \frac{1}{z - a} \\
&= \frac{1}{z} \cdot \frac{1}{1 - a/z}
\end{aligned}
$$

$$= \frac{1}{z} \sum_{n=0}^{\infty} \left(\frac{a}{z}\right)^n \quad \text{(geometric series!)}$$

$$= \sum_{n=1}^{\infty} \frac{a^{n-1}}{z^n}.$$

The geometric series we have written converges, provided $|z| > |a|$. It is now possible to pick out the coefficients of z^{-n} and identify them as the terms of the sequence u_n. Doing this we see that

$$u_n = \begin{cases} 0 & \text{if } n = 0, \\ a^{n-1} & \text{if } n > 0. \end{cases}$$

The strategy of using algebra to form a geometric series can be used endlessly to find inverse z-transforms. Here is a variation on the same theme.

Example: Inversion by long division. Very often, the power series in z^{-1} needed to determine an inverse z-transform can be found simply by long division. For example, given the z-transform

$$U(z) = \frac{z^2 + 2z}{z^2 - 2z + 1},$$

it can be expanded in powers of z^{-1} by long division to give

$$U(z) = 1 + \frac{4}{z} + \frac{7}{z^2} + \frac{10}{z^3} + \cdots + \frac{3n+1}{z^n} + \cdots.$$

The coefficients in this power series give the inverse z-transform $\mathcal{Z}^{-1}\{U(z)\}_n = u_n = 3n + 1$.

Example: Inversion by partial fractions. Another powerful tool for finding inverse z-transforms (just as it is for inverting Laplace transforms) is partial fraction decomposition. For example, the function

$$U(z) = \frac{z + 6}{z^2 - 4}$$

can be written in partial fractions as

$$U(z) = \frac{2}{z - 2} - \frac{1}{z + 2}.$$

Now the result of a previous example can be used to conclude that

$$\mathcal{Z}^{-1}\{U(z)\}_n = u_n = \begin{cases} 0 & \text{if } n = 0, \\ 2^n - (-2)^{n-1} & \text{if } n > 0. \end{cases}$$

Before turning to some important properties of the z-transform, we will provide a short table of z-transforms, some of which have already been derived as examples, others of which can be found in the problem section. Table 8.1 shows the input sequence u_n, the corresponding z-transform, and the region of the complex plane in which the transform is valid.

The z-transform, like all the transforms we have seen, has a variety of useful properties. Only two properties will concern us, and we enumerate those now.

TABLE 8.1
A short table of z-transforms.

u_n	$U(z)$	Valid for
1	$\frac{z}{z-1}$	$\|z\| > 1$
$\delta(n)$	1	all z
a^n	$\frac{z}{z-a}$	$\|z\| > \|a\|$
$\cos n\theta$	$\frac{z-\cos\theta}{z-2\cos\theta+z^{-1}}$	$\|z\| > 1$
$\sin n\theta$	$\frac{\sin\theta}{z-2\cos\theta+z^{-1}}$	$\|z\| > 1$
$\begin{cases} 0 & \text{if } n=0, \\ a^{n-1} & \text{if } n>0. \end{cases}$	$\frac{1}{z-a}$	$\|z\| > \|a\|$
$\begin{cases} 0 & \text{if } n=0 \text{ or } n \text{ odd}, \\ (-1)^{\frac{n}{2}+1}a^{n-2} & \text{if } n \text{ even}. \end{cases}$	$\frac{1}{z^2+a^2}$	$\|z\| > \|a\|$
$\begin{cases} 0 & \text{if } n \text{ even}, \\ (-1)^{\frac{n-1}{2}}a^{n-1} & \text{if } n \text{ odd}. \end{cases}$	$\frac{z}{z^2+a^2}$	$\|z\| > \|a\|$

1. **Linearity.** The z-transform is a linear operator, which means that if a and b are constants and u_n and v_n are sequences, then

$$\mathcal{Z}\{au_n + bv_n\} = a\mathcal{Z}\{u_n\} + b\mathcal{Z}\{v_n\}.$$

In words, the z-transform of a sum of sequences is the sum of the z-transforms, and the z-transform of a constant times a sequence is that constant times the z-transform (problem 173).

2. **Shift property.** As with the DFT, the shift property of the z-transform is extremely important, particularly for the solution of difference equations. It is not difficult to derive this property, so let's do it. Let u_{n+1} denote the sequence

$$\{u_1, u_2, u_3, \ldots\},$$

which is produced by shifting the sequence u_n to the left one place. Then

$$
\begin{aligned}
\mathcal{Z}\{u_{n+1}\} &= \sum_{n=0}^{\infty} u_{n+1} z^{-n} \\
&= z \sum_{n=0}^{\infty} u_{n+1} z^{-n-1} \\
&= z \sum_{n=1}^{\infty} u_n z^{-n} \\
&= z \left(-u_0 + \sum_{n=0}^{\infty} u_n z^{-n} \right) \\
&= -zu_0 + zU(z).
\end{aligned}
$$

We have used $U(z)$ to denote the z-transform of the original sequence u_n. We see that the the z-transform of the shifted sequence is related directly to the z-transform of the original sequence *and* to the initial term of the original sequence

u_0. This property should be compared to the property for the Laplace transform of the derivative of a function; the resemblance is not an accident!

A similar calculation can be used (problem 174) to show that the z-transform of the k-fold shifted sequence u_{n+k} is given by

$$\mathcal{Z}\{u_{n+k}\} = z^k U(z) - z^k u_0 - z^{k-1} u_1 - \cdots - z u_{k-1},$$

where k is any positive integer.

Solution of Initial Value Problems

In the first section of Chapter 7 the fundamental distinctions between boundary value problems and initial value problems were described, and then the DFT was used to solve boundary value problems. With the shift property in hand, we can now investigate how the z-transform is used to solve initial value problems. We will proceed by example and consider the second-order difference equation initial value problem

$$u_{n+2} - u_{n-1} + 6u_n = \sin\left(\frac{\pi n}{2}\right)$$

for $n \geq 2$, subject to the initial conditions $u_0 = 0$ and $u_1 = 3$. The goal is to find the sequence u_n that satisfies both of the initial conditions (for $n = 0$ and $n = 1$) *and* the difference equation (for $n \geq 2$). The problem is called an initial value problem because the *initial* two terms of the unknown sequence are given, and the difference equation governs the evolution of the system for all other values of n.

The solution begins by taking the z-transform of both sides of the difference equation:

$$\mathcal{Z}\{u_{n+2}\} - \mathcal{Z}\{u_{n+1}\} + 6\mathcal{Z}\{u_n\} = \mathcal{Z}\left\{\sin\left(\frac{\pi n}{2}\right)\right\}.$$

The fact that the z-transform is a linear operator is essential in this step. Now a liberal use of the shift property leads to

$$(z^2 - z - 6)U(z) - z^2 u_0 - z u_1 = \mathcal{Z}\left\{\sin\left(\frac{\pi n}{2}\right)\right\}.$$

The two given initial conditions fit the needs of the shift property perfectly. Letting $u_0 = 0, u_1 = 3$ and using the z-transform of the sine sequence derived earlier, we can write

$$(z^2 - z - 6)U(z) - 3z = \frac{1}{z + z^{-1}}.$$

As with all transform techniques, the most immediate task is to solve for the *transform* of the unknown sequence; in this case, we must solve for $U(z)$. Notice that this is an algebraic problem, and its solution is

$$U(z) = \frac{3z^3 + 4z}{(z^2 + 1)(z^2 - z - 6)}.$$

Having found the z-transform of the solution, the actual solution is only an inverse z-transform away. This looks like a grim task; however, the tricks learned in the previous examples will serve us well. First, a partial fraction decomposition of $U(z)$ leads us to

$$U(z) = \frac{31}{50} \cdot \frac{1}{z - 3} - \frac{16}{25} \cdot \frac{1}{z + 2} - \frac{1}{50} \cdot \frac{7z + 1}{z^2 + 1}.$$

The first two terms in the partial fraction representation have inverses that we have already encountered. The third term is also a familiar z-transform whose inverse can be found in problem 176 and Table 8.1. Combining the three inverse z-transforms gives us the solution

$$u_n = \frac{31}{50}3^n - \frac{16}{25}(-2)^n - \frac{1}{50}\begin{cases} (-1)^{\frac{n}{2}-1} & \text{if } n \text{ is even,} \\ 7(-1)^{\frac{n-1}{2}} & \text{if } n \text{ is odd.} \end{cases}$$

While this expression is rather cumbersome, it can be verified that it produces the correct initial conditions ($u_0 = 0$ and $u_1 = 3$). Furthermore, it generates the sequence $\{0, 3, 3, 20, 48, \ldots\}$, which is precisely the sequence that results if the original difference equation is evaluated recursively. The benefit of the z-transform solution is that it provides a single analytical expression that can be used to find u_n for *any* values of $n \geq 0$. The method of solution used for this initial value problem is perfectly general: it can be applied to any constant coefficient difference equation with initial conditions, of any order, homogeneous or nonhomogeneous, and it provides a solution at least up to the final inversion step. If the inverse z-transform step cannot be done analytically, then numerical methods must be used. This brings us to the point of this discussion: the relationship between the z-transform and the DFT.

The z-Transform and the DFT

We have spent several pages introducing the z-transform, its properties, and its use for the solution of initial value problems. All of this work has been analytical. It is now time to investigate numerical methods that must be used when the z-transform or its inverse cannot be determined exactly. This leads directly to the DFT.

Let's begin with an observation: for all of the examples of z-transforms done earlier, and for all of the z-transforms that appear in Table 8.1, there is a condition of the form $|z| > R_0$ that gives the region of validity of the z-transform in the complex plane. In other words, each z-transform is defined *outside* of a circle of radius R_0 in the complex plane. Therefore, we will take the definition of the z-transform

$$U(z) = \sum_{n=0}^{\infty} u_n z^{-n} \quad \text{for} \quad |z| > R_0,$$

and evaluate it on a circle $\mathcal{C}$ of radius $R > R_0$. Recall that a circle of radius R in the complex plane can be parameterized as

$$z = R(\cos\theta + i\sin\theta) = Re^{i\theta}$$

for $0 \leq \theta < 2\pi$, where z is a point on the circle, and the angle θ is the parameter. This gives us the representation

$$U(z) = \sum_{n=0}^{\infty} u_n (Re^{i\theta})^{-n}.$$

We must now think about approximations to $U(z)$ that might be implemented on a computer. Said slightly differently, we must now imagine how this continuous z-transform can be made discrete. To begin with, we will be able to compute $U(z)$

only at a finite number of points. So, we will choose N equally spaced sample points on the circle $\mathcal{C}$ and denote them

$$z_k = Re^{i\theta_k} \quad \text{where} \quad \theta_k = \frac{2\pi k}{N} \quad \text{and} \quad k = 0 : N - 1.$$

The other aspect of the z-transform that is not discrete is the infinite sum. A discrete form of the z-transform must involve a finite sum. Since we have chosen N sample points z_k, it stands to reason that we should take N terms of the z-transform sum. If we incorporate both of these discretizations (samples of z and the finite sum) into the definition of the z-transform, we have

$$
\begin{aligned}
U(z_k) &\approx \sum_{n=0}^{N-1} u_n z_k^{-n} \\
&= \sum_{n=0}^{N-1} u_n \left(Re^{i2\pi k/N} \right)^{-n} \\
&= \underbrace{\sum_{n=0}^{N-1} (u_n R^{-n}) e^{-i2\pi nk/N}}_{\text{a DFT!}}.
\end{aligned}
$$

Hopefully the conclusion of this little argument is clear. If we use our usual notation and let $U_k = U(z_k)$, we can write that

$$U_k \approx N\mathcal{D}_N \left\{ R^{-n} u_n \right\}_k \tag{8.4}$$

for $k = 0 : N - 1$. In words, the z-transform of the (infinite) sequence u_n can be approximated by applying the DFT to the auxiliary sequence $R^{-n} u_n$, where $n = 0 : N - 1$. The only condition on R is that it must exceed the radius of the critical circle R_0. Therefore, R must be regarded as a parameter in this method precisely in the manner that the approximate inverse of the Laplace transform involved the parameter c.

The relationship between the z-transform and the DFT turns out to be remarkably straightforward. How do the respective inverse transforms come together? There are at least two ways to display the connection. The first is to formally invert the relationship (8.4). Taking inverse DFTs of both sides gives

$$\mathcal{D}^{-1} \left\{ U_k \right\}_n \approx N R^{-n} u_n.$$

Therefore,

$$u_n \approx \frac{R^n}{N} \mathcal{D}^{-1} \left\{ U_k \right\}_n.$$

This argument displays the relationship between the inverse z-transform and the DFT, but it does not quite give the entire account for a simple reason: we have yet to see the exact inverse z-transform! In all of the preceding examples, we conspired to find inverse z-transforms by devious means such as long division, geometric series, or partial fractions. Not once did we use an inversion "formula." Therefore, we will sketch the final scene in broad strokes since it requires an excursion into complex variables.

The z-transform definition

$$U(z) = \sum_{n=0}^{\infty} u_n z^{-n} \quad \text{for} \quad |z| > R_0$$

really says that the sequence u_n consists of the coefficients of the Laurent[3] series for $U(z)$ in the region $|z| > R_0$. These coefficients are readily found by an application of the theory of residues, and the result is that

$$u_n = \frac{1}{2\pi i} \oint_C U(z) z^{n-1} dz,$$

where C can be taken as any circle with radius $R > R_0$. With this exact inversion formula, we can find a discrete version of the inverse z-transform. As outlined in problem 180, the contour integral can be approximated by summing the integrand at the same N points on the circle C,

$$z_k = R e^{i 2\pi k / N},$$

that we used earlier. If this discretization is carried out carefully, then indeed we discover that

$$u_n \approx \frac{R^n}{N} \mathcal{D}^{-1} \{U_k\}_n$$

for $n = 0 : N - 1$. The relationship between the inverse z-transform and the inverse DFT transform can then be expressed as

$$\mathcal{Z}^{-1}\{U_k\}_n \approx \frac{R^n}{N} \mathcal{D}^{-1} \{U_k\}_n$$

for $n = 0 : N - 1$. As in the inversion of the Laplace transform, the parameter R appears and must be determined experimentally.

8.4. The Chebyshev Transform

Did you ever look at the multiple angle formulas for the cosine function and wonder about the patterns in the coefficients? If not, let's do it now. Here are the first few formulas:

$$
\begin{aligned}
\cos\theta &= \cos\theta, \\
\cos 2\theta &= 2\cos^2\theta - 1, \\
\cos 3\theta &= 4\cos^3\theta - 3\cos\theta, \\
\cos 4\theta &= 8\cos^4\theta - 8\cos^2\theta + 1, \\
\cos 5\theta &= 16\cos^5\theta - 20\cos^3\theta + 5\cos\theta, \\
\vdots &= \vdots
\end{aligned}
$$

The list can be continued indefinitely, but the first few entries lead to some immediate observations. Notice that $\cos n\theta$ can be expressed as a polynomial of degree n in

---

[3]French analyst PIERRE ALPHONSE LAURENT (1813–1854) is best known for his generalization of Taylor series in the complex plane.

$\cos \theta$. If n is even, that polynomial consists entirely of even powers of $\cos \theta$; if n is odd, the polynomial consists of odd powers only. In both cases, the coefficients of the polynomial alternate signs, and the leading coefficient in the polynomial for $\cos n\theta$ is 2^{n-1}. These properties and many others hold for all positive integers n, and they have fascinated mathematicians for centuries. The most eminent person to study these polynomials was the nineteenth century Russian mathematician P. L. Chebyshev[4] (variations on this name include Tchebycheff), whose name has been given to them.

Let's rewrite these polynomials with the notational simplification that $x = \cos \theta$. We will also denote the nth polynomial in the list $T_n(x)$ with the agreement that $T_0(x) = 1$. The list now reads

$$
\begin{aligned}
T_0(x) &= 1, \\
T_1(x) &= x, \\
T_2(x) &= 2x^2 - 1, \\
T_3(x) &= 4x^3 - 3x, \\
T_4(x) &= 8x^4 - 8x^2 + 1, \\
T_5(x) &= 16x^5 - 20x^3 + 5x, \\
\vdots &= \vdots
\end{aligned}
$$

We have listed the first six **Chebyshev polynomials**. They may be characterized very simply for any nonnegative integer n in the following way:

$$ T_n(x) = \cos n\theta \quad \text{where} \quad \theta = \cos^{-1} x, $$

or more simply as

$$ T_n(x) = \cos(n \cos^{-1} x) $$

for $n = 0, 1, 2, \ldots$. This set of polynomials has been studied extensively for the past 150 years, and its properties have been uncovered and recorded. The polynomials arise in a startling variety of seemingly disparate subjects, from approximation theory to algebra and number theory, and undoubtedly there are connections with other disciplines that have yet to be discovered. Our goal is to relate the Chebyshev polynomials to the DFT, but along the way we will explore their properties and learn a bit more about them.

Let's survey some of the more useful and frequently encountered properties of the Chebyshev polynomials. The simpler demonstrations will be left as exercises with hints; the deeper results (and there are many more of them) are accompanied by references (a good general reference is [12]).

1. **Polynomial properties.** As mentioned, T_n is an nth-degree polynomial; T_n is an even function when n is even and an odd function when n is odd. The graphs of the first few Chebyshev polynomials are shown in Figure 8.2.

---

[4]PAFNUTI LIWOWICH CHEBYSHEV (1821–1894) was associated with the University of Petrograd for much of his life. His name appears in many different branches of mathematics, perhaps most notably in number theory and probability. While everyone agrees upon his importance in mathematical history, almost nobody agrees on the spelling of his name, which appears in many different forms. Indeed, the controversy inspired Phillip J. Davis to write a book, *The Thread, a Mathematical Yarn*, in which the name Pafnuti Liwowich Chebyshev is the central, unifying theme.

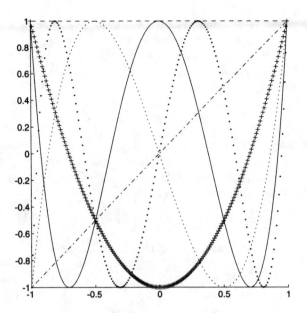

FIG. 8.2. *The figure shows the graphs of the Chebyshev polynomials $T_1, \ldots, T_5$ on the interval $[-1, 1]$. The polynomials can be identified by counting the zero crossings: T_n has n zero crossings. Note that the even- (odd-) order polynomials are even (odd) functions, that T_n has all n of its zeros on $(-1, 1)$, and that T_n has $n + 1$ extreme values on $[-1, 1]$.*

2. **Zeros.** The n zeros of T_n (points at which $T_n(x) = 0$) are real and lie in the interval $(-1, 1)$. From the definition, $T_n(x) = \cos(n \cos^{-1} x)$, they are easily shown to be (problem 184)

$$\xi_j = \cos\left(\frac{2j - 1}{n} \frac{\pi}{2}\right)$$

for $j = 1 : n$.

3. **Extreme values.** On the interval $[-1, 1]$, $|T_n(x)| \leq 1$ and T_n attains its extreme values of ± 1 at the $n + 1$ points (problem 185)

$$\eta_j = \cos\left(\frac{\pi j}{n}\right)$$

for $j = 0 : n$.

4. **Multiplicative property.** For $m > n \geq 0$

$$T_m T_n = \frac{1}{2}(T_{m+n} + T_{m-n})$$

(problem 186).

5. **Semigroup property.** For $m \geq 0$ and $n \geq 0$

$$T_m(T_n(x)) = T_{mn}(x)$$

(problem 187).

6. **Minimax property.** The normalized Chebyshev polynomials are formed by dividing through by the leading coefficient 2^{n-1} giving the **monic** polynomials $\tilde{T}_n(x) = 2^{1-n}T_n(x)$ with leading coefficient one. An extremely important property of the Chebyshev polynomials is that on the interval $[-1, 1]$, among all nth-degree polynomials p_n,

$$||p_n||_\infty \geq ||\tilde{T}_n||_\infty = 2^{1-n}$$

for $n > 0$, where $||f||_\infty = \max_{-1 \leq x \leq 1}|f(x)|$. In other words, among all nth-degree polynomials, $\tilde{T}_n$ has the smallest maximum absolute value on $[-1, 1]$. This "best minimax" property [118] is the basis of many approximation methods that involve Chebyshev polynomials.

7. **Recurrence relation.** The trigonometric identity

$$\cos n\theta \cos(n-2)\theta = 2\cos\theta\cos(n-1)\theta$$

can be used directly to show (problem 188) that the Chebyshev polynomials satisfy the recurrence relation

$$T_n(x) = 2xT_{n-1}(x) - T_{n-2}(x)$$

for $n = 2, 3, 4, \ldots$.

8. **Differential equation.** Computing $T_n'(x)$ and $T_n''(x)$ (problem 189) shows that T_n satisfies the second-order differential equation

$$(1-x^2)y''(x) - xy'(x) + n^2y(x) = 0$$

for $n = 0, 1, 2, \ldots$.

9. **Orthogonality.** Undoubtedly, much of the utility of Chebyshev polynomials arises from the fact that they comprise a set of **orthogonal polynomials**. Here is the crucial orthogonality property: For nonnegative integers m and k

$$\int_{-1}^{1} \frac{T_m(x)T_k(x)}{\sqrt{1-x^2}}\,dx = \begin{cases} 0 & \text{if } m \neq k, \\ \pi/2 & \text{if } m = k \neq 0, \\ \pi & \text{if } m = k = 0. \end{cases}$$

This property says that the polynomials T_n are orthogonal on the interval $[-1, 1]$ with respect to the weight function $(1-x^2)^{-1/2}$. The proof is direct and worth reviewing, since it reveals the important connection between the orthogonality of the Chebyshev polynomials and the orthogonality of the cosine functions that we have already studied. If we start with the orthogonality integral above and make the change of variables $x = \cos\theta$, we see that

$$\int_{-1}^{1} \frac{T_m(x)T_k(x)}{\sqrt{1-x^2}}\,dx = \int_0^\pi \cos m\theta \cos k\theta\,d\theta = \begin{cases} 0 & \text{if } m \neq k, \\ \pi/2 & \text{if } m = k \neq 0. \end{cases}$$

If $m = n = 0$ the value of the integral is π.

10. **Representation of polynomials.** The orthogonality of the Chebyshev polynomials allows them to be used in the representation of other functions. This

representation is exact and finite when the other functions are polynomials. The process is familiar and instructive and will be used again. Let p_N be a polynomial of degree N. We seek coefficients c_k such that

$$p_N(x) = \frac{c_0}{2} + c_1 T_1(x) + \cdots + c_N T_N(x) = \sum_{k=0}^{N} {}' c_k T_k(x), \qquad (8.5)$$

where we have used the notation Σ' to designate a sum in which the first term is weighted by one-half. In a method analogous to the determination of Fourier coefficients, we multiply both sides of the representation (8.5) by an arbitrary T_m, where $0 \le m \le N$, *and* by the weight function $(1 - x^2)^{-1/2}$. We then integrate over $[-1, 1]$, to discover that

$$\int_{-1}^{1} \frac{T_m(x) p_N(x)}{\sqrt{1 - x^2}}\, dx = \sum_{k=0}^{N} {}' c_k \underbrace{\int_{-1}^{1} \frac{T_m(x) T_k(x)}{\sqrt{1 - x^2}}\, dx}_{\frac{\pi}{2} \delta_{mk}}.$$

As indicated, the orthogonality of the T_n's means that only the $k = m$ term of the sum survives, leaving

$$c_k = \frac{2}{\pi} \int_{-1}^{1} \frac{T_k(x) p_N(x)}{\sqrt{1 - x^2}}\, dx$$

for $k = 0 : N$. Note that the factor of $1/2$ on the c_0 term accounts for the special case of $m = k = 0$ in the orthogonality property. (See problems 190 and 197.)

11. **Minimum least squares property.** The orthogonality can be used to show (problem 191) that of all nth-degree polynomials p_n, the normalized polynomial $\check{T}_n$ minimizes the quantity

$$\int_{-1}^{1} \frac{(p_n(x))^2}{\sqrt{1 - x^2}}\, dx.$$

12. **Least squares representations of functions.** We have seen that polynomials can be represented exactly by a finite linear combination of Chebyshev polynomials. What about the representation of arbitrary continuous functions f on $[-1, 1]$? This leads us to consider an expansion of the form

$$f(x) = \sum_{k=0}^{\infty} {}' c_k T_k(x) \qquad (8.6)$$

consisting of an infinite series of Chebyshev polynomials. We may proceed formally as in the polynomial case, using the orthogonality property, to find that the coefficients c_k in this expansion are given by

$$c_k = \frac{2}{\pi} \int_{-1}^{1} \frac{T_k(x) f(x)}{\sqrt{1 - x^2}}\, dx \qquad (8.7)$$

for $k = 0, 1, 2, \ldots$. Having determined the coefficients c_k, the critical questions are whether the series $\sum_k c_k T_k$ converges to f and in what sense. The answers are known and we can investigate further.

In practice one would compute only a partial sum of the expansion (8.6). Therefore, we will denote the Nth partial sum

$$s_N(x) = \sum_{k=0}^{N}{}' c_k T_k(x).$$

Note that s_N is a polynomial of degree N. It can be shown [118] that of all Nth-degree polynomials, s_N (with coefficients given by (8.7)) is the **least squares approximation** to f on $[-1, 1]$ with respect to the weight function $(1-x^2)^{-1/2}$. To state this carefully, we must define the weighted two-norm of a function f as

$$\|f\|_2 = \left\{ \int_{-1}^{1} \frac{(f(x))^2}{\sqrt{1-x^2}} dx \right\}^{1/2}.$$

Then s_N satisfies

$$\|f - s_N\|_2 \le \|f - p_N\|_2$$

for every Nth-degree polynomial p_N (with equality only if $p_N = s_N$).

The convergence theory for Chebyshev expansions is well developed [12], [69], [118]. Mean square convergence (convergence in the two-norm) can be assessed by first using the orthogonality to deduce that (problem 192)

$$\|f - s_N\|_2^2 = \sum_{k=N+1}^{\infty} c_k^2.$$

Thus, the rate of convergence in the 2-norm can be estimated knowing the rate of decay of the coefficients c_k which is determined by the smoothness of f on $[-1, 1]$ (in much the same way that the decay rate of Fourier coefficients was determined in Chapter 6).

There are also rather deep and recent results on the pointwise error in Chebyshev expansions. Rivlin [118] has shown that if $p_N^*(x)$ is the best uniform (or minimax) approximation to f on $[-1, 1]$ (minimizing $\|f - p_N^*\|_\infty$ over all Nth-degree polynomials) then

$$\|f - p_N^*\|_\infty \le \|f - s_N\|_\infty < 4 \left(1 + \frac{\ln N}{\pi^2} \right) \|f - p_N\|_\infty,$$

where s_N is the least squares approximation to f. A practical interpretation of this result is that "since $4(1 + \ln N/\pi^2) < 10$ for $N < 2,688,000$, the Chebyshev series is within a decimal place of the minimax approximation for all such polynomial approximations" [69]. This confirms the experience of many practitioners that Chebyshev expansions give very accurate and rapidly converging approximations to functions.

A few of the important properties of the Chebyshev polynomials have been displayed in this brief tour. We now move toward the practical matter of computing the coefficients c_k in Chebyshev expansions and showing how the DFT appears rather miraculously. We will proceed adroitly on two avenues: the first is continuous, the second is discrete.

Consider once again the representation

$$f(x) = \sum_{k=0}^{\infty}{'} c_k T_k(x)$$

for a continuous function f on the interval $[-1, 1]$. If we let $x = \cos\theta$, let $g(\theta) = f(\cos\theta)$ and recall that $T_k(x) = T_k(\cos\theta) = \cos k\theta$; then we may write

$$g(\theta) = f(\cos\theta) = \sum_{k=0}^{\infty}{'} c_k \cos k\theta,$$

where $0 \le \theta \le \pi$. We recognize this representation as the Fourier cosine series for $g(\theta)$ on the interval $[0, \pi]$. This suggests a computational strategy for approximating the coefficients $c_0, c_1, \dots, c_N$; it consists of the following two steps:

- Sample the given function f on the interval $[-1, 1]$ at the $N + 1$ points $x_n = \cos(n\pi/N)$, where $n = 0 : N$. This gives the samples $g_n = f(x_n) = f(\cos(n\pi/N))$ for $n = 0 : N$.

- Apply the discrete cosine transform (DCT) (preferably in the form of an FFT) to the sequence g_n to obtain the coefficients F_k which are approximations to the exact coefficients c_k. They are given explicitly by

$$F_k = \frac{1}{N}\left[g_0 + 2\sum_{n=1}^{N-1} g_n \cos\left(\frac{\pi n k}{N}\right) + g_N \cos(\pi k) \right]$$

for $k = 0 : N$.

This argument exhibits the connection between the Chebyshev coefficients and the DCT quite clearly.

Let's now reach the same destination along the discrete pathway. It turns out that like the complex exponential, the Chebyshev polynomials have both continuous and discrete orthogonality properties. However, there are some unexpected developments that arise with the discrete orthogonality. The first twist is that there are actually *two* different discrete orthogonality properties. Furthermore, the relevant discrete orthogonality properties use the extreme points of the polynomials. Recall that we used $\eta_n = \cos(\pi n/N)$ to denote the extreme points of T_N where $n = 0 : N$. With these two clues in mind it is not difficult to use the orthogonality of the DCT to show the following two discrete orthogonality properties (problem 193), where $0 \le j, k, n \le N$.

▶ **Discrete Orthogonality Property 1 (with respect to grid points)** ◀

$$\sum_{k=0}^{N}{''} T_k(\eta_j) T_k(\eta_n) = \begin{cases} 0 & \text{if } j \ne n, \\ N/2 & \text{if } j = n \ne 0, N, \\ N & \text{if } j = n = 0, N. \end{cases}$$

▶ **Discrete Orthogonality Property 2 (with respect to degree)** ◀

$$\sum_{n=0}^{N}{''} T_j(\eta_n) T_k(\eta_n) = \begin{cases} 0 & \text{if } j \ne k, \\ N/2 & \text{if } j = k \ne 0, N, \\ N & \text{if } j = k = 0, N. \end{cases}$$

The notation Σ'' indicates a sum whose first and last terms are weighted by one-half. The terminology is critical: in the first orthogonality property, the value of the sum depends on the indices of the grid points (j and n), whereas in the second property the value of the sum depends on the degree of the polynomials (j and k). We might add that the surprises with discrete orthogonality do not end here. There are actually two more orthogonality relations for the Chebyshev polynomials that use the zeros, rather than the extreme points, as grid points (problems 195 and 208).

We may now proceed in a way that mimics the computation of the coefficients in the continuous Chebyshev expansion. We look for a representation of a given function f at the points η_n that has the form

$$f(\eta_n) = \sum_{j=0}^{N}{}'' F_j T_j(\eta_n) \tag{8.8}$$

for $n = 0 : N$. Multiplying both sides of this representation by the arbitrary polynomial $T_k(\eta_n)$, where $k = 0 : N$, and summing over the points η_n for $n = 0 : N$ we find that

$$\sum_{n=0}^{N}{}'' T_k(\eta_n) f(\eta_n) = \sum_{j=0}^{N}{}'' F_j \underbrace{\sum_{n=0}^{N}{}'' T_j(\eta_n) T_k(\eta_n)}_{\substack{\frac{N}{2}\delta(j-k) \\ \text{if } 0 < j, k < N}}$$

for $k = 0 : N$.

Now the discrete orthogonality enters in a predictable way. Using the second discrete orthogonality property (with respect to degree), we see that if $k \neq 0$ or N, then the inner sum has a value of $N/2$ when $j = k$ and vanishes otherwise. This says that

$$F_k = \frac{2}{N} \sum_{n=0}^{N}{}'' T_k(\eta_n) f(\eta_n)$$

for $k = 1 : N - 1$. However, $T_k(\eta_n) = T_k(\cos(\pi n/N)) = \cos(\pi n k/N)$. Therefore, letting $g_n = f(\eta_n)$, the coefficients F_k are given by

$$F_k = \frac{2}{N} \sum_{n=0}^{N}{}'' g_n \cos\left(\frac{\pi n k}{N}\right)$$

for $k = 1 : N - 1$. A similar argument reveals that for $k = 0$ and $k = N$

$$F_0 = \frac{1}{N} \sum_{n=0}^{N}{}'' g_n \quad \text{and} \quad F_N = \frac{1}{N} \sum_{n=0}^{N}{}'' g_n \cos(\pi n).$$

However, since F_0 and F_N are weighted by $1/2$ in the representation (8.8), we can use a single definition for all of the F_k's; it is

$$F_k = \frac{2}{N} \sum_{n=0}^{N}{}'' g_n \cos\left(\frac{\pi n k}{N}\right)$$

for $k = 0 : N$. Once again, this expression should be recognized as the discrete cosine transform of the sequence $g_n = f(\eta_n)$. This leads us to the conclusion that was reached earlier on the continuous pathway:

The coefficients of the N-term Chebyshev expansion for a function f on the interval $[-1, 1]$ can be approximated by applying the N-point DCT to the samples

$$g_n = f(\cos(\pi n/N))$$

for $n = 0 : N$.

This argument gives us the forward discrete Chebyshev transform.

▶ **Forward Discrete Chebyshev Transform** ◀

$$F_k = \frac{2}{N} \sum_{n=0}^{N} {}'' f(\eta_n) T_k(\eta_n)$$

$$= \frac{2}{N} \sum_{n=0}^{N} {}'' g_n \cos\left(\frac{\pi n k}{N}\right)$$

for $k = 0 : N$.

We can now apply a similar argument and use the first discrete orthogonality property (with respect to the grid points) to establish the inverse transform (problem 194).

▶ **Inverse Discrete Chebyshev Transform** ◀

$$f(\eta_n) = \sum_{j=0}^{N} {}'' F_j T_j(\eta_n)$$

for $n = 0 : N$.

Despite all of the meandering that we have done through properties of the Chebyshev polynomials, this final result is the intended destination. It shows clearly the relationship among Chebyshev polynomials, the Fourier cosine series, and the DCT. It also confirms that there is a fast method for computing Chebyshev coefficients, namely the FFT implementation of the DCT. Let's solidify these ideas with a numerical example.

Example: Chebyshev coefficients and expansion. Consider the function $f(x) = |x|$ on the interval $[-1, 1]$. While this is not a profound choice of a test function, it does have the virtue that the coefficients in its Chebyshev expansion can be found analytically. Verify that (problem 198) this function has the expansion

$$|x| = \frac{2}{\pi} + \frac{4}{\pi} \sum_{k=1}^{\infty} \frac{(-1)^{k-1}}{4k^2 - 1} T_{2k}(x).$$

Note that $c_{2k+1} = 0$ because f is an even function.

If we sample f at the points $x_n = \cos(n\pi/N)$ and apply the DCT, approximations to the coefficients c_k can be computed. As a simple example, approximations to c_k, computed for $N = 8$ and $N = 16$, are shown in Table 8.2.

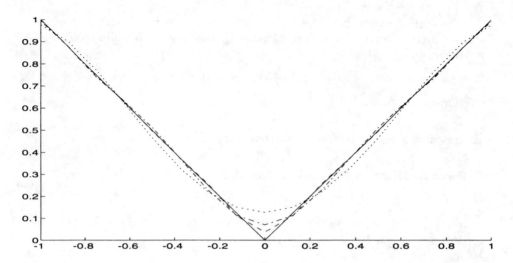

FIG. 8.3. *Given the exact coefficients in the Chebyshev expansion of $f(x) = |x|$, the function can be reconstructed from its partial sums. The figure shows the function f (solid curve) and the partial sums of the Chebyshev expansion for $N = 4$ (dotted curve), $N = 8$ (dashed and dotted curve), and $N = 16$ (dashed curve).*

TABLE 8.2

Approximations to Chebyshev coefficients for $f(x) = |x|$.

k	c_k (Exact)	F_k for $N = 8$	F_k for $N = 16$
0	1.27(0)	1.26(0)	1.27(0)
2	4.24(−1)	4.41(−1)	4.29(−1)
4	−8.49(−2)	−1.04(−1)	−8.91(−2)
6	3.64(−2)	5.87(−2)	4.08(−2)
8	−2.02(−2)	−4.97(−2)	−2.49(−2)
10	1.29(−2)	−	1.79(−2)
12	−8.90(−2)	−	−1.44(−2)
14	6.53(−3)	−	1.28(−2)
16	−4.99(−3)	−	−1.23(−2)

$c_{2k+1} = F_{2k+1} = 0$ identically for all N.

The numerical results show that the approximations converge to the exact values of c_k with increasing N (the continuity of f and its periodic extension might suggest that the errors should decrease as N^{-2}, and indeed they do). In keeping with approximations related to the DFT, we also see a typical degradation of the approximations for the higher frequency coefficients.

The reconstruction of f from its coefficients can also be observed. Figure 8.3 shows the partial sums of the Chebyshev expansion of $f(x) = |x|$ (using the exact values of c_k) for $N = 4, 8, 16$. Clearly, very few terms are needed to obtain a good representation of this function. Not surprisingly, the poorest fit occurs near the discontinuity in the derivative of f.

Theory and applications of Chebyshev polynomials extend far beyond the brief glimpse we have provided; see the list of references for additional excitement. We close on a note of intrigue with a picture (Figure 8.4) of the so-called **white curves** of the Chebyshev polynomials, the geometry of the which is elaborated in Rivlin [118].

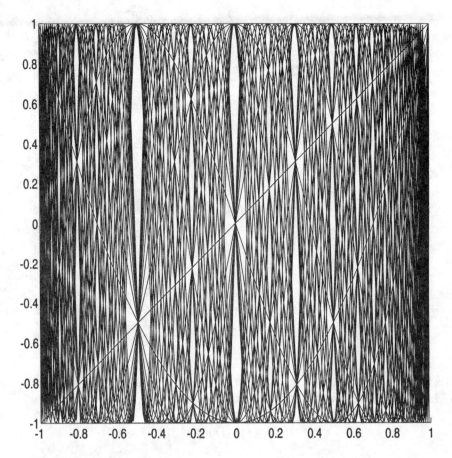

FIG. 8.4. *The first 30 Chebyshev polynomials, when graphed on the square $-1 \leq x \leq 1, -1 \leq y \leq 1$, reveal striking and unexpected patterns known as white curves.*

8.5. Orthogonal Polynomial Transforms

The subject of orthogonal polynomials, which carries many august names, is one of the most elegant and tantalizing areas of mathematics. Orthogonal polynomials arise in a variety of different applications, including approximation theory and the solution of differential equations. Not surprisingly, there are some important computational issues associated with their use. In this section, we will show how any system of orthogonal polynomials can generate a discrete transform pair. The discussion will use the Legendre[5] polynomials as a specific case to illustrate the procedure that can be used for any set of orthogonal polynomials. We will begin with a few essential words about orthogonal polynomials. The wealth of rich classical theory that will unfortunately, but necessarily, be omitted from this discussion can be found in many sources [32], [81], [106], [145].

We will let p_n denote the nth polynomial of a system of orthogonal polynomials

---

[5] ADRIEN MARIE LEGENDRE (1752–1833) is ranked with Laplace and Lagrange as one of the greatest analysts of eighteenth and nineteenth century Europe. He held university positions in Paris, as well as several minor governmental posts. Legendre made significant contributions to the theory of elliptic integrals and number theory. His study of spherical harmonics led to Legendre's differential equation.

where $n = 0, 1, 2, \ldots$. The degree of p_n is n, and we will always use a_n to denote the leading coefficient (coefficient of x^n) of p_n. Every system of polynomials is defined on a specific interval $[a, b]$ with respect to a positive weight function w. The orthogonality property satisfied by the set $\{p_n\}$ has the form

$$\int_a^b w(x)p_j(x)p_k(x)dx = d_k^2 \delta(j - k)$$

for $j, k = 0, 1, 2, \ldots$, where

$$d_k^2 = \int_a^b w(x)p_k^2(x)dx > 0.$$

For example, the set of Chebyshev polynomials just studied satisfies the orthogonality property

$$\int_{-1}^1 \frac{T_j(x)T_k(x)}{\sqrt{1 - x^2}} dx = \begin{cases} 0 & \text{if } j \neq k, \\ \pi/2 & \text{if } j = k \neq 0, \\ \pi & \text{if } j = k = 0, \end{cases}$$

with $[a, b] = [-1, 1]$, $w(x) = (1 - x^2)^{-1/2}$, $d_k^2 = \pi/2$ for $k \geq 1$, and $d_0^2 = \pi$. The Legendre polynomials, which we will investigate shortly, satisfy the orthogonality property

$$\int_{-1}^1 P_j(x)P_k(x)dx = \frac{2}{2k + 1}\delta(j - k)$$

with $[a, b] = [-1, 1]$, $w(x) = 1$, and $d_k^2 = 2/(2k + 1)$. As a slightly different example, the Laguerre polynomials have the orthogonality property

$$\int_0^\infty e^{-x} L_j(x)L_k(x)dx = \delta(j - k),$$

where $[a, b) = [0, \infty)$, $w(x) = e^{-x}$, and $d_k^2 = 1$. The literally endless properties (for example, recurrence relations, associated differential equations, generating functions, properties of zeros) of these and many other systems of orthogonal polynomials are tabulated in a variety of handbooks [3].

Of particular relevance to this discussion is the problem of representing fairly arbitrary functions as expansions of orthogonal polynomials. Given a function f defined on an interval $[a, b]$, we ask if coefficients c_k can be found such that

$$f(x) = \sum_{k=0}^\infty c_k p_k(x), \tag{8.9}$$

where $\{p_k\}$ is a system of orthogonal polynomials defined on the interval $[a, b]$. We can proceed formally, just as we have done many times already. Multiplying both sides of this representation by $w(x)p_j(x)$, integrating over $[a, b]$, and appealing to the orthogonality property leads immediately to the coefficients in the form (problem 199)

$$c_k = \frac{1}{d_k^2} \int_a^b w(x)f(x)p_k(x)dx \tag{8.10}$$

for $k = 0, 1, 2, \ldots$. The question of convergence of expansions of this form is of course quite important. The theory is well developed and, not surprisingly, parallels the convergence theory for Fourier series [81], [145].

We can already anticipate how discrete transforms based on orthogonal polynomials will be formulated. The process of computing the coefficients c_k of a given function f can be regarded as a *continuous* forward transform that might be denoted $c_k = \mathcal{P}\{f(x)\}$. The process of reconstructing a function from a set of coefficients can be viewed as an inverse transform that we might denote $f(x) = \mathcal{P}^{-1}\{c_k\}$. For example, the continuous Legendre transform pair for a function f has the form

$$c_k = \mathcal{P}\{f(x)\} = \frac{2k+1}{2} \int_{-1}^{1} f(x) P_k(x) dx \quad \text{(forward transform)}$$

and

$$f(x) = \mathcal{P}^{-1}\{c_k\} = \sum_{k=0}^{\infty} c_k P_k(x) \quad \text{(inverse transform)}.$$

The task is to find discrete versions of this continuous transform pair. However, some properties of orthogonal polynomials will be needed first. A fact of great importance in this business is that systems of orthogonal polynomials have *two* discrete orthogonality properties. We will need to develop both properties *and* avoid confusing them.

In order to proceed, a few properties of the zeros of orthogonal polynomials need to be stated. Any system of orthogonal polynomials $\{p_n\}$ on the interval $[a, b]$ has the property that the nth polynomial has precisely n zeros on (a, b). We will denote the n zeros of p_n by ξ_m where $m = 1, \ldots, n$. An interesting property (not of immediate use for our present purposes) is that the zeros of any two consecutive polynomials in a set of orthogonal polynomials are interlaced. This can be seen in the graphs of several Legendre polynomials shown in Figure 8.5. The zeros of the orthogonal polynomials will play a fundamental role in all that follows.

We may now begin the quest for discrete orthogonal polynomial transforms. Having seen how the DFT and the discrete Chebyshev transform were developed, we can now make a reasonable conjecture that will also prove to be correct. To make a discrete forward transform from a continuous forward transform (8.10), we might anticipate that the integral defining the coefficients c_k will need to be approximated by a sum:

$$c_k = \frac{1}{d_k^2} \int_a^b w(x) f(x) p_k(x) dx \quad \text{becomes} \quad c_k \approx \frac{1}{d_k^2} \sum_{n=1}^{N} \alpha_n f(x_n) p_k(\xi_n),$$

where the α_n's and x_n's are weights and nodes in a chosen quadrature rule. To make a discrete inverse transform from (8.10), we might anticipate that the series representation for f will need to be truncated and evaluated at certain grid points:

$$f(x) = \sum_{k=0}^{\infty} c_k p_k(x) \quad \text{becomes} \quad f(x_n) \approx \sum_{k=0}^{N-1} c_k p_k(x_n).$$

Let's first consider the forward transform and the approximation of the integral for the expansion coefficients c_k. One might appeal to any number of quadrature (integration) rules to approximate this integral. But if there are orthogonal polynomials in the picture (as there are here), then there is one technique that begs to be used, and that is **Gaussian quadrature**. A quick synopsis of Gaussian quadrature goes

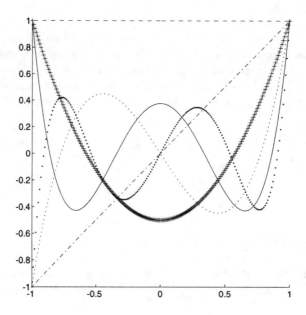

FIG. 8.5. *The first six Legendre polynomials $P_0, \ldots, P_5$ are shown in the figure on their natural interval $[-1, 1]$. Note that P_n has n zeros and that the zeros of two consecutive polynomials are interlaced.*

something like this. The integral $\int_a^b g(x)w(x)dx$ can be approximated by a rule of the form

$$\int_a^b g(x)w(x)dx \approx \sum_{n=1}^{N} \alpha_n g(\xi_n),$$

where the ξ_n's are the zeros of the orthogonal polynomial p_N on the interval $[a, b]$ with respect to the weight function w. The weights α_n are generally tabulated or can be determined. This rule has the remarkable property that it is exact for integrands $g(x) = x^p$ that are polynomials of degree $p = 0 : 2N - 1$. It achieves this super-accuracy by allowing *both* the weights α_n *and* the nodes ξ_n (a total of $2N$ parameters) to be chosen. It is a direct, but rather cumbersome task to show that the weights of any Gaussian quadrature rule are given by [80], [106]

$$\alpha_n = \frac{a_N}{a_{N-1}} \frac{d_{N-1}^2}{p_N'(\xi_n)p_{N-1}(\xi_n)}$$

for $n = 0, 1, 2, \ldots$, where a_N and a_{N-1} are the leading coefficients of p_N and p_{N-1}, respectively.

If we now use this quadrature rule to approximate the expansion coefficients c_k, we find that

$$c_k \approx \frac{1}{d_k^2} \sum_{n=1}^{N} \alpha_n f(\xi_n) p_k(\xi_n) = \frac{a_N}{a_{N-1}} \frac{d_{N-1}^2}{d_k^2} \sum_{n=1}^{N} \frac{f(\xi_n) p_k(\xi_n)}{p_N'(\xi_n) p_{N-1}(\xi_n)}$$

for $k = 0, 1, 2, \ldots$. With these thoughts in mind, it is now possible to propose a discrete transform pair associated with the orthogonal polynomials $\{p_k\}$. Let's adopt our usual notation and let $f_n = f(\xi_n)$ be the samples of the function f that comprise the input sequence where $n = 1 : N$. We will also let F_k denote the sequence of transform coefficients that approximate $c_0, \ldots, c_{N-1}$. Then here is our proposed discrete transform pair.

▶ **Discrete Orthogonal Polynomial Transform Pair** ◀

$$F_k = \frac{a_N}{a_{N-1}} \frac{d^2_{N-1}}{d^2_k} \sum_{n=1}^{N} \frac{f_n p_k(\xi_n)}{p'_N(\xi_n) p_{N-1}(\xi_n)} \quad \text{(forward transform)} \tag{8.11}$$

and

$$f_n = \sum_{k=0}^{N-1} F_k p_k(x) \quad \text{(inverse transform)}, \tag{8.12}$$

where $k = 0 : N - 1$ in the forward transform and $k = 1 : N$ in the inverse transform.

So far this is all a conjecture, but hopefully a rather convincing one. Now the task is to show that these two relations between the sequences f_n and F_k really do constitute a transform pair. In principle, this should be an easy matter: one could substitute (8.11) into (8.12) (or visa versa) and check for an identity. Unfortunately, as we have seen repeatedly in the past, this procedure ultimately relies on a (discrete) orthogonality property. Therefore, in order to establish this transform pair, we must ask about the discrete orthogonality of this system. The answer is both curious and unexpected.

The investigation into discrete orthogonality of orthogonal polynomials requires a special result called the **Christoffel[6]–Darboux[7] formula**. This formula is not that difficult to derive, but it would take us further afield than necessary [81], [145]. Therefore, we will simply state the fact that if $\{p_k\}$ is a system of orthogonal polynomials on (a, b), then

$$\sum_{k=0}^{N-1} \frac{p_k(x) p_k(y)}{d^2_k} = \frac{a_{N-1}}{a_N d^2_{N-1}} \frac{p_N(x) p_{N-1}(y) - p_{N-1}(x) p_N(y)}{x - y},$$

where $x \neq y$ are any two points on the interval (a, b). When $x = y$, it follows (by L'Hôpital's rule) that

$$\sum_{k=0}^{N-1} \frac{p^2_k(x)}{d^2_k} = \frac{a_{N-1}}{a_N d^2_{N-1}} p'_N(x) p_{N-1}(x).$$

Consider what happens if we let x and y be two zeros of p_N in the Christoffel–Darboux formula. With $x = \xi_n$ and $y = \xi_m$, noting that $p_N(\xi_m) = p_N(\xi_n) = 0$, we find the first discrete orthogonality property.

▶ **Discrete Orthogonality Property 1 (with respect to grid points)** ◀

$$\sum_{k=0}^{N-1} \frac{p_k(\xi_m) p_k(\xi_n)}{d^2_k} = \delta(m - n) \frac{a_{N-1}}{a_N d^2_{N-1}} p'_N(\xi_n) p_{N-1}(\xi_n) \tag{8.13}$$

---

[6] ELWIN BRUNO CHRISTOFFEL (1829–1900) was a professor of mathematics in both Zurich and Strasbourg. He is known for his research on potential theory, minimal surfaces, and general curved surfaces.

[7] The work of JEAN GASTON DARBOUX (1842–1917) in geometry and analysis was well known through both his teaching and his books. He was the secretary of the Paris Academy of Sciences and co-founder of the *Bulletin des sciences mathématiques et astronomiques*.

for $m, n = 1 : N$.

The sum of products of the first N polynomials evaluated at zeros of p_N vanishes unless the polynomials are evaluated at the same zero of p_N. This *is* an orthogonality property, although it is not analogous to the continuous orthogonality properties that we have already seen. The sum is on the order of the polynomials, not on the grid points, as is usually the case. We will call this property **orthogonality with respect to the grid points** since the $\delta(m-n)$ term operates on the indices of the grid points.

We can put this property to use immediately in verifying the inverse transform of our alleged transform pair. We begin with the forward transform (8.11)

$$F_k = \frac{a_N}{a_{N-1}} \frac{d_{N-1}^2}{d_k^2} \sum_{m=1}^{N} \frac{f_m p_k(\xi_m)}{p_N'(\xi_m) p_{N-1}(\xi_m)},$$

and solve for one of the terms of the input sequence f_n. Multiplying both sides of this relation by $p_k(\xi_n)$ and summing over the degrees of the polynomials ($k = 0 : N - 1$) results in

$$
\begin{aligned}
\sum_{k=0}^{N-1} F_k p_k(\xi_n) &= \sum_{k=0}^{N-1} \frac{a_N}{a_{N-1}} \frac{d_{N-1}^2}{d_k^2} \sum_{m=1}^{N} \frac{f_m p_k(\xi_m) p_k(\xi_n)}{p_N'(\xi_m) p_{N-1}(\xi_m)} \\
&= \frac{a_N d_{N-1}^2}{a_{N-1}} \sum_{m=1}^{N} \frac{f_m}{p_N'(\xi_m) p_{N-1}(\xi_m)} \underbrace{\sum_{k=0}^{N-1} \frac{p_k(\xi_m) p_k(\xi_n)}{d_k^2}}_{\text{use orthogonality}} \\
&= \frac{a_N d_{N-1}^2}{a_{N-1}} \sum_{m=1}^{N} \left(\frac{f_m}{p_N'(\xi_m) p_{N-1}(\xi_m)} \right. \\
&\qquad \left. \times\, \delta(m-n) \frac{a_{N-1}}{a_N d_{N-1}^2} p_N'(\xi_n) p_{N-1}(\xi_n) \right) \\
&= f_n,
\end{aligned}
$$

for $n = 1 : N$. As shown by the underbrace in the second line, the inner sum (on k) can be simplified by the first orthogonality relation, which in turn allows a single term of the outer sum (on m) to survive. The outcome is the inverse transform of our alleged transform pair.

Recall that with the DFT a single orthogonality property sufficed to derive both the forward and inverse transform. In the orthogonal polynomial case, an attempt to use the same orthogonality property to "go in the opposite direction" (start with the inverse transform and derive the forward transform) leads nowhere. A second (dual) orthogonality property is needed. We could resort to another long-winded derivation of this second property or make a very astute observation.

Let's write the first orthogonality property in the form

$$\sum_{k=0}^{N-1} \frac{p_k(\xi_m) p_k(\xi_n)}{d_k^2} = \delta(m-n) \lambda_n^2,$$

where

$$\lambda_n^2 = \frac{a_{N-1} p_N'(\xi_n) p_{N-1}(\xi_n)}{(a_N d_{N-1}^2)}$$

just stands for the constants multiplying $\delta(m - n)$. This sum can now be written

$$\sum_{k=0}^{N-1} \frac{p_k(\xi_m)}{\lambda_n d_k} \frac{p_k(\xi_n)}{\lambda_n d_k} = \sum_{k=0}^{N-1} c_{km} c_{kn} = \delta(m - n),$$

where we have let $c_{kn} = p_k(\xi_n)/(d_k \lambda_n)$. If we now view the c_{kn}'s as the elements of an $N \times N$ matrix C, then this equation can be written in matrix form as $C^T C = I$. Two important consequences follow from this fact: first, $C^T = C^{-1}$, which says that C is an orthogonal matrix (reflecting the first orthogonality property), and second, $CC^T = I$ (problem 196). If we now write out the components of the matrix equation $CC^T = I$, we see that

$$\sum_{n=1}^{N} c_{jn} c_{kn} = \delta(j - k).$$

Replacing the c_{kn}'s by what they really represent, we have

$$\sum_{n=1}^{N} \frac{p_j(\xi_n)}{d_j \lambda_n} \frac{p_k(\xi_n)}{d_k \lambda_n} = \delta(j - k).$$

And finally, replacing λ_n, a brand new orthogonality property appears.

▶ **Discrete Orthogonality Property 2 (with respect to degree)** ◀

$$\sum_{n=1}^{N} \frac{p_j(\xi_n) p_k(\xi_n)}{p'_N(\xi_n) p_{N-1}(\xi_n)} = \delta(j - k) \frac{a_{N-1} d_k^2}{a_N d_{N-1}^2} \tag{8.14}$$

for $j, k = 0 : N - 1$. This orthogonality property *does* parallel the continuous orthogonality properties since the sum is over the grid points and the "zero-nonzero switch" is with respect to the degree of the polynomials.

We may now take the final step and verify the forward transform of the proposed transform pair. Beginning with the alleged inverse transform (8.12)

$$f_n = \sum_{j=0}^{N-1} F_j p_j(\xi_n)$$

for $n = 1 : N$, we multiply both sides by $p_k(\xi_n)/p'_N(\xi_n) p_{N-1}(\xi_n)$ and sum over the grid points ξ_n for $n = 1 : N$. This leads to the following line of thought, which inevitably uses the second discrete orthogonality property:

$$\sum_{n=1}^{N} \frac{f_n p_k(\xi_n)}{p'_N(\xi_n) p_{N-1}(\xi_n)} = \sum_{n=1}^{N} \frac{p_k(\xi_n)}{p'_N(\xi_n) p_{N-1}(\xi_n)} \sum_{j=0}^{N-1} F_j p_j(\xi_n)$$

$$= \sum_{j=0}^{N-1} F_j \underbrace{\sum_{n=1}^{N} \frac{p_j(\xi_n) p_k(\xi_n)}{p'_N(\xi_n) p_{N-1}(\xi_n)}}_{\text{use orthogonality}}$$

$$= \sum_{j=0}^{N-1} F_j \delta(j - k) \frac{a_{N-1} d_k^2}{a_N d_{N-1}^2}$$

$$= F_k \frac{a_{N-1} d_k^2}{a_N d_{N-1}^2}.$$

Therefore,

$$F_k = \frac{a_N}{a_{N-1}} \frac{d_{N-1}^2}{d_k^2} \sum_{n=1}^{N} \frac{f_n p_k(\xi_n)}{p_N'(\xi_n) p_{N-1}(\xi_n)}$$

for $k = 0 : N - 1$, which is the proposed forward transform (8.11) given above.

In summary, we have established a discrete transform pair based on a system of orthogonal polynomials. It is a curious state of affairs that two different orthogonality properties are needed to establish a complete transform pair. At this point we could proceed to derive transform pairs for any of the many orthogonal polynomial systems that populate the mathematical universe. We will be content with just one such exercise, and it is a transform that arises frequently in practice. Let's consider the set of Legendre polynomials and its discrete transform.

As mentioned earlier, the Legendre polynomials $\{P_n\}$ reside on the interval $[-1, 1]$ and use the weight function $w(x) = 1$. Here are the first few Legendre polynomials with their graphs shown in Figure 8.5:

$$
\begin{aligned}
P_0(x) &= 1, \\
P_1(x) &= x, \\
P_2(x) &= \frac{3}{2}x^2 - \frac{1}{2}, \\
P_3(x) &= \frac{5}{2}x^3 - \frac{3}{2}, \\
P_4(x) &= \frac{35}{8}x^4 - \frac{15}{4}x^2 + \frac{3}{8}.
\end{aligned}
$$

We will cite some of the more interesting and useful properties of the Legendre polynomials, some of which are needed to develop the discrete transform. The full development of these properties can be found in many sources [81], [145]; some will be elaborated in the problems.

1. **Degree.** The polynomial P_n has degree n and is even/odd when n is even/odd.

2. **Zeros.** The polynomial P_n has n (real) zeros on the interval $(-1, 1)$. If n is odd, then $P_n(0) = 0$.

3. **Orthogonality.** We reiterate the (continuous) orthogonality property of the Legendre polynomials:

$$\int_{-1}^{1} P_j(x) P_k(x) dx = \frac{2}{2k + 1} \delta(j - k)$$

for $j, k = 0, 1, 2, \ldots$. This means that the scaling factors d_k^2 are given by

$$d_k^2 = \int_{-1}^{1} P_k^2(x) dx = \frac{2}{2k + 1}.$$

4. **Rodrigues' formula.** There are several ways to define and generate Legendre polynomials. Many treatments begin with the definition known as Rodrigues'[8]

---

[8] We know only that OLINDE RODRIGUES (1794–1851) was an economist and reformer who evidently was sufficiently familiar with the work of Legendre to derive the formula for the Legendre polynomials that bears his name.

formula and all other properties follow (problem 201):

$$P_n(x) = \frac{1}{2^n n!} \frac{d^n}{dx^n}(x^2 - 1)^n$$

for $n = 0, 1, 2, \ldots$.

5. **Recurrence relations.** The Legendre polynomials satisfy several recurrence relations that relate two or three consecutive polynomials and/or their derivatives. The fundamental three-term recurrence relation is

$$(n + 1)P_{n+1}(x) = (2n + 1)xP_n(x) - nP_{n-1}(x) \qquad (8.15)$$

for $n = 1, 2, 3, \ldots$. Given P_0 and P_1, this relation can be used to generate the Legendre polynomials. Two other recurrence relations that will be needed are

$$nP_n(x) - xP_n'(x) + P_{n-1}'(x) = 0 \qquad (8.16)$$

and

$$P_n'(x) - nP_{n-1}(x) - xP_{n-1}'(x) = 0. \qquad (8.17)$$

The relation that we will use for the discrete Legendre transform results if we multiply (8.16) by x and subtract the resulting equation from (8.17). We find (problem 202) that

$$nxP_n(x) + (1 - x^2)P_n'(x) - nP_{n-1}(x) = 0 \qquad (8.18)$$

for $n = 1, 2, 3, \ldots$.

6. **Differential equation.** The nth Legendre polynomial P_n satisfies the differential equation (problem 203)

$$(1 - x^2)y''(x) - 2xy'(x) + n(n + 1)y(x) = 0.$$

7. **Ratio of leading coefficients.** Letting a_n be the leading coefficient in P_n, it can be shown that

$$\frac{a_n}{a_{n-1}} = \frac{2n - 1}{n}.$$

With these properties in hand, we may now deduce the discrete orthogonality properties and the discrete transform pair for Legendre polynomials. Let's begin with the orthogonality properties. The first discrete orthogonality property on N points (8.13) was found to be

$$\sum_{k=0}^{N-1} \frac{P_k(\xi_m)P_k(\xi_n)}{d_k^2} = \delta(m - n)\frac{a_{N-1}}{a_N d_{N-1}^2}P_N'(\xi_n)P_{N-1}(\xi_n),$$

where the ξ_n's are the zeros of P_N. We can now tailor this general property to the Legendre polynomials. The scaling factors d_k^2 and d_{N-1}^2 are given in property 3 above, and the ratio of leading coefficients, a_{N-1}/a_N, is given in property 7. The term $P_N'(\xi_n)$ can also be replaced by a nonderivative term by letting $n = N$ and $x = \xi_n$ in (8.18) and noting that $P_N(\xi_n) = 0$. After some rearranging we see that

$$P_N'(\xi_n) = \frac{NP_{N-1}(\xi_n)}{1 - \xi_n^2}.$$

Assembling all of these pieces, we arrive at the first orthogonality property.

▶ **Discrete Orthogonality Property 1 for Legendre Polynomials** ◀

$$\sum_{k=0}^{N-1}(2k+1)P_k(\xi_m)P_k(\xi_n) = \delta(m-n)\frac{N^2P_{N-1}^2(\xi_n)}{1-\xi_n^2} \tag{8.19}$$

for $m, n = 1 : N$. Notice that this is an orthogonality property with respect to the grid points, which are the zeros of P_N.

We can carry out a very similar substitution process (problem 205) with the second discrete orthogonality property (8.14) and deduce the corresponding property for Legendre polynomials.

▶ **Discrete Orthogonality Property 2 for Legendre Polynomials** ◀

$$\sum_{n=1}^{N}\frac{P_j(\xi_n)P_k(\xi_n)(1-\xi_n^2)}{P_{N-1}^2(\xi_n)} = \delta(j-k)\frac{N^2}{2k+1} \tag{8.20}$$

for $j, k = 0 : N - 1$. This second orthogonality property is with respect to the degree of the polynomials. These orthogonality properties can now be used to derive the discrete Legendre transform pair *or* we can appeal directly to the general transform pair (8.11) and (8.12). By either path, we are led to the forward N-point transform (problem 206).

▶ **Forward Discrete Legendre Transform** ◀

$$F_k = \frac{2k+1}{N^2}\sum_{n=1}^{N}\frac{f_n(1-\xi_n^2)}{P_{N-1}(\xi_n)}P_k(\xi_n)$$

for $k = 0 : N - 1$, where the input sequence $f_n = f(\xi_n)$ is the given function f sampled at the zeros of P_N. Going in the opposite direction, given a set of coefficients F_k, the function f can be reconstructed at the zeros of P_N using the inverse discrete transform.

▶ **Inverse Discrete Legendre Transform** ◀

$$f_n = \sum_{k=0}^{N-1} F_k P_k(\xi_n)$$

for $n = 1 : N$.

Since it is a somewhat tangential topic, this excursion into the realm of other discrete transforms has probably lasted long enough. In closing, we mention that the account of orthogonal polynomial transforms given in this chapter is incomplete in more than one aspect. There are still many discrete transforms that are not included in the orthogonal polynomial framework given here. We mention only the Walsh–

Hadamard[9], the Hilbert[10], the discrete Bessel[11] (or Hankel[12]) [83], [84], and the Kohonen–Loewe transforms at the head of the list of omissions. The book by Elliot and Rao [57] gives a good account of these and many more discrete transforms.

Another issue of extreme importance that is under active investigation is the matter of *fast* discrete transforms. For all of the discrete transforms mentioned, there is the attendant question of whether the N-point transform can be computed in FFT time (roughly $N \log N$ operations) rather than matrix-vector multiply time (roughly N^2 operations). In addition to the DFT, the discrete Chebyshev transform and the Walsh–Hadamard transforms have fast versions because of their kinship with the DFT. However, the search for fast algorithms for other discrete transforms is a wide open and tempting quest. It appears that wavelet methods have recently led to a fast Legendre transform [10], and there may be extensions to other discrete transforms.

8.6. The Hartley Transform

Some Background

It seems appropriate to include the Hartley transform in this chapter because it is intertwined so closely with the Fourier transform both mathematically and historically. It is of mathematical interest because it resembles the Fourier transform and shares many analogous properties. Computationally, it is also compelling since, in its discrete form, it may be a legitmate alternative to the DFT, with alleged advantages. Historically, the story of the Hartley transform is relatively brief, but it parallels that of the Fourier transform on a much-compressed scale. It made its first official appearance in a 1942 paper by Ralph V. L. Hartley[13] in the *Proceedings of the Institute of Radio Engineers* [72]. The need to sample signals and approximate the continuous transform on computers led inevitably to the discrete form of the Hartley transform (DHT). Like the DFT, the DHT is a matrix-vector product, and it requires laborious calculations, particularly for long input sequences. The final step occurred in 1984 when Ronald N. Bracewell announced (and patented) the fast Hartley transform (FHT). This algorithm achieves its speed in much the same way as the FFT and computes the DHT in FFT time (order $N \log N$). It also led to claims that "for every application of the Fourier transform there is an application of the Hartley transform," and that the FFT has been "made obsolete by the Hartley formalism" [14].

In the last ten years there has been much animated discussion both on and off the record about the relative merits of the FFT and the FHT, with claims of superiority

---

[9] JACQUES HADAMARD (1865–1963) was a French mathematician who is best known for his proof of the prime number theorem (on the density of primes) in 1896. He also wrote on the psychology of mathematical creativity.

[10] Born in Königsberg in 1862, DAVID HILBERT founded the formalist school of mathematical thought in an attempt to axiomatize mathematics. His famous 23 problems, proposed at the International Mathematical Congress in Paris in 1900, set the agenda for twentieth century mathematics. He was a professor of mathematics at Göttingen from 1895 to 1930, and died there in 1943.

[11] F. W. BESSEL (1784–1846) was a friend of Gauss and a well-known astronomer. He is best known among mathematicians and engineers for the second-order differential equation, and the functions satisfying it, that bear his name.

[12] HERMANN HANKEL (1839–1873) was a German historian of mathematics. Although his name is associated with an integral transform, he appears to have done most of his work in algebra.

[13] As a researcher at Western Electric, RALPH V. L. HARTLEY worked on the design of receivers for transatlantic radiotelephones. During World War I, he proposed a theory for the perception of sound by the human ear and brain. His early work in information theory led to Hartley's Law and was recognized when the fundamental unit of information was named the Hartley.

on both sides. In this section we will stop short of jumping into the FFT/FHT fray and try to keep the discussion as nonpartisan as possible. There is plenty to say if we just present the essentials of the DHT and point out its remarkable similarities to the DFT. We will collect the arguments for and against the FHT as an alternative to the FFT. If a final judgment is necessary, we will leave it to the reader!

The Hartley Transform

We will begin is with the definition of the continuous Hartley transform, essentially in the form given by Hartley in his 1942 paper. The input to the Hartley transform is a real-valued function h defined on $(-\infty, \infty)$ that satisfies the condition $\int_{-\infty}^{\infty} |h(x)| dx < \infty$. Although the input may be a time-dependent signal, we will maintain our convention of using x as the independent variable. The Hartley transform operates on the input h and returns another function H of the frequency variable ω. The kernel of the Hartley transform is the combination

$$\operatorname{cas} x \equiv \cos x + \sin x,$$

which clearly resembles the kernel of the Fourier transform $(\cos x + i \sin x)$ except that it is real-valued. In its most commodious form, the forward Hartley transform is given as follows.

▶ **Forward Hartley Transform** ◀

$$H(\omega) = \int_{-\infty}^{\infty} h(x)\operatorname{cas}(2\pi\omega x)dx. \tag{8.21}$$

Given the transform H, the original input can be recovered using the inverse Hartley transform.

▶ **Inverse Hartley Transform** ◀

$$h(x) = \int_{-\infty}^{\infty} H(\omega)\operatorname{cas}(2\pi\omega x)d\omega. \tag{8.22}$$

Expressions (8.21) and (8.22) form the continuous Hartley transform pair. Two important properties can be gleaned immediately: the transform involves only real-valued quantities, and the Hartley transform is its own inverse.

Given the definition of the transform pair, one may deduce the relationship between the Fourier and Hartley transforms. As with any function, the even and odd parts of the Hartley transform are given by

$$E(\omega) = \frac{H(\omega) + H(-\omega)}{2} = \int_{-\infty}^{\infty} h(x) \cos(2\pi\omega x)dx$$

and

$$O(\omega) = \frac{H(\omega) - H(-\omega)}{2} = \int_{-\infty}^{\infty} h(x) \sin(2\pi\omega x)dx.$$

It is now easily shown (problem 209) that the Fourier transform of h,

$$F(\omega) = \int_{-\infty}^{\infty} h(x)e^{-i2\pi\omega x}dx,$$

is related to the Hartley transform by the following identities.

▶ **Hartley → Fourier** ◀

$$F(\omega) = E(\omega) - iO(\omega).$$

Conversely the Hartley transform can be obtained from the Fourier transform with the following relationship.

▶ **Fourier → Hartley** ◀

$$H(\omega) = \text{Re}\left\{F(\omega)\right\} - \text{Im}\left\{F(\omega)\right\}.$$

Before turning to the discrete transform, there is another Fourier–Hartley link that should be mentioned since it bears on the controversy between the two formalisms. One of the most common reasons for computing the Fourier transform of an input signal is to determine two quantities: its **power spectrum**

$$P(\omega) = |F(\omega)|^2 = |\text{Re}\left\{F(\omega)\right\}|^2 + |\text{Im}\left\{F(\omega)\right\}|^2$$

and its **phase**

$$\phi(\omega) = \arctan\left(\frac{\text{Im}\left\{F(\omega)\right\}}{\text{Re}\left\{F(\omega)\right\}}\right).$$

A point frequently cited by Hartley proponents is that these two important quantities are easily obtained from the Hartley transform by the relations (problem 210)

$$P(\omega) = \frac{H(\omega)^2 + H(-\omega)^2}{2} \quad \text{and} \quad \phi(\omega) = \arctan\left(\frac{H(-\omega) - H(\omega)}{H(\omega) + H(-\omega)}\right).$$

The Discrete Hartley Transform (DHT)

As mentioned earlier, the necessity of sampling input signals and approximating the Hartley transform numerically led naturally to the discrete version of the Hartley transform. The framework for the DHT is absolutely identical to that used throughout this book for the DFT. An input signal h is sampled at N equally spaced points of an interval $[0, A]$ in the spatial (or time) domain. We will denote these samples h_n, where $n = 0 : N - 1$. We have followed the most common convention of using indices in the range $0 : N - 1$, but the periodicity of the Hartley kernel allows other choices such as $1 : N$ or $-N/2 + 1 : N/2$.

The output of the DHT is a real-valued sequence H_k given by the following relation.

▶ **Forward Discrete Hartley Transform** ◀

$$H_k = \frac{1}{N} \sum_{n=0}^{N-1} h_n \text{cas}\left(\frac{2\pi nk}{N}\right) \tag{8.23}$$

for $k = 0 : N - 1$. Notice that the sequence H_k defined in this way is periodic with period N. Furthermore, the kth Hartley coefficient should be interpreted as the weight associated with the kth mode. Several representative modes of the DHT are shown in Figure 8.6. Notice that each mode is a linear combination of a sine and cosine,

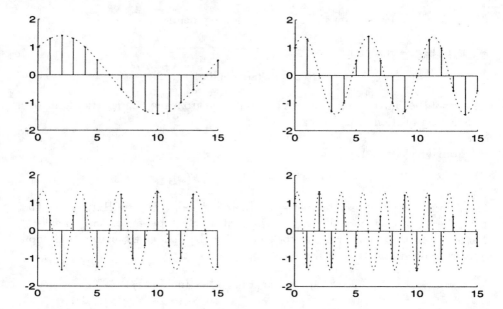

FIG. 8.6. *Modes of the discrete Hartley transform,* $\mathrm{cas}(2\pi nk/N)$, *with frequencies (clockwise from upper left)* $k = 1, 3, 9,$ *and* 5 *on a grid with* $N = 16$ *points are shown. Each mode is a single shifted sine (or cosine) wave.*

and can be viewed as a single shifted sine or cosine mode. With this choice of indices $(n, k = 0 : N - 1)$, the highest frequency occurs at $k = N/2$, while the kth and $(N - k)$th frequencies are the same. The reciprocity relations of the DFT carry over entirely to the DHT.

As with the DFT, the inverse DHT follows with the aid of orthogonality relations. The usual arsenal of trigonometric identities and/or geometric series can be used (problem 211) to show that

$$\sum_{n=0}^{N-1} \mathrm{cas}\left(\frac{2\pi nk}{N}\right) \mathrm{cas}\left(\frac{2\pi nj}{N}\right) = \begin{cases} N & \text{if } j - k = 0 \text{ or a multiple of } N, \\ 0 & \text{otherwise.} \end{cases}$$

It is now possible (problem 213) to deduce the inverse of the relationship given in (8.23).

► **Inverse Discrete Hartley Transform** ◄

$$h_n = \sum_{k=0}^{N-1} H_k \mathrm{cas}\left(\frac{2\pi nk}{N}\right) \tag{8.24}$$

for $n = 0 : N - 1$. The sequence h_n defined by the inverse DHT is real and periodic with period N. However, the most significant property of this transform pair (particularly in the proselytism of the DHT) is the fact that the DHT is its own inverse. In other words, the same algorithm (computer program) can be used for both the forward and inverse transform. This cannot be said of the DFT, in which the forward transform of a real sequence is a conjugate even sequence, which necessitates a distinct algorithm for the inverse transform.

Now the properties of the DHT follow in great profusion, most of them with palpable similarity to the familiar DFT properties. We will highlight a few of these features and leave many more to the exercises. The relationships between the DFT and DHT are of particular interest. Arguing as we did with the continuous transforms, the even and odd parts of the DHT are given by

$$E_k = \frac{H_k + H_{N-k}}{2} \quad \text{and} \quad O_k = \frac{H_k - H_{N-k}}{2}.$$

Notice that the periodicity of the sequence H_k allows us to replace H_{-k} by H_{N-k}. If we let F_k denote the DFT of the real input sequence h_n, it is a straighforward calculation to show that

$$F_k = E_k - iO_k \quad \text{and} \quad H_k = \text{Re}\{F_k\} - \text{Im}\{F_k\},$$

where in both relations $k = 0 : N - 1$. It is useful and reassuring to write these relations in a slightly different way. Recall that since the input sequence h_n is assumed to be real, the DFT coefficients have the conjugate even symmetry $F_k = F_{N-k}^*$. This says that there are N independent real quantities in the DFT sequence: the real and imaginary parts of $F_1, \ldots, F_{\frac{N}{2}-1}$ plus F_0 and $F_{\frac{N}{2}}$, which are both real. Therefore, the DHT $\rightarrow$ DFT relations can be found easily (problem 212).

▶ **DHT → DFT** ◀

$$\text{Re}\{F_k\} = \frac{H_k + H_{N-k}}{2} \quad \text{and} \quad \text{Im}\{F_k\} = -\frac{H_k - H_{N-k}}{2} \qquad (8.25)$$

for $k = 0 : N/2$. We can also use the conjugate even symmetry of the DFT sequence to convert the DFT to DHT.

▶ **DFT → DHT** ◀

$$H_k = \text{Re}\{F_k\} - \text{Im}\{F_k\} \quad \text{and} \quad H_{N-k} = \text{Re}\{F_k\} + \text{Im}\{F_k\} \qquad (8.26)$$

for $k = 0 : N/2$. Note that these relations imply $F_0 = H_0$ and $F_{\frac{N}{2}} = H_{\frac{N}{2}}$. The second pair shows clearly that given a real input sequence of length N, the DFT and the DHT are both sets of N distinct real quantities that are intimately related.

A few properties of the DHT are worth listing, particularly those with obvious DFT parallels. In all cases the proofs of these properties follow directly from the definition (8.23) and are relegated to the problem section. The operational notation $\mathcal{H}$ is used to denote the DHT; that is, $H_k = \mathcal{H}\{h_n\}_k$.

1. **Periodicity.** If the N-point sequences h_n and H_k are related by the DHT transform pair (8.23) and (8.24), then $h_n = h_{n\pm N}$ and $H_k = H_{k\pm N}$.

2. **Reversal.** $\mathcal{H}\{h_{-n}\}_k = H_{-k}$.

3. **Even sequences.** If h_n is an even sequence ($h_{N-n} = h_n$), then H_k is even.

4. **Odd sequences.** If h_n is an odd sequence ($h_{N-n} = -h_n$), then H_k is odd.

5. **Shift property.**

$$\mathcal{H}\{h_{n-p}\}_k = \cos\left(\frac{2\pi pk}{N}\right) H_k + \sin\left(\frac{2\pi pk}{N}\right) H_{N-k}.$$

6. Convolution. Letting $G_k = \mathcal{H}\{g_n\}_k$,

$$\mathcal{H}\{(h * g)_n\}_k = \frac{N}{2}(H_k G_k - H_{N-k} G_{N-k} + H_k G_{N-k} + H_{N-k} G_k),$$

where $(h * g)_n = \sum_{j=0}^{N-1} h_j g_{n-j}$ is the cyclic convolution of the two sequences h_n and g_n.

For the details on these and other properties, plus an excursion into two-dimensional DHTs, filtering with the DHT and matrix formulations of the DHT, the reader is directed to Bracewell's testament, *The Hartley Transform* [14].

In closing we shall attempt to summarize the current state of affairs regarding the DFT and the DHT. We must look at the issue from mathematical, physical, and computational perspectives (and even then we will have undoubtedly committed some oversimplifications).

Mathematically, there is a near equivalence between the two discrete transforms when the input consists of N real numbers. The two sets of transform coefficients consist of N distinct real numbers that are simply related. Furthermore, quantities of physical interest, such as the power and phase, are easily computed from either set of coefficients. From this perspective, there is little reason to prefer one transform over another.

In terms of physical applicability, Nature appears to have a predilection for the Fourier transform. It is the Fourier transform that appears as the analog output of optical and diffraction devices; it is the Fourier transform that arises most naturally in classical models of physical phenomena such as diffusion and wave propagation. From this perspective, a reasonable conclusion is that if a particular problem actually calls for the Hartley transform (or a set of Hartley coefficients), then the DHT should be used. In the vast preponderance of problems in which the Fourier transform (or a set of DFT coefficients) is needed, the DFT should be used. So far, these conclusions are hardly profound or unexpected.

The question remains: is there anything to be gained by using the DHT to compute the DFT? (The opposite question does not seem to attract as much attention.) It is this computational issue that has drawn the lines between the DFT and the DHT camps. As mentioned earlier, there are now fast versions of both the DFT and the DHT, which reduce the computation of these transforms from an $O(N^2)$ to an $O(N \log N)$ chore. If a comparison is made between the FHT and the *complex* form of the FFT (which is occasionally done), then there is a factor-of-two advantage in favor of the FHT (in arithmetic operations). As is well known (see Chapter 4), there are compact symmetric versions of the FFT that exploit the symmetries of a real input sequence and reduce the computational effort by that same factor of two. Therefore, if the appropriate comparison between the DHT and the *real* FFT is made, the two algorithms have virtually identical operation counts (one count favors the compact FFT by $N - 2$ additions). The difference in operation counts is far outweighed by practical considerations such as hardware features, data handling, overall architecture (serial vs. vector vs. multiprocessor), as well as the efficiency of software implementation.

Let's prolong the discussion a bit more and agree to call the performance issue a toss-up; that is, in a fixed computer environment, the computation of the Hartley coefficients by the FHT is as efficient as the computation of Fourier coefficients by the FFT. Then two additional factors should be cited. If the FHT is used to compute

Fourier coefficients, then an extra postprocessing step is needed (given by (8.25)). While there is minimal arithmetic in this step, it does require an additional pass through the data, which will almost certainly give the advantage to the FFT. On the other hand, the FHT has the clear advantage that only one code (program or microprocessor) is needed to do both the forward and inverse DHT. The real FFT needs one code for the forward transform and a separate (but equally efficient) code for the inverse transform. If one is designing chips for satellite guidance systems, this may be a deciding factor; for supercomputer software packages, it probably is not. On that note of equivocation we will close this brief glance at the Hartley transform, and indeed this chapter on related transforms.

Notes

One of the most ardent champions of the fast Hartley transform was the late Oscar Buneman. His papers [23], [24] contain valuable reading on further refinements and implementations of the FHT, as well as the extension to multidimensional DHTs. Further deliberations on the uses and abuses of the DHT, as well as comments on the FFT vs. FHT issue can be found in [15], [53], [71], [100], [123], and [127]. Attempts to determine the current status of FHT codes were unsuccessful. Bracewell's book [14] contains BASIC programs for various versions of the FHT. Each program carries the citation *Copyright* 1985 *The Board of Trustees of the Leland Stanford Junior University*. The extent to which this caption limits the use and reproduction of FHT codes is not known.

8.7. Problems

Laplace Transforms

172. Inverse Laplace transforms. Use the inversion method described in the text to approximate the inverse Laplace transform of the following functions of $s = c + i\omega$. The exact inverse is also given.

(a) $F(s) = 1/(s - 3)$ where $c = \operatorname{Re}\{s\} > 3$ has the inverse $f(t) = e^{3t}$.

(b) $F(s) = s/(s^2 + 16)$ where $c = \operatorname{Re}\{s\} > 0$ has the inverse $f(t) = \cos 4t$.

(c) $F(s) = 1/s^2$ where $c = \operatorname{Re}\{s\} > 0$ has the inverse $f(t) = t$.

In each case experiment with the number of sample points N, the length of the sampling interval Ω, and the free parameter c to determine how the errors behave. For fixed values of N and Ω, try to determine experimentally the optimal value of c.

z-Transforms

173. Linearity of the z-transform. Show that if a and b are constants and if u_n and v_n are sequences, then

$$\mathcal{Z}\{au_n + bv_n\} = a\mathcal{Z}\{u_n\} + b\mathcal{Z}\{v_n\}.$$

174. Shift property.   Show that if u_{n+k} is the sequence obtained from u_n by shifting it to the left by k places, then

$$\mathcal{Z}\{u_{n+k}\} = z^k U(z) - z^k u_0 - z^{k-1} u_1 - \cdots - z u_{k-1},$$

where k is any positive integer.

175. Finding z-transforms.   Find the z-transform of the following sequences u_n where $n = 0, 1, 2, 3, \ldots$. Indicate the values of z for which the transforms are valid.

   (a) $u_n = e^{-an}$ where $a > 0$ is any real number.

   (b) $u_n = \frac{1}{2}(1 + (-1)^n) = \begin{cases} 1 & \text{if } n \text{ is even,} \\ 0 & \text{if } n \text{ is odd.} \end{cases}$

   (c) $u_n = (-1)^n$.

   (d) $u_n = \sin n\theta$, where θ is any real number.

   (e) $\{u_n\} = \{0, 0, 2, 0, 0, 2, 0, 0, 2, \ldots\}$.

   (f) $u_n = \begin{cases} 0 & \text{if } n = 0, \\ 3(\frac{1}{2})^{n-1} & \text{if } n \geq 1. \end{cases}$

176. Two special inverses.   Verify the last two transform pairs of Table 8.1:

$$\mathcal{Z}^{-1}\left\{\frac{1}{z^2 + a^2}\right\} = \begin{cases} 0 & \text{if } n = 0 \text{ or } n \text{ odd,} \\ (-1)^{\frac{n}{2}+1} a^{n-2} & \text{if } n \text{ even,} \end{cases}$$

and

$$\mathcal{Z}^{-1}\left\{\frac{z}{z^2 + a^2}\right\} = \begin{cases} 0 & \text{if } n \text{ even,} \\ (-1)^{\frac{n-1}{2}} a^{n-1} & \text{if } n \text{ odd.} \end{cases}$$

What are the regions of validity for these transforms?

177. Inverse z-transforms.   Find the sequences that are the inverse z-transforms of the following functions.

   (a) $U(z) = \frac{1}{z+3}$.

   (b) $U(z) = \frac{z^2 + 4z - 2}{z^2 + z}$.

   (c) $U(z) = \frac{1}{z^2 - 4}$.

   (d) $U(z) = \frac{z^2 + 9}{z^2 - 4}$.

   (e) $U(z) = \frac{z}{z^2 - 4}$.

   (f) $U(z) = z^{-p}$ where p is a positive integer.

178. Solving initial value problems.   Use the z-transform to solve the following initial value problems.

   (a) $u_{n+1} + 4u_n = 0$ for $n \geq 0$ with $u_0 = 2$.

   (b) $u_{n+1} - 3u_n = 6$ for $n \geq 0$ with $u_0 = 0$.

   (c) $u_{n+2} - 9u_n = 0$ for $n \geq 0$ with $u_0 = 0$ and $u_1 = 2$.

(d) $u_{n+2} - 2u_{n+1} - 15u_n = 2^n$ for $n \geq 0$ with $u_0 = u_1 = 0$.

179. Fibonacci sequence. One of the most celebrated sequences in mathematics is the Fibonacci sequence, which arises in many diverse applications. It can be generated from the initial value problem

$$u_{n+2} = u_{n+1} + u_n$$

for $n = 0, 1, 2, 3, \ldots$, with $u_0 = 1$ and $u_1 = 2$. Use the z-transform to solve this initial value problem and generate the Fibonacci sequence.

180. Inverse z-transform from the definition. Begin with the inversion formula for the z-transform

$$u_n = \frac{1}{2\pi i} \oint_C U(z) z^{n-1} dz$$

for $n = 0, 1, 2, 3, \ldots$, where C is a circle of radius R centered at the origin. Approximate this contour integral by a sum of integrand values at the points $z_k = Re^{i2\pi k/N}$, where $k = 0 : N - 1$, and show that

$$u_n \approx \frac{R^n}{N} \mathcal{D}^{-1} \{U_k\}_n$$

for $n = 0 : N - 1$.

181. Laplace transform to z-transform. Begin with the definition of the Laplace transform

$$\tilde{U}(s) = \int_0^\infty u(t) e^{-st} dt,$$

and make the change of variables $z = e^s$ (where z and s are complex). Let u_n be samples of $u(t)$ at the points $t_n = n$ for $n = 0, 1, 2, 3, \ldots$ to derive the z-transform $U(z)$ as an approximation to the Laplace transform $\tilde{U}(s)$. If the Laplace transform is valid for $\text{Re}\{s\} > s_0$, for what values of z is the corresponding z-transform valid?

182. The convolution theorem. It should come as no surprise that the z-transform has a convolution theorem. Prove that

$$\mathcal{Z}^{-1}\{U(z)V(z)\} = u_n * v_n = \sum_{k=0}^n = u_k v_{n-k}.$$

183. Inverse z-transforms with nonsimple poles. The problem of inverting a z-transform $U(z)$ that has multiple roots in the denominator is more challenging. It can be done either by long division or by using the convolution theorem. For example, to invert $U(z) = (z - 2)^{-2}$, one may let $F(z) = G(z) = (z - 2)^{-1}$ and note that $f_n = g_n = 2^{n-1}$ for $n \geq 1$ and $f_0 = g_0 = 0$. The convolution theorem can then be used. Find the inverse z-transforms of the following functions.

(a) $U(z) = \dfrac{1}{(z - 2)^2}$,   (b) $U(z) = \dfrac{z}{(z - 2)^2}$.

Chebyshev Transforms

184. Zeros of the Chebyshev polynomials.   Show that the zeros of the nth Chebyshev polynomial T_n are given by

$$\xi_j = \cos\left(\frac{2j-1}{n}\frac{\pi}{2}\right)$$

for $j = 1 : n$.

185. Extreme values.   Show that $|T_n(x)| \le 1$ on $[-1, 1]$ for $n = 0, 1, 2, \ldots$, and that the extreme values of T_n occur at the points

$$\eta_j = \cos\left(\frac{\pi j}{n}\right)$$

for $j = 0 : n$.

186. Multiplicative property.   Use the definition of the Chebyshev polynomials to show that

$$T_m T_n = \frac{1}{2}(T_{m+n} + T_{m-n}).$$

187. Semigroup property.   Use the definition of the Chebyshev polynomials to show that

$$T_m(T_n(x)) = T_{mn}(x).$$

188. Recurrence relation.   Prove the recurrence relation

$$T_n(x) = 2xT_{n-1}(x) - T_{n-2}(x)$$

for $n = 2, 3, 4, \ldots$.

189. Differential equation.   Compute $T_n'(x)$ and $T_n''(x)$ in terms of the variable $\theta = \cos^{-1} x$ to show that T_n satisfies the differential equation

$$(1 - x^2)y''(x) - xy'(x) + n^2 y(x) = 0$$

for $n = 0, 1, 2, \ldots$.

190. Representation of polynomials.   Show that the polynomial

$$p(x) = 1 + x + x^2 + x^3 + x^4 + x^5$$

has the representation

$$p(x) = \frac{15}{8} + \frac{19}{8}T_1(x) + T_2(x) + \frac{9}{16}T_3(x) + \frac{1}{8}T_4(x) + \frac{1}{16}T_5(x).$$

Show further (in a process called **economization**) that the truncated polynomials

$$p_4(x) = \frac{15}{8} + \frac{19}{8}T_1(x) + T_2(x) + \frac{9}{16}T_3(x) + \frac{1}{8}T_4(x)$$

and

$$p_3(x) = \frac{15}{8} + \frac{19}{8}T_1(x) + T_2(x) + \frac{9}{16}T_3(x)$$

give approximations to p that satisfy

$$|p(x) - p_4(x)| \leq \frac{1}{16} \quad \text{and} \quad |p(x) - p_3(x)| \leq \frac{3}{16}$$

for $-1 \leq x \leq 1$.

191. Least squares property. Show that of all nth-degree polynomials p_n, the normalized Chebyshev polynomial $\tilde{T}_n$ minimizes

$$\int_{-1}^{1} \frac{(p_n(x))^2}{\sqrt{1 - x^2}} dx.$$

(Hint: Expand an arbitrary p_n as a sum of Chebyshev polynomials.)

192. Error in truncating a Chebyshev expansion. Use the orthogonality of the Chebyshev polynomials to show that if the expansion

$$f(x) = \sum_{k=0}^{\infty}{}' c_k T_k(x)$$

is truncated after $N + 1$ terms, the resulting partial sum s_N satisfies

$$\|f - s_N\|_2^2 = \sum_{k=N+1}^{\infty} c_k^2.$$

193. Discrete orthogonality. Assuming that $0 \leq j, k \leq N$, prove the discrete orthogonality properties with respect to *both* the grid points and the degree of the polynomials

$$\sum_{k=0}^{N}{}'' T_k(\eta_j) T_k(\eta_n) = \begin{cases} 0 & \text{if } j \neq n, \\ N/2 & \text{if } j = n \neq 0, N \\ N & \text{if } j = n = 0, N, \end{cases} \quad (\#1 \text{ with respect to grid points}),$$

and

$$\sum_{n=0}^{N}{}'' T_j(\eta_n) T_k(\eta_n) = \begin{cases} 0 & \text{if } j \neq k, \\ N/2 & \text{if } j = k \neq 0, N \\ N & \text{if } j = k = 0, N, \end{cases} \quad (\#2 \text{ with respect to degree}),$$

where $\eta_n = \cos(\pi j/N)$ are the x-coordinates of the extreme points of T_N. (Hint: Use the discrete orthogonality of the cosine.)

194. Chebyshev series synthesis. Assume that the coefficients $c_0, c_1, \ldots, c_N$ of the expansion

$$f(x) = \sum_{k=0}^{\infty}{}' c_k T_k(x)$$

are given. Show how the inverse DCT can be used to approximate $f(x)$ at selected points of $[-1, 1]$.

195. Discrete orthogonality on the zeros. The Chebyshev polynomials also have a discrete orthogonality property that uses the zeros of the polynomials as grid points. Let ξ_n be the zeros of T_N for $n = 1 : N$. Show that for $0 \le j, k \le N - 1$ the following relations hold:

$$\sum_{n=1}^{N} T_j(\xi_n) T_k(\xi_n) = \begin{cases} 0 & \text{if } j \ne k, \\ N/2 & \text{if } j = k \ne 0, \\ N & \text{if } j = k = 0. \end{cases}$$

(Hint: Use the discrete orthogonality of the cosine.) Note that this property is with respect to the degree of the polynomials in analogy to the discrete orthogonality property 2 of the text.

196. A matrix property.   Recall that a matrix identity was used to derive the second discrete orthogonality property from the first. Show that if an $N \times N$ matrix C satisfies $C^T C = I$, then $CC^T = I$.

197. Representation of polynomials.   Use the discrete orthogonality on zeros (problem 195) to show that any polynomial p of degree $N - 1$ can be expressed in the form

$$p(x) = \sum_{k=0}^{N-1} {}' c_k T_k(x),$$

where

$$c_k = \frac{2}{N} \sum_{n=1}^{N} p(\xi_n) T_k(\xi_n)$$

for $k = 0 : N - 1$.

198. Representations of functions.   Verify the following Chebyshev expansions analytically or with a symbolic algebra package:

$$\text{sgn}(x) = \frac{4}{\pi} \sum_{k=1}^{\infty} \frac{(-1)^{k-1}}{2k - 1} T_{2k-1}(x) \quad \text{and} \quad |x| = \frac{2}{\pi} + \frac{4}{\pi} \sum_{k=1}^{\infty} \frac{(-1)^{k-1}}{4k^2 - 1} T_{2k}(x).$$

In each case

  (a) Approximate the coefficients in the expansion using the forward discrete Chebyshev transform and compare the computed results to the exact coefficients. How do the errors vary with N?

  (b) Use the exact coefficients $c_0, \ldots, c_N$ as input to the inverse discrete Chebyshev transform and compute approximations to the function values at selected points of $[-1, 1]$. Monitor the errors as they vary with N.

Legendre Transforms

199. Legendre coefficients.   Use the orthogonality of the Legendre polynomials to show that the coefficients in the expansion

$$f(x) = \sum_{k=0}^{\infty} c_k P_k(x)$$

on $[-1, 1]$ are given by

$$c_k = \frac{2k + 1}{2} \int_{-1}^{1} f(x) P_k(x) dx$$

for $k = 0, 1, 2, \ldots$.

200. Legendre expansions for polynomials.

(a) Find the representation of $f(x) = 3x^3 - 2x$ in terms of Legendre polynomials.

(b) Show that any polynomial of degree n can be represented exactly by a linear combination of $P_0, \ldots, P_n$.

(c) Show that P_n is orthogonal to all polynomials of degree less than n.

201. Rodrigues' formula.   Use Rodrigues' formula

$$P_n(x) = \frac{1}{2^n n!} \frac{d^n}{dx^n} (x^2 - 1)^n$$

for $n = 0, 1, 2, \ldots$ to determine P_0, P_1, P_2, and P_3.

202. Recurrence relations.   Combine recurrence relations (8.16) and (8.17) to obtain the relation

$$nx P_n(x) + (1 - x^2) P_n'(x) - n P_{n-1}(x) = 0$$

for $n = 1, 2, 3, \ldots$, which is needed for the discrete orthogonality properties. Simplify this relation when $x = \xi_k$, a zero of P_n.

203. Legendre's differential equation.   Eliminate the term P_{n-1}' between the recurrence relations (8.16) and (8.17) to derive the differential equation

$$(1 - x^2) y''(x) - 2xy'(x) + n(n+1)y(x) = 0$$

satisfied by P_n.

204.   Some discrete Legendre transforms.   Let $\mathcal{P}\{f_n\}_k$ denote the kth component of the N-point discrete Legendre transform of the sequence f_n. Show that

(a) $\mathcal{P}\{1\}_k = \delta(k)$.

(b) $\mathcal{P}\{\xi_n\}_k = \delta(k - 1)$ where ξ_n are the zeros of P_N.

(c) $\mathcal{P}\{\xi_n g(\xi_n)\}_k = \dfrac{k+1}{2k+3} G_{k+1} + \dfrac{k}{2k-1} G_{k-1}$, where g is an arbitrary function and $g_k = \mathcal{P}\{g_n\}_k$. (Hint: Use the recurrence relation (8.15).)

205. Discrete orthogonality property 2 for Legendre polynomials.   Derive the second discrete orthogonality property for Legendre polynomials (8.20) from the general property (8.14).

206. Discrete Legendre transform pair   Derive the discrete Legendre transform pair

$$F_k = \frac{2k+1}{N^2} \sum_{n=1}^{N} \frac{f_n(1 - \xi_n^2)}{P_{N-1}(\xi_n)} P_k(\xi_n)$$

for $k = 0 : N - 1$ and

$$f_n = \sum_{k=0}^{N-1} F_k P_k(\xi_n)$$

for $n = 1 : N$ by

(a) appealing to the general transform pair (8.11) and (8.12), and

(b) using the discrete orthogonality properties of the Legendre polynomials (noting that a different property must be used for the forward and inverse transforms).

207. Legendre expansions.   For each of the following functions on $[-1, 1]$ compute as many coefficients c_k in the expansion $\sum_{k=0}^{\infty} c_k P_k(x)$ as possible (using either analytical methods or a symbolic algebra package).

(a) $f(x) = \sqrt{\dfrac{1-x}{2}}$,   (b) $f(x) = 1 - |x|$,   (c) $f(x) = \begin{cases} 1 & \text{if } 0 < x < 1, \\ 0 & \text{if } -1 < x < 0. \end{cases}$

Then

(a) Approximate these same coefficients using an N-point discrete Legendre transform and compare the results to the exact coefficients c_k for various values of N.

(b) Use the exact coefficients $c_0, \ldots, c_{N-1}$ as input to the inverse N-point discrete Legendre transform to approximate $f(\xi_n)$ and compare the results to the original functions.

208. Open questions about discrete orthogonality.   The discrete Chebyshev transform was presented separately in this chapter because it is related so closely to the discrete cosine transform. However, it could also be developed within the general framework of discrete orthogonal polynomial transforms.

(a) Find the two discrete orthogonality properties of the Chebyshev polynomials from the general orthogonality properties (8.13) and (8.14). Note that these two properties use the *zeros*, not the extreme points, of the polynomials as grid points. Conclude that there are now *four* different discrete orthogonality relations for the Chebyshev polynomials: two that use the extreme points as grid points and two that use the zeros as grid points; in each case, we have property 1 (with respect to the grid points), and property 2 (with respect to the degree of the polynomials).

(b) Show that the discrete orthogonality property with respect to degree derived in part (a) of this problem (using the zeros as grid points) corresponds to the orthogonality property derived above in problem 195. (Hint: You will need the identity that $T_N'(\xi_n) = N U_{N-1}(\xi_n)$ where $U_n = \sin(n\theta)/\sin\theta$ is the nth Chebyshev polynomial of the second kind).

(c) Derive the discrete Chebyshev transform pair that uses the zeros of the polynomials as grid points. How does it compare to the first transform pair?

(d) Since the Chebyshev polynomials have four different discrete orthogonality properties, do *all* orthogonal polynomials have orthogonality properties that use the extreme points (instead of the zeros) as grid points? If so, are there two such properties: property 1 with respect to the grid points, and property 2 with respect to the degree of the polynomials?

The Hartley Transform

209. Continuous Hartley and Fourier transforms.   Given a function defined on $(-\infty, \infty)$, show that its Fourier transform F is related to its Hartley transform H by

$$H = \text{Re}\left\{F\right\} - \text{Im}\left\{F\right\} \quad \text{and} \quad F = E - iO,$$

where E and O are the even and odd parts of the Hartley transform, respectively. Show from this representation of F that it has the conjugate even symmetry $F(-\omega) = F(\omega)^*$.

210. Power spectrum and phase.   Show that the power spectrum and phase of a function can be obtained from the continuous Hartley transform by the relations

$$P(\omega) = \frac{H(\omega)^2 + H(-\omega)^2}{2} \quad \text{and} \quad \phi(\omega) = \arctan\left(\frac{H(-\omega) - H(\omega)}{H(\omega) + H(-\omega)}\right).$$

211. Orthogonality.   Prove the orthogonality property of the DHT kernel

$$\sum_{n=0}^{N-1} \text{cas}\left(\frac{2\pi nk}{N}\right) \text{cas}\left(\frac{2\pi nj}{N}\right) = \begin{cases} N & \text{if } j - k = 0 \text{ or a multiple of } N, \\ 0 & \text{otherwise.} \end{cases}$$

212. DFT to DHT and back.   Verify the DHT $\rightarrow$ DFT relationships

$$\text{Re}\left\{F_k\right\} = \frac{H_k + H_{N-k}}{2} \quad \text{and} \quad \text{Im}\left\{F_k\right\} = -\frac{H_k - H_{N-k}}{2}$$

for $k = 0 : N/2$, and the DFT $\rightarrow$ DHT relationships

$$H_k = \text{Re}\left\{F_k\right\} - \text{Im}\left\{F_k\right\} \quad \text{and} \quad H_{N-k} = \text{Re}\left\{F_k\right\} + \text{Im}\left\{F_k\right\}$$

for $k = 0 : N/2$.

213. Inverse DHT.   Using the orthogonality of the cas functions, prove that the DHT is its own inverse up to a multiplicative constant.

214. DHT properties.   Verify the following properties of the DHT.

(a) **Reversal.** $\mathcal{H}\{h_{-n}\}_k = H_{-k}$.

(b) **Even sequences.** If h_n is an even sequence ($h_{N-n} = h_n$), then H_k is even.

(c) **Odd sequences.** If h_n is an odd sequence ($h_{N-n} = -h_n$), then H_k is odd.

(d) **Shift property.**

$$\mathcal{H}\{h_{n-p}\}_k = \cos\left(\frac{2\pi pk}{N}\right) H_k + \sin\left(\frac{2\pi pk}{N}\right) H_{N-k}.$$

(e) **Convolution.** Letting $G_k = \mathcal{H}\{g_n\}_k$,

$$\mathcal{H}\{(h * g)_n\}_k = \frac{N}{2}(H_k G_k - H_{N-k}G_{N-k} + H_k G_{N-k} + H_{N-k}G_k),$$

where $(h*g)_n = \sum_{j=0}^{N-1} h_j g_{n-j}$ is the cyclic convolution of the two sequences h_n and g_n.

215. Further properties. Given an input sequence h_n, let F_k and H_k be its DFT and DHT, respectively. Show that the following relationships are true.

(a) **Sum of sequence (DC component):**

$$\sum_{n=0}^{N-1} h_n = NF_0 = NH_0.$$

(b) **First value:**

$$h_0 = \sum_{k=0}^{N-1} F_k = \sum_{k=0}^{N-1} H_k.$$

(c) **Parseval's theorem:**

$$\sum_{n=0}^{N-1} |h_n|^2 = N \sum_{k=0}^{N-1} |F_k|^2 = N \sum_{k=0}^{N-1} |H_k|^2.$$

216. Convolution and symmetry. Show that if either of the two sequences h_n or g_n in a cyclic convolution is even, then the convolution theorem for the DHT simplifies considerably and becomes

$$\mathcal{H}\{(h * g)_n\}_k = NH_kG_k.$$

217. Generalized Hartley kernels. The function $\operatorname{cas} x = \cos x + \sin x$ is a sine function shifted by $\pi/4$ radians. A more general transform may be devised by using the kernel $\sqrt{2}\sin(x+\phi)$, where ϕ is any angle other than a multiple of $\pi/2$. Show that if this kernel is used in the forward transform, the kernel for the inverse transform is $\sqrt{\cot\phi}\sin x + \sqrt{\tan\phi}\cos x$ [14].

Chapter **9**

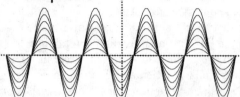

Quadrature and the DFT

It [Fourier series] reveals the transcendence of analysis over geometrical perception. It signalizes the flight of human intellect beyond the bounds of the senses.
– Edward B. Van Vleck, 1914

9.1.   Introduction

The theory and techniques of **numerical integration**, or **quadrature**, comprise one of the truly venerable areas of numerical analysis. Its history reflects contributions from some of the greatest mathematicians of the past three hundred years. The subject is very tangible since most methods are ultimately designed for the practical problem of approximating the area of regions under curves. And yet, as we will see, there are some subtleties in the subject as well. It is not surprising that eventually the DFT and quadrature should meet. After all, the DFT is an approximation to Fourier coefficients or Fourier transforms, both of which are defined in terms of integrals. The design of quadrature rules for oscillatory integrands (particularly Fourier integrals) goes back *at least* to the work of Filon in 1928 [58], and the literature is filled with contributions to the subject throughout the intervening years [22], [45], [80]. During the 1970s, the subject of the DFT as a quadrature rule was revisited in a flurry of correspondences [1], [2], [56], [105], that provided provocative numerical evidence. The subject remains important today, as practitioners look for the most accurate and efficient methods for approximating Fourier integrals [95]. The goal of this section is to explore the fundamental connections between the DFT and quadrature rules, to resolve a few subtle points, and to provide a survey of the work that has been done over many years.

In previous chapters we considered how the DFT can be used to approximate both Fourier coefficients and Fourier transforms. Relying heavily on the Poisson Summation Formula, it was possible to develop accurate error bounds for these approximations. In the process, we were actually using the DFT as a quadrature rule to approximate definite integrals; yet, ironically, the notion of quadrature errors was never mentioned. In this chapter we will establish the DFT as a *bona fide* quadrature rule; in fact, as we have seen, it is essentially the familiar trapezoid rule. From this vantage point, it will be possible to gain complementary insights into the DFT and its errors. The DFT error bounds that arise in the quadrature setting rely on the powerful Euler–Maclaurin Summation Formula. These error bounds are stronger than the results of the previous chapters in some cases, and weaker in others. We will also look at how higher-order (more accurate) quadrature rules can be related to the DFT. There are cases in which higher-order methods provide better approximations to Fourier coefficients, and other situations in which the DFT appears to be nearly optimal. As usual, the journey is as rewarding as the final results, so let us begin.

9.2.   The DFT and the Trapezoid Rule

For ease of exposition we use the interval $[-\pi, \pi]$, although everything that is said and done can be applied to any finite interval. We will consider the problem of computing the Fourier coefficients of a function f that is defined on the interval $[-\pi, \pi]$. As we have already seen, these coefficients are given by

$$c_k = \frac{1}{2\pi} \int_{-\pi}^{\pi} f(x)e^{-ikx}dx \quad \text{for} \quad k = 0, \pm 1, \pm 2 \ldots$$

(corresponding to $A = 2\pi$). Note that the problem of approximating $\int_{-\pi}^{\pi} f(x)dx$ is just the special case of approximating $2\pi c_0$. Recall how the trapezoid rule is used to approximate such an integral. The interval $[-\pi, \pi]$ is first divided into N subintervals

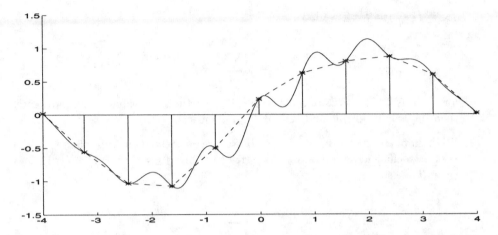

FIG. 9.1. *The trapezoid rule for $\int_{-\pi}^{\pi} g(x)dx$ results when the integrand g is replaced by a piecewise linear function over N subintervals of $[-\pi, \pi]$. The area under the piecewise linear curve is the sum of areas of trapezoids.*

of width $2\pi/N$. To use fairly standard notation, we will now let h rather than Δx denote the width of these subintervals. The resulting grid points will be denoted $x_n = nh$, where $n = -N/2 : N/2$, with $x_{\pm\frac{N}{2}} = \pm\pi$. For the moment, let the integrand be denoted $g(x) = f(x)e^{-ikx}$. Then the trapezoid rule approximation to $\int_{-\pi}^{\pi} g(x)dx$ with N subintervals (Figure 9.1) results when the integrand g is approximated by a piecewise linear function (connected straight line segments). The area of the resulting N trapezoids is

$$T_N\{g\} = \frac{2\pi}{N}\left[\frac{1}{2}g(x_{-\frac{N}{2}}) + \sum_{-\frac{N}{2}+1}^{\frac{N}{2}-1} g(x_n) + \frac{1}{2}g(x_{\frac{N}{2}})\right],$$

where we have introduced the notation $T_N\{g\}$ to stand for the N-point trapezoid rule approximation with g as an integrand.

One small rearrangement of this expression will be very instructive. Letting $g(x) = f(x)e^{-ikx}$ and $x_n = 2\pi n/N$, we can write

$$\begin{aligned}
c_k &\approx \frac{1}{2\pi}T_N\{f(x)e^{-ikx}\} \\
&= \frac{1}{2\pi}\frac{2\pi}{N}\left[\frac{1}{2}f(-\pi)\cos(\pi k) + \sum_{-\frac{N}{2}+1}^{\frac{N}{2}-1} f(x_n)e^{-ikx_n} + \frac{1}{2}f(\pi)\cos(\pi k)\right] \\
&= \frac{\cos(\pi k)}{2N}f(-\pi) + \frac{1}{N}\sum_{-\frac{N}{2}+1}^{\frac{N}{2}-1} f(x_n)e^{-i\frac{2\pi nk}{N}} + \frac{\cos(\pi k)}{2N}f(\pi).
\end{aligned}$$

This last expression begins to resemble the DFT. Two cases now present themselves.

1. If the given function f satisfies the condition that $f(-\pi) = f(\pi)$, then the first and last terms of the rule may be combined. Letting $f_n = f(x_n)$ and

$f_{\frac{N}{2}} = f(-\pi) = f(\pi)$, we have

$$\frac{1}{2\pi}T_N\{f(x)e^{-ikx}\} = \frac{1}{N}\sum_{n=-\frac{N}{2}+1}^{\frac{N}{2}} f_n e^{-i2\pi nk/N} = \mathcal{D}\{f_n\}_k$$

for $k = -N/2+1 : N/2$. In other words, if f has the same value at the endpoints, then the trapezoid rule and the DFT are identical.

2. What happens if $f(-\pi) \neq f(\pi)$? If we are interested in approximating the Fourier coefficients c_k, then we can agree to define $f_{\frac{N}{2}} = \frac{1}{2}(f(-\pi) + f(\pi))$, and we *still* have

$$\frac{1}{2\pi}T_N\{f(x)e^{-ikx}\} = \frac{1}{N}\sum_{n=-\frac{N}{2}+1}^{\frac{N}{2}} f_n e^{-i2\pi nk/N} = \mathcal{D}\{f_n\}_k$$

for $k = -N/2+1 : N/2$. Notice that this "agreement" about $f_{\frac{N}{2}}$ is not arbitrary. Letting $f_{\frac{N}{2}}$ be the average of the endpoint values is precisely the requirement (AVED) that we have used consistently to define the input to the DFT.

In either case, we see that the trapezoid rule applied to $f(x)e^{-ikx}$ gives the same result as the DFT of the sequence f_n, provided that average values are used at endpoints and discontinuities in defining f_n. For this reason we will identify the trapezoid rule with the DFT for the remainder of this chapter.

There is a wealth of theory about the error in quadrature rules such as the trapezoid rule. It seems reasonable to appeal to those results to estimate the error in the trapezoid rule (DFT) approximation to Fourier coefficients. The most familiar statement about the error in the trapezoid rule can be found in every numerical analysis book (e.g., [22], [45], [80], [166]), and for our particular problem it looks like this.

THEOREM 9.1. ERROR IN THE TRAPEZOID RULE. *Let $g \in C^2[-\pi, \pi]$ (that is, g has at least two continuous derivatives on $[-\pi, \pi]$). Let $T_N\{g\}$ be the trapezoid rule approximation to $\int_{-\pi}^{\pi} g(x)dx$ using N uniform subintervals. Then the error in the trapezoid rule approximation is*

$$E_N \equiv T_N\{g\} - \int_{-\pi}^{\pi} g(x)dx = \frac{\pi}{6}\left(\frac{2\pi}{N}\right)^2 g''(\xi) = \frac{\pi}{6}h^2 g''(\xi)$$

where $-\pi < \xi < \pi$.

The most significant message in this result is that the error in the trapezoid rule decreases as h^2 (equivalently as N^{-2}), meaning that if h is halved (or N is doubled) we can expect a roughly fourfold decrease in the error. We will observe this pattern in many examples. This theorem will now be used to estimate the error in trapezoid rule approximations to the Fourier coefficients of f. On the face of it, this theorem seems to say that if $f \in C^2[-\pi, \pi]$, then the error should decrease as N^{-2} as the number of grid points N increases. We might suspect already that this result does not tell the whole story, since we know that there are situations in which the error in the DFT decreases more rapidly than this; indeed, there are instances in which the DFT is exact. Let's investigate a little further. In our particular case, $g(x) = f(x)e^{-ikx}$, and it follows that

$$g''(x) = e^{-ikx}(f''(x) - 2ikf'(x) - k^2 f(x)).$$

Since the exact location of the mystery point ξ is not known, it is difficult to evaluate $g''(\xi)$. It is customary at this point to settle for an error bound which takes the form

$$|E_N| \leq \frac{\pi}{6} \left(\frac{2\pi}{N}\right)^2 \sup_{(-\pi,\pi)} [|f''(x)| + 2|kf'(x)| + k^2|f(x)|].$$

In practice, the Fourier coefficients c_k would be approximated for $k = -N/2 + 1 : N/2$. Notice that for values of k near $\pm N/2$ this bound decreases very slowly with respect to N. In fact, for $k = N/2$ the bound is essentially independent of N. This is not consistent with observations that were made in Chapter 6, in which we found uniform bounds for the error for *all* $k = -N/2 + 1 : N/2$. The conclusion is that the standard trapezoid rule error theorem, when applied to the DFT, can be misleading. As we will soon see, it does not account for the possible continuity of higher derivatives of f; equally important, it does not reflect the periodicity of f and its derivatives which is often present in the calculation of Fourier coefficients.

In order to forge ahead with this question, we need a more powerful tool, and fortunately it exists in the form of the celebrated **Euler–Maclaurin[1] Summation Formula**. This remarkable result can be regarded, among other things, as a generalization of the trapezoid rule error theorem given above. In a form best suited to our purposes, it can be stated in the following theorem.

THEOREM 9.2. EULER–MACLAURIN SUMMATION FORMULA. *Let $g \in C^{2p+2}[-\pi, \pi]$ for $p \geq 1$, and let $T_N\{g\}$ be the trapezoid rule approximation to $\int_{-\pi}^{\pi} g(x)dx$ with N uniform subintervals. Then the error in this approximation is*

$$\begin{aligned} E_N &= T_N\{g\} - \int_{-\pi}^{\pi} g(x)dx \\ &= \sum_{m=1}^{p} \left\{ \left(\frac{2\pi}{N}\right)^{2m} \frac{B_{2m}}{(2m)!} \left[g^{(2m-1)}(\pi) - g^{(2m-1)}(-\pi)\right] \right. \\ &\qquad \left. + 2\pi \left(\frac{2\pi}{N}\right)^{2p+2} \frac{B_{2p+2}}{(2p+2)!} g^{(2p+2)}(\xi) \right\}, \end{aligned}$$

where $-\pi < \xi < \pi$, and B_n is the nth Bernoulli number (to be discussed shortly).

A proof of this theorem is not central to our present purposes, and a variety of proofs are easily found [22], [45], [166]. First let's interpret the result and then put it to use. We see that the theorem gives a representation for E_N, the error in the trapezoid rule approximation on the interval $[-\pi, \pi]$. In contrast to Theorem 9.1, the Euler–Maclaurin Formula expresses the error as a sum of terms plus one final term that involves the "last" continuous derivative of the integrand. Each successive term reflects additional smoothness of g and can be regarded as another correction to the error given in Theorem 9.1. We see that this error expansion accounts for the continuity of the higher derivatives of g as well as the endpoint values of g and its

---

[1] COLIN MACLAURIN (1698–1746) became a professor in Aberdeen, Scotland at the age of nineteen and succeeded James Gregory at the University of Edinburgh eight years later. He met Newton in 1719 and published the *Theory of Fluxions* elaborating Newton's calculus. The special case of Taylor's series known today as the Maclaurin series was never claimed by Maclaurin, who quite properly cited Brook Taylor and James Stirling for its discovery. Maclaurin and Euler independently published the Euler–Maclaurin Formula in about 1737 by generalizing a previous result of Gregory.

derivatives. The constants B_{2m} that appear are the **Bernoulli**[2] **numbers,** a few of which appear in Table 9.1. Note that although the terms of this sequence initially decrease in magnitude, they ultimately increase in magnitude very swiftly, a fact of impending importance (see problem 231).

TABLE 9.1
A few Bernoulli numbers.

n	0	1	2	4	6	8	10	12	14	16
B_n	1	$-\frac{1}{2}$	$\frac{1}{6}$	$-\frac{1}{30}$	$\frac{1}{42}$	$-\frac{1}{30}$	$\frac{5}{66}$	$-\frac{691}{2730}$	$\frac{7}{6}$	$-\frac{3617}{510}$

$B_1 = -1/2$, $B_{2m+1} = 0$ for $m \geq 1$.

It is instructive to consider some specific cases of the Euler–Maclaurin error expansion, and then generalizations will follow easily. Assume for the moment that we wish to approximate the Fourier coefficients of a function $f \in C^4[-\pi, \pi]$ using the trapezoid rule (DFT). Once again, let the full integrand be $g(x) = f(x)e^{-ikx}$, and consider the case in which $f(-\pi) \neq f(\pi)$. This condition will generally imply that $g'(-\pi) \neq g'(\pi)$ (problem 221). As a consequence, the Euler–Maclaurin formula with $p = 1$ tells us that

$$E_N = \frac{1}{12}\left(\frac{2\pi}{N}\right)^2 (g'(\pi) - g'(-\pi)) - \frac{\pi}{360}\left(\frac{2\pi}{N}\right)^4 g^{(iv)}(\xi). \qquad (9.1)$$

This is essentially the result given by Theorem 9.1, since by the mean value theorem $g'(\pi) - g'(-\pi) = 2\pi g''(\xi)$. Therefore, in this particular case, we expect the errors in the trapezoid rule to decrease as N^{-2}. Let's see how well this error estimate works.

Example: No endpoint agreement. Consider the problem of approximating the Fourier coefficients c_k of the function $f(x) = \sin((3x - \pi)/4)$ on $[-\pi, \pi]$. Although f is infinitely differentiable, $f(-\pi) \neq f(\pi)$, and we can expect Theorem 9.1 to apply to the trapezoid rule (DFT) errors. Table 9.2 shows the errors in the approximations to c_2, c_8, and c_{16} for several values of N. In all three cases, both the real and imaginary parts of the errors decrease almost precisely by a factor of four when N is doubled, *once N is sufficiently large.* The error can decrease more slowly when $k \approx N/2$; therefore, the coefficients c_8 and c_{16} require larger values of N before the N^{-2} behavior is observed. This behavior should be seen in light of the discussion in Chapter 6: as we saw there, the errors in approximating the Fourier coefficients of a function with different endpoint values decrease as N^{-1} for all $k = -N/2 + 1 : N/2$.

TABLE 9.2
Errors in trapezoid rule (DFT) approximations to
Fourier coefficients c_2, c_8, and c_{16} of $f(x) = \sin((3x - \pi)/4)$.

N	E_N for c_2	E_N for c_8	E_N for c_{16}
8	$7.0(-3) + 1.7(-2)i$	$-2.1(-1) + 2.0(-2)i$	$-2.1(-1) + 1.0(-2)i$
16	$1.6(-3) + 4.1(-3)i$	$2.8(-3) + 2.0(-2)i$	$-2.1(-1) + 1.0(-2)i$
32	$3.9(-4) + 1.0(-3)i$	$4.4(-4) + 4.3(-3)i$	$6.9(-4) + 1.0(-2)i$
64	$9.6(-5) + 2.6(-4)i$	$1.0(-4) + 1.0(-3)i$	$1.0(-4) + 2.1(-3)i$
128	$2.4(-5) + 6.4(-5)i$	$2.4(-5) + 2.6(-4)i$	$2.5(-5) + 5.1(-4)i$

---

[2] JAKOB BERNOULLI (1654–1705) was one member (together with his brother Johann and nephew Daniel) of a family of prolific Swiss mathematicians. Jakob occupied the chair at the University of Basel from 1687 until his death and did fundamental work in the calculus of variations, combinatorics, and probability.

Now let's assume that $f \in C^4[-\pi, \pi]$ has the additional properties $f(-\pi) = f(\pi)$ and $f'(-\pi) = f'(\pi)$. It then follows (problem 221) that $g'(-\pi) = g'(\pi)$ and the first term in Euler–Maclaurin error expression (9.1) vanishes. This leaves us with an error

$$E_N = -\frac{\pi}{360} \left(\frac{2\pi}{N}\right)^4 g^{(iv)}(\xi).$$

At first glance we seem to have an error term that decreases as N^{-4} with increasing N. Therefore, the Euler–Maclaurin formula predicts a faster convergence rate than Theorem 9.1 because the smoothness and endpoint properties of f have been recognized. And this faster convergence rate can be observed, provided that the frequency k is small. However, recalling that $g(x) = f(x)e^{-ikx}$, the derivative $g^{(iv)}$ has a term proportional to k^4. This means that for frequencies with k approaching $\pm N/2$, the error may decrease very slowly with N, and for the $k = N/2$ coefficient, the error may be insensitive to increases in N. These predictions should be compared to the error results that were derived using the Poisson Summation Formula. For the particular case at hand, $f^{(p)}(-\pi) = f^{(p)}(\pi)$ for $p = 0, 1$, Theorem 6.3 for periodic, non-band-limited functions asserts that the error in the DFT decreases as N^{-3} *for all* $k = -N/2 : N/2 - 1$. Therefore, the Euler–Maclaurin error bound predicts a faster rate of convergence as N increases, provided $k \ll N/2$. For $k \approx \pm N/2$, the error bound given in Chapter 6 is stronger. A short numerical experiment will be useful.

Example: Endpoint agreement of f and f'. Consider the function $f(x) = (x - 1)(\pi^2 - x^2)^2$, which has the property $f(-\pi) = f(\pi)$ and $f'(-\pi) = f'(\pi)$. Table 9.3 shows the trapezoid rule (DFT) errors in approximating c_2 and c_8 on $[-\pi, \pi]$. To highlight the rates at which the errors decrease, the "Factor" columns give the amount by which the real and imaginary parts of the error are reduced from their values in the previous row.

TABLE 9.3
Errors in trapezoid rule (DFT) approximations to
Fourier coefficients c_2 and c_8 of $f(x) = (x - 1)(\pi^2 - x^2)^2$.

N	E_N for $k = 2$	Factor	E_N for $k = 8$	Factor
8	$2.2(-2) - 2.9(-1)i$	–	$6.0(-3) - 1.5(-1)i$	–
16	$9.2(-4) - 1.6(-2)i$	2.3,18	$6.0(-3) - 1.5(-1)i$	8.5,1.0
32	$5.1(-5) - 1.0(-3)i$	18,16	$8.6(-5) - 4.8(-3)i$	7.0,3.1
64	$3.1(-6) - 6.1(-5)i$	16,16	$3.6(-6) - 2.6(-4)i$	24,18
128	$1.9(-7) - 3.8(-6)i$	16,16	$2.0(-7) - 1.5(-5)i$	18,17

The pattern in the error reduction as N increases is quite striking. For the approximations to c_2, the errors decrease by a factor of 16 almost immediately each time N is doubled. For the higher frequency coefficient c_8, larger values of N are needed before that same reduction rate is observed, all of which confirms the error bounds of Theorem 9.2.

Clearly, this pattern can be extended to higher orders of smoothness ($p > 1$). In general, if $g \in C^{2p+2}[-\pi, \pi]$ and g and its odd derivatives through order $2p - 1$ have equal values at the endpoints, then the error in using the trapezoid rule with N subintervals on $[-\pi, \pi]$ is given by

$$E_N = 2\pi \left(\frac{2\pi}{N}\right)^{2p+2} \frac{B_{2p+2}}{(2p + 2)!} g^{(2p+2)}(\xi)$$

where $-\pi < \xi < \pi$. For the problem of computing Fourier coefficients ($g(x) = f(x)e^{-ikx}$), the Euler–Maclaurin formula suggests that the error decreases as N^{-2p-2} as N increases. This rate of error reduction will generally be observed for the low frequency coefficients in which the term k^{2p+2}, arising from $g^{(2p+2)}$, does not compete with N^{2p+2}. This estimate is a stronger bound than that given by Theorem 6.3. However, for the high frequency coefficients (k near $\pm N/2$), the Euler–Maclaurin bound allows the possibility that the error could decrease very slowly with increasing N, and the error bound of Theorem 6.3 could become stronger. The way in which the two error results, one based on the Poisson Summation Formula and the other based on the Euler–Maclaurin Summation Formula, complement each other, is summarized in Table 9.4. In all cases, the function f is assumed to have as many continuous derivatives on $[-\pi, \pi]$ as needed. The Fourier coefficients c_k for $k = -N/2 + 1 : N/2$ are assumed to be approximated by an N-point DFT (trapezoid rule). The table shows how the k and N dependence of the error bound, as predicted by the two theories, varies with the behavior of f at the endpoints.

TABLE 9.4
Comparison of error bounds: Poisson summation
(Theorem 6.3) versus Euler–Maclaurin summation (Theorem 9.2).

Endpoint properties	Poisson Summation Formula (Theorem 6.3)	Euler–Maclaurin Formula (Theorem 9.2)
$f(-\pi) \neq f(\pi)$	N^{-1}	$(k/N)^2$
$f(-\pi) = f(\pi)$	N^{-2}	$(k/N)^2$ for cosine coeffs
		$(k/N)^4$ for sine coeffs
$f(-\pi) = f(\pi)$ $f'(-\pi) = f'(\pi)$	N^{-3}	$(k/N)^4$
$f^{(p)}(-\pi) = f^{(p)}(\pi)$ $p = 0, 1, 2$	N^{-4}	$(k/N)^4$ for cosine coeffs $(k/N)^6$ for sine coeffs
$f^{(p)}(-\pi) = f^{(p)}(\pi)$ $p = 0, 1, 2, 3$	N^{-5}	$(k/N)^6$

The Euler–Maclaurin error expansion also exposes an interesting and often-observed difference between approximating sine and cosine coefficients. Assume that $f \in C^4[-\pi, \pi]$ and that $f(-\pi) = f(\pi)$. Then the Euler–Maclaurin formula (9.1) predicts that the error in approximating the sine coefficients is

$$
\begin{aligned}
E_N &= T_N - \int_{-\pi}^{\pi} f(x) \sin kx \, dx \\
&= \frac{1}{12} \left(\frac{2\pi}{N} \right)^2 (g'(\pi) - g'(-\pi)) - \frac{\pi}{360} \left(\frac{2\pi}{N} \right)^4 g^{(iv)}(\xi)
\end{aligned}
$$

where $-\pi < \xi < \pi$, and where $g(x) = f(x)\sin(kx)$. It is easily shown (problem 221) that the difference $g'(\pi) - g'(-\pi)$ vanishes for the sine coefficient, but not for the cosine coefficient. This means that one could expect errors in the approximations to the sine coefficients to decrease as N^{-4}, while the usual N^{-2} convergence rate still governs the cosine coefficients. Of course, this rule extends to higher orders of smoothness, as shown in Table 9.4. This effect is easily observed.

Example: Improved sine coefficients. We will approximate the Fourier coefficients c_2 and c_8 of $f(x) = (x-1)(x^2 - \pi^2)$ on the interval $[-\pi, \pi]$ with N-point DFTs. This function satisfies $f(-\pi) = f(\pi)$. The results tabulated in Table 9.5 show the real and imaginary parts of the errors together with the factor by which they are reduced from the previous approximation. With a few anomalies along the way, the real part of the error (corresponding to the cosine coefficients) decreases by a factor of roughly four each time that N is doubled according to the N^{-2} dependence. The imaginary part of the errors (corresponding to the sine coefficients) decreases 16-fold when N is doubled, reflecting an N^{-4} dependence. In both cases, k is sufficiently small that the predicted dependence on N is actually observed for relatively small values of N. As expected, the approximations to the higher frequency coefficient are slightly less accurate.

TABLE 9.5

Errors in trapezoid rule (DFT) approximations to
Fourier coefficients c_2 and c_8 of $f(x) = (x-1)(x^2 - \pi^2)$.

N	E_N for $k = 2$	Factor	E_N for $k = 8$	Factor
8	$-1.2(-1) - 2.3(-2)i$	–	$6.5(0) - 1.2(-2)i$	–
16	$-2.7(-2) - 1.2(-3)i$	4.4,19	$-4.6(-2) - 1.2(-2)i$	14,1.0
32	$-6.4(-3) - 7.5(-5)i$	4.2,16	$-7.3(-3) - 3.6(-4)i$	6.3,33
64	$-1.6(-3) - 4.7(-6)i$	4.0,16	$-1.7(-3) - 1.9(-5)i$	4.3,19
128	$-4.0(-4) - 2.9(-7)i$	4.0,16	$-4.0(-4) - 1.1(-6)i$	4.2,17

The Euler–Maclaurin formula leads to some intriguing and subtle questions when we consider functions that are infinitely differentiable ($C^\infty[-\pi, \pi]$) with all derivatives 2π-periodic. One might be tempted to argue that in this case, all of the terms of the error expansion vanish and the trapezoid rule (DFT) should be exact for computing the Fourier coefficients of such functions. One need look no further than the following simple example [166] to see that there is a fallacy lurking. Surely $f(x) = \cos 4x$ is infinitely differentiable and all of its derivatives are 2π-periodic. Yet, in computing its Fourier coefficient c_0 exactly and by a trapezoid rule with $N = 4$ points, we see that

$$c_0 = \int_{-\pi}^{\pi} \cos 4x \, dx = 0$$

while

$$T_4\{\cos 4x\} = \frac{\pi}{2}\left(\frac{1}{2} + 1 + 1 + \frac{1}{2}\right) = 2\pi.$$

Clearly, the trapezoid rule is not exact. You are quite right if you claim that aliasing is the explanation for this error and argue that the error vanishes provided that N is large enough to resolve all of the modes of the integrand. Indeed, $T_N\{\cos 4x\} = 0$, provided that $N \geq 5$. But how does this failure of the trapezoid rule show up in the Euler–Maclaurin error bound, which *seems* to predict that the trapezoid rule should be exact? The answer is subtle and worth investigating.

Suppose that $g \in C^\infty[-\pi, \pi]$, and all of its derivatives are 2π-periodic. Let's look at the trapezoid rule approximations to $\int_{-\pi}^{\pi} g(x)dx$. Since all derivatives of g have equal values at $\pm\pi$, the Euler–Maclaurin error expansion says that

$$E_N = 2\pi \left(\frac{2\pi}{N}\right)^p \frac{B_p}{p!} g^{(p)}(\xi_p) \tag{9.2}$$

for arbitrarily large values of p, where $-\pi < \xi_p < \pi$. The critical question in determining whether or not the trapezoid rule is exact becomes the question of how this expression for E_N behaves as $p \to \infty$ (for fixed values of N). We will need some facts about the *asymptotic* behavior of the Bernoulli numbers B_p as $p \to \infty$. It is known [3] that

$$B_p \leq \frac{4p!}{(2\pi)^p} \quad \text{for} \quad p \geq 2 \quad \text{and} \quad B_p \sim \frac{2p!}{(2\pi)^p} \quad \text{as} \quad p \to \infty,$$

where we write $A \sim B$ as $p \to \infty$ if the ratio A/B approaches 1 as $p \to \infty$. Using the first of these relations in (9.2), we find that

$$|E_N| \leq 8\pi N^{-p} \max_{(-\pi,\pi)} |g^{(p)}(\xi_p)|. \tag{9.3}$$

Of course, this relationship is not too useful without an estimate of how $g^{(p)}$ behaves as p increases. Therefore, let us consider the example above, in which $g(x) = \cos kx$ and k is a positive integer. It follows that $|g^{(p)}(x)| \leq k^p$, and we can use (9.3) to deduce that

$$|E_N| \leq 8\pi \left(\frac{k}{N}\right)^p.$$

From this we may conclude that if $|k| < N$ then $\lim_{p\to\infty} |E_N| = 0$ and the trapezoid rule is exact. In the above example with $k = N = 4$, $\lim_{p\to\infty} |E_N| \neq 0$, and we expect the trapezoid rule to be in error. This result does not preclude the possibility that the trapezoid rule may be exact for some values of $k > N$. In fact (problem 222), if N is fixed, then $T_N\{\cos kx\}$ is exact as long as k is not a multiple of N. For example, $T_4\{\cos 7x\} = 0$, which is exact.

This may appear to contradict what was learned in Chapter 6 about the DFT for periodic band-limited functions. But notice that $\int_{-\pi}^{\pi} \cos(kx)dx$ can be viewed as the Fourier coefficient c_k for $f(x) = 1$. Chapter 6 taught us that the N-point DFT is exact in computing c_k, provided that $|k| \leq N/2$, and indeed it is in this case.

Example: Periodic C^∞ functions. The plot thickens when we consider 2π-periodic, C^∞ functions other than simple sines and cosines. Davis and Rabinowitz [45] suggest the function $g(x) = (1 + \sigma \sin mx)^{-1}$, which belongs to $C^\infty[-\pi, \pi]$ and has 2π-periodic derivatives of all orders when m is an integer (Figure 9.2). It can be shown that if $|\sigma| < 1$, then

$$\int_{-\pi}^{\pi} \frac{dx}{1 + \sigma \sin mx} = \frac{2\pi}{\sqrt{1 - \sigma^2}}$$

independent of m. (This result is most easily obtained using complex variables and the theory of residues.)

In order to form an error bound from the Euler–Maclaurin expansion, we must estimate $g^{(p)}$, particularly for large values of p. This is a rather cumbersome calculation (although $g^{(p)}$ appears *not* to be bounded as $p \to \infty$), so we turn directly to the numerical evidence as given in Table 9.6.

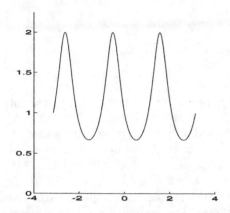 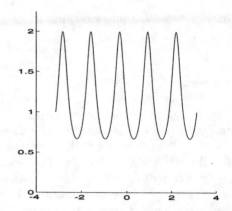

FIG. 9.2.   *The function* $g(x) = (1 + \sigma \sin mx)^{-1}$ *is shown using* $\sigma = .5$, *with* $m = 3$ *(left) and* $m = 5$ *(right). It is infinitely differentiable and all of its derivatives are* 2π*-periodic. The trapezoid rule applied to* $\int_{-\pi}^{\pi} g(x)dx$ *is extremely accurate for even moderate values of* N.

TABLE 9.6

Errors in trapezoid rule approximations to
$\int_{-\pi}^{\pi} (1 + \sigma \sin mx)^{-1} dx = 2\pi(1 - \sigma^2)^{-1/2}$, *with* $\sigma = 0.5$.

	$N = 8$	$N = 16$	$N = 32$	$N = 64$	$N = 128$
$m = 3$	3.9(−4)	1.0(−8)	0	0	0
$m = 5$	3.9(−4)	1.0(−8)	0	0	0
$m = 8$	−9.7(−1)	−9.7(−1)	7.5(−2)	3.9(−4)	1.0(−8)
$m = 17$	3.9(−4)	1.0(−8)	0	0	0
$m = 25$	3.9(−4)	1.0(−8)	0	0	0
$m = 40$	−9.7(−1)	−9.7(−1)	7.5(−2)	3.8(−4)	1.0(−8)

$a(-n)$ *means* $a \times 10^{-n}$, *while* 0 *means the error is* $< 10^{-10}$.

There are patterns and irregularities in these errors that are rather mysterious. Most importantly, note that for many values of m the trapezoid rule is exact (error denoted 0) even for small values of N. Surely this reflects the fact that derivatives of g of all orders are periodic. However, two choices of m (and undoubtedly there are others) show relatively slow decay of errors as N increases. This fact requires further investigation, and we cannot offer a full explanation at this time.

As the two previous examples indicate, the trapezoid rule can behave in unexpected ways for C^∞ functions: occasionally, it is more accurate than the theory predicts, which is always a welcome occurrence. It would be nice to say something conclusive about when the trapezoid rule is exact, but a general statement is elusive. This is partly due to the presence of the intermediate point ξ in the error term of the Euler–Maclaurin expansion, which can always conspire to make the error much smaller than the bounds suggest. Nevertheless, here is a modest statement about when the trapezoid rule is exact for C^∞ functions.

THEOREM 9.3.   ZERO ERROR IN THE TRAPEZOID RULE. *Assume* $f \in C^\infty[-\pi, \pi]$ *and that* f *and all of its derivatives are* 2π*-periodic. Let* $|f^{(p)}(x)| \le \alpha^p$ *on* $[-\pi, \pi]$ *for some* $\alpha > 0$ *and for all* $p \ge 0$. *Then the trapezoid rule (DFT) is exact when used to* *approximate the Fourier coefficient*

$$c_k = \frac{1}{2\pi} \int_{-\pi}^{\pi} f(x)e^{-ikx}dx,$$

provided that $\alpha < |k| < N$.

This theorem gives sufficient conditions for the trapezoid rule to be exact, but the conditions appear to be rather weak. As the above examples illustrate, the trapezoid rule can be unexpectedly accurate or virtually exact in situations that do not meet the conditions of this theorem. The proof of the theorem collects several of the observations already made in the chapter, and is left as an exercise (problem 225).

Before parting with the relationship between the trapezoid rule and the DFT, we will consider one final matter. It merits attention because it pertains to the persistent issue of proper treatment of endpoints. The endpoint question appears in a slightly different form as a technique known as **subtracting the linear trend**. This idea, attributed to Lanczos [1], [91], is developed fully in problems 219 and 220; we will sketch the essentials here. Assume that the Fourier coefficients of f are to be approximated on $[-\pi, \pi]$, and that although f is continuous on $[-\pi, \pi]$, it does not satisfy $f(-\pi) = f(\pi)$. Then the DFT will see a discontinuity in the input, and by the error results of Chapter 6, we expect the error to decrease rather slowly as N^{-1}. Alternatively, by the Euler–Maclaurin error bound we expect errors that decrease as $(k/N)^2$. We now define the linear function

$$\ell(x) = \frac{(x+\pi)f(\pi) - (x-\pi)f(-\pi)}{2\pi},$$

which satisfies $\ell(-\pi) = f(-\pi)$ and $\ell(\pi) = f(\pi)$. It follows that the auxiliary function $\phi(x) = f(x) - \ell(x)$ has the property that $\phi(-\pi) = \phi(\pi) = 0$, and hence the DFT applied to ϕ should have errors that are smaller than when the DFT is applied directly to f. Furthermore, the Fourier coefficients of ℓ can be determined exactly, which allows the DFT of f to be recovered. This technique results in approximations that converge more rapidly to the Fourier coefficients of f as N increases. Let's see how well it works.

Example: Subtracting the linear trend. We return to the function $f(x) = \sin((3x - \pi)/4)$ considered in an earlier example (see Table 9.2). This function has the property that $f(-\pi) \neq f(\pi)$, and as we saw, the DFT approximations converge to the Fourier coefficients c_k at a rate proportional to $(k/N)^2$. As shown in Figure 9.3, if the function $\ell(x) = (x + \pi)/2\pi$ is subtracted from f, the resulting function $\phi = f - \ell$ satisfies $\phi(-\pi) = \phi(\pi) = 0$. Table 9.7 shows the errors that result when the DFT with the linear trend subtracted is used to approximate three Fourier coefficients of f.

TABLE 9.7

Errors in approximations to Fourier coefficients c_2, c_8, and c_{16} of $f(x) = \sin((3x + \pi)/4)$ with linear trend subtracted.

N	E_N for c_2	E_N for c_8	E_N for c_{16}
8	$7.0(-3) - 3.5(-4)i$	$-2.1(-1) - 1.8(-4)i$	$-2.1(-1) - 2.2(-5)i$
16	$1.6(-3) - 1.9(-5)i$	$2.8(-3) - 1.8(-4)i$	$-2.1(-1) - 2.2(-5)i$
32	$3.9(-4) - 1.1(-6)i$	$4.4(-4) - 5.4(-6)i$	$6.9(-4) - 2.2(-5)i$
64	$9.6(-5) - 7.0(-8)i$	$1.0(-4) - 2.9(-7)i$	$1.0(-4) - 6.8(-7)i$
128	$2.4(-5) - 4.3(-9)i$	$2.4(-5) - 1.8(-8)i$	$2.5(-5) - 3.6(-8)i$
RF	4.0,16	4.2,16	4.0,19

RF = Reduction Factor = ratio of errors in last two rows.

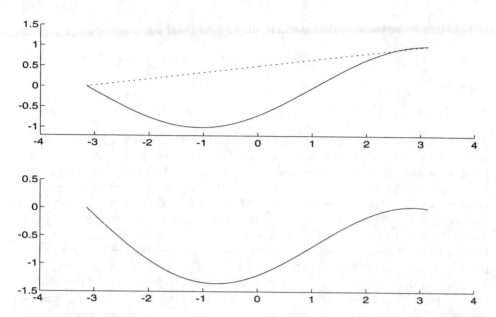

FIG. 9.3. *The top figure is a graph of the function $f(x) = \sin((3x - \pi)/4)$ on the interval $[-\pi, \pi]$. If the function $\ell(x) = (x + \pi)/2\pi - 1$ (dotted line in top figure) is subtracted from f, the resulting function ϕ satisfies $\phi(-\pi) = \phi(\pi) = 0$ (bottom figure). The DFT converges more quickly as N increases when applied to ϕ, rather than f, and produces better approximations to the sine coefficients.*

This table should be compared to Table 9.2, in which the DFT errors *without* subtracting the linear trend are recorded. The approximations to the real part (cosine coefficients) are identical in both tables; in other words, subtracting the linear trend offers no improvement in the cosine coefficients, and the errors in these approximations decrease as $(k/N)^2$, as indicated by the reduction factor. However, the errors in the approximations to the sine coefficients (the imaginary parts of the errors) are improved significantly when the linear trend is subtracted showing a convergence rate proportional to $(k/N)^4$. We also see the typically slower convergence for the larger values of k. The explanation for this occurrence is pursued in problem 221.

9.3. Higher-Order Quadrature Rules

Up until now we have focused on the trapezoid rule because of its virtual identity with the DFT. However, a few words should be offered about the use of other quadrature rules to approximate Fourier integrals. As mentioned earlier, **Filon's rule** [45], [58] dates from 1928 and is designed for Fourier integrals. The method is most easily described as it applies to the integrals

$$\int_{-\pi}^{\pi} f(x) \sin kx\, dx \quad \text{and} \quad \int_{-\pi}^{\pi} f(x) \cos kx\, dx.$$

The interval $[-\pi, \pi]$ is divided into $2N$ subintervals of equal width h, and over each pair of subintervals (taking groups of three points) the function f is approximated by a quadratic interpolating polynomial. The products of these quadratics with $\sin kx$

and $\cos kx$ can be integrated analytically to give a rule of the form

$$\int_{-\pi}^{\pi} f(x)\sin kx dx \approx h[-\alpha\cos(\pi k)(f(\pi) - f(-\pi)) + \beta S_N + \gamma S_N']$$

and

$$\int_{-\pi}^{\pi} f(x)\cos kx dx \approx h[\beta C_N + \gamma C_N'].$$

The constants α, β, γ are easily computed, and the terms S_N, S_N', C_N, and C_N' are simple sums, with $N \pm 1$ terms, involving $f(x)\sin kx$ and $f(x)\cos kx$ evaluated at the grid points. The method is easy to implement and has an error proportional to h^4 provided $kh << 1$. It can also be generalized to higher-order interpolating polynomials [56]. Not surprisingly, the method performs best when f is smooth and when k is small. In practice, $N \approx 10k$ (ten points per period) seems to give good results, although larger values of N must be used if f is also oscillatory. However, there is an interesting connection between Filon's rule and the DFT. The sums S_N, S_N', C_N, and C_N' are essentially trapezoid rule sums on either the even or odd points. Therefore, if a full set of N Fourier coefficients is needed, these sums (which comprise most of the computational effort) can be done using the DFT, which, in practice, means the FFT. As we will show momentarily, Filon's rule gives approximations to Fourier integrals which are generally as accurate as DFT approximations. Thus, when used in conjunction with the FFT, it presents a very efficient method.

The other direction in which one might look for more accurate approximations to Fourier integrals is toward even higher-order quadrature rules. In fact, there is a temptingly simple way in which higher-order quadrature rules might be married to the DFT (and the FFT). Suppose that a particular quadrature rule has the form

$$\int_{-\pi}^{\pi} g(x)dx \approx h \sum_{n=-\frac{N}{2}+1}^{\frac{N}{2}} \alpha_n g(x_n),$$

where the x_n's are uniformly spaced points on $[-\pi, \pi]$, h is the grid spacing, and the α_n's are known weights. To approximate the Fourier coefficients of a function f using this rule, we have that

$$\frac{1}{2\pi}\int_{-\pi}^{\pi} f(x)e^{-ikx}dx \approx \frac{h}{2\pi} \sum_{n=-\frac{N}{2}+1}^{\frac{N}{2}} \alpha_n f(x_n)\omega_N^{-nk} = \mathcal{D}\{\alpha_n f_n\}_k.$$

It appears possible to obtain more accurate (higher-order) approximations simply by applying the DFT to a slightly modified input sequence ($\alpha_n f_n$ rather than f_n), at virtually no extra cost (problem 224). Indeed, this approach has been used in practice with many well-known quadrature rules including Filon's rule. *When* this maneuver works, it provides an effective method for computing Fourier coefficients. However, it must be used with some care: as is well known, some families of quadrature rules (for example, **Newton[3]–Cotes[4] rules**) actually become less accurate as the order increases.

---

[3]ISAAC NEWTON was born in Lincolnshire, England in 1642. At the age of 18, he attended Cambridge University, where he met the well-known professor of Greek and mathematics Isaac Barrow. Newton conceived of his method of fluxions, which led to the calculus in 1665, while working on the quadrature of curves. The *Philosophae Naturalis Principia Mathematica*, which contains his calculus and the theory of gravitation, appeared in 1687. Newton was 85 years old when he died.

[4]ROGER COTES was born in 1682 and, as a professor of astronomy and natural philosophy

We will illustrate some general conclusions about higher-order methods with the well-known **Simpson's[5] rule**. There are several ways to derive Simpson's rule; the most common approach replaces the integrand f by quadratic interpolating polynomials on pairs of subintervals. However, there is another way to arrive at Simpson's rule, and it is so useful that it merits a quick glimpse. We begin with the trapezoid rule and assume that it is applied to the integral

$$I = \int_{-\pi}^{\pi} g(x)dx$$

with a grid spacing of $h = 2\pi/N$. We will assume that the function g has several continuous derivatives, but for the moment no special endpoint conditions. As we have seen, the error in this trapezoid rule approximation can be expressed as

$$I - T_N\{g\} = c_1 h^2 + c_2 h^4 + c_3 h^6 + \ldots, \qquad (9.4)$$

where the constants c_i are known from the Euler–Maclaurin formula. In other words, the error has an expansion in even powers of h that can be continued to h^{2p} provided $g^{(2p)}$ is continuous on $[-\pi, \pi]$.

The goal is to obtain a better approximation: one whose error is proportional not to h^2, but to a higher power of h. Toward this end, imagine that a second trapezoid rule approximation to I is computed with $2N$ subintervals and a grid spacing of $h/2$. We could then write that

$$I - T_{2N}\{g\} = c_1 \left(\frac{h}{2}\right)^2 + c_2 \left(\frac{h}{2}\right)^4 + c_3 \left(\frac{h}{2}\right)^6 + \ldots. \qquad (9.5)$$

The question is whether these two trapezoid rule approximations (and their error expansions) can be combined to give a higher-order method. Indeed they can: if the first trapezoid rule (9.4) is subtracted from four times the second (9.5), we find that

$$I - \underbrace{\left[\frac{4T_{2N} - T_N}{3}\right]}_{S_{2N}} = -\frac{3}{4}c_2 h^4 - \frac{15}{16}c_3 h^6 - \ldots.$$

We see that a simple combination of the two trapezoid rules, which we have denoted S_{2N}, differs from the exact value of the integral I by an amount that depends upon h^4. The goal has been accomplished: we have a new higher-order rule called S_{2N}, and it can be obtained with virtually no work once the two trapezoid rule approximations have been computed. It may not be obvious, but the new rule S_{2N} is just Simpson's rule based on $2N$ uniform subintervals (problem 223). This is the first step of a process called **Richardson[6] extrapolation** [117] or simply **extrapolation**.

But why stop here? We see that the error in the rule S_{2N} also has an expansion in powers of h, assuming that g is sufficiently differentiable. Therefore, another

at Cambridge, assisted Newton with the second edition of his *Principia*. Cotes created many computational tables and died at the age of 33.

[5] THOMAS SIMPSON (1710–1761) was a self-taught English mathematican who is also known for discoveries in plane geometry and trigonometry. There is evidence suggesting that Simpson's rule was known to Cavalieri in 1639 and to Gregory in about 1670.

[6] LEWIS FRY RICHARDSON (1881–1953) was an eccentric and visionary English meteorologist who conceived of numerical weather forecasting in the 1920s. He also carried out experimental work (measuring coastlines) that led to the formulation of fractal geometry.

extrapolation step may be done to eliminate the leading h^4 error term. The result is a method that involves two Simpson's rule approximations with a leading error term proportional to h^6. We will not carry this entire procedure to its conclusion (problem 223) except to say that the repeated use of extrapolation leads to a systematic quadrature method called **Romberg integration** [22], [80], [120]. Perhaps even more important is the fact that the idea of extrapolation needn't be confined to the trapezoid rule. It can be used with other quadrature rules (for example, Filon's rule [56]), it can be used to improve approximations to derivatives, and it finds use in the numerical solution of differential equations.

But we have diverged a bit. What does this say about DFT approximations to Fourier coefficients? There are two situations to discuss. Letting $g(x) = f(x)e^{-ikx}$, we can apply the above remarks to the computation of the Fourier coefficients of f on $[-\pi, \pi]$. If f has $2p + 2$ continuous derivatives on $[-\pi, \pi]$, but no special endpoint properties, then extrapolation may be carried out (p times in fact) to provide improved approximations over the original trapezoid rule (DFT) approximations. Practical experience suggests that the work required to do one or two extrapolations is justified by the improvement in accuracy. Often, lack of smoothness of f obviates the value of further extrapolations. A simple example demonstrates the effectiveness of extrapolation.

Example: Extrapolation in quadrature. The function $f(x) = x^2 - \pi^2$ is real and even, therefore its Fourier coefficients are also real and even. Since $f(-\pi) = f(\pi)$, but $f'(-\pi) \neq f'(\pi)$, we expect the errors in the DFT approximations to the Fourier coefficients c_k to decrease as N^{-2} for $k << N/2$. Therefore, extrapolation should offer some improvements at little additional cost. Table 9.8 shows the errors in the approximation to c_2.

TABLE 9.8
Errors in approximations to the Fourier coefficient c_2 of
$f(x) = x^2 - \pi^2$ by extrapolation from the DFT.

N	E_N for T_N	E_N for first extrap.	E_N for second extrap.	E_N for third extrap.
16	9.3(−1)	−	−	−
32	1.1(−1)	−1.6(−1)	−	−
64	1.2(−2)	−2.0(−2)	−1.0(−2)	−
128	2.6(−3)	−5.9(−4)	−7.2(−4)	9.0(−4)
256	6.4(−4)	−3.2(−5)	−5.6(−6)	−5.7(−7)
512	1.6(−4)	−1.9(−6)	−7.6(−8)	−1.2(−8)
RF	4	16	73	475

RF = Reduction Factor = ratio of errors in last two rows.

The errors in the trapezoid rule (DFT) approximations (second column) decrease by a factor of four each time N is doubled, abiding by the expected N^{-2} convergence rate. The errors in the first extrapolation (Simpson's rule in the third column) show a strict reduction by a factor of 16 each time N is doubled, conforming to the N^{-4} convergence rate. The second and third extrapolations show reductions of factors exceeding 64 and 256 each time N is doubled, according to the expected N^{-6} and N^{-8} convergence rates. The RF row gives the actual factor by which the error is reduced between $N = 256$ and $N = 512$. This is a convincing demonstration of the remarkable effectiveness of extrapolation applied to quadrature.

The other situation to consider is that in which f does have some special periodicity or endpoint properties. We will now assume that f and its first $2p + 2$ derivatives are 2π-periodic and continuous. This means that $g(x) = f(x)e^{-ikx}$ also has this degree of smoothness and periodicity. The following conclusions also apply to nonperiodic functions whose odd derivatives happen to agree at the endpoints (problem 230). If the trapezoid rule (DFT) with N subintervals is used to approximate the Fourier coefficients of f, then as we have seen from the Euler–Maclaurin Summation Formula, the error in the approximation is given by

$$E_N = c_p h^{2p+2} g^{(2p+2)}(\xi)$$

where $-\pi < \xi < \pi$, and c_p is a known constant given by the Euler–Maclaurin remainder term. Suppose we now wish to obtain a better approximation by extrapolation (which is equivalent to using a higher-order method). The error in the DFT no longer has an expansion in powers of h, since the periodicity of f and its derivatives has reduced this expansion to a single term. Without an error expansion, extrapolation cannot be done, or more precisely, extrapolation will not lead to more accurate approximations. Of course, Simpson's rule or extrapolation may still be applied, but in general, we should expect no improvement over the "superconvergent" approximations given by the trapezoid rule (DFT). The trapezoid rule applied to periodic integrands is exceptionally accurate, and the resulting approximations have as much accuracy as the smoothness of the integrand allows.

One last numerical experiment will allow us to demonstrate many of the foregoing remarks. We close this section with a case study of a family of functions that has been used often to demonstrate the performance of the DFT as a quadrature rule [1], [45].

Example: A comparison of all methods. We will consider the problem of approximating the Fourier coefficients c_k of the function $f(x) = x \cos k_0 x$ on the interval $[0, 2\pi]$ where k_0 is an integer. The exact values of the coefficients c_k can be determined for any integers k and k_0; note that neither the real nor the imaginary part of c_k vanishes identically. With these exact values of the coefficients, the errors in various approximations can be computed easily. Specifically, we will investigate the performance of the trapezoid rule (DFT) with and without the linear trend subtracted (denoted $\mathcal{D}_N^*$ and $\mathcal{D}_N$, respectively), Simpson's rule (denoted S_N), and Filon's rule (denoted F_N) for various values of k, k_0, and N. The errors in these approximations are given in Table 9.9. When errors are less that 10^{-15}, as is the case with many of the approximations to the cosine coefficient, the error is given a value of zero.

Many of the conclusions of this section are demonstrated quite convincingly in this table. As mentioned, the first three methods produce very accurate approximations to the cosine coefficients

$$\operatorname{Re}\{c_k\} = \frac{1}{2\pi} \int_0^{2\pi} x \cos(k_0 x) \cos(kx) dx.$$

A quick calculation (problem 226) reveals that the integrand $g(x) = x \cos(k_0 x) \cos(kx)$, while not periodic, satisfies the condition that $g^{(2p-1)}(0) = g^{(2p-1)}(2\pi)$ for $p \geq 1$ provided k and k_0 are integers. The Euler–Maclaurin formula suggests that the trapezoid rule may be extraordinarily accurate in this case, and indeed it is virtually exact. We see that $\mathcal{D}_N^*$ and S_N also inherit this superaccuracy, but there is no reason to incur the extra expense of these rules. Depending on the choice of k and k_0, F_N can also be extremely accurate.

TABLE 9.9
Errors in approximations to Fourier coefficients of $f(x) = x \cos k_0 x$;
$\mathcal{D}^*$ and $\mathcal{D}$ = DFT with and without linear trends subtracted, respectively;
S_N = Simpson's rule, while F_N = Filon's rule.

k_0	k	N	Error in $\mathcal{D}_N$	Error in $\mathcal{D}_N^*$	Error in S_N	Error in F_N
1	1	32	$2.0(-2)i$	$3.9(-5)i$	$-2.1(-4)i$	$2.4(-4) - 2.7(-4)i$
–	–	64	$5.0(-3)i$	$2.4(-6)i$	$-1.3(-5)i$	$1.5(-5) - 1.7(-5)i$
–	–	128	$1.3(-3)i$	$1.5(-7)i$	$-8.1(-7)i$	$9.5(-7) - 1.0(-6)i$
–	–	256	$3.2(-4)i$	$9.5(-9)i$	$-5.1(-8)i$	$6.0(-8) - 6.6(-8)i$
–	–	512	$7.9(-5)i$	$5.9(-10)i$	$-3.2(-9)i$	$3.7(-9) - 4.1(-9)i$

k_0	k	N	Error in $\mathcal{D}_N$	Error in $\mathcal{D}_N^*$	Error in S_N	Error in F_N
1	20	32	$5.7(-1)i$	$3.7(-3)i$	$1.1(0)i$	$7.5(-5)i$
–	–	64	$1.1(-1)i$	$6.7(-5)i$	$-4.7(-2)i$	$9.6(-7)i$
–	–	128	$2.6(-2)i$	$3.3(-6)i$	$-1.8(-3)i$	$2.0(-8)i$
–	–	256	$6.3(-3)i$	$1.9(-7)i$	$-1.0(-4)i$	$-1.8(-0)i$
–	–	512	$1.6(-3)i$	$1.2(-8)i$	$-6.4(-6)i$	$-1.2(-10)i$

k_0	k	N	Error in $\mathcal{D}_N$	Error in $\mathcal{D}_N^*$	Error in S_N	Error in F_N
20	1	32	$5.6(-2)i$	$3.5(-2)i$	$-9.1(-2)i$	$8.9(-2)i$
–	–	64	$6.2(-3)i$	$1.1(-3)i$	$-1.0(-2)i$	$1.0(-2)i$
–	–	128	$1.3(-3)i$	$6.3(-5)i$	$-3.0(-4)i$	$3.0(-4)i$
–	–	256	$3.2(-4)i$	$3.8(-6)i$	$-1.6(-5)i$	$1.6(-5)i$
–	–	512	$7.9(-5)i$	$2.4(-7)i$	$-9.6(-7)i$	$9.6(-7)i$

k	k_0	N	Error in $\mathcal{D}_N$	Error in $\mathcal{D}_N^*$	Error in S_N	Error in F_N
20	20	32	$-2.3(-1)i$	$8.1(-1)i$	$-3.3(-1)i$	$8.3(0) + 3.0(-1)i$
–	–	64	$1.4(-1)i$	$3.4(-2)i$	$2.6(-1)i$	$1.6(0) - 3.3(-2)i$
–	–	128	$2.7(-2)i$	$1.4(-3)i$	$-1.1(-2)i$	$1.4(-1) + 9.3(-2)i$
–	–	256	$6.4(-3)i$	$7.8(-5)i$	$-4.6(-4)i$	$9.3(-3) + 5.4(-4)i$
–	–	512	$1.6(-3)i$	$4.8(-6)i$	$-2.6(-5)i$	$5.9(-4) + 3.3(-5)i$

No error shown means the error is $< 10^{-15}$, and $a(-n)$ means $a \times 10^{-n}$.

Turning to the sine coefficients

$$\text{Im}\,\{c_k\} = \frac{1}{2\pi} \int_0^{2\pi} x \cos(k_0 x) \sin(kx)\, dx,$$

we see the errors in the DFT approximations (indicated $\mathcal{D}_N$) decreasing at the predictable rate of N^{-2} as N is increased. When the linear trend is subtracted (indicated by $\mathcal{D}_N^*$), the accuracy of the approximations is improved momentously, exceeding that of the Simpson's rule approximations (S_N). The errors in both $\mathcal{D}_N^*$ and S_N decrease as N^{-4} once N is sufficiently large relative to k and k_0. Filon's rule performs extremely well in all cases. The errors in all of its approximations decrease strictly in an N^{-4} fashion except in the two cases in which the method is exact in approximating cosine coefficients. With all of the methods, less accurate approximations are obtained (for the same values of N) for both higher frequency coefficients (k large) and for more oscillatory integrands (k_0 large). It is remarkable that overall the DFT with the linear trend subtracted ($\mathcal{D}_N^*$) is comparable, if not superior, to the higher-order and more expensive methods.

It would be nice to conclude with some sweeping statements about the best quadrature method for computing Fourier integrals. Unfortunately, such a statement most likely does not exist in any generality, in part because of the capricious nature of the error terms. We have seen that, for the problem of approximating a set of N Fourier coefficients of a function f, the DFT can be identified with the trapezoid rule. If f has both periodicity and smoothness of some of its derivatives, then the DFT gives approximations that are extremely accurate. In the absence of special endpoint conditions, the techniques of subtracting the linear trend and extrapolation can be used to improve DFT approximations. Since the DFT is generally implemented via the FFT, any methods based on the trapezoid rule will always be very efficient. Numerical evidence confirms that Filon's rule, particularly with extrapolation, also provides accurate approximations; when joined with the FFT it also becomes a competitive method.

9.4.   Problems

218. Finite sums from the Euler–Maclaurin Summation Formula.   The first two terms ($p = 1$) of the Euler–Maclaurin Summation Formula for an arbitrary finite interval $[a, b]$ are given by

$$T_N\{g\} - \int_a^b g(x)dx = \frac{h^2}{12}(g'(b) - g'(a)) - (b - a)\frac{h^4}{720}g^{(iv)}(\xi)$$

for $a < \xi < b$, where $h = (b - a)/N$ and g is at least four times continuously differentiable. Use this expression with $a = 0$, $b = N$, and special choices of g to evaluate the following finite sums.

$$\text{(a)} \sum_{n=1}^N n = \frac{N(N - 1)}{2}, \qquad \text{(b)} \sum_{n=1}^N n^2 = \frac{N(N + 1)(2N + 1)}{6},$$

$$\text{(c)} \sum_{n=1}^N n^3 = \frac{N^2(N + 1)^2}{4}.$$

Find a closed form expression for $\sum_{n=1}^N \sqrt{n}$.

219. Subtracting the linear trend.   Carry out the full derivation of the method of subtracting out the linear trend.

(a) Let f be a continuous function on $[-\pi, \pi]$, define

$$\ell(x) = \frac{(x + \pi)f(\pi) - (x - \pi)f(-\pi)}{2\pi},$$

and let $\phi(x) = f(x) - \ell(x)$. Also let f_n and ϕ_n be samples of f and ϕ at the N equally spaced grid points of $[-\pi, \pi]$. Then, for $k = -N/2 + 1 : N/2$, show that approximations to the Fourier coefficients c_k are given by

$$c_k \approx \mathcal{D}\{f_n\}_k = \mathcal{D}\{\phi_n\}_k + \frac{i\cos(k\pi)}{k}\left(\frac{f(\pi) - f(-\pi)}{2\pi}\right)$$

for $k \neq 0$ and

$$c_0 \approx \mathcal{D}\{f_n\}_0 = \mathcal{D}\{\phi_n\}_0 + \frac{1}{2}(f(-\pi) + f(\pi))$$

for $k = 0$.

(b) Argue that since ϕ is at least continuous on $[-\pi, \pi]$ and $\phi(-\pi) = \phi(\pi)$, the error in the DFT of ϕ_n should decrease at least as rapidly as the error in the DFT of f_n.

(c) Why does this technique improve approximations to the sine coefficients (imaginary part of c_k), but not to the cosine coefficients?

220. Subtracting the linear trend calculation. Carry out the strategy of the previous problem to approximate the Fourier coefficients c_k of the function

$$f(x) = \begin{cases} 0 & \text{for} \quad -\pi < x \le 0, \\ x & \text{for} \quad 0 \le x < \pi. \end{cases}$$

Compute approximations to the same coefficients using the DFT without the linear trend technique with several increasing values of N. Comment on the improvements that you observe in the approximations to both the cosine coefficients ($\text{Re}\{c_k\}$) and the sine coefficients ($\text{Im}\{c_k\}$).

221. Effect of endpoint conditions. Assume that f has as many continuous derivatives as necessary. Investigate the effect of various endpoint conditions on the error in the trapezoid rule approximations to the Fourier coefficients of f on $[-\pi, \pi]$. Use the Euler–Maclaurin error term and let $g(x) = f(x)e^{-ikx}$.

(a) Show that if $f(\pi) \ne f(-\pi)$ then in general $g'(\pi) \ne g'(-\pi)$ and the errors in the trapezoid rule approximations decrease as N^{-2} with increasing N, provided $k << N$.

(b) Verify the claim made in the text that if $f(\pi) = f(-\pi)$ then the error in the DFT approximations to the sine coefficients of f decreases as N^{-4} while the error in the approximations to the cosine coefficients decreases as N^{-2}, provided $k << N$.

(c) Show that if $f(\pi) = f(-\pi)$ and $f'(\pi) = f'(-\pi)$, then the errors in the trapezoid rule approximations decrease as N^{-4} with increasing N, provided $k << N$.

222. Trapezoid rule and harmonics. Consider using the trapezoid rule (DFT) to approximate the following integrals that involve trigonometric functions on $[-\pi, \pi]$.

(a) Verify that the N-point trapezoid rule approximation to $\int_{-\pi}^{\pi} \cos kx \, dx$ is exact, provided that $|k| < N$. Assuming that N is fixed, is it exact for any values of $|k| \ge N$?

(b) Verify that for approximations to $\int_{-\pi}^{\pi} \cos kx \, dx$, the error term

$$E_N = 2\pi \left(\frac{2\pi}{N} \right)^p \frac{B_p}{p!} f^{(p)}(\xi)$$

approaches zero as $p \to \infty$, provided $|k| < N$. Use the properties of B_p given in the text.

(c) Suppose that the trapezoid rule (DFT) is used to approximate the Fourier cosine coefficients of the function $f(x) = \cos k_0 x$ on the interval $[-\pi, \pi]$ where $k_0 \ne 0$ is an integer:

$$c_k = \frac{1}{2\pi} \int_{-\pi}^{\pi} \cos(k_0 x) \cos(kx) \, dx.$$

For fixed values of k_0 and N, for what values of k is the trapezoid rule exact?

223. Simpson's rule by extrapolation. Using $N = 4$, show that Simpson's rule using $2N$ points is given by

$$S_{2N} = \frac{4T_{2N} - T_N}{3},$$

where T_{2N} and T_N are the trapezoid rules using $2N$ and N points, respectively. Show that if another extrapolation is performed using S_N and S_{2N}, then another quadrature rule,

$$R_{2N} = \frac{16S_{2N} - S_N}{15},$$

is obtained with an error proportional to h^6.

224. Simpson's rule directly. As it is usually applied directly, Simpson's rule using $2N$ points is given by

$$\int_{-\pi}^{\pi} g(x)dx \approx \frac{h}{3}(g(x_{-N})4g(x_{-N+1}) + 2g(x_{-N+2})$$

$$+ \cdots + 2g(x_{N-2}) + 4g(x_{N-1}) + g(x_N)),$$

where $h = 2\pi/N$ and the uniformly spaced grid points are given by $x_n = nh$ for $n = -N : N$.

(a) Describe how Simpson's rule could be implemented directly (not through extrapolation) on the integrand $g(x) = f(x)e^{-ikx}$ by expressing it in terms of DFTs of modified input sequences.

(b) Assume that the DFT is implemented via the FFT. For the sake of simplicity assume that the FFT of a sequence of length N costs $N \log N$ arithmetic operations (the base of the logarithm is not important here). Compare the cost of implementing Simpson's rule with $2N$ points (i) directly as described in part (a), and (ii) by computing the trapezoid rule approximations T_N and T_{2N} and using extrapolation. Assume that FFTs are used for all of the quadrature rules.

225. Zero error DFT. Work out the details of the proof of Theorem 9.3, stating that if $f \in C^\infty[-\pi, \pi]$ and has 2π-periodic derivatives with $|f^{(p)}(x)| \leq \alpha^p$, then the trapezoid rule (DFT) is exact in computing the Fourier coefficients c_k provided that $\alpha < |k| < N$. (Suggestion: Start with the general error expression (9.3) and compute $g^{(p)}(x)$ where $g(x) = f(x)e^{-ikx}$. Use the geometric series to bound the sum that arises in the expression for $g^{(p)}$. Note the conditions on N, k, α needed for $\lim_{p\to\infty} |E_N| = 0$.)

226. Endpoint conditions. Verify that the function

$$g(x) = x\cos(k_0 x)\cos(kx)$$

satisfies $g^{(2p-1)}(2\pi) - g^{(2p-1)}(0) = 0$ when k and k_0 are integers, and thus the leading terms of the Euler–Maclaurin error expansion vanish. (Suggestion: Use

$$2\cos ax \cos bx = \cos(a+b)x + \cos(a-b)x \quad \text{and} \quad h(x) = x\cos(mx).$$

Then compute $h^{(2p-1)}(2\pi) - h^{(2p-1)}(0)$.)

227. Comparing theories. Consider the function $f(x) = (\pi^2 - x^2)^2$ on the interval $[-\pi, \pi]$. The DFT is to be used to approximate the Fourier coefficients of f. Using the theory of Chapter 6 (based on the Poisson Summation Formula) and the theory of this chapter (based on the Euler–Maclaurin Summation Formula), how do you predict that the error in these approximations should decrease as N is increased?

228. What does it cost? Assume that for a particular integral, the error in the trapezoid rule, Filon's rule, and Simpson's rule decreases as N^{-2}, N^{-4}, and N^{-4}, respectively (which is the general behavior of these methods).

(a) By what factor do you need to increase the number of grid points in each method to achieve a reduction in the error by a factor of 64?

(b) Neglecting "lower-order amounts of work," assume that each of the three methods above can be implemented on N points with a single FFT that costs $N \log N$ arithmetic operations. What is the increase in computational effort that is needed to achieve a 64-fold reduction in the error in each method?

229. Is T_N exact? The functions $f(x) = e^{\pm \sin x}$ and $f(x) = e^{\pm \cos x}$ are infinitely differentiable, and they and their derivatives are 2π-periodic. Furthermore,

$$\int_{-\pi}^{\pi} f(x)dx = 2\pi \sum_{m=0}^{\infty} \frac{1}{(m!2^m)^2} \approx 7.95492652101\ldots$$

for all four functions! Investigate the performance of the trapezoid rule in approximating the value of I. Does the theory predict that the trapezoid rule is exact? Does the numerical evidence support the theory? How well does the trapezoid rule (DFT) perform in computing the Fourier coefficients of f?

230. More vanishing derivatives. The function $f(x) = e^{x^2(1-x)^2}$, while not periodic, has the property that its odd derivatives vanish at $x = 0$ and $x = 1$. Does the theory predict that the trapezoid rule is exact in approximating $\int_0^1 f(x)dx$? Carry out the numerical experiments for $N = 2^m, m = 3, \ldots, 10$ and compare the numerical results to the theory. Other functions with the property that its odd derivatives vanish at the endpoints are [105]

$$f(x) = e^{-ax^2} \cos kx, \quad f(x) = xe^{-ax^2} \sin kx, \quad f(x) = \frac{x \sin kx}{(x^2 + a^2)^2},$$

where $a > 0$, $k \in \mathbf{Z}$, all with respect to the interval $[0, 2\pi]$. Investigate the performance of the trapezoid rule with these integrands.

231. Bernoulli numbers. Show that the Bernoulli numbers are the coefficients in the Taylor series for the function

$$f(x) = \frac{x}{e^x - 1}$$

about $x = 0$. Confirm that the series converges for $|x| < 2\pi$.

Chapter **10**

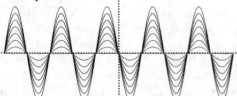

The Fast Fourier Transform

Truly that method greatly reduces the tediousness of mechanical calculations; success will teach the one who tries it.
– Carl Friedrich Gauss, ca 1805

10.1.   Introduction

Until now we have explored the uses and applications of the DFT assuming that its actual calculation could always be carried out in a handy black box. If nothing else, that black box can contain the explicit definition of the DFT in terms of a matrix-vector product. However, we have occasionally hinted that there are better ways to do it. Therefore, in this chapter we will look inside of that particular black box known as the **fast Fourier transform (FFT)**. For the initiated, this treatment will seem rather cursory; for those who view the FFT with awe and wonder (as we all should), the chapter may provide some guidance for further study; and for those who would prefer to leave the FFT inside of the black box, you have reached the end of this book!

This chapter should be regarded as a high altitude reconnaisance of a broad and complex landscape. One quickly realizes that the FFT is not a single algorithm, but a large and still proliferating family of methods, all designed to compute the various forms of the DFT with remarkable efficiency.

One reason that the FFT literature is both rich and occasionally impermeable is that there are several different frameworks in which FFTs may be developed and presented. Some FFTs arise quite naturally in one setting, but are difficult to formulate in another. Other FFTs have clearly equivalent expressions in several frameworks. As we all know, many attempts to understand the FFT have ended in a maze of signal flow diagrams or in a cloud of subscripted subscripts. However, we are also fortunate that the subject abounds with excellent presentations, in many mathematical languages and with many different perspectives. It would be foolish to duplicate and impossible to improve upon these existing treatments. We mention specifically the superb books of Brigham [20], [21], which for 20 years have surely offered revelations about the FFT to more people than any other source.

Let's begin with some history. The FFT was unveiled by a pair of IBM researchers named John Tukey and James Cooley in 1965. One might be tempted to think that their four-page paper "An Algorithm for the Machine Calculation of Complex Fourier Series" [42] was the first and last word on the subject, but it was neither. It certainly was not the last word: in the years since 1965, the original Cooley–Tukey method has been seized by practitioners from many different fields, refined for countless applications, and modified for a variety of computer architectures. The size of the "industry" spawned by that single paper is reflected by the fact that it has been cited 2047 times between 1965 and 1991 in scientific journals from *Psychophysiology* to *Econometrics* to *X-Ray Spectroscopy*. There can be little doubt that the idea presented by Cooley and Tukey in 1965 changed the face of signal processing, geophysics, statistics, computational mathematics, and the world around us forever [35].

On the other hand, the 1965 paper was not the first word on the subject either. In that publication, Cooley and Tukey cited the 1958 paper by I. Good [68], which is generally regarded as the origin of the prime factor algorithm (PFA) FFT. However, there is a much longer FFT lineage, not known to Cooley and Tukey in 1965, that was revealed shortly thereafter by Rudnick [121] and Cooley, Lewis, and Welch [40]. Those studies uncovered the crux of the FFT in a 1942 paper by Danielson and Lanczos [44], which itself was inspired by work of C. D. T. Runge[1] in 1903 and 1905 [122]. Those interested in tracing this thread further into the past must read

---

[1] Born in Bremen, Germany in 1856, CARL DAVIS TOLME RUNGE studied mathematics in Munich and Berlin. His interests turned to physics, and his most notable work was in the field of spectroscopy. He was a professor at Göttingen until he died in 1927.

the fascinating investigation by Heideman, Johnson, and Burrus [74], in which the authors document several nineteenth century methods for the efficient computation of Fourier coefficients. Of these methods, only the one proposed by Gauss in 1805 [61] (published posthumously in 1866) is a genuine FFT. Ironically, references to Gauss' method have appeared at least twice in the literature in the past 100 years: once in 1904 in H. Burkhardt's encyclopedia of mathematics [25] and once again in 1977 in H. H. Goldstine's history of numerical analysis [67].

The goal of this chapter is to describe several different avenues along which the FFT may be approached. The various FFT paths we will travel in this chapter are

- splitting methods,

- index expansions (one- $\rightarrow$ multi-dimensional transforms),

- matrix factorizations,

- prime factor and convolution methods.

The chapter will conclude with some remarks about related issues such as FFT performance, computing symmetric DFTs, and implementation of FFTs on advanced architectures. There is a lot to say, and we have given ourselves little space, so let's begin.

10.2. Splitting Methods

If you had ten minutes and a single page to explain how the FFT works, the splitting method would be the best approach to use. Let us recall that the goal is to compute the DFT of a (possibly complex) sequence x_n which has length N and is assumed to have a period of N. As we have noted many times, there are several commonly used forms of the DFT. The FFT can be developed for any of these forms, but a specific choice must be made for the sake of exposition. For the duration of this chapter we will use the definition in which both indices n and k belong to the set $\{0, \ldots, N-1\}$. We will use the definition

$$X_k = \sum_{n=0}^{N-1} x_n \omega_N^{-nk} \tag{10.1}$$

for $k = 0 : N - 1$, where $\omega_N = e^{i2\pi/N}$. We will dispense with the scaling factor of $1/N$, which can always be included at the end of the calculation if it is needed at all.

Both history and pedagogy agree that in developing the FFT, it is best to begin with the case $N = 2^M$, where M is a natural number; this is called the radix-2 case. If we now split x_n into its even and odd subsequences by letting $y_n = x_{2n}$ and $z_n = x_{2n+1}$, and then substitute these subsequences into the definition (10.1), we find that

$$X_k = \sum_{n=0}^{\frac{N}{2}-1} y_n \omega_N^{-2nk} + z_n \omega_N^{-(2n+1)k} \tag{10.2}$$

for $k = 0 : N - 1$. We now use a simple but crucial symmetry property of the complex exponential, which might be identified as the linchpin of the whole FFT enterprise: it is that

$$\omega_N^{-2nk} = \omega_{N/2}^{-nk} \quad \text{or equivalently} \quad e^{-i4\pi nk/N} = e^{-i2\pi nk/(N/2)}.$$

(More generally, we can always replace ω_N^{-pq} by $\omega_{N/q}^{-p}$, where p and q are real numbers (see problem 232).) Using this property, expression (10.2) can be written as

$$X_k = \underbrace{\sum_{n=0}^{\frac{N}{2}-1} y_n \omega_{N/2}^{-nk}}_{\text{a DFT of length } \frac{N}{2}} + \omega_N^{-k} \underbrace{\sum_{n=0}^{\frac{N}{2}-1} z_n \omega_{N/2}^{-nk}}_{\text{a DFT of length } \frac{N}{2}} \quad.$$

And now, if we stand back, we see that the original DFT has been expressed as a simple combination of the DFT of the sequence y_n and the DFT of the sequence z_n, both of length $N/2$. Keeping with our convention, we will call these half-length DFTs Y_k and Z_k, respectively. By letting $k = 0 : N/2 - 1$, we can write

$$\begin{aligned} X_k &= Y_k + \omega_N^{-k} Z_k, \\ X_{k+\frac{N}{2}} &= Y_{k+\frac{N}{2}} + \omega_N^{-(k+\frac{N}{2})} Z_{k+\frac{N}{2}}. \end{aligned}$$

We now note that $\omega_N^{-\frac{N}{2}} = -1$, and that the sequences Y_k and Z_k have a period of $N/2$ to conclude that

$$X_k = Y_k + \omega_N^{-k} Z_k, \tag{10.3}$$

$$X_{k+\frac{N}{2}} = Y_k - \omega_N^{-k} Z_k \tag{10.4}$$

for $k = 0 : N/2 - 1$, which accounts for all N DFT coefficients X_k. Expressions (10.4) are often called **combine formulas** or **butterfly relations**. They give a recipe for combining two DFTs of length $N/2$ (corresponding to the even and odd subsequences of the original sequence) to form the DFT of the original sequence.

It is worthwhile to pause and note the savings that have already been achieved. Computing the sequence X_k explicitly from its definition (10.1) requires approximately N^2 complex multiplications and N^2 complex additions. On the other hand, if the splitting method is used and if the sequences Y_k and Z_k are computed as matrix-vector products, then the DFT requires approximately $2(N/2)^2 = N^2/2$ multiplications and $N^2/2$ additions to compute Y_k and Z_k plus an extra $N/2$ multiplications and N additions for the butterfly relations. We see that a DFT computed by one splitting of the input sequence requires roughly $N^2/2$ multiplications and $N^2/2$ additions, which is a factor-of-two savings. It is also worth noting that this single application of the splitting method is Procedure 2 of Chapter 4.

However, we are not finished. We just assumed that the DFT sequences Y_k and Z_k would be computed as matrix-vector products; of course, they needn't be. A full FFT algorithm results when the splitting idea is applied to the computation of Y_k and Z_k. In a divide-and-conquer spirit, it is then repeated on their subsequences, and so on, for $M = \log_2 N$ steps. Eventually, the original problem of computing a DFT of length N has been replaced by the problem of computing N DFTs of length 1. At this point there are really no DFTs left to be done, since the DFT of a sequence of length 1 is itself.

The procedure we have just described is shown schematically in Figure 10.1 and is one of the fundamental FFT algorithms. It is often described as **decimation-in-time**, since it involves the splitting of the input (or time) sequence. This FFT is very similar to the one described by Cooley and Tukey in their 1965 paper (although it was derived differently there). The method consists of two stages:

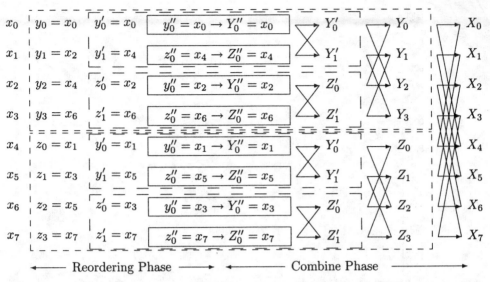

$$\longleftarrow \text{Reordering Phase} \longrightarrow \longleftarrow \text{Combine Phase} \longrightarrow$$

FIG. 10.1. *The Cooley–Tukey FFT, shown here for a sequence of length $N = 8$, takes place in two phases. In the reordering phase, the input sequence x_n is successively split in an odd/even fashion until N sequences of length 1 remain (shown inside the solid boxes). The combine phase begins with N trivial DFTs of length 1 (also shown inside the solid boxes). Then the DFT sequences of length 1 are combined in pairs to form DFTs of length 2 (shown inside the coarse dashed boxes). Then DFT sequences of length 2 are combined in pairs to form two DFTs of length 4 (shown inside the fine dashed boxes and denoted Y_k and Z_k). Finally, the two DFTs of length 4 are combined to form the desired DFT X_k, of length 8. All of the combine steps use the Cooley–Tukey butterfly relations and are shown as webs of arrows.*

- a **reordering** stage (often called **bit-reversal**), in which the input sequence is successively split into even and odd subsequences; and

- a **combine** stage, in which the butterfly relations (10.3) and (10.4) are used to combine sequences of length 1 into sequences of length 2, then sequences of length 2 into sequences of length 4, and so on, until the final transform sequence X_k is formed from two sequences of length $N/2$.

Now let's count complex arithmetic operations with the help of Figure 10.1. The combine stage consists of $M = \log_2 N$ steps, each of which involves $N/2$ butterflies, each of which requires one complex multiplication and two complex additions. This gives a total operation count for an N-point complex DFT of

$$N \log_2 N \quad \text{complex additions} \quad \text{and} \quad \frac{N}{2} \log_2 N \quad \text{complex multiplications.}$$

Even this tally can be improved if one accounts for multiplications by ± 1 and $\pm i$ (which needn't be done). The point of overwhelming practical importance is that the FFT provides a factor of $N/\log N$ in computational savings over the evaluation of the DFT by its definition. As shown in Figure 10.2, this factor itself increases with N.

Note that the FFT just described can be done **in-place**, meaning that the computation can be done without additional storage arrays. The method does have the bothersome reordering stage, but, as we will see, this is often unnecessary or FFTs

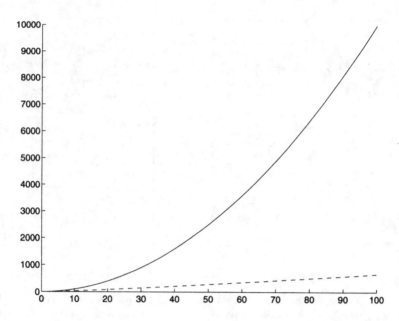

FIG. 10.2. *A comparison of the computational complexities of the DFT (solid line) and the FFT (dashed line) is shown graphically. Plotted is the floating-point operation count as a function of N. The DFT is an $O(N^2)$ operation, while the FFT has a complexity of $O(N \log_2 N)$, assuming that $N = 2^p$. Similar savings are produced for other values of N as well.*

can be found that avoid it. It should also be noted that the splitting algorithm can be generalized to the radix-r case for sequences of length $N = r^M$, where $r > 1$ and M are natural numbers, provided one has efficient ways to evaluate r-point DFTs (see problems 233 and 234). With a bit more ingenuity, the method can be extended to the **mixed radix** case for sequences of length $N = r_1^{M_1} \cdots r_L^{M_L}$, where $r_i > 1$ and M_i and L are natural numbers. A quick physical interpretation of the splitting method might go as follows. The even subsequence y_n is a sampling of the original sequence x_n. The odd subsequence z_n is also a sampling of x_n, shifted by one grid point with respect to the first sampling. The DFT of the original sequence is just an average of the DFTs of the two subsequences. Since the second sample is shifted relative to the first, the shift theorem of Chapter 3 requires that the rotation ω_N^{-k} multiply the DFT of the odd subsequence [19].

With more brevity, we will now outline how another canonical FFT can be derived by the splitting method. As before, we begin with an input sequence x_n of length N, but this time it is split into its first-half and last-half subsequences x_n and $x_{n+N/2}$. We make these replacements in the DFT definition (10.1) to find that

$$X_k = \sum_{n=0}^{\frac{N}{2}-1} x_n \omega_N^{-nk} + x_{n+\frac{N}{2}} \omega_N^{-(n+N/2)k} = \sum_{n=0}^{\frac{N}{2}-1} \left[x_n + (-1)^k x_{n+\frac{N}{2}} \right] \omega_{N/2}^{-nk/2}, \quad (10.5)$$

where $k = 0 : N - 1$. Once again, the crucial properties $\omega_N^{-N/2} = -1$ and $\omega_N^{-nk} = \omega_{N/2}^{-nk/2}$ have been used.

We are now headed down a different road, and a new orientation is needed. Replacing k by $2k$ in (10.5) tells us that the even coefficients of the DFT are given by

$$X_{2k} = \sum_{n=0}^{\frac{N}{2}-1} \left(x_n + x_{n+\frac{N}{2}} \right) \omega_{N/2}^{-nk}, \tag{10.6}$$

whereas if k is replaced by $2k + 1$, we find the odd coefficients given by

$$X_{2k+1} = \sum_{n=0}^{\frac{N}{2}-1} \left[(x_n - x_{n+\frac{N}{2}})\omega_N^{-n} \right] \omega_{N/2}^{-nk}. \tag{10.7}$$

The expressions (10.6) and (10.7) can be used for $k = 0 : N/2 - 1$ to generate all N values $X_0, \ldots, X_{N-1}$.

We see that both the even and odd terms of X_k can be found by performing DFTs of length $N/2$ on simple combinations of the subsequences x_n and $x_{n+N/2}$. These simple combinations are both subsequences of length $N/2$, and we call them y_n and z_n, respectively. They are defined by another set of butterfly relations, namely

$$y_n = x_n + x_{n+\frac{N}{2}}, \tag{10.8}$$

$$z_n = (x_n - x_{n+\frac{N}{2}})\omega_N^{-n} \tag{10.9}$$

for $n = 0 : N/2 - 1$. Once again, the task of computing a DFT of length N has been replaced by the task of computing the DFTs of shorter sequences, in this case y_n and z_n, both of length $N/2$. As with the previous FFT we do not stop here. The same strategy is applied to compute the DFTs of y_n and z_n, then again to compute the resulting DFTs of length $N/4$, until eventually (recalling that N is a power of two), the job has been reduced to computing N DFTs of length 1, which is trivial. Therefore, the only work involves preparing the input sequences for each successive DFT, which can be done by applying the butterfly relations (10.8) and (10.9) with the appropriate value of N.

This procedure leads to another quite different FFT which is illustrated schematically in Figure 10.3. It is the algorithm proposed by Gentleman and Sande in 1966 [64]. Notice that in contrast to the Cooley–Tukey FFT (which requires a reordering of the input sequence, but computes the transform coefficients in natural order), this FFT takes the input sequence in natural order, but produces the transform coefficients in scrambled (bit-reversed) order. For this reason, this FFT is often described as **decimation-in-frequency**. In contrast to the Cooley–Tukey FFT, the combine stage occurs first and the reordering stage occurs last. This FFT can also be done in-place, it has the same computational cost as the Cooley–Tukey FFT, and it can be extended to more general values of N (see problem 236).

In both of these FFTs, the scrambling of the input or output sequence is an annoyance that can be avoided. By a clever rearrangement of the intermediate sequences, it is actually possible to devise "self-sorting" FFTs that accept a naturally ordered input sequence and produce a naturally ordered output sequence: unfortunately, the price is an extra storage array. These FFTs are generally associated with the name Stockham [38], although they are special cases of a family of FFTs proposed by Glassman [66]. Another rearrangement of the Cooley–Tukey FFT produces an FFT in which the length of the inner loops remains constant ($N/2$). The advantage of this

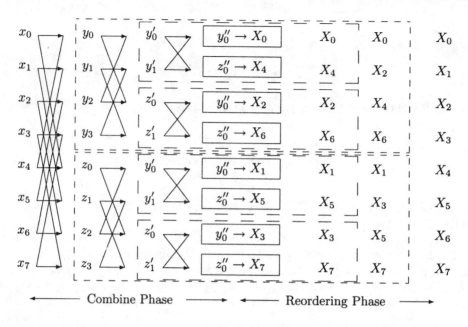

◄——— Combine Phase ———►   ◄——— Reordering Phase ———►

FIG. 10.3. *The Gentleman–Sande FFT, shown here for a sequence of length $N = 8$, also takes place in two phases. The input sequence x_n goes into the combine phase in natural order. After the first combine step two subsequences of length 4, y_n and z_n (shown inside the fine dashed boxes), are produced whose DFTs are the even and odd terms of the desired DFT, X_{2k} and X_{2k+1}, respectively. After the second combine step, four subsequences of length 2 are produced (shown inside the coarse dashed boxes) whose DFTs are $\{X_0, X_4\}, \{X_2, X_6\}, \{X_1, X_5\}, \{X_3, X_7\}$, respectively. After the third combine step, eight subsequences of length 1 are produced whose DFTs can be found trivially (solid boxes). These three combine steps result in the desired DFT X_k in scrambled (bit-reversed) order. The reordering phase recovers the DFT in its natural order. All combine steps use the Gentleman– Sande butterfly relations (shown as webs of arrows).*

FFT for parallel computation was realized in a prescient 1968 paper by Pease [112]. Much more recently, the seemingly inescapable trade-off between reordering and extra storage has been overcome, but only within the prime factor setting (to be discussed shortly).

It should also be mentioned that in many important FFT applications (such as computing convolutions and solving boundary value problems), forward DFTs are ultimately followed by inverse DFTs. Therefore, if one is willing to do computations in the transform domain with respect to bit-reversed indices (and these are often very simple computations), it is possible to use a decimation-in-frequency FFT for the forward transforms and a decimation-in-time FFT for the inverse transforms, thereby avoiding a reordering of either the input or output data.

Finally, we mention that all FFTs, including those presented in this section, can be adapted to compute inverse DFTs (IDFTs). Recall that the DFT and the IDFT differ only by a multiplicative factor of $1/N$ and the replacement of ω by ω^{-1}. Therefore, the simplest way to produce an inverse FFT is to replace all references to ω in an FFT program by ω^{-1} (equivalently, negate all references to angles). The scaling by $1/N$ can be handled after the transform is done, if necessary. However, inverse FFTs can be created in other ways as well (see problems 237 and 238).

10.3. Index Expansions

In this section we give just a hint of another approach that can be used to derive both of the FFTs of the previous section plus many more. It also has historical significance because it is the technique used by Cooley and Tukey. We assume that the length of the input sequence N can be factored as $N = pq$. The indices n and k that appear in the definition of the DFT may now be represented in the forms

$$n = n_1 p + n_0 \quad \text{for} \quad n_0 = 0 : p - 1, \; n_1 = 0 : q - 1,$$
$$k = k_1 q + k_0 \quad \text{for} \quad k_0 = 0 : q - 1, \; k_1 = 0 : p - 1.$$

To reflect these **index expansions**, we now write x_n and X_k as $x(n_1, n_0)$ and $X(k_1, k_0)$, and substitute into the DFT definition (10.1). We immediately find that

$$X(k_1, k_0) = \sum_{n_0=0}^{p-1} \sum_{n_1=0}^{q-1} x(n_1, n_0) \omega_N^{-(n_1 p + n_0)(k_1 q + k_0)}$$

for $k_0 = 0 : q - 1$, $k_1 = 0 : p - 1$. The exponent can now be expanded and simplified noting that $\omega_N^{-pq} = \omega_N^{-N} = 1$ and $\omega_N^{-n_1 k_0 p} = \omega_q^{-n_1 k_0}$. This leads to

$$X(k_1, k_0) = \sum_{n_0=0}^{p-1} \omega_N^{-n_0(k_1 q + k_0)} \underbrace{\sum_{n_1=0}^{q-1} x(n_1, n_0) \omega_q^{-n_1 k_0}}_{p \text{ DFTs of length } q}$$

for $k_0 = 0 : q - 1$, $k_1 = 0 : p - 1$. This expression requires some inspection. The inner sum on n_1 can be identified as p DFTs of length q of sequences consisting of every pth term of the input sequence x_n. Since this inner sum depends directly on k_0 and n_0, we will denote it $Y(k_0, n_0)$. This allows us to write that

$$X(k_1, k_0) = \sum_{n_0=0}^{p-1} Y(k_0, n_0) \omega_N^{-n_0(k_1 q + k_0)} \tag{10.10}$$

for $k_0 = 0 : q - 1$, and $k_1 = 0 : p - 1$. We must now simplify the exponent of ω_N by noting that

$$\omega_N^{-n_0(k_1 q + k_0)} = \omega_p^{-n_0 k_1} \omega_N^{-n_0 k_0}.$$

With this observation we have that

$$X(k_1, k_0) = \sum_{n_0=0}^{p-1} \{ Y(k_0, n_0) \omega_N^{-n_0 k_0} \} \omega_p^{-n_0 k_1} \tag{10.11}$$

$$\underbrace{\qquad\qquad\qquad\qquad\qquad\qquad\qquad}_{q \text{ DFTs of length } p}$$

for $k_0 = 0 : q - 1$ and $k_1 = 0, p - 1$. This last expression may be interpreted as q DFTs of length p of sequences derived from the intermediate sequences $Y(k_0, n_0)$.

Now let's step away from the forest far enough to make an extremely valuable observation, one that will lead to many more FFTs. At least conceptually, we see that a single DFT of length $N = pq$ may be cast as a two-dimensional DFT of size $p \times q$. To be sure, there are some complications, such as the appearance of the **"twiddle**

factors" $\omega_N^{-n_0 k_0}$ in the second set of DFTs. Nevertheless, this interpretation is very important. It may be illuminated by Figure 10.4, which shows the two-dimensional computational array in which the FFT takes place. Note the arrangement of the four indices. The input sequence x_n is ordered *by rows* in this array. The inner sum (p DFTs of length q) corresponds to transforming the columns of the array, since with n_0 fixed, the input sequences for these DFTs consist of every pth element of the original sequence. The outer sum corresponds to transforming in the direction of the rows of the array (with k_0 fixed).

If FFTs for sequences of length p and q are already available, then this single factorization would result in a complete FFT for sequences of length N. Otherwise, the procedure can be repeated for subsequent factorizations of p and q, which results in the conversion of a single DFT into a multidimensional DFT. For example, if $N = pqr$, the N-point DFT could be expressed as a three-dimensional ($p \times q \times r$) DFT. If p and q are themselves prime or relatively prime, there are still other possibilities that will be mentioned later.

The computational savings with this single factoring of N should be noted. If the DFTs of length p and q are done as matrix-vector products, then q DFTs of length p require roughly qp^2 complex multiplications and additions, while p DFTs of length q require roughly pq^2 complex multiplications and additions. Thus, the DFT of length N requires

$$qp^2 + pq^2 = pq(p + q) = N(p + q)$$

complex multiplications and additions compared to N^2 operations to compute the DFT by the definition. If this process is repeated for the DFTs of length p and q, then it can be shown that an $N \log N$ operation count results for the entire DFT. We mention an interesting variation on index expansions due to de Boor [46], who interprets the FFT as nested polynomials "with a twist."

The decimation-in-time FFT presented in the previous section may be viewed in this framework by letting $p = 2$ and $q = N/2$. We then have that

$$X(k_1, k_0) = \sum_{n_0=0}^{1} \omega_N^{-n_0(k_1 q + k_0)} \sum_{n_1=0}^{\frac{N}{2}-1} x(n_1, n_0) \omega_{\frac{N}{2}}^{-n_1 k_0}$$

for $k_0 = 0 : N/2 - 1$, $k_1 = 0 : 1$. The inner sum consists of two DFTs of length $N/2$, one transforming the even terms of the input sequence, the other transforming the odd terms of the input sequence. This reflects the odd/even rearrangement of the input sequence that arose in the splitting method. The outer sum can be identified as the two-term butterfly relations (10.3) and (10.4) that appeared in the splitting approach. Thus, one factoring of N ($N = 2 \times N/2$) corresponds to one step of the splitting method. As with the splitting method, a full FFT results when the factoring is next applied to the DFTs of length $N/2$.

Hopefully this glimpse of the index expansion idea at least suggests how it can lead to many more FFT algorithms. Just within the example presented above, it is possible to reverse the order of the two sums, which leads to a decimation-in-frequency FFT (see problem 239). However, there is much more generality: if N is composite and consists of M different factors (counting multiplicity), then the DFT can be expressed in terms of M nested sums, and each index, n and k, has M terms in its expansion. Now the M nested sums can be rearranged in $M!$ different ways, leading to as many different FFTs, some of which are useful, most of which are not. Fortunately this sudden proliferation of FFTs can be organized greatly by the next perspective.

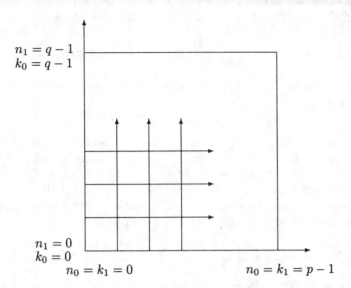

FIG. 10.4. *A single DFT of length $N = pq$ may be viewed as p DFTs of length q followed by q DFTs of length p that take place in the two-dimensional $(p \times q)$ array shown above. The first wave of DFTs transforms along the columns of the array; the second wave of DFTs transforms along the rows of the array. The indices n_0 and k_1 increase along the rows of the array, while the indices n_1 and k_0 increase along the columns of the array.*

10.4. Matrix Factorizations

We will now survey a very powerful and encompassing setting in which FFTs can be derived and understood. This approach has been promoted very convincingly in a recent book by Van Loan [155], which deserves to be recognized as the next in a series of definitive books about the FFT. The book succeeds in expressing most known FFTs as matrix factorizations of the DFT matrix $\mathbf{W}_N$, thus unifying and organizing the entire FFT menagerie. The idea of expressing the FFT as a factorization of $\mathbf{W}_N$ first appears in the late 1960's in papers by W. M. Gentleman [63] and F. Theilheimer [151], and was elaborated upon frequently thereafter. Hopefully our glancing coverage of this approach will suggest its generality.

In order to interpret the DFT as a matrix-vector product, we will think of the periodic input and output sequences of the DFT as the column N-vectors

$$\mathbf{x} = (x_0, x_1, \ldots, x_{N-1})^T \quad \text{and} \quad \mathbf{X} = (X_0, X_1, \ldots, X_{N-1})^T.$$

This enables us to write the DFT in the form $\mathbf{X} = \mathbf{W}_N \mathbf{x}$. To be quite specific, consider the case of a DFT of length $N = 16$ which is to be computed by the Cooley–Tukey (decimation-in-time) FFT. Recall the butterfly relations (10.3) and (10.4) that indicate how two DFTs, Y_k and Z_k, of length $N/2 = 8$ can be combined to form X_k, the DFT of length $N = 16$:

$$X_k = Y_k + \omega_N^{-k} Z_k, \tag{10.12}$$

$$X_{k+N/2} = Y_k - \omega_N^{-k} Z_k \quad \text{for} \quad k = 0:7. \tag{10.13}$$

Letting $\mathbf{Y}$ and $\mathbf{Z}$ be the eight-vectors corresponding to the sequences Y_k and Z_k, we can write the butterfly relations (10.12) and (10.13) in the form

$$\mathbf{X} = \mathbf{W}_{16}\mathbf{x} = \begin{pmatrix} \mathbf{I}_8 & \mathbf{\Omega}_8 \\ \mathbf{I}_8 & -\mathbf{\Omega}_8 \end{pmatrix} \begin{pmatrix} \mathbf{Y} \\ \mathbf{Z} \end{pmatrix} \equiv \mathbf{B}_{16} \begin{pmatrix} \mathbf{Y} \\ \mathbf{Z} \end{pmatrix},$$

where in all that follows $\mathbf{I}_m$ is the identity matrix of order m and $\mathbf{\Omega}_m$ is the $(m \times m)$ diagonal matrix

$$\mathbf{\Omega}_m = \mathrm{diag}\,(1, \omega_{2m}^{-1}, \omega_{2m}^{-2}, \ldots, \omega_{2m}^{-(m-1)}).$$

The matrix used for this combine step is called a **butterfly matrix** and we have denoted it $\mathbf{B}_{16}$. It should be verified (problem 241) that this product is exactly the eight pairs of butterfly relations given in (10.12) and (10.13).

Now note that $\mathbf{Y}$ and $\mathbf{Z}$ are themselves DFTs (of length $N = 8$) of the even and odd subsequences of $\mathbf{x}$, which we will call $\mathbf{x}_{\text{even}}$ and $\mathbf{x}_{\text{odd}}$. Therefore, similar butterfly matrices may be used to compute them. For example,

$$\mathbf{Y} = \mathbf{W}_8\mathbf{x}_{\text{even}} = \begin{pmatrix} \mathbf{I}_4 & \mathbf{\Omega}_4 \\ \mathbf{I}_4 & -\mathbf{\Omega}_4 \end{pmatrix} \begin{pmatrix} \mathbf{Y}' \\ \mathbf{Z}' \end{pmatrix} \equiv \mathbf{B}_8 \begin{pmatrix} \mathbf{Y}' \\ \mathbf{Z}' \end{pmatrix},$$

where $\mathbf{Y}'$ and $\mathbf{Z}'$ denote the DFTs (of length $N = 4$) of the even and odd subsequences of $\mathbf{x}_{\text{even}}$. A similar expression may be written for the transform of $\mathbf{x}_{\text{odd}}$:

$$\mathbf{Z} = \mathbf{W}_8\mathbf{x}_{\text{odd}} = \begin{pmatrix} \mathbf{I}_4 & \mathbf{\Omega}_4 \\ \mathbf{I}_4 & -\mathbf{\Omega}_4 \end{pmatrix} \begin{pmatrix} \mathbf{Y}'' \\ \mathbf{Z}'' \end{pmatrix} \equiv \mathbf{B}_8 \begin{pmatrix} \mathbf{Y}'' \\ \mathbf{Z}'' \end{pmatrix},$$

where $\mathbf{Y}''$ and $\mathbf{Z}''$ are the DFTs (of length $N = 4$) of the even and odd subsequences of $\mathbf{x}_{\text{odd}}$.

Before we become strangled by our notation, we should find a pattern. So far we have that

$$\mathbf{X} = \mathbf{W}_{16}\mathbf{x} = \mathbf{B}_{16} \begin{pmatrix} \mathbf{B}_8 & \mathbf{0} \\ \mathbf{0} & \mathbf{B}_8 \end{pmatrix} \begin{pmatrix} \mathbf{Y}' \\ \mathbf{Z}' \\ \mathbf{Y}'' \\ \mathbf{Z}'' \end{pmatrix}.$$

If we were to continue in this manner and express the remaining DFTs of length $N = 4$ and $N = 2$ in terms of the butterfly matrices $\mathbf{B}_4$ and $\mathbf{B}_2$, the result would be

$$\mathbf{X} = \mathbf{W}_{16}\mathbf{x} = \mathbf{B}_{16} \begin{pmatrix} \mathbf{B}_8 & \mathbf{0} \\ \mathbf{0} & \mathbf{B}_8 \end{pmatrix} \begin{pmatrix} \mathbf{B}_4 & \mathbf{0} & \mathbf{0} & \mathbf{0} \\ \mathbf{0} & \mathbf{B}_4 & \mathbf{0} & \mathbf{0} \\ \mathbf{0} & \mathbf{0} & \mathbf{B}_4 & \mathbf{0} \\ \mathbf{0} & \mathbf{0} & \mathbf{0} & \mathbf{B}_4 \end{pmatrix}$$

$$\times \begin{pmatrix} \mathbf{B}_2 & 0 & 0 & 0 & 0 & 0 & 0 & 0 \\ 0 & \mathbf{B}_2 & 0 & 0 & 0 & 0 & 0 & 0 \\ 0 & 0 & \mathbf{B}_2 & 0 & 0 & 0 & 0 & 0 \\ 0 & 0 & 0 & \mathbf{B}_2 & 0 & 0 & 0 & 0 \\ 0 & 0 & 0 & 0 & \mathbf{B}_2 & 0 & 0 & 0 \\ 0 & 0 & 0 & 0 & 0 & \mathbf{B}_2 & 0 & 0 \\ 0 & 0 & 0 & 0 & 0 & 0 & \mathbf{B}_2 & 0 \\ 0 & 0 & 0 & 0 & 0 & 0 & 0 & \mathbf{B}_2 \end{pmatrix} \mathbf{P}^T\mathbf{x}. \qquad (10.14)$$

We see that the DFT matrix $\mathbf{W}_{16}$ has been expressed as a product of four matrices consisting of the butterfly matrices $\mathbf{B}_m$ for $m = 2, 4, 8, 16$. At each step, the matrix $\mathbf{B}_m$

holds the butterfly relations for combining two DFTs of length $m/2$ into one DFT of length m. Notice that as this factorization progresses, the rearrangements of the input sequence "pile up" at the right end of the product. Therefore, this product of matrices operates not on the input sequence $\mathbf{x}$, but on a permuted input sequence $\mathbf{P}^T\mathbf{x}$, which is the scrambled sequence that we have already encountered in the decimation-in-time FFT. (The matrix $\mathbf{P}$ is a **permutation matrix** created by performing successive odd/even rearrangements of the columns of the identity matrix $\mathbf{I}_{16}$; premultiplying $\mathbf{x}$ by $\mathbf{P}^T$ rearranges the components of $\mathbf{x}$ in the same manner.) This formulation clearly shows the two stages of this particular FFT: the input sequence is first reordered by the permutation matrix $\mathbf{P}^T$, and then it passes through $\log_2 N$ combine stages.

We have written the matrix factorization in this rather cumbersome way to motivate the next step, which expresses it much more compactly. Recall that the **Kronecker product** of a $p \times q$ matrix $\mathbf{A}$ and an $r \times s$ matrix $\mathbf{B}$ is the $pr \times qs$ matrix

$$\mathbf{A} \otimes \mathbf{B} = \begin{pmatrix} a_{11}\mathbf{B} & \cdots & a_{1q}\mathbf{B} \\ \vdots & \ddots & \vdots \\ a_{p1}\mathbf{B} & \cdots & a_{pq}\mathbf{B} \end{pmatrix},$$

where the a_{ij}'s are the elements of $\mathbf{A}$. We see that the factorization in (10.14) can be simplified nicely by the Kronecker product to give

$$\mathbf{X} = \mathbf{W}_{16}\mathbf{x} = (\mathbf{I}_1 \otimes \mathbf{B}_{16})(\mathbf{I}_2 \otimes \mathbf{B}_8)(\mathbf{I}_4 \otimes \mathbf{B}_4)(\mathbf{I}_8 \otimes \mathbf{B}_2)\mathbf{P}^T\mathbf{x}.$$

In general, the Cooley–Tukey FFT for $N = 2^M$ can be expressed as

$$\mathbf{X} = \mathbf{W}_N\mathbf{x} = \mathbf{A}_M\mathbf{A}_{M-1}\cdots\mathbf{A}_2\mathbf{A}_1\mathbf{P}^T\mathbf{x},$$

where $\mathbf{A}_k = \mathbf{I}_r \otimes \mathbf{B}_m, m = 2^k$, and $rm = N$ for $k = 1 : M$. As before, we have let

$$\mathbf{B}_{2m} = \begin{pmatrix} \mathbf{I}_m & \mathbf{\Omega}_m \\ \mathbf{I}_m & -\mathbf{\Omega}_m \end{pmatrix} \quad \text{and} \quad \mathbf{\Omega}_m = \text{diag}\,(1, \omega_{2m}^{-1}, \omega_{2m}^{-2}, \ldots, \omega_{2m}^{-(m-1)}).$$

The matrix factorization approach gives a concise representation of the Cooley–Tukey FFT. Now, how does it generalize? First, the matrix factorization idea can be applied directly to the radix-r case (see problem 245). A greater challenge is to extend it to the mixed-radix case in which N is a product of powers of different integers; but it can be done. Furthermore, noting that the DFT matrix $\mathbf{W}_N$ is symmetric, it follows that

$$\mathbf{W}_N^T = \left(\mathbf{A}_M\mathbf{A}_{M-1}\cdots\mathbf{A}_2\mathbf{A}_1\mathbf{P}^T\right)^T = \mathbf{P}\mathbf{A}_1^T\mathbf{A}_2^T\cdots\mathbf{A}_{M-1}^T\mathbf{A}_M^T$$

is also a factorization that corresponds to an FFT. This transpose factorization, in which the combine stage appears first and the reordering stage appears last is the decimation-in-frequency FFT of Gentleman and Sande that we have already encountered. Notice that inverse DFT algorithms could also be obtained by taking the inverse of the above factorizations.

But there is yet another level of generality. Instead of applying the permutation matrix P before or after the combine phase, it is possible to distribute it throughout the combine phase. The factorization

$$\mathbf{W}_N^T = \mathbf{A}_M\mathbf{P}_M\mathbf{A}_{M-1}\mathbf{P}_{M-1}\cdots\mathbf{A}_2\mathbf{P}_2\mathbf{A}_1\mathbf{P}_1,$$

where the $\mathbf{P}_i$'s are permutation matrices, is an FFT in which a (usually simple) reordering of the data occurs before each butterfly step. Clearly, many possible FFTs could result from this strategy. Among the few useful ones are the Stockham self-sorting FFTs and the Pease FFT mentioned earlier. We have still not exhausted the scope of the matrix factorization approach, and it will reappear to simplify the discussion of the next section.

We close the discussion on FFTs in the general Cooley–Tukey mode by mentioning a clever and effective variation on the methods discussed heretofore. Revealed first in a paper by Duhamel and Hollmann [52] in 1984, the **split-radix FFT** can be derived within any of the above frameworks and it is fairly easy to describe. In the radix-2 case, the split-radix method proceeds by splitting the original sequence of length N into one sequence of length $N/2$ and *two* more sequences of length $N/4$. It turns out that a splitting of this form requires less arithmetic than either a regular radix-2 or radix-4 splitting. The strategy is then repeated on subsequences to produce a competitive, if not superior, FFT. The split-radix idea can also be generalized to other radices [157] and to symmetric sequences [51], [126].

10.5.   Prime Factor and Convolution Methods

The FFTs discussed so far represent conceptually about half of all known FFT algorithms. In this section we attempt to summarize briefly the ideas that lead to the "other half." This task is more difficult since the foundations of prime factor algorithms (PFAs) are closely related to a strategy of evaluating DFTs through convolution. In fact, the two approaches can be combined within the same FFT, which further blurs the distinction between them. Nevertheless, a coarse survey is worth attempting in order to complete the FFT picture.

The idea underlying prime factor FFTs is generally traced to the 1958 paper by Good [68]. The essential ideas were also published by L. H. Thomas [152] in 1963. Good's 1958 paper is concerned primarily with a problem in the design of experiments. Almost as an afterthought, the author appends "some analogous short cuts in practical Fourier analysis" with the hope that the work "may be of some interest and practical value!" The prime factor idea was dormant for almost 20 years after the publication of Good's paper. Its revival is usually attributed to the 1977 paper by Kolba and Parks [86], with subsequent work by Winograd [164], Burrus [26], Burrus and Eschenbacher [27], and Temperton [147] leading to the ultimate acceptance of prime factor algorithms as competitive, and in some cases, preferable, FFTs [85], [148], [150].

Perhaps the simplest way to reach the crux of the prime factor idea is to return to the decimation-in-time FFT of length $N = pq$ as it was presented in the index expansion setting. Recall that the N-point DFT can be viewed as two waves of DFTs:

1. p DFTs of length q:

$$Y(k_0, n_0) = \sum_{n_1=0}^{q-1} x(n_1, n_0)\omega_q^{-n_1 k_0}$$

   for $n_0 = 0 : p - 1$, $k_0 = 0 : q - 1$, and $k_1 = 0 : p - 1$.

2. q DFTs of length p:

$$X(k_1, k_0) = \sum_{n_0=0}^{p-1} \{Y(k_0, n_0)\omega_N^{-n_0 k_0}\}\omega_p^{-n_0 k_1}$$

for $k_0 = 0 : q - 1$ and $k_1 = 0 : p - 1$.

With just a bit of a leap, it is possible to rewrite this two-step process in the matrix form

$$\mathbf{X} = \mathbf{W}_N \mathbf{x} = (\mathbf{W}_p \otimes \mathbf{I}_q)\, \Omega\, (\mathbf{I}_p \otimes \mathbf{W}_q) \mathbf{P}^T \mathbf{x},$$

where $\mathbf{P}^T$ represents the reordering of the input sequence, $(\mathbf{I}_p \otimes \mathbf{W}_q)$ represents the p transforms of length q, and $(\mathbf{W}_p \otimes \mathbf{I}_q)$ represents the q transforms of length p. The important issue is the diagonal $N \times N$ matrix Ω that holds the "twiddle factors" (powers of ω) that must be applied between the two waves of transforms. *If these twiddle factors could be eliminated*, then we would have, by a property of the Kronecker product $((\mathbf{A} \otimes \mathbf{B})(\mathbf{C} \otimes \mathbf{D}) = (\mathbf{AC}) \otimes (\mathbf{BD}))$, that

$$\mathbf{W}_N P = (\mathbf{W}_p \otimes \mathbf{I}_q)(\mathbf{I}_p \otimes \mathbf{W}_q) = \mathbf{W}_p \otimes \mathbf{W}_q.$$

(We have also used the fact that a permutation matrix $\mathbf{P}$ satisfies $\mathbf{P}^{-1} = \mathbf{P}^T$.) This would result in the reduction of a DFT of length N to a *genuine* two-dimensional DFT of size $p \times q$ with no intermediate "twiddling" required. And this is the motivation of the prime factor movement.

The underlying task becomes one of finding orderings of the input and output sequences, represented by permutation matrices $\mathbf{C}$ and $\mathbf{R}$, respectively, such that

$$\mathbf{C}^T \mathbf{W}_N R = \mathbf{W}_p \otimes \mathbf{W}_q.$$

If this task can be accomplished, then the DFT can be written as

$$\underbrace{\mathbf{C}^T \mathbf{X}}_{\substack{\text{reordered} \\ \text{output}}} = \mathbf{C}^T \mathbf{W}_N \mathbf{x} = (\mathbf{W}_p \otimes \mathbf{W}_q)\, \underbrace{\mathbf{R}^T \mathbf{x}}_{\substack{\text{reordered} \\ \text{input}}}.$$

The quest for these orderings requires a refreshing excursion into elementary number theory, and while it is entirely worthwhile, we can only summarize the salient results [98], [107]. The conclusion is that if p and q are relatively prime (have no common divisors other than 1), then such orderings can be found. The orderings are accomplished through mappings $\mathcal{M}$ that take a single index array of length $N = pq$ into a $p \times q$ index array:

$$\mathcal{M} : \{0, 1, 2, \ldots, N - 1\} \rightarrow \{0, 1, 2, \ldots, p - 1\} \times \{0, 1, 2, \ldots, q - 1\}.$$

In the original prime factor scheme proposed by Good, the ordering of the input sequence, $\mathbf{R}^T$, was given by the **Ruritanian map**, while the reordering of the output sequence, $\mathbf{C}^T$, was given by the **Chinese Remainder Theorem (CRT) map**.

As an example, consider the case in which $N = 3 \cdot 5$, for which we need reorderings $\mathbf{C}^T$ and $\mathbf{R}^T$ such that

$$\mathbf{C}^T \mathbf{X} = (\mathbf{W}_5 \otimes \mathbf{W}_3) \mathbf{R}^T \mathbf{x} = (\mathbf{W}_5 \otimes \mathbf{I}_3)(\mathbf{I}_5 \otimes \mathbf{W}_3) \mathbf{R}^T \mathbf{x}.$$

The reordered vectors $\mathbf{R}^T \mathbf{x}$ and $\mathbf{C}^T \mathbf{X}$ may be represented as 3×5 arrays. For this particular case, the appropriate reorderings are given by

$$\mathbf{R}^T \mathbf{x} = \begin{pmatrix} x_0 & x_3 & x_6 & x_9 & x_{12} \\ x_5 & x_8 & x_{11} & x_{14} & x_2 \\ x_{10} & x_{13} & x_1 & x_4 & x_7 \end{pmatrix}$$

and

$$\mathbf{C}^T\mathbf{X} = \begin{pmatrix} X_0 & X_6 & X_{12} & X_3 & X_9 \\ X_{10} & X_1 & X_7 & X_{13} & X_4 \\ X_5 & X_{11} & X_2 & X_8 & X_{14} \end{pmatrix}.$$

The prime factor FFT now proceeds in the following steps.

1. Form the array $\mathbf{R}^T\mathbf{x}$ from the input sequence.

2. Apply $(\mathbf{I}_5 \otimes \mathbf{W}_3)$, which means performing three-point DFTs of the columns of $\mathbf{R}^T\mathbf{x}$ (overwriting the original array).

3. Apply $(\mathbf{W}_5 \otimes \mathbf{I}_3)$, which means performing five-point DFTs of the rows of the array produced in step 2 (overwriting the previous array).

4. Recover the components of $\mathbf{X}$ from the array $\mathbf{C}^T\mathbf{X}$.

The prime factor algorithm can seem like an improbable sleight-of-hand until it is actually worked out in detail. This opportunity is provided in problems 243 and 244.

We hasten to add that the prime factor idea can be extended to cases in which N has several factors which are mutually prime. Prime factor FFTs are competitive because they avoid computations with "twiddle factors," because there are no index calculations or data movement during the actual combine phase, and because very efficient "small N DFT modules" can be designed to perform the smaller DFTs that arise. A selection of these specially crafted and streamlined modules is usually included with prime factor programs. More recently, by using more general maps or "rotated DFTs" [27], [149], PFAs have been discovered that are *both* in-place (require no additional storage arrays) and in-order (avoid reordering of the input and output sequences).

We now turn to the question of designing DFT modules for small values of N, such as those required by the prime factor FFTs. This leads to at least two more interesting paths. The first is a variation that has proven to be extremely effective in the design of FFTs. In his work on the complexity of FFT and convolution algorithms, Winograd [164] realized the possibility of expressing a DFT matrix as a product of three matrices, $\mathbf{W}_N = \mathbf{ABC}$, such that the elements of $\mathbf{A}$ and $\mathbf{C}$ are either zero or ± 1, and the diagonal matrix $\mathbf{B}$ has elements that are either real or purely imaginary. It can be proved that such a factoring minimizes the number of multiplications required in the FFT. Winograd factorizations exist for many small prime-order DFTs and can be incorporated into prime factor algorithms to evaluate the small N DFTs that arise.

Another path takes us a bit further afield, but it is worth the diversion. As we have seen, there is an intimate connection between DFTs and the convolution of two sequences. The connection is usually considered to be a one-way street: if an FFT is used to evaluate DFTs, then convolutions may be evaluated very efficiently by way of the convolution theorem. However, there is a history of ideas that "go in the opposite direction." In a 1970 paper, Bluestein [11] made the observation that writing

$$(k - n)^2 = k^2 - 2kn + n^2$$

allows us to rewrite the DFT of order N as

$$X_k = \sum_{n=0}^{N-1} x_n \omega_N^{-nk} = \sum_{n=0}^{N-1} x_n \omega_{2N}^{-2nk} = \omega_{2N}^{-k^2} \sum_{n=0}^{N-1} \{x_n \omega_{2N}^{-n^2}\} \omega_{2N}^{(k-n)^2}.$$

The sum on the right side may be recognized as the cyclic convolution of two sequences $\{x_n\omega_{2N}^{-n^2}\}$ and $\{\omega_{2N}^{n^2}\}$. In other words,

$$\omega_{2N}^{k^2}X_k = \{x_n\omega_{2N}^{-n^2}\} * \{\omega_{2N}^{n^2}\}.$$

While this maneuver does not lead directly to an efficient FFT algorithm, it does prove that a DFT of *any* order can be computed in $O(N \log N)$ operations by evaluating Bluestein's convolution (possibly embedded in a larger convolution) by FFTs.

The Bluestein approach is actually predated by the quite different convolution method of Rader [113], who observed that if $N = p$ is an odd prime, then there is a number-theoretic mapping that prescribes a row and column permutation of the DFT matrix that puts it in circulant form. Since a circulant matrix may be identified with convolution, well-known fast convolution methods may be applied. The critical mapping in the Rader FFT relies upon the notion of **primitive roots** of prime numbers, and it can be extended to cases in which $N = p^k$, where p is an odd prime and $k > 1$ is a natural number. The Rader factorization, when combined with fast convolution, has been used to design very efficient small prime-order DFT modules that are often integrated with prime factor FFTs.

10.6. FFT Performance

The preceding tour of the FFT has been breathless and brief, but perhaps it provides an impression of the richness and breadth of the subject. In this final section, we will discuss FFT performance and address the inevitable question of which is the best FFT. Not surprisingly, the answer to this question is "it depends!" In defending this nonanswer, we will cite several factors that determine FFT performance and insure that there will always be open problems for FFT users and architects.

1. **Symmetries.** It was recognized from the start (even in Good's 1958 paper) that if the input sequence has certain symmetries (such as real, real/even, or real/odd), then corresponding savings in storage and computation can be realized in the FFT. As mentioned in Chapter 4, sequences with symmetries can be transformed with savings by pre- and postprocessing algorithms. An alternative approach is to incorporate the symmetry directly into the FFT itself. This idea is generally attributed to Edson [9], who proposed an FFT for real sequences that has roughly half of the computational and storage costs of the complex FFT. The same strategy has been extended via the splitting method to design compact symmetric FFTs for even, odd, quarter-wave even and quarter-wave odd sequences that have one-fourth the computation and storage costs of complex FFTs [18], [141]. Symmetric FFTs have also been devised for prime factor FFTs [109] (an unfinished task). Clearly, the exploitation of symmetries improves FFT performance and leads to new and specialized FFT variations.

2. **Multiple transforms.** In many applications, FFTs do not come one at a time, but rather in herds. This generally occurs when problems are posed in more than one spatial dimension (for example, filtering of two-dimensional images or solutions of boundary value problems in a three-dimensional region). As we saw in Chapter 4, the DFT of a two-dimensional array of data is done by overwriting each row of the array with its DFT, and then computing the DFT of each resulting column. Clearly, a wave of multiple DFTs must be done in each

direction. There is a wealth of literature about how new FFTs can be designed and how existing FFTs can be adapted to economize the computation of multiple FFTs. There are situations in which it is advantageous to convert M sequences of length N into a supersequence of length MN and then perform N steps of a "truncated FFT" on the supersequence [136]. Regardless of how multiple FFTs are done, very often the *average* performance, measured in computation time per FFT, can be significantly improved with multiple FFTs. Needless to say, computer architecture bears heavily on this issue.

3. **Overhead.** By *overhead* we mean less-tangible, and often neglected costs, that impact the comparision of FFTs. These include all computational factors beyond the actual execution of the butterfly relations. Examples of overhead are additional storage arrays, computation and storage of powers of ω (see Van Loan [155] for excellent discussions of different ways to compute powers of ω), permutations and movement of data, and the arithmetic and storage necessitated by index mappings. It is difficult to compare FFTs accurately unless the playing field is leveled with respect to these factors.

4. **Hybrid strategies.** The pigeon-hole approach of this chapter suggests that different FFTs must never be seen together. In fact, the mixing of FFT methods within larger FFTs can often be very effective and leads to a nearly countless family of hybrid FFTs. One example may suffice. Assume that an FFT is to be applied to a sequence of length $N = 296000 = 2^5 \cdot 5^3 \cdot 7^2$. One approach is to let $N = r_1 r_2 r_3$ where $r_1 = 2^5, r_2 = 5^3$, and $r_3 = 7^2$, and call on a mixed-radix Cooley–Tukey FFT. On the other hand, it would be possible to let $r_1 = 5 \cdot 7 \cdot 8, r_2 = 4 \cdot 5 \cdot 7$, and $r_3 = 5$ (each r_i consists of mutually prime factors), use prime factor FFTs on each of the transforms of length r_i, and then combine them in a mixed-radix Cooley–Tukey FFT. It would be difficult to enumerate, analyze, or compare the various hybrid FFT options for even a single large value of N.

5. **Architecture.** FFTs reside on "hard-wired" microchips inside the navigational systems of space craft; they are performed by experimental codes for massively parallel distributed memory computers; they can be called "on-line" in micro-computer/workstation environments; and they can be found in software libraries for highly vectorized supercomputers [7], [59], [87], [136] and parallel computers [18]. More recently, symmetric FFTs have been developed for parallel comput-ers [16], [75]. Therefore, it goes without saying that FFT performance depends on hardware and architecture issues. With the explosive progress of computing technology, there is an attendant need to tailor FFTs to new advanced architec-tures. Some existing FFTs fit new architectures ideally (for example, the radix-2 Cooley–Tukey FFT on a hypercube architecture [18]); other architectures may demand entirely new FFTs. (Occasionally these "new" FFTs may actually be old FFTs reincarnated; for example, Bluestein's algorithm, which is not efficient on scalar computers, may be the *only* efficient way of computing certain FFTs on a hypercube [139].) This factor alone insures that there will never be a single *best* FFT; the answer depends upon the arena in which the race is run.

6. **Software.** FFTs abound in software for everything from microcomputers to su-percomputers. No single package will conform to all applications and computing

environments. Those interested in reliable and widely used mainframe FFT software might begin with *FFTPACK* [140] and its vectorized version *VFFTPACK*.

10.7. Notes

Those readers interested in the historical development of the FFT should locate a 1967 volume (AU-15) of the *IEEE Transactions on Audio and Electroacoustics*, which is a collection of early papers on the recently revealed FFT. Volume AU-17 of the same journal (1969) contains the proceedings of the *Arden House Workshop of FFT Processing*, which also is of historical interest [39]. Extensive FFT bibliographies can be found in Brigham [21], Van Loan [155], and Heideman and Burrus [73].

10.8. Problems

232. Crucial properties of ω_N^{nk}. Letting $\omega_N = e^{i2\pi/N}$ for any natural number N, prove that (i) $\omega_N^N = 1$, (ii) $\omega_N^{N/2} = -1$, (iii) $\omega_N^{N/4} = i$, (iv) $\omega_N^p = \omega_{N/p}$, and (v) $\omega_N^{-pq} = \omega_{N/p}^{-q}$.

233. Radix-3 FFT. With $N = 3^M$, derive the butterfly relations for the Cooley–Tukey decimation-in-time FFT using the splitting method.

234. Radix-4 FFT. With $N = 4^M$, derive the butterfly relations for the Cooley–Tukey decimation-in-time FFT using the splitting method.

235. Real operation counts. Use the results of the chapter and the previous two problems to find the real operation counts for N-point decimation-in-time radix-2, radix-3, and radix-4 FFTs. Assume that a complex multiplication requires two real additions and four real multiplications. Assume that a radix-r FFT consists of $\log_r N$ stages, each of which has N/r butterflies. Neglect all multiplications by ± 1 and avoid all duplicate multiplications. Express all operation counts in the form $CN\log_2 N$, where terms proportional to N are neglected. Verify that the entries in Table 10.1 give the appropriate values of C.

TABLE 10.1
*Real operation counts for radix-2,3,4 FFTs
of length N; values of C in $CN\log_2 N$.*

	$N = 2^M$	$N = 3^M$	$N = 4^M$
Real +	3.00	$4\log_3 2 \approx 2.52$	$11/4 = 2.75$
Real ×	2.00	$16/3\log_3 2 \approx 3.26$	$3/2 = 1.5$
Total	5.00	5.78	4.25

Comment on the relative costs of these FFTs. How would you design an FFT for a sequence of length $N = 256$?

236. Another radix-3 FFT. Use the splitting method to find the butterfly relations for the Gentleman–Sande decimation-in-frequency FFT for the case $N = 3^M$.

237. Inverse FFTs. Here is a way to devise inverse FFTs. Begin with the decimation-in-frequency butterfly relations (10.8) and (10.9) in the form

$$X_{2k} = \mathcal{D}_{\frac{N}{2}}\{x_n + x_{n+\frac{N}{2}}\}_k,$$
$$X_{2k+1} = \mathcal{D}_{\frac{N}{2}}\{(x_n + x_{n+\frac{N}{2}})\, \omega_N^{-n}\}_k$$

for $k = 0 : N/2 - 1$, where $\mathcal{D}_p$ represents the DFT of length p. Equivalently, we have

$$\mathcal{D}_{\frac{N}{2}}^{-1}\{X_{2k}\}_n = x_n + x_{n+\frac{N}{2}},$$
$$\mathcal{D}_{\frac{N}{2}}^{-1}\{X_{2k+1}\}_n = (x_n - x_{n+\frac{N}{2}})\, \omega_N^{-n}$$

for $n = 0 : N/2 - 1$, where $\mathcal{D}_p^{-1}$ represents the inverse DFT of length p. Now formally invert the second pair of relations (solve for x_n and $x_{n+N/2}$) and show that the result is the decimation-in-time butterfly relations (10.3) and (10.4) with ω replaced by ω^{-1} and with a scaling factor of $1/2$. Argue that these relations describe how to combine two IDFTs of length $N/2$ to form an IDFT of length N.

238. Another inverse FFT.   Beginning with the butterfly relations for the decimation-in-time FFT (10.3) and (10.4), follow the procedure of the preceding problem to obtain a decimation-in-frequency inverse FFT.

239. The Gentleman–Sande FFT by index expansions.   Follow the index expansion method outlined in the text with $p = N/2$ and $q = 2$ to derive the Gentleman–Sande butterfly relations (10.8) and (10.9) for $N = 2^M$. The outcome depends on the ordering of the two nested sums.

240. An $N = 16$ FFT by index expansions.   Letting $N = 4 \cdot 4$, express the 16-point DFT as two nested 4-point DFTs. Arrange the two nested sums so that both sets of 4-point DFTs require *no* multiplications.

241. Matrix factorization.   Consider the $N = 16$ decimation-in-time FFT. Show that the butterfly relations for the first splitting can be expressed in the form

$$\mathbf{X} = \mathbf{W}_{16}\mathbf{x} = \begin{pmatrix} \mathbf{I}_8 & \mathbf{\Omega}_8 \\ \mathbf{I}_8 & -\mathbf{\Omega}_8 \end{pmatrix} \begin{pmatrix} \mathbf{Y} \\ \mathbf{Z} \end{pmatrix} \equiv \mathbf{B}_{16}\begin{pmatrix} \mathbf{Y} \\ \mathbf{Z} \end{pmatrix},$$

where $\mathbf{I}_m$ is the identity matrix of order m and $\mathbf{\Omega}_m$ is the $(m \times m)$ diagonal matrix

$$\mathbf{\Omega}_m = \mathrm{diag}\,(1, \omega_{2m}^{-1}, \omega_{2m}^{-2}, \ldots, \omega_{2m}^{-(m-1)}).$$

242. An $N = 6$ FFT by index expansions.   Letting $N = 3 \cdot 2$ use the index expansion method to derive an FFT for $N = 6$. Note that depending upon how the two nested sums are arranged, it is possible to produce either a decimation-in-time or decimation-in-frequency FFT.

243. An $N = 6$ prime factor FFT.   In the previous problem, you most likely encountered "twiddle factors" (extra multiplications by powers of ω). These terms can be eliminated by a careful reordering of the input and output sequences. Consider the case in which $N = pq$, $p = 3$, and $q = 2$ and proceed as follows.

  (a) Let $\langle m \rangle_p$ (said "m mod p") denote the remainder in dividing m by p. Check that the **Chinese Remainder Theorem map** $\mathcal{C}$ and the **Ruritanian map** $\mathcal{R}$ given by

$$\mathcal{C}(k) = (\langle k \rangle_3, \langle k \rangle_2) \quad \text{and} \quad \mathcal{R}(n) = (\langle 2n \rangle_3, \langle n \rangle_2)$$

when applied to the terms of the sequence $\mathbf{s} = (0, 1, 2, 3, 4, 5)^T$ produce the following reorderings of $\mathbf{s}$:

$$\mathbf{C}^T\mathbf{s} = \begin{pmatrix} 0 & 3 \\ 4 & 1 \\ 2 & 5 \end{pmatrix} \quad \text{and} \quad \mathbf{R}^T\mathbf{s} = \begin{pmatrix} 0 & 3 \\ 2 & 5 \\ 4 & 1 \end{pmatrix}$$

respectively. (Note that row indices vary from $0 \to 2$ and column indices vary from $0 \to 1$.)

(b) Now reorder the input sequence $\mathbf{x}$ so it has the form $\mathbf{R}^T\mathbf{x}$ and reorder the output sequence $\mathbf{X}$ so it has the form $\mathbf{C}^T\mathbf{X}$. Verify that by taking three-point DFTs of the columns of $\mathbf{R}^T\mathbf{x}$ (overwriting the original array) and then taking two-point DFTs of the columns of the resulting array, the transform coefficients X_k appear in the array $\mathbf{C}^T\mathbf{X}$.

(c) Why does it work? Note that $\mathcal{C}$ and $\mathcal{R}$ are one-to-one and onto mappings. Verify that the mappings

$$\hat{k} = \mathcal{C}^{-1}(k_1, k_0) = \langle 4k_1 + 3k_0 \rangle_6 \quad \text{and} \quad \hat{n} = \mathcal{R}^{-1}(n_1, n_0) = \langle 2n_1 + 3n_0 \rangle_6$$

are the inverses of $\mathcal{C}$ and $\mathcal{R}$.

(d) Write out the definition of the six-point DFT as a pair of nested sums, replacing the x_n by $x(\mathcal{R}(n))$, X_k by $X(\mathcal{C}(k))$, n by $\hat{n}$, and k by $\hat{k}$. Verify that the "twiddle factors" do not appear, and that the six-point DFT becomes a genuine two-dimensional (3×2) DFT.

244. An $N = 15$ prime factor FFT. Carry out the procedure of the preceding problem in the case $N = pq, p = 3$, and $q = 5$. Show that the maps

$$\mathcal{C}(k) = (\langle k \rangle_3, \langle k \rangle_5) \quad \text{and} \quad \mathcal{R}(n) = (\langle 2n \rangle_3, \langle 2n \rangle_5)$$

produce the orderings $\mathbf{C}^T\mathbf{X}$ and $\mathbf{R}^T\mathbf{x}$ given in Section 10.5. Show that when the inverse maps

$$\hat{k} = \mathcal{C}^{-1}(k_1, k_0) = \langle 10k_1 + 6k_0 \rangle_{15} \quad \text{and} \quad \hat{n} = \mathcal{R}^{-1}(n_1, n_0) = \langle 5n_1 + 3n_0 \rangle_{15}$$

are used in the definition of the DFT, a genuine two-dimensional (3×5) DFT results.

245. Matrix factorization for radix-3. Show that the Cooley–Tukey decimation-in-time FFT for $N = 3^4 = 81$ can be expressed as

$$\mathbf{X} = \mathbf{W}_{81}\mathbf{x} = (\mathbf{I}_1 \otimes \mathbf{B}_{81})(\mathbf{I}_3 \otimes \mathbf{B}_{27})(\mathbf{I}_9 \otimes \mathbf{B}_9)(\mathbf{I}_{27} \otimes \mathbf{B}_3)\mathbf{P}^T\mathbf{x}.$$

In analogy with the radix-2 case, find the block entries of the butterfly matrices $\mathbf{B}_m$ in terms of the diagonal matrices $\mathbf{\Omega}_m$ and the identity matrices $\mathbf{I}_m$.

Appendix

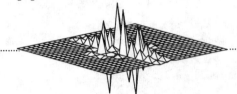

Table of DFTs

In the following table we have collected several analytical DFTs. A few words of explanation are needed. Each DFT entry is arranged as follows.

Discrete input name	$f_n, \ n \in \mathcal{N}$	
Graph of f_n	$F_k, \ k \in \mathcal{N}$	Graph of F_k
	$\|c_k - F_k\|, \ k \in \mathcal{N}$	
Continuum input name	$f(x), \ x \in I$	
Graph of $f(x)$	$c_k, \ k \in \mathbf{Z}$	Graph of $\|c_k - F_k\|$
	Comments	$\max \|c_k - F_k\|$

The first column has two boxes. The upper box gives the name of the input, below which are graphs of the real and imaginary parts of the discrete input sequence. The lower box contains the name of the continuum input, and the corresponding continuum input graphs. The middle column has six boxes containing, in order from top to bottom, the formula of the input sequence f_n; the analytical N-point DFT output F_k; a measure of the difference $\|c_k - F_k\|$, where c_k is the kth Fourier coefficient; the formula of the continuum input function $f(x)$; the formula for c_k; an entry for comments, perhaps the most important of which is the AVED warning. This means that *average values at endpoints and discontinuities* must be used if the correct DFT is to be computed. The third column consists of two boxes. The upper box displays graphically the real and imaginary parts of the DFT. The lower box gives the maximum error $\max \|c_k - F_k\|$, and displays graphically the error $\|c_k - F_k\|$ for a small (24-point) example.

Unless otherwise noted, the function is assumed to be sampled on the interval $[-A/2, A/2]$. The difference $\|c_k - F_k\|$ is generally in the form CN^{-p} for some constant C and some positive integer p, which should be interpreted in an asymptotic sense for $N \to \infty$; in other words, if $\|F_k - c_k\| = CN^{-p}$, then $\lim_{N \to \infty} \left| \frac{F_k - c_k}{N^p} \right| = C$. While this measure is different than the pointwise errors that were derived in Chapter 6, it does agree with those estimates in its dependence on N.

The following notational conventions hold throughout *The Table of DFTs*:

$$\mathcal{N} = \left\{ -\frac{N}{2} + 1, \ldots, \frac{N}{2} \right\}, \quad I = \left[-\frac{A}{2}, \frac{A}{2} \right], \quad \mathbf{Z} = \{0, \pm 1, \pm 2, \ldots\};$$

$$\omega_N = e^{i\frac{2\pi}{N}}, \quad \theta_k = \frac{2\pi k}{N}, \quad \delta(k) = \begin{cases} 1 & \text{if} \quad k = 0, \\ 0 & \text{if} \quad k \neq 0; \end{cases}$$

$$\hat{\delta}_N(k) = \begin{cases} 1 & \text{if} \quad k = 0 \text{ or a multiple of } N, \\ 0 & \text{otherwise.} \end{cases}$$

AVED = average values at endpoints and discontinuities,
C is a constant independent of k and N.

TABLE OF DFTS

Discrete input name	$f_n, \ n \in \mathcal{N}$	
Graph of f_n	$F_k, \ k \in \mathcal{N}$	Graph of F_k
	$|c_k - F_k|, \ k \in \mathcal{N}$	
Continuum input name	$f(x), \ x \in I$	
Graph of $f(x)$	$c_k, \ k \in \mathbf{Z}$	Graph of $|c_k - F_k|$
	Comments	$\max |c_k - F_k|$
1. Impulse	$\delta(n - n_0)$	$F_k:$
$\mathcal{R}:$	$\frac{1}{N}\omega_N^{-n_0 k}$	$\mathcal{R}:$
$\mathcal{I}:$	Exact	$\mathcal{I}:$
1. None	–	
	–	
	$n_0 \in \mathcal{N}$	
2a. Paired impulses	$\frac{1}{2}(\delta(n - n_0) + \delta(n + n_0))$	$F_k:$
$\mathcal{R}:$	$\frac{1}{N}\cos(\frac{2\pi n_0 k}{N})$	$\mathcal{R}:$
$\mathcal{I}:$	Exact	$\mathcal{I}:$
2a. None	–	
	–	
	$n_0 \in \mathcal{N}$	

2b. Paired impulses $\mathcal{R}:$ $\mathcal{I}:$	$\frac{1}{2}(\delta(n+n_0) - \delta(n-n_0))$ $\frac{i}{N}\sin(\frac{2\pi n_0 k}{N})$ Exact	$F_k:$ $\mathcal{R}:$ $\mathcal{I}:$
2b. None	$-$ $-$ $n_0 \in \mathcal{N}$	
3. Complex harmonic $\mathcal{R}:$ $\mathcal{I}:$	$\omega_N^{nk_0}$ $\hat{\delta}_N(k-k_0)$ (periodic) Exact	F_k and $c_k:$ $\mathcal{R}:$ $\mathcal{I}:$
3. Complex harmonic $\mathcal{R}:$ $\mathcal{I}:$	$e^{i\frac{2\pi k_0 x}{A}}$ $\delta(k-k_0)$ $k_0 \in \mathbf{Z}$	 $\max|c_k - F_k| = 0$
3a. Constant $\mathcal{R}:$ $\mathcal{I}:$	1 $\hat{\delta}_N(k)$ (periodic) Exact	F_k and $c_k:$ $\mathcal{R}:$ $\mathcal{I}:$
3a. Constant $\mathcal{R}:$ $\mathcal{I}:$	1 $\delta(k)$ Case 3: $k_0 = 0$	 $\max|c_k - F_k| = 0$
4a. Cosine harmonic $\mathcal{R}:$ $\mathcal{I}:$	$\cos(\frac{2\pi k_0 n}{N})$ $\frac{1}{2}(\hat{\delta}_N(k-k_0) + \hat{\delta}_N(k+k_0))$ Exact	F_k and $c_k:$ $\mathcal{R}:$ $\mathcal{I}:$
4a. Cosine harmonic $\mathcal{R}:$ $\mathcal{I}:$	$\cos(\frac{2\pi k_0 x}{A})$ $\frac{1}{2}(\delta(k-k_0) + \delta(k+k_0))$ $k_0 \in \mathbf{Z}$	 $\max|c_k - F_k| = 0$

4b. Critical mode $\mathcal{R}:$ $\mathcal{I}:$	$\cos(\pi n) = (-1)^n$ $\hat{\delta}_N(k - \tfrac{N}{2})$ Exact	F_k and c_k: $\mathcal{R}:$ $\mathcal{I}:$
4b. Critical mode $\mathcal{R}:$ $\mathcal{I}:$	$\cos(\tfrac{\pi N x}{A})$ $\tfrac{1}{2}(\delta(k - \tfrac{N}{2}) + \delta(k + \tfrac{N}{2}))$ Case 4a: $k_0 = \tfrac{N}{2}$	 $\max|c_k - F_k| = 0$
4c. Sine harmonic $\mathcal{R}:$ $\mathcal{I}:$	$\sin(\tfrac{2\pi k_0 n}{N})$ $\tfrac{i}{2}(\hat{\delta}_N(k + k_0) - \hat{\delta}_N(k - k_0))$ Exact	F_k and c_k: $\mathcal{R}:$ $\mathcal{I}:$
4c. Sine harmonic $\mathcal{R}:$ $\mathcal{I}:$	$\sin(\tfrac{2\pi k_0 x}{A})$ $\tfrac{i}{2}(\delta(k + k_0) - \delta(k - k_0))$ $k_0 \in \mathbf{Z}$	 $\max|c_k - F_k| = 0$
5. Complex wave $\mathcal{R}:$ $\mathcal{I}:$	$\omega_N^{nk_0}$ $\dfrac{\sin(\pi(k-k_0))\sin(2\pi(k-k_0)/N)}{2N\sin^2(\pi(k-k_0)/N)}$ $\sim C(k - k_0)N^{-2}$	F_k and c_k: $\mathcal{R}:$ $\mathcal{I}:$
5. Complex wave $\mathcal{R}:$ $\mathcal{I}:$	$e^{i\tfrac{2\pi k_0 x}{A}}$ $\dfrac{\sin(\pi(k-k_0))}{\pi(k-k_0)}$ AVED, $k_0 \notin \mathbf{Z}$	$|c_k - F_k|,\quad k_0 = 2.4$ $\max \approx 10^{-2},\ N = 24$
6a. Cosine $\mathcal{R}:$ $\mathcal{I}:$	$\cos(\tfrac{\pi k_0 n}{N})$ $\dfrac{\cos(\pi k)\sin(\pi k_0/2)}{4N}$ $\times \left(\dfrac{\sin\theta_+}{\sin^2\theta_+/2} - \dfrac{\sin\theta_-}{\sin^2\theta_-/2}\right)$ $\sim C k_0 N^{-2}$	F_k and c_k: $\mathcal{R}:$ $\mathcal{I}:$
6a. Cosine $\mathcal{R}:$ $\mathcal{I}:$	$\cos(\tfrac{\pi k_0 x}{A})$ $-\dfrac{2k_0\cos(\pi k)\sin(\pi k_0/2)}{\pi(4k^2 - k_0^2)}$ $k_0 \notin \mathbf{Z},\ \theta_\pm = \tfrac{\pi(2k\pm k_0)}{N}$	$|c_k - F_k|,\quad k_0 = 2.4$ $\max \approx 10^{-3},\ N = 24$

6b. Half cosine	$\cos(\frac{\pi n}{N})$	F_k and c_k:
$\mathcal{R}$:	$\frac{\cos(\pi k)}{4N}\left(\frac{\sin\theta_+}{\sin^2\theta_+/2} - \frac{\sin\theta_-}{\sin^2\theta_-/2}\right)$	$\mathcal{R}$:
$\mathcal{I}$:	$\sim CN^{-2}$	$\mathcal{I}$:
6b. Half cosine	$\cos(\pi x)$ on $(-\frac{1}{2}, \frac{1}{2}]$	$\|c_k - F_k\|$
$\mathcal{R}$:	$-\frac{2\cos(\pi k)}{\pi(4k^2-1)}$	
$\mathcal{I}$:	Case 6a: $\theta_\pm = \frac{\pi(2k\pm 1)}{N}$, $k_0 = 1, A = 1$	max $\approx 10^{-3}$, $N = 24$
6c. Sine	$\sin(\frac{\pi k_0 n}{N})$	F_k and c_k:
$\mathcal{R}$:	$i\frac{\cos(\pi k)\sin(\pi k_0/2)}{4N}$ $\times\left(\frac{\sin\theta_+}{\sin^2\theta_+/2} + \frac{\sin\theta_-}{\sin^2\theta_-/2}\right)$	$\mathcal{R}$:
$\mathcal{I}$:	$\sim CkN^{-2}$	$\mathcal{I}$:
6c. Sine	$\sin(\frac{\pi k_0 x}{A})$	$\|c_k - F_k\|$, $k_0 = 2.4$
$\mathcal{R}$:	$i\frac{4k\cos(\pi k)\sin(\pi k_0/2)}{\pi(4k^2-k_0^2)}$	
$\mathcal{I}$:	AVED, $k_0 \notin \mathbf{Z}, \theta_\pm = \frac{\pi(2k\pm k_0)}{N}$	max $\approx 10^{-2}$, $N = 24$
6d. Even sine	$\|\sin(2\pi n/N)\|$	F_k and c_k:
$\mathcal{R}$:	$\frac{1+\cos(\pi k)}{4N}\left(\frac{\sin\theta_+}{\sin^2\theta_+/2} - \frac{\sin\theta_-}{\sin^2\theta_-/2}\right)$	$\mathcal{R}$:
$\mathcal{I}$:	$\sim CN^{-2}$	$\mathcal{I}$:
6d. Even sine	$\|\sin(2\pi x/A)\|$	$\|c_k - F_k\|$
$\mathcal{R}$:	$\frac{1}{\pi}\frac{1+\cos(\pi k)}{1-k^2}$	
$\mathcal{I}$:	$\theta_\pm = \frac{2\pi}{N}(k\pm 1), c_{\pm 1} = F_{\pm 1} = 0$	max $\approx 10^{-3}$, $N = 24$
7. Linear	n/N	F_k and c_k:
$\mathcal{R}$:	$F_0 = 0, F_k = \frac{i\cos(\pi k)\sin\theta_k}{4N\sin^2\theta_k/2}$	$\mathcal{R}$:
$\mathcal{I}$:	$\sim CkN^{-2}$	$\mathcal{I}$:
7. Linear	x/A	$\|c_k - F_k\|$
$\mathcal{R}$:	$c_0 = 0, c_k = \frac{i\cos(\pi k)}{2\pi k}$	
$\mathcal{I}$:	AVED, $f_{-\frac{N}{2}} = 0$	max $\approx 10^{-2}$, $N = 24$

8. Triangular wave	$1 - 2|n|/N$	F_k and c_k:
$\mathcal{R}$:	$F_0 = \frac{1}{2}, F_k = \frac{1-\cos(\pi k)}{N^2 \sin^2 \theta_k/2}$	$\mathcal{R}$:
$\mathcal{I}$:	$\sim CN^{-2}$	$\mathcal{I}$:
8. Triangular wave	$1 - 2|x|/A$	$|c_k - F_k|$
$\mathcal{R}$:	$c_0 = \frac{1}{2}, c_k = \frac{1-\cos(\pi k)}{\pi^2 k^2}$	
$\mathcal{I}$:		max $\approx 10^{-3}$, $N = 24$
9. Rectangular wave	$\begin{cases} -1, & -N/2 < n < 0 \\ 1, & 0 < n < N/2 \end{cases}$	F_k and c_k:
$\mathcal{R}$:	$F_0 = 0, F_k = i\frac{(\cos(\pi k)-1)\sin\theta_k}{2N \sin^2 \theta_k/2}$	$\mathcal{R}$:
$\mathcal{I}$:	$\sim CkN^{-2}$	$\mathcal{I}$:
9. Rectangular wave	$\begin{cases} -1, & -A/2 < x < 0 \\ 1, & 0 < x < A/2 \end{cases}$	$|c_k - F_k|$
$\mathcal{R}$:	$c_0 = 0, c_k = \frac{i(\cos(\pi k)-1)}{\pi k}$	
$\mathcal{I}$:	AVED, $f_n = 0$ for $n = 0, \pm N/2$	max $\approx 10^{-2}$, $N = 24$
10. Square pulse	$\begin{cases} 1, & |n| < M/2 \\ 0, & M/2 < |n| < N/2 \end{cases}$	F_k and c_k:
$\mathcal{R}$:	$F_0 = \frac{M}{N}, F_k = \frac{\sin(\pi k M/N)\sin\theta_k}{2N \sin^2 \theta_k/2}$	$\mathcal{R}$:
$\mathcal{I}$:	$\sim CkN^{-2}$	$\mathcal{I}$:
10. Square pulse	$\begin{cases} 1, & |x| < a/2 \\ 0, & a/2 < |x| < A/2 \end{cases}$	$|c_k - F_k|$
$\mathcal{R}$:	$c_0 = \frac{a}{A}, c_k = \frac{\sin(\pi k a/A)}{\pi k}$	
$\mathcal{I}$:	AVED, $0 < \frac{a}{A} = \frac{M}{N} < 1$	max $\approx 10^{-2}$, $N = 24$
10a. Square pulse	$\begin{cases} 1, & |n| < N/4 \\ 0, & N/4 < |n| < N/2 \end{cases}$	F_k and c_k:
$\mathcal{R}$:	$F_0 = \frac{1}{2}, F_k = \frac{\sin(\pi k/2)\sin\theta_k}{2N \sin^2 \theta_k/2}$	$\mathcal{R}$:
$\mathcal{I}$:	$\sim CkN^{-2}$	$\mathcal{I}$:
10a. Square pulse	$\begin{cases} 1, & |x| < 1/2 \\ 0, & 1/2 < |x| < 1 \end{cases}$	$|c_k - F_k|$
$\mathcal{R}$:	$c_0 = \frac{1}{2}, c_k = \frac{\sin(\pi k/2)}{\pi k}$	
$\mathcal{I}$:	AVED, Case 10: $A = 2a = 1$	max $\approx 10^{-2}$, $N = 24$

11. Exponential	$e^{-aAn/N},\ 0 \le n \le N-1$	F_k and c_k:
$\mathcal{R}:$	$\dfrac{\sigma(1-e_N^2)-i2\sigma e_N \sin\theta_k}{2N(1-2e_N\cos\theta_k+e_N^2)}$	$\mathcal{R}:$
$\mathcal{I}:$	$\sim CN^{-1}$	$\mathcal{I}:$
11. Exponential	$e^{-ax},\ 0 < x < A$	$|c_k - F_k|,\ a=2,\ A=3$
$\mathcal{R}:$	$\dfrac{\sigma(aA-i2\pi k)}{a^2A^2+4\pi^2k^2}$	
$\mathcal{I}:$	AVED, $\sigma = 1 - e^{-aA},\ e_N = e^{-aA/N}$	$\max \approx 10^{-2},\ N=24$
12. Even exponential	$e^{-aA|n|/N}$	F_k and c_k:
$\mathcal{R}:$	$\dfrac{(1-e_N^2)(1-e_2\cos(\pi k))}{N(1-2e_N\cos\theta_k+e_N^2)}$	$\mathcal{R}:$
$\mathcal{I}:$	$\sim CN^{-2}$	$\mathcal{I}:$
12. Even exponential	$e^{-a|x|},\ |x| < A/2$	$|c_k - F_k|,\ a=2,\ A=1$
$\mathcal{R}:$	$\dfrac{2aA(1-e_2\cos(\pi k))}{a^2A^2+4\pi^2k^2}$	
$\mathcal{I}:$	$e_2 = e^{-aA/2},\ e_N = e^{-aA/N}$	$\max \approx 10^{-3},\ N=24$
13. Odd exponential	$(n/|n|)e^{-aA|n|/N}$	F_k and c_k:
$\mathcal{R}:$	$i\dfrac{2(e_N \sin\theta_k)(e_2\cos(\pi k)-1)}{N(1-2e_N\cos\theta_k+e_N^2)}$	$\mathcal{R}:$
$\mathcal{I}:$	$\sim CN^{-1}$	$\mathcal{I}:$
13. Odd exponential	$(x/|x|)e^{-a|x|},\ |x| < A/2$	$|c_k - F_k|,\ a=2,\ A=1$
$\mathcal{R}:$	$i\dfrac{4\pi k(e_2\cos(\pi k)-1)}{a^2A^2+4\pi^2k^2}$	
$\mathcal{I}:$	AVED, $e_2 = e^{-aA/2},\ e_N = e^{-aA/N}$	$\max \approx 10^{-2},\ N=24$
14. Linear/exponential	$(nA/N)e^{-aA|n|/N}$	F_k and c_k:
$\mathcal{R}:$	$i\dfrac{2A\sin\theta_k(\sigma_k(e_N^3-e_N)+\cdots}{N^2(e_N^4-4\cos\theta_k(e_N^3+e_N)+\cdots}$ $\dfrac{\cdots+Ne_2e_N\cos(\pi k)(1-e_N\cos\theta_k))}{\cdots+2e_N^2(\cos(2\theta_k)+2)+1)}$	$\mathcal{R}:$
$\mathcal{I}:$	$\sim CN^{-1}$	$\mathcal{I}:$
14. Linear/exponential	$xe^{-a|x|}$	$|c_k - F_k|,\ a=2,\ A=3$
$\mathcal{R}:$	$i\dfrac{2\pi kA}{d}(e_2\cos(\pi k)+$ $\dfrac{4aA}{d}(e_2\cos(\pi k) - 1))$	
$\mathcal{I}:$	$\sigma_k = 1 + (\frac{N}{2}-1)e_2\cos(\pi k)$ $e_2 = e^{-aA/2},\ e_N = e^{-aA/N}$ $d = a^2A^2 + 4\pi^2k^2,\qquad$ AVED	$\max \approx 10^{-3},\ N=24$

15a. Cosine/exponential	$e^{-aA\|n\|/N}\cos(\pi n/N)$	F_k and c_k:
$\mathcal{R}:$	$\dfrac{(1-e_N^2)+2e_2e_N\cos(\pi k)\sin\theta_+}{2N(1-2e_N\cos\theta_+ +e_N^2)}$ $+\dfrac{(1-e_N^2)-2e_2e_N\cos(\pi k)\sin\theta_-}{2N(1-2e_N\cos\theta_- +e_N^2)}$	$\mathcal{R}:$
$\mathcal{I}:$	$\sim CN^{-2}$	$\mathcal{I}:$

15a. Cosine/exponential	$e^{-a\|x\|}\cos(\pi x/A)$	$\|c_k-F_k\|$, $a=2$, $A=3$
$\mathcal{R}:$	$\dfrac{e_2k_+\cos(\pi k)+a}{A(k_+^2+a^2)}-\dfrac{e_2k_-\cos(\pi k)-a}{A(k_-^2+a^2)}$	
$\mathcal{I}:$	$e_2=e^{-aA/2},\qquad e_N=e^{-aA/N}$ $\theta_\pm=\dfrac{\pi}{N}(2k\pm1),\quad k_\pm=\dfrac{\pi}{A}(2k\pm1)$ AVED	$\max\approx10^{-3}$, $N=24$

15b. Sine/exponential	$e^{-aA\|n\|/N}\sin(\pi n/N)$	F_k and c_k:
$\mathcal{R}:$	$i\left(\dfrac{1-e_N\cos\theta_+ +e_2e_N\cos(\pi k)\sin\theta_+}{N(1-2e_N\cos\theta_+ +e_N^2)}\right.$ $\left.-\dfrac{1-e_N\cos\theta_- -e_2e_N\cos(\pi k)\sin\theta_-}{N(1-2e_N\cos\theta_- +e_N^2)}\right)$	$\mathcal{R}:$
$\mathcal{I}:$	$\sim CN^{-1}$	$\mathcal{I}:$

15b. Sine/exponential	$e^{-a\|x\|}\sin(\pi x/A)$	$\|c_k-F_k\|$, $a=2$, $A=3$
$\mathcal{R}:$	$\dfrac{ie_2k_+\sin((2k+1)\pi/2))+a}{A(k_+^2+a^2)}$ $-\dfrac{ie_2k_-\sin((2k-1)\pi/2))+a}{A(k_-^2+a^2)}$	
$\mathcal{I}:$	$e_2=e^{-aA/2},\qquad e_N=e^{-aA/N}$ $\theta_\pm=\dfrac{\pi}{N}(2k\pm1),\quad k_\pm=\dfrac{\pi}{A}(2k\pm1)$ AVED	$\max\approx10^{-3}$, $N=24$

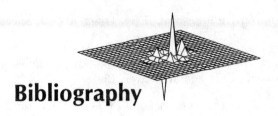

Bibliography

[1] F. ABRAMOVICI, *The accurate calculation of Fourier integrals by the fast Fourier transform technique*, J. Comp. Phys., 11 (1973), pp. 28–37.

[2] ——, *Letter: The accuracy of finite Fourier transforms*, J. Comp. Phys., 17 (1975), pp. 446–449.

[3] M. ABRAMOWITZ AND I. STEGUN, *Handbook of Mathematical Functions*, Dover, New York, 1972.

[4] J. ARSAC, *Fourier Transforms and the Theory of Distributions*, Prentice-Hall, Englewood Cliffs, NJ, 1966.

[5] L. AUSLANDER AND M. SCHENEFELT, *Fourier transforms that respect crystallographic symmetries*, IBM J. Res. and Dev., 31 (1987), pp. 213–223.

[6] L. AUSLANDER AND R. TOLIMIERI, *Is computing with the finite Fourier transform pure or applied mathematics?*, Bull. Amer. Math Soc., 1 (1979), pp. 847–897.

[7] D. BAILEY, *A high peformance FFT algorithm for vector supercomputers*, Internat. J. Supercomputer Applications, 2 (1988), pp. 82–87.

[8] D. BAILEY AND P. SWARZTRAUBER, *A fast method for the numerical evaluation of continuous Fourier and Laplace transforms*, SIAM J. Sci. Comput., 15 (1994), pp. 1105–1110.

[9] G. BERGLAND, *A fast Fourier transform algorithm for real-valued series*, Comm. ACM, 11 (1968), pp. 703–710.

[10] G. BEYLKIN AND M. BREWSTER, *Fast numerical algorithms using wavelet bases on the interval*, 1993, in progress.

[11] L. BLUESTEIN, *A linear filtering approach to the computation of the discrete Fourier transform*, IEEE Trans. Audio and Electroacoustics, AU–18 (1970), pp. 451–455.

[12] J. BOYD, *Chebyshev and Fourier Spectral Methods*, Springer-Verlag, Berlin, 1989.

[13] R. BRACEWELL, *The Fourier Transform and Its Applications*, McGraw-Hill, New York, 1978.

[14] ——, *The Hartley Transform*, Oxford University Press, U.K., 1986.

[15] R. BRACEWELL, *Assessing the Hartley transform*, IEEE Trans. Acoust. Speech Signal Process., ASSP-38 (1990), pp. 2174–2176.

[16] B. BRADFORD, *Fast Fourier Transforms for Direct Solution of Poisson's Equation*, PhD thesis, University of Colorado at Denver, 1991.

[17] A. BRASS AND G. PAWLEY, *Two and three dimensional FFTs on highly parallel computers*, Parallel Computing, 3 (1986), pp. 167–184.

[18] W. BRIGGS, L. HART, R. SWEET, AND A. O'GALLAGHER, *Multiprocessor FFT methods*, SIAM J. Sci. Stat. Comput., 8 (1987), pp. 27–42.

[19] W. L. BRIGGS, *Further symmetries of in-place FFTs*, SIAM J. Sci. Stat. Comput., 8 (1987), pp. 644–655.

[20] E. O. BRIGHAM, *The Fast Fourier Transform*, Prentice-Hall, Englewood Cliffs, NJ, 1974.

[21] ——, *The Fast Fourier Transform and Its Applications*, Prentice-Hall, Englewood Cliffs, NJ, 1988.

[22] R. BULIRSCH AND J. STOER, *Introduction to Numerical Analysis*, Springer-Verlag, Berlin, 1980.

[23] O. BUNEMAN, *Conversion of FFT to fast Hartley transforms*, SIAM J. Sci. Stat. Comput., 7 (1986), pp. 624–638.

[24] ——, *Multidimensional Hartley transforms*, Proc. IEEE, 75 (1987), p. 267.

[25] H. BURKHARDT, *Encyklopädie der mathematischen wissenschaften*, 1899–1916.

[26] C. BURRUS, *A new prime factor FFT algorithm*, in Proc. 1981 IEEE ICASSP, 1981, pp. 335–338.

[27] C. BURRUS AND P. ESCHENBACHER, *An in-place in-order prime factor FFT algorithm*, IEEE Trans. Acoust. Speech Signal Process., 29 (1981), pp. 806–817.

[28] C. BURRUS AND T. PARKS, *DFT/FFT and Convolution Algorithms*, John Wiley, New York, 1985.

[29] B. BUZBEE, G. GOLUB, AND C. NIELSON, *On direct methods for solving Poisson's equation*, SIAM J. Numer. Anal., 7 (1971), pp. 627–656.

[30] H. S. CARSLAW, *Introduction to the Theory of Fourier's Series and Integrals*, Third Ed., Dover, New York, 1930.

[31] B. CHAR, *MAPLE V Library Reference Manual*, Springer-Verlag, Berlin, 1991.

[32] T. CHIHARA, *An Introduction to Orthogonal Polynomials*, Gordon and Breach, New York, 1975.

[33] C. CHU, *The Fast Fourier Transform on Hypercube Parallel Computers*, PhD thesis, Cornell University, Ithaca, NY, 1988.

[34] R. V. CHURCHILL, *Fourier Series and Boundary Value Problems*, McGraw-Hill, New York, 1941.

[35] B. CIPRA, *The FFT: Making technology fly*, SIAM News, May 1993.

[36] V. ČIŽEK, *Discrete Fourier Transforms and Their Applications*, Adam Hilger, Bristol, England, 1986.

[37] R. CLARKE, *Transform Coding of Images*, Academic Press, New York, 1985.

[38] W. COCHRANE, *What is the fast Fourier transform?*, IEEE Trans. Audio and Electroacoustics, AU–15 (1967), pp. 45–55.

[39] J. COOLEY, R. GARWIN, C. RADER, B. BOGERT, AND T. STOCKHAM, *The 1968 Arden House workshop on fast Fourier transform processing*, IEEE Trans. Audio and Electroacoustics, AU–17 (1969), pp. 66–76.

[40] J. COOLEY, P. LEWIS, AND P. WELCH, *Historical notes on the fast Fourier transform*, IEEE Trans. Audio and Electroacoustics, AU–15 (1967), pp. 76–79.

[41] ——, *The fast Fourier transform algorithm: Programming considerations in the calculation of sine, cosine and Laplace transforms*, J. Sound Vibration, 12 (1970), pp. 315–337.

[42] J. COOLEY AND J. TUKEY, *An algorithm for the machine calculation of complex Fourier series*, Math. Comp., 19 (1965), pp. 297–301.

[43] R. COURANT AND D. HILBERT, *Methods of Mathematical Physics*, Vol. 1, Interscience Publishers, New York, 1937. Original in German in 1924.

[44] G. DANIELSON AND C. LANCZOS, *Some improvements in practical Fourier analysis and their application to x-ray scattering from liquids*, J. Franklin Inst., 233 (1942), pp. 365–380, 435–452.

[45] P. DAVIS AND P. RABINOWITZ, *Numerical Integration*, Blaisdell, New York, 1967.

[46] C. DE BOOR, *FFT as nested multiplication with a twist*, SIAM J. Sci. Stat. Comput., 1 (1980), pp. 173–178.

[47] S. R. DEANS, *The Radon Transform and Some of Its Applications*, John Wiley, New York, 1983.

[48] M. B. DOBRIN, *Introduction to Geophysical Prospecting*, Third ed., McGraw-Hill, New York, 1976.

[49] J. DOLLIMORE, *Some algorithms for use with the fast Fourier transform*, J. Inst. Math. Appl., 12 (1973), pp. 115–117.

[50] H. DUBNER AND J. ABATE, *Numerical inversion of Laplace transforms*, J. Assoc. Comput. Mach, 15 (1968), p. 115.

[51] P. DUHAMEL, *Implementation of the split radix FFT algorithms for complex, real, and real-symmetric data*, IEEE Trans. Acoust. Speech Signal Process., ASSP–34 (1986), pp. 285–295.

[52] P. DUHAMEL AND H. HOLLMANN, *Split radix FFT algorithms*, Electron. Lett., 20 (1984), pp. 14–16.

[53] P. DUHAMEL AND M. VETTERLI, *Improved Fourier and Hartley algorithms: Application to cyclic convolution of real data*, IEEE Trans. Acoust. Speech Signal Process., ASSP-35 (1987), pp. 818–824.

[54] A. DUTT AND V. ROKHLIN, *Fast Fourier transforms for nonequispaced data*, SIAM J. Sci. Comput., 14 (1993), pp. 1368–1393.

[55] H. DYM AND H. MCKEAN, *Fourier Series and Integrals*, Academic Press, Orlando, FL, 1972.

[56] B. EINARSSON, *Letter: Use of Richardson extrapolation for the numerical calculation of Fourier transforms*, J. Comp. Phys., 21 (1976), pp. 365–370.

[57] D. F. ELLIOT AND K. R. RAO, *Fast Transforms, Algorithms, Analysis and Applications*, Academic Press, Orlando, FL, 1982.

[58] L. FILON, *On a quadrature formula for trigonometric integrals*, Proc. Royal Soc. Edinburgh, 49 (1928), pp. 38–47.

[59] B. FORNBERG, *A vector implementation of the fast Fourier transform*, Math. Comp., 36 (1981), pp. 189–191.

[60] J. FOURIER, *Œuvres de Fourier: Théorie analytique de la chaleur*, Vol. 1, Gauthier-Villars et Fils, Paris, 1888.

[61] C. GAUSS, *Theoria interpolationis methodo nova tractata*. Carl Friedrich Gauss Werke, Band 3, Königlichen Gesellschaft der Wissenschaften, Göttingen, 1866.

[62] I. M. GEL'FAND, G. E. SHILOV, AND N. Y. VILENKIN, *Generalized Functions*, Vol. 5, Academic Press, New York, 1966.

[63] W. GENTLEMAN, *Matrix multiplication and fast Fourier transforms*, Bell Systems Tech. J., 47 (1968), pp. 1099–1103.

[64] W. GENTLEMAN AND G. SANDE, *Fast Fourier transforms for fun and profit*, Proc. 1966 Fall Joint Computer Conf. AFIPS, 29 (1966), pp. 563–578.

[65] W. M. GENTLEMAN, *Implementing Clenshaw-Curtis quadrature II: Computing the cosine transform*, Comm. ACM, 15 (1972), pp. 343–346.

[66] J. GLASSMAN, *A generalization of the fast Fourier transform*, IEEE Trans. Comp., C-19 (1970), pp. 105–116.

[67] H. H. GOLDSTINE, *A History of Numerical Analysis from the 16th through the 19th Century*, Springer-Verlag, Berlin, 1977.

[68] I. GOOD, *The interaction algorithm and practical Fourier analysis*, J. Royal Stat. Soc. Series B, 20 (1958), pp. 361–372.

[69] D. GOTTLIEB AND S. A. ORSZAG, *Numerical Analysis of Spectral Methods*, Society for Industrial and Applied Mathematics, Philadelphia, 1977.

[70] R. HAMMING, *Digital Filters*, Second ed., Prentice-Hall, Englewood Cliffs, NJ, 1983.

[71] H. HAO AND R. BRACEWELL, *A three-dimensional DFT algorithm using the fast Hartley transform*, Proc. IEEE, 75 (1987), p. 264.

[72] R. HARTLEY, *A more symmetrical Fourier analysis applied to transmission problems*, Proc. Inst. Radio Engrg., 30 (1942), pp. 144–150.

[73] M. HEIDEMAN AND C. BURRUS, *A bibliography of fast transform and convolution algorithms*, Department of Electrical Engineering Technical Report 8402, Rice University, Houston, TX, 1984.

[74] M. HEIDEMAN, D. JOHNSON, AND C. BURRUS, *Gauss and the history of the fast Fourier transform*, Arch. Hist. Exact Sciences, 34 (1985), pp. 265–277.

[75] V. E. HENSON, *Parallel compact symmetric FFTs*, in Vector and Parallel Computing: Issues in Applied Research and Development, J. Dongarra, I. Duff, and S. M. Patrick Gaffney, eds., Chichester, England, 1989, Ellis Horwood.

[76] ——, *Fourier Methods of Image Reconstruction*, PhD thesis, University of Colorado at Denver, 1990.

[77] ——, *DFTs on irregular grids: The anterpolated DFT*. Technical Report NPS-MA-92-006, Naval Postgraduate School, Monterey, CA, 1992.

[78] G. T. HERMAN, *Image Reconstruction from Projections*, Academic Press, Orlando, FL, 1980.

[79] R. HOCKNEY, *A fast direct solution of Poisson's equation using Fourier analysis*, J. Assoc. Comput. Mach., 12 (1965), pp. 95–113.

[80] E. ISAACSON AND H. KELLER, *Analysis of Numerical Methods*, John Wiley, New York, 1966.

[81] D. JACKSON, *Fourier Series and Orthogonal Polynomials*, Mathematical Association of America, Washington, D.C., 1941.

[82] A. JERRI, *The Shannon sampling thoerem: Its various extensions and applications: A tutorial review*, Proc. IEEE, 65 (1977), pp. 1565–1596.

[83] ——, *An extended Poisson sum formula for the generalized integral transforms and aliasing error bound for the sampling theorem*, J. Appl. Anal, 26 (1988), pp. 199–221.

[84] ——, *Integral and Discrete Transforms with Applications and Error Analysis*, Marcel Dekker, New York, 1992.

[85] H. JOHNSON AND C. BURRUS, *An in-place in-order radix-2 FFT*, Proc. IEEE ICASSP, San Diego, 1984, p. 28A.2.

[86] D. KOLBA AND T. PARKS, *A prime factor FFT algorithm using high speed convolution*, IEEE Trans. Acoust. Speech Signal Process., ASSP-25 (1977), pp. 281–294.

[87] D. KORN AND J. LAMBIOTTE, *Computing the fast Fourier transform on a vector computer*, Math. Comp., 33 (1979), pp. 977–992.

[88] T. KÖRNER, *Fourier Analysis*, Cambridge University Press, U.K., 1988.

[89] H. KREISS AND J. OLIGER, *Stability of the Fourier method*, SIAM J. Numer. Anal., 16 (1979), pp. 421–433.

[90] J. LAGRANGE, *Recherches sur la nature et la propagation du son*. Miscellanea Taurinensia (Mélanges de Turin), Vol. I, Nos. I–X, 1759, pp. 1–112. Reprinted in Oeuvres de Lagrange, Vol. I, J. A. Serret, ed., Paris, 1876, pp.39–148.

[91] C. LANCZOS, *Applied Analysis*, Prentice-Hall, Englewood Cliffs, NJ, 1956.

[92] R. M. LEWITT, *Reconstruction algorithms: Transform methods*, Proc. IEEE, 71 (1983), pp. 390–408.

[93] M. LIGHTHILL, *Introduction to Fourier Analysis and Generalised Functions*, Cambridge University Press, U.K., 1958.

[94] E. LINFOOT, *A sufficiency condition for Poisson's formula*, J. London Math. Soc., 4 (1928), pp. 54–61.

[95] J. LYNESS, *Some quadrature rules for finite trigonometric and related integrals*, in Numerical Integration: Recent Developments and Software Applications, P. Keast and G. Fairweather, eds., D. Reidel, Dordrecht, the Netherlands, 1987 pp. 17–34.

[96] *MATHCAD User's Guide*, MathSoft, Cambridge, MA, 1991.

[97] *MATLAB Reference Guide*, The Math Works, Natick, MA, 1992.

[98] J. MCCLELLAN AND C. RADER, *Number Theory in Digital Signal Processing*, Prentice-Hall, Englewood Cliffs, NJ, 1979.

[99] J. H. MCCLELLAN AND T. W. PARKS, *Eigenvalue and eigenvector decomposition of the discrete Fourier transform*, IEEE Trans. Audio and Electroacoustics, 20 (1972), pp. 66–74.

[100] H. MECKELBURG AND D. LIPKA, *Fast Hartley transform algorithm*, Electron. Lett., 21 (1985), pp. 341–43.

[101] L. MORDELL, *Poisson's summation formula and the Riemann zeta function*, J. London Math. Soc., 4 (1928), pp. 285–291.

[102] F. NATTERER, *Fourier reconstruction in tomography*, Numer. Math., 47 (1985), pp. 343–353.

[103] ——, *Efficient evaluation of oversampled functions*, J. Comp. Appl. Math., 14 (1986), pp. 303–309.

[104] ——, *The Mathematics of Computerized Tomography*, John Wiley, New York, 1986.

[105] K.-C. NG, *Letter: On the accuracy of numerical Fourier transforms*, J. Comp. Phys., 16 (1973), pp. 396–400.

[106] A. NIKIFOROV, S. SUSLOV, AND V. UVAROV, *Classical Orthogonal Polynomials of a Discrete Variable*, Springer-Verlag, Berlin, 1991.

[107] H. NUSSBAUMER, *FFT and Convolution Algorithms*, Springer-Verlag, Berlin, 1982.

[108] A. OPPENHEIMER AND R. SCHAFER, *Digital Signal Processing*, Prentice-Hall, Englewood Cliffs, NJ, 1975.

[109] J. OTTO, *Symmetric prime factor fast Fourier transform algorithms*, SIAM J. Sci. Stat. Comput., 10 (1989), pp. 419–431.

[110] R. PALEY AND N. WIENER, *Fourier Transforms in the Complex Plane*, American Mathematical Society, Providence, RI, 1934.

[111] A. PAPOULIS, *The Fourier Integral and Its Applications*, McGraw-Hill, New York, 1962.

[112] M. PEASE, *An adaptation of the fast Fourier transform for parallel processing*, J. Assoc. Comput. Mach., 15 (1968), pp. 252–264.

[113] C. RADER, *Discrete Fourier transforms when the number of data samples is prime*, Proc. IEEE, 5 (1968), pp. 1107–1108.

[114] J. RADON, *Uber die bestimmung von funktionen durch ihre integralwerte langs gewisser mannigfaltigkeiten*, Berichte Sachsische Akademie der Wissenschaften, Leipzig, Math.-Phys., Kl., 69 (1917), pp. 262–267.

[115] G. RAISBECK, *The order of magnitude of the Fourier coefficients in functions having isolated singularities*, Amer. Math. Monthly, (1955), pp. 149–154.

[116] K. RAO AND P. YIP, *The Discrete Cosine Transform: Algorithms, Advantages and Applications*, Academic Press, Orlando, FL, 1990.

[117] L. RICHARDSON, *The deferred approach to the limit*, Phil. Trans. Royal Soc., 226 (1927), p. 300.

[118] T. RIVLIN, *Chebyshev Polynomials*, John Wiley, New York, 1990.

[119] V. ROKHLIN, *A fast algorithm for the discrete Laplace transformation*, Research Report YALEU/DCS/RR-509, Yale University, New Haven, CT, January 1987.

[120] W. ROMBERG, *Vereinfachte numerische integration*, Norske Vid. Slesk. Forh. Trondheim, 28 (1955), pp. 30–36.

[121] P. RUDNICK, *Note on the calculation of Fourier series*, Math. Comp., 20 (1966), pp. 429–430.

[122] C. RUNGE, *Über die zerlegung einer empirischen funktion in sinuswellen*, Z. Math. Phys., 52 (1905), pp. 117–123.

[123] R. SAATCILAR, S. ERGINTAV, AND N. CANITEZ, *The use of the Hartley transform in geophysical applications*, Geophysics, 55 (1990), pp. 1488–1495.

[124] C. E. SHANNON, *Communication in the presence of noise*, Proc. IRE, 37 (1949), pp. 10–21.

[125] L. A. SHEPP AND B. F. LOGAN, *The Fourier reconstruction of a head section*, IEEE Trans. Nucl. Sci., NS–21 (1974), pp. 21–43.

[126] H. SORENESEN, M. HEIDEMAN, AND C. BURRUS, *On calculating the split-radix FFT*, IEEE Trans. Acoust. Speech Signal Process., ASSP–34 (1986), pp. 152–156.

[127] H. SORENSEN, D. JONES, C. BURRUS, AND M. HEIDEMAN, *On computing the discrete Hartley transform*, IEEE Trans. Acoust. Speech Signal Process., ASSP–33 (1985), pp. 1231–1238.

[128] I. STAKGOLD, *Green's functions and boundary value problems*, John Wiley, New York, 1979.

[129] H. STARK, J. WOODS, I. PAUL, AND R. HINGORANI, *Direct Fourier reconstruction in computer tomography*, IEEE Trans. Acoust. Speech Signal Process., ASSP–29 (1981), pp. 237–244.

[130] ——, *An investigation of computerized tomography by direct Fourier reconstruction and optimum interpolation*, IEEE Trans. Biomedical Engrg., BME–28 (1981), pp. 496–505.

[131] F. STENGER, *Numerical methods based on sinc and analytic functions*, Springer-Verlag, New York, 1993.

[132] D. STONE AND G. CLARKE, *In situ measurements of basal water quality as an indicator of the character of subglacial drainage systems*, Hydrological Processes, (1994). to appear.

[133] P. SWARZTRAUBER, *A direct method for the discrete solution of separable elliptic equations*, SIAM J. Numer. Anal., 11 (1974), pp. 1136–1150.

[134] ——, *The methods of cyclic reduction, Fourier analysis, and cyclic reduction-Fourier analysis for the discrete solution of Poisson's equation on a rectangle*, SIAM Review, 19 (1977), pp. 490–501.

[135] ——, *Algorithm 541: Efficient fortran subprograms for the solution of separable elliptic partial differential equations*, ACM Trans. Math. Software, 5 (1979), pp. 352–364.

[136] ——, *Vectorizing the FFTs*, in Parallel Computations, G. Rodrigue, ed., New York, 1982, Academic Press.

[137] ——, *Fast Poisson solvers*, in Studies in Numerical Analysis, G. Golub, ed., Washington, D.C., 1984, Mathematical Association of America.

[138] ——, *Multiprocessor FFTs*, Parallel Comput., 5 (1987), pp. 197–210.

[139] P. SWARZTRAUBER, R. A. SWEET, W. L. BRIGGS, V. E. HENSON, AND J. OTTO, *Bluestein's FFT for arbitrary n on the hypercube*, Parallel Comput., 17 (1991), pp. 607–617.

[140] P. N. SWARZTRAUBER, *FFTPACK, a package of fortran subprograms for the fast Fourier transform of periodic and other symmetric sequences*, 1985. Available from NETLIB. Send email to netlib@ornl.gov.

[141] ——, *Symmetric FFTs*, Math. Comp., 47 (1986), pp. 323–346.

[142] R. SWEET, *Direct methods for the solution of Poisson's equation on a staggered grid*, J. Comp. Phys., 12 (1973), pp. 422–428.

[143] ——, *Crayfishpak: A vectorized fortran package to solve Helmholtz equations*, in Recent Developments in Numerical Methods and Software for ODE/ADE/PDEs, G. Byrne and W. Schiesser, eds., World Scientific, Singapore, 1992, pp. 37–54.

[144] R. SWEET AND U. SCHUMANN, *Fast Fourier transforms for direct solution of Poisson's equation with staggered boundary conditions*, J. Comp. Phys., 75 (1988), pp. 123–137.

[145] G. SZEGÖ, *Orthogonal Polynomials*, Fourth ed., American Mathematical Society, Providence, RI, 1975.

[146] W. M. TELFORD, L. P. GELDART, R. E. SHERIFF, AND D. A. KEYS, *Applied Geophysics*, Cambridge University Press, U.K., 1976.

[147] C. TEMPERTON, *A note on prime factor FFT algorithms*, J. Comp. Phys., 52 (1983), pp. 198–204.

[148] ——, *Self-sorting mixed radix fast Fourier transforms*, J. Comp. Phys., 52 (1983), pp. 1–23.

[149] ——, *Implementation of a self-sorting in-place prime factor FFT algorithm*, J. Comp. Phys, 58 (1985), pp. 283–299.

[150] ——, *Self-sorting in-place fast Fourier transforms*, SIAM J. Sci. Comput., 12 (1991), pp. 808–823.

[151] F. THEILHEIMER, *A matrix version of the fast Fourier transform*, IEEE Trans. Audio and Electroacoustics, AU–17 (1969), pp. 158–161.

[152] L. THOMAS, *Using a computer to solve problems in physics*, in Applications of Digital Computers, Ginn, Boston, MA, 1963.

[153] G. P. TOLSTOV, *Fourier Series*, Prentice-Hall, Englewood Cliffs, NJ, 1962. Reprinted by Dover, New York, 1976.

[154] C. TRUESDELL, *The Tragicomical History of Thermodynamics: 1822–1854*, Springer-Verlag, Berlin, 1980.

[155] C. VAN LOAN, *Computational Frameworks for the Fast Fourier Transform*, Society for Industrial and Applied Mathematics, Philadelphia, 1992.

[156] E. B. VAN VLECK, *The influence of Fourier series upon the development of mathematics*, Science, 39 (1914), pp. 113–124.

[157] M. VETTERLI AND P. DUHAMEL, *Split radix algorithms for length p^m DFTs*, IEEE Trans. Acoust. Speech Signal Process., ASSP–34 (1989), pp. 57–64.

[158] J. S. WALKER, *Fourier Analysis*, Oxford University Press, New York, 1988.

[159] ——, *Fast Fourier Transforms*, CRC Press, Boca Raton, FL, 1991.

[160] P. WALKER, *The Theory of Fourier Series and Integrals*, John Wiley, New York, 1986.

[161] W. WEEKS, *Numerical inversion of Laplace transforms*, J. Assoc. Comput. Mach., 13 (1966), p. 419.

[162] E. WHITTAKER AND G. ROBINSON, *The Calculus of Observation*, Blackie and Sons, London, 1924.

[163] O. WING, *An efficient method of numerical inversion of Laplace transforms*, Arch. Elect. Comp., 2 (1967), p. 153.

[164] S. WINOGRAD, *On computing the discrete Fourier transform*, Math. Comp., 32 (1978), pp. 175–199.

[165] S. WOLFRAM, *Mathematica: A System for Doing Mathematics by Computer*, Addison-Wesley, Boston, 1988.

[166] D. M. YOUNG AND R. T. GREGORY, *A Survey of Numerical Mathematics*, Addison-Wesley, Boston, 1972.

Index